Also in the Variorum Collected Studies Series:

DAVID A. KING
Astrolabes from Medieval Europe

DAVID A. KING
Astronomy in the Service of Islam

JULIO SAMSÓ
Astronomy and Astrology in al-Andalus and the Maghrib

J.L. MANCHA
Studies in Medieval Astronomy and Optics

GAD FREUDENTHAL
Science in the Medieval Hebrew and Arabic Traditions

PAUL KUNITZSCH
Stars and Numbers
Astronomy and Mathematics in the Medieval Arab and Western Worlds

RAYMOND MERCIER
Studies on the Transmission of Medieval Mathematical Astronomy

DONALD R. HILL
Studies in Medieval Islamic Technology
From Philo to al-Jazari – from Alexandria to Diyar Bakr

EDWARD S. KENNEDY
Astronomy and Astrology in the Medieval Islamic World

RICHARD LORCH
Arabic Mathematical Sciences
Instruments, Texts and Transmission

A.I. SABRA
Optics, Astronomy and Logic
Studies in Arabic Science and Philosophy

ROSHDI RASHED
Optique et mathématiques
Recherches sur l'histoire de la Pensée Scientifique en Arabe

VARIORUM COLLECTED STUDIES SERIES

Islamic Astronomy and Geography

To Julio Samsó and
the Barcelona school,
مع فائق تقديري

David A. King

Islamic Astronomy and Geography

Routledge
Taylor & Francis Group
LONDON AND NEW YORK

First published 2012 by Ashgate Publishing

Published 2016 by Routledge
2 Park Square, Milton Park, Abingdon, Oxon OX14 4RN
52 Vanderbilt Avenue, New York, NY 10017

Routledge is an imprint of the Taylor & Francis Group, an informa business

British Library Cataloguing in Publication Data
King, David A.
Islamic astronomy and geography.
– (Variorum collected studies series ; CS1009)
1. Islamic astronomy – History. 2. Astronomy – Mathematics – History.
3. Astronomical geography – History.
I. Title II. Series
520.9'1767–dc23

ISBN 978-1-4094-4201-1

Library of Congress Control Number: 2012931346

ISBN 13: 978-1-4094-4201-1 (hbk)

VARIORUM COLLECTED STUDIES SERIES CS1009

CONTENTS

This volume contains xlii + 376 pages

PREFACE

This volume supplements my three previous Variorum volumes on the history of Islamic astronomy: *Islamic Mathematical Astronomy*, *Islamic Astronomical Instruments*, and *Astronomy in the Service of Islam*.[1] The contents include an introductory essay on Islamic astronomy, general overviews of instruments and manuscript illustrations, overviews of astronomical activity in specific geographical regions, and a detailed study of one aspect of Islamic astrology. Also included are an overview of 'sacred geography', that is, the notion of the world centred on the Kaʿba in Mecca, and the determination of the sacred direction (*qibla*) towards Mecca by mathematical means one the one hand and the methods of folk astronomy on the other. The results of this dichotomy of approaches to the qibla problem are illustrated for one particular region, Central Asia, by an 11th-century treatise. The volume concludes with three studies treating of geodetic measurements, mathematical cartography, and geographical coordinates. The last two of these three papers deal with subjects that are the standard fare of the medieval scientists, but whose treatment in the medieval sources under discussion is no longer in keeping with the high standards of earlier Islamic science.

I 'Islamic astronomy'. The 1996 volume *Astronomy before the Telescope* provides useful and reliable introductions to Greek, Indian, Byzantine and medieval European astronomy, and a lot more. This essay attempts to provide

---

[1] The following abbreviations are used in this preface:

DSB	*Dictionary of Scientific Biography*, 15 vols., New York, 1970–1978.
EI_2	*Encyclopaedia of Islam*, new edn., 12 vols., Leiden, 1960–2005.
SATMI	*In Synchrony with the Heavens – Studies in Astronomical Timekeeping and Instrumentation in Medieval Islamic Civilization*, 2 vols., I: *The Call of the Muezzin – Studies I–IX*, II: *Instruments of Mass Calculation – Studies X–XVIII*, Leiden, 2004–2005.
Studies A	*Islamic Mathematical Astronomy*, Variorum, 1986/93.
Studies B	*Islamic Astronomical Instruments*, Variorum, 1987/95.
Studies C	*Astronomy in the Service of Islam*, Variorum, 1993.
Studies D	*Astrolabes from Medieval Europe*, Ashgate-Variorum, 2011.
World-Maps	*World-Maps for Finding the Direction and Distance to Mecca – Innovation and Tradition in Islamic Science*, Leiden & London, 1999.
Suhayl	*Suhayl – Journal for the History of the Exact and Natural Sciences in Islamic Civilisation* (Barcelona) (usd.proves.ub.edu/suhayl/).

an overview of Muslim activity in this field. Meanwhile, Benno van Dalen is continuing the research of the doyen of our field, the late Ted Kennedy, on the Islamic astronomical handbooks with tables known as *zījes* (Kennedy published his classic survey in 1956[2]); Benno aims to publish a new survey of these works, with tables of contents and basic parameters. For an interim report on these *zījes*, and other tables not in *zījes*, see my 2001 paper co-authored with Julio Samsó and Bernard Goldstein.[3] Also, detailed studies of astronomical timekeeping and instrumentation are now published (*SATMI*, vols. 1 and 2). One aspect of the Muslim encounter with the astronomical sciences that is lacking from my overview is astrology, which was of prime importance in the history of Islamic civilization.

II 'From inscriptions to context: some Islamic astronomical instruments and their secrets'. This paper was prepared especially for an audience of Islamicists; in it I tried to stress the value of instruments as historical sources in their own right. Most Islamic instruments are signed and dated, and their provenance is clear. For the three Mecca-centred world-maps see *World-Maps* and *SATMI*, vol. 1, study VIIc. On instrumentation in general see *SATMI*, vol. 2, where most of the instruments mentioned in this paper are described in detail, and a list of Islamic astrolabes, ordered chronologically by region, is presented. A most useful contribution to the context of Islamic instrumentation has been published by François Charette.[4] In contrast with Islamic instruments, the majority of medieval European astrolabes are unsigned and undated. See *Studies D* for some detailed studies of various historically significant medieval European pieces and an ordered list of all known astrolabes from that milieu.

III 'Some illustrations in Islamic scientific manuscripts ...'. These go far beyond the illustrations one expects in mathematical and astronomical manuscripts and constitute a very useful means of presenting our subject to a modern audience.

IV–VI: Over the years I have tried to stress the fact that from the 11th century onwards, that is, after the time covered by Fuat Sezgin in his monumental *Geschichte des arabischen Schrifttums*, Islamic astronomy developed within a

---

[2] 'A survey of Islamic astronomical tables', *Transactions of the American Philosophical Society*, 46:2 (1956), repr. 1989. For Kennedy's bibliography see now *Suhayl* 9 (2009/10), pp. 185–214, available on the Internet.

[3] 'Astronomical handbooks and tables from the Islamic world (750–1900): An interim report', *Suhayl* 2 (2001), pp. 9–105, available on the Internet.

[4] François Charette, 'The Locales of Islamic Astronomical Instrumentation', *History of Science* 44 (2006), pp. 123–38. See also n. 9 below.

series of regional schools. The only surveys of the history of astronomy in Islamic Iran are two early essays by Ted Kennedy.[5] My surveys of astronomy in Fatimid and Mamluk Egypt and the Maghrib are reprinted here. A bio-bibliographical survey of astronomy in medieval Yemen was published in book form in 1983,[6] and a brief survey of astronomy in medieval Syria in a 1993 exhibition catalogue.[7] For astronomy in al-Andalus see the various publications of the Barcelona school.[8]

IV 'Aspects of Fatimid astronomy'. This was prepared for a conference on Fatimid Egypt, mainly for historians of art and architecture. It summarizes my work on Ibn Yūnus and his milieu; for more on Ibn Yūnus see my 1972 doctoral dissertation and the extensive article in the *Dictionary of Scientific Biography*. I used the occasion to show the importance of astronomical sources for understanding the orientation of mosques and cities, in this case, Cairo. For more on the Cairo ventilators see now *SATMI*, VIIb. This paper is dedicated to the memory of Felicitas Jaritz, one of the very few historians of Islamic architecture to realize the importance of orientations, in gratitude for many fruitful exchanges in Cairo during the 1970s.

V 'Mamluk astronomy and the institution of the *muwaqqit*'. This is a summary of the 1983 overview 'Mamluk astronomy' (repr. in *Studies A*, III) and the 1995 study 'On the role of the muezzin and muwaqqit in medieval Islamic societies' (repr. in *SATMI*, vol. 1, V). Our knowledge of this subject took a major step forward with the publication of François Charette's study of the illustrated treatise on over 100 instruments by Najm al-Dīn al-Miṣrī, a work that I

---

5 'The Exact Sciences in Iran under the Saljuqs and Mongols' and 'The Exact Sciences in Iran under the Timurids', in *The Cambridge History of Iran,* V (1968), pp. 659–79, and VI (1986): pp. 568–80.

6 *Mathematical Astronomy in Medieval Yemen – A Bio-Bibliographical Survey* (Publications of the American Research Center in Egypt), Malibu, Ca.: Undena, 1983.

7 'L'astronomie en Syrie à l'époque islamique', in *Syrie - Mémoire et civilisation*, Sophie Cluzan *et al.*, eds., Paris: Institut du monde arabe, 1993, pp. 386–95, 432–43 and 480.

8 See www.ub.edu/arab/publicacions.html, and especially the new edition of Julio Samsó, *Las ciencias de los antiguos en al-Andalus*. Madrid: Mapfre, 1992/2011, and the same author's *Islamic Astronomy and Medieval Spain*, Aldershot: Variorum, 1994; *idem*, *Astronomy and Astrology in al-Andalus and the Maghrib*, Aldershot: Ashgate-Variorum, 2007; and *idem*, *Astrometeorologia y astrología medievales*, Barcelona: Universitat de Barcelona, 2008. On the Hyderabad manuscript see Àngel Mestres, 'Maghribī Astronomy in the 13th Century: a Description of Manuscript Hyderabad Andra Pradesh State Library 298', in Josep Casulleras and Julio Samsó, eds., *From Baghdad to Barcelona – Studies in the Islamic Exact Sciences in Honour of Prof. Juan Vernet*, 2 vols., Barcelona: Instituto 'Millás Vallicrosa' de Historia de la Ciencia Árabe, 1996, I, pp. 383–443; and *idem*, *Materials andalusins en el Zīj d'Ibn Isḥāq al-Tūnisī* (edited text and tables, with introduction and commentary in English), doctoral thesis, University of Barcelona, 2000. See also n. 12 below.

had previously, and also in this study, attributed to his contemporary from Aleppo, Ibn al-Sarrāj.[9] One of the reasons for my incorrect attribution was the fact that the anonymous author presents only tables for latitude 36°; I have since realized that this latitude was chosen not because the author worked in Aleppo but because 36° is the latitude of the middle of the fourth of the seven climates, that is, the middle of the inhabited earth. In other words, the tables of Najm al-Dīn for latitude 36° were for didactic rather than practical purposes. Another important contribution has been made by Sonja Brentjes, exposing significant biographical information about the Mamluk astronomers and their affiliations.[10] This study of mine was dedicated to the memory of my friend Salah Abdel-Khalek.

VI 'On the history of astronomy in the medieval Maghrib'. The earliest version of this paper was published in Tunis in 1990; this version appeared in Casablanca in 2007. In it, I try to deal with all aspects of astronomy, at both levels, the mathematical and the folk astronomical traditions, and give instrumentation the place it deserves. The inspiration for this study was the rediscovery of the unique Hyderabad manuscript of the astronomical handbook (*zīj*) of the 13th-century Tunisian astronomer Ibn Isḥāq, the most significant astronomer of the Maghrib over many centuries. Given the prolific output of the Barcelona school on Andalusi and Maghribi astronomy, any survey such as this will be quickly outdated. The reader is referred to the works of Julio Samsó, Àngel Mestres – for a study of the materials in the Hyderabad manuscript – and Mònica Rius and their colleagues, which contain updated information on various aspects of this subject.[11]

VII 'A Hellenistic astrological table ...'. Few tables can have had such a chequered career as Vettius Valens' little table for calculating the duration of human life. We see how a concept – here in the form of a table and accompanying text – could be revived and could thrive centuries after it was first introduced in the Islamic world. Discussed here are all of the known occurrences of the table, mostly from Mamluk and Ottoman Egypt, as well as one from the medieval Maghrib. In none of the sources is there any indication that the table and its underlying theory are absurd. This paper was dedicated to the late David Pingree,

---

9 François Charette, *Mathematical Instrumentation in 14th-Century Egypt and Syria – The illustrated treatise of Najm al-Dīn al-Miṣrī* (Islamic Philosophy Theology and Science - Texts and Studies, vol. LI), Leiden & Boston: Brill, 2003.

10 Sonja Brentjes, 'Shams al-Dīn al-Sakhāwī on muwaqqits, mu'adhdhins, and teachers of various astronomical disciplines in Mamluk cities of the fifteenth century', in Emilia Calvo, Mercè Comes, Roser Puig and Mònica Rius, eds., *A Shared Legacy – Islamic Science East and West – Homage to Professor J.M. Millàs Vallicrosa*, Barcelona: Universitat de Barcelona, 2008, pp. 129–50.

11 See nn. 8 above and 12 below.

to whom I have felt indebted ever since he read my doctoral thesis with a red pen in hand.

VIII 'The sacred geography of Islam'. The idea of the astronomically-aligned Kaʿba in Mecca as the physical focus of worship in the Muslim world inspired a great deal of activity amongst medieval scholars. First, the *fuqahāʾ*, that is, the scholars of the sacred law, and the specialists in folk astronomy developed their own solutions to the qibla problem. The book on this subject started in New York in the 1980s remains unfinished, although it is hoped that a paper on the sources for Islamic sacred geography can be submitted to *Suhayl*.

The study continues with a discussion of the ways in which Muslim astronomers calculated the sacred direction, featuring trigonometric formulae, tables and maps. For more information see *Studies C* and *World-Maps*.

Finally, these two approaches resulted in different directions being proposed for one and the same locality by different interest groups. I use the examples of Cordova, Cairo and Samarqand, for each of which we have adequate textual material to illustrate a spectrum of qibla values that were accepted in these cities. Those values not based on calculation are found in treatises on the sacred law or on folk astronomy. Mònica Rius and Petra Schmidl have published detailed analyses of medieval texts on the qibla in al-Andalus and the Yemen, respectively.[12] For the situation in Central Asia see IX below. The orientation of religious architecture can only be understood with reference to such medieval sources.

IX 'Al-Bazdawī on the qibla in early Islamic Transoxania'. The source published here is a legal treatise dealing with the qibla by an 11th-century Ḥanafī *qāḍī* in Samarqand. The author mentions directions based on tradition, others based on folk astronomy, and also directions calculated using geographical coordinates. The study, originally dedicated to Dick Frank, is revived here in the unrealistic hope that it might attract a historian of Islamic architecture. It has been left just as it was published some 30 years ago, as a tribute to the early days of the *Journal for the History of Arabic Science*. In those days there were great hopes for the newly-founded Institute for the History of Arabic Science. As I write this in October 2012, the future of Aleppo and the Institute, indeed of Syria itself, is in the balance.

---

[12] Mònica Rius, *La Alquibla en al-Andalus y al-Maǧrib al-Aqsà*, Barcelona: Institut 'Millás Vallicrosa' de Història de la Ciència Àrab, 2000; and Petra Schmidl, *Volkstümliche Astronomie im islamischen Mittelalter – Zur Bestimmung der Gebetszeiten und der Qibla bei al-Aṣbaḥī, Ibn Raḥīq und al-Fārisī*, 2 vols. (Islamic Philosophy, Theology and Science, Texts and Studies, ed. Hans Daiber, vol. LXVIII), Leiden & Boston, MA: Brill, 2008.

X: 'Too many cooks ... : a newly-rediscovered account of the earliest Muslim geodetic measurements'. It is doubtful whether any scientific experiment ever suffered so much from contemporaneous or later documentation as the geodetic measurements conducted from Baghdad in the early 9th century for the Caliph al-Ma'mūn. A manuscript preserved in Princeton (see also XI) contains a copy of a hitherto-unknown report by the judge entrusted by the Caliph to oversee the measurements, a task that was clearly beyond him.

XI 'A world-map in the tradition of al-Bīrūnī and al-Khāzinī'. Serious Islamic world-maps were fitted with proper grids for longitude and latitude and had cities marked according to contemporaneous geographical data. Most of these maps have disappeared without trace, except for the various lists of coordinates of cities that underlay some of them. A very corrupt world-map preserved in a Princeton manuscript (see also X) contains evidence that it reflects a work of the great scientist al-Bīrūnī, albeit using the derivative geographical data of al-Khāzinī (on which see *World-Maps*). The map well reflects the decline in mathematical cartography that occurred in the central parts of the Islamic world. My paper was dedicated to my friend Hossam Elkhadem.

XII 'Mathematical geography in 15th-century Egypt'. Continued uncritical copying of geographic coordinates without any means of checking their accuracy could lead to prominent scholars recording information of little or no scientific value. Here we are witnesses to an unhappy end to serious work in one aspect of the Muslim scientific endeavour, the persons involved being two of the leading astronomers of 15th-century Cairo. This paper was originally dedicated to my colleague in Frankfurt, Hans Daiber.

It is fitting that these studies should conclude with two which clearly indicate the decline of scientific activity in medieval Islamic society. I am often asked why Muslim activity in the sciences declined. The first thing to clarify is that Muslim activity in the sciences, at least in astronomy and mathematics, continued into the 19th century. Once this is realized one can begin to understand that Muslim scientists after the 15th century, with few exceptions, simply dealt with the same old problems of ancient and medieval astronomy and mathematics, citing the same old authorities and delving into abridgements of, and commentaries and super-commentaries on their works. No new questions were posed since there was, in most places in the Islamic world, no access to any findings based on the telescope. We are dealing with 'astronomy before the telescope', a very apt title to the volume in which my 1996 overview of Islamic astronomy – the first study in this book – appeared.

In a 2004 study entitled 'Reflections on some new studies on applied science in Islamic societies', now available on the Internet, I outlined the problems of the diffusion of our current knowledge of the history of Islamic science in the modern Muslim world. I suggested that, in particular, my surveys of the ways in which Muslims have for 1400 years determined the direction of Mecca and regulated the times of their prayers (*World-Maps* and *In Synchrony with the Heavens*) will probably not have any influence in the Muslim world. The same is true, of course, of my four Variorum volumes on the history of Islamic astronomy, and, indeed, for the entire corpus of modern studies on the history of Islamic science.

It gives me pleasure to thank Lindsay Farthing of Ashgate Variorum for converting a pile of papers and Word files into the present volume.

I conclude by expressing my gratitude to John Smedley, the editor of the Variorum Collected Studies Series, featuring over a thousand volumes. He and his press have thereby made a monumental contribution to scholarship. I owe him my gratitude for being generous and patient since the mid 1980s. Perhaps the best way to thank him is to promise him not to submit any more materials for another volume.

DAVID A. KING

Frankfurt am Main
October 2012

ACKNOWLEDGEMENTS

For permission to reprint the various articles in this volume I thank: the British Museum Press, London (I); Garnet, Reading (II); State University of New York Press, Albany, N.Y., and The Library of Congress, Washington, D.C. (III); Presses de l'Université de Paris-Sorbonne, Paris (IV); Cambridge University Press (V); Professor Mohammed Abattouy (VI); Elsevier, Oxford (VIII); the Institute for the History of Arabic Science, Aleppo (IX); Prof. Julio Samsó, editor, *Suhayl – Journal for the History of the Exact and Natural Sciences in Islamic Civilisation*, Barcelona (X); and Archives et bibliothèques de Belgique / Archief- en Bibliotheekwezen in België, Royal Library of Belgium, Brussels (XI).

Koninklike Brill NV of Leiden, who have published close to 3,000 pages of my books, gave permission for including articles VII and XII but wanted to charge a substantial fee for reproduction. These pieces have therefore been republished here by Variorum in a new format.

PUBLISHER'S NOTE

The articles in this volume, as in all others in the Variorum Collected Studies Series, have not been given a new, continuous pagination. In order to avoid confusion, and to facilitate their use where these same studies have been referred to elsewhere, the original pagination has been maintained wherever possible.

Each article has been given a Roman number in order of appearance, as listed in the Contents. This number is repeated on each page and is quoted in the index entries.

David A. King
List of Publications
October, 2012

Notes: Titles of books are printed bold. References of the form **X-n** indicate that an article has been reprinted as no. n in vol. X of the five Variorum volumes:

A	*Islamic Mathematical Astronomy* (1986/1993, see nos. 79/132)
B	*Islamic Astronomical Instruments* (1987/1995, see nos. 87/163)
C	*Astronomy in the Service of Islam* (1993, see no. 131)
D	*Astrolabes from Medieval Europe* (2011, see no. 255)
E	*Islamic Astronomy and Geography* (2012, see no. 260)

Some other works are available in new versions in:

SATMI *In Synchrony with the Heavens – Studies in Astronomical Timekeeping and Instrumentation in Medieval Islamic Civilization* (2 vols., 2004–05, see nos. 230–231)

Articles published in ***Suhayl*** are available on the Internet at www.ub.edu/arab/suhayl/.

Various articles are reprinted in ***Islam and Science***: Muzaffar Iqbal, ed., *Islam and Science: Historic and Contemporary Perspectives*, 4 vols., esp. III: *New Perspectives on the History of Islamic Science*; and IV: *Studies in the Making of Islamic Science: Knowledge in Motion*, Farnham: Ashgate, 2012.

The signs + and ++ indicate that the work in question was translated by Kurt Maier or the late Wolf-Dieter Wagner, respectively. An asterisk is used for further works not in English that were translated by others.

1972

1 *The Astronomical Works of Ibn Yūnus*, Ph.D. dissertation, Yale University, Department of Near Eastern Languages and Literatures, 1972. [Available from ProQuest.com (formerly University Microfilms, Ann Arbor, Mich.), no. 7229740.]

2 'The ᶜAbd al-A'imma astrolabe forgeries' (with Owen Gingerich & George Saliba), *Journal for the History of Astronomy* 3 (1972), pp. 188–98. [Repr. in **B-VI**.]

1973

3 'al-Khalīlī's auxiliary tables for solving problems of spherical astronomy', *Journal for the History of Astronomy* 4 (1973), pp. 99–110. [Repr. in **A-XI**; see now ***SATMI*****, I–II**.]

4 'Ibn Yūnus' *Very Useful Tables* for reckoning time by the sun', *Archive for History of Exact Science* 10 (1973), pp. 342–94. [Repr. in **A-IX**; see now ***SATMI*****, I–II**.]

5 A review of Bernard R. Goldstein, *al-Biṭrūjī: On the Principles of Astronomy*, New Haven, Conn., & London, 1971, in *Journal of the American Oriental Society* 93 (1973), pp. 566–7.

6 A review of Ahmed Saidan, *Arabic Arithmetic: The Arithmetic of Abū al-Wafā' al-Būzajānī* [in Arabic], Amman, n.d. [1972?], in *ISIS* 64 (1973), pp. 123–5.

1974

7 'A double-argument table for the lunar equation attributed to Ibn Yūnus', *Centaurus* 18 (1974), pp. 129–46. [Repr. in **A-V**.]

8 'On medieval Islamic multiplication tables', *Historia Mathematica* 1 (1974), pp. 317–23. [Repr. in **A-XIV**; see also no. 32.]

9 'Smithsonian Institution Project in Medieval Islamic Astronomy', *Historia Mathematica* 1 (1974), pp. 183–4.

10 'An analog computer for solving problems of spherical astronomy: The *shakkāzīya* quadrant of Jamāl al-Dīn al-Māridīnī', *Archives Internationales d'Histoire des Sciences* 24 (1974), pp. 219–42. [Repr. in **B-X**.]

11 A review of Edward S. Kennedy & David Pingree, *The Astrological History of Māshā'allāh*, Cambridge, Mass., 1971, in *Journal of Near Eastern Studies* 33 (1974), pp. 158–60.

12 A review of Edward S. Kennedy, *A Commentary upon al-Bīrūnī's Taḥdīd [nihāyāt] al-amākin*, Beirut, 1973, in *Centaurus* 19 (1974), pp. 320–23.

1975

13 'al-Khalīlī's qibla table', *Journal of Near Eastern Studies* 34 (1975), pp. 81–122. [Repr. in **A-XIII**.]

14 'On the astronomical tables of the Islamic Middle Ages', *Studia Copernicana* 13 (1975), pp. 37–56. [Repr. in **A-II**.]

15 'Astronomical timekeeping (*ᶜilm al-mīqāt*) in medieval Islam', *Actes du XXIXe Congrès International des Orientalistes*, Paris: L'Asiathèque, 1975, II:2, pp. 86–90.

16 'Ibn al-Shāṭir' in *Dictionary of Scientific Biography*, vol. XII, New York: Charles Scribner's Sons, 1975, pp. 357–64.

17 'Medieval mechanical devices', an essay review of Donald R. Hill, *The Book of Knowledge of Ingenious Mechanical Devices*, Dordrecht & Boston, Mass., 1974, *History of Science* 13 (1975), pp. 284–9. [Repr. in **B-XX**.]

1976

18 'Ibn Yūnus' in *Dictionary of Scientific Biography*, vol. XIV, New York: Charles Scribner's Sons, 1976, pp. 574–80.

19 A review of Saleh Ahmed & Rushdi Rashed, *Al-Bāhir en algèbre d'as-Samaw'al*, Damascus, 1972, in *ISIS* 67 (1976), pp. 307–8.

1977

20 'A fourteenth-century Tunisian sundial for regulating the times of Muslim prayer', in Walter G. Saltzer & Yasukatsu Maeyama, eds., *ΠΡΙΣΜΑΤΑ*: *Naturwissenschafts-geschichtliche Studien – Festschrift für Willy Hartner*, Wiesbaden: Franz Steiner, 1977, pp. 187–202. [Repr. in **B-XVIII**; see now ***SATMI***, **IV**.]

21 'Ibn al-Shāṭir's *Ṣandūq al-yawāqīt*: An astronomical compendium' (with Louis Janin), *Journal for the History of Arabic Science* 1 (1977), pp. 187–256. [Repr. in **B-XII**.]

22 A review of Bernard R. Goldstein, *The Astronomical Tables of Levi ben Gerson*, Hamden, Conn., 1974, in *ISIS* 68 (1977), pp. 476–7.

23 A review of Donald R. Hill, *On the Construction of Water-Clocks; Kitāb Arshimīdas fī ʿamal al-binkāmāt*, London, 1976, in *History of Science* 15 (1977), pp. 295–8. [Repr. in **B-XXI**.]

1978

24 'Astronomical timekeeping in fourteenth-century Syria', *Proceedings of the First International Symposium for the History of Arabic Science (Aleppo, 1976)*, 2 vols., Aleppo: Institute for the History of Arabic Science, 1978, I, pp. 391–415 (Arabic), and II, pp. 75–84 (English). [Repr. in **A-X**; see now ***SATMI*, I–II**.]

25 *Project in Medieval Islamic Astronomy – A Progress Report with Bibliography*, Cairo: American Research Center in Egypt (Project Report No. 1), Jan. 1978.

26 'Three sundials from Islamic Andalusia', *Journal for the History of Arabic Science* 2 (1978), pp. 358–92. [Repr. in **B-XV**.]

27 'Notes on the astrolabist Nasṭūlus/Basṭūlus', *Archives Internationales d'Histoire des Sciences* 28 (1978), pp. 115–18. [Repr. in **B-IV**; see also no. 63.]

28 'Le cadran solaire de la mosquée d'Ibn Ṭūlūn au Caire' (with Louis Janin), *Journal for the History of Arabic Science* 2 (1978), pp. 331–57. [Repr. in **B-XVI**.]

29 'al-Khalīlī' in *Dictionary of Scientific Biography*, vol. XV, Supp. I, New York: Charles Scribner's Sons, 1978, pp. 259–61.

30 a/b 'Islamic mathematics and astronomy', an essay review of the chapters on astronomy and mathematics in Seyyed Hossein Nasr, *Islamic Science: An Illustrated Study*, London, 1976, in *Journal for the History of Astronomy* 9 (1978), pp. 212–19, repr. in *Bibliotheca Orientalis* 35 (1978), pp. 339–43. [The latter version is repr. in **A-XVII**.]

1979

31 'Report on a field-trip to India, September-October, 1978', *Newsletter of the American Research Center in Egypt*, no. 108 (Spring, 1979), pp. 21–4.

32 'Supplementary notes on medieval Islamic multiplication tables', *Historia Mathematica* 6 (1979), pp. 405–17. [A supplement to no. 8; repr. in **A-XV**.]

33 'On the early history of the universal astrolabe in Islamic astronomy and the origin of the term *shakkāzīya* in medieval scientific Arabic', *Journal for the History of Arabic Science* 3 (1979), pp. 244–57. [Repr. in **B-VII**.]

34 'Ibn Yūnus and the pendulum: A history of errors', *Archives Internationales d'Histoire des Sciences* 29 (1979), pp. 35–52. [Repr. in **B-XIX** (abridged).]

35 'Mathematical astronomy in medieval Yemen', *Arabian Studies* 5 (1979), pp. 61–5. [Repr. in **A-IV**; see no. 58.]

36 'Astronomical timekeeping in Ottoman Turkey', *Proceedings of the International Symposium on the Observatories in Islam, 19-23 Sept., 1977*, Istanbul: Millî Eğitim Basımevi, 1980, pp. 245–69. [Repr. in **A-XII**.]

37 'A classification of Islamic astronomical literature and the present state of research on the manuscript sources', *Proceedings of the International Symposium on the Observatories in Islam, 19-23 Sept., 1977*, Istanbul: Millî Eğitim Basımevi, 1980, pp. 169–80.

38 'The sundial on the West Wall of the Madrasa of Sultan Qāytbāy in Jerusalem' (with Archibald G. Walls), *art and architecture research papers* 15 (July, 1979), pp. 16–21. [Repr. in **B-XVII**.]

39 'Ḳibla. ii. Astronomical aspects' [sacred direction], in *The Encyclopaedia of Islam*, new edition, vol. V, fascs. 79–80, Leiden: E.J. Brill, 1979, pp. 83–8. [Repr. in **C-IX**.]

40 'On the sources for the study of early Islamic mathematics', an essay review of Fuat Sezgin, *Geschichte des arabischen Schrifttums*, V: Mathematik, Leiden: E.J. Brill, 1974, in *Journal of the American Oriental Society* 99 (1979), pp. 450–59.

41 A review of Ali Abdallah Daffa, *The Muslim Contribution to Mathematics*, London & Atlantic Highlands, N.J., 1977, in *History of Science* 17 (1979), pp. 295–6. [Repr. in **A-XVIII**.]

42 A review of William Brice, Colin Imber & Richard Lorch, *The Dā'ire-yi Mu'addel of Seydī Alī Re'īs*, Manchester, 1976, in *Journal for the History of Astronomy* 10 (1979), pp. 51–3. [Repr. in **B-XIII**.]

1980

43 'New light on the *Zīj al-ṣafā'iḥ* of Abū Jaᶜfar al-Khāzin', *Centaurus* 23 (1980), pp. 105–17. [Repr. in **B-XI**.]

44 'The exact sciences in medieval Islam: Some remarks on the present state of research', *Bulletin of the Middle East Studies Association of North America* 4 (1980), pp. 10–26. [Repr. in **A-I** (abridged).]

45 'A handlist of the Arabic and Persian astronomical manuscripts in the Maharaja Mansingh II Library in Jaipur', *Journal for the History of Arabic Science* 4 (1980), pp. 81–6. [Repr. in **A-XVI**.]

46 'Ibn al-Majdī's tables for calculating ephemerides' (with E.S. Kennedy), *Journal for the History of Arabic Science* 4 (1980), pp. 48–68. [Repr. in **A-VI**.]

1981

47 ***A Catalogue of the Scientific Manuscripts in the Egyptian National Library*** **[in Arabic], vol. 1: A critical handlist of the scientific collections – Indexes of copyists and owners, Cairo: General Egyptian Book Organization, 1981, xviii + 781 pp.** [See nos. 77 and 78.]

Review:
James Weinberger in *Notes of the Middle East Librarians Association*, 28 (1983), pp. 28–30.

48 'On the origin of the astrolabe according to the medieval Arabic sources', *Journal for the History of Arabic Science* 5 (1981), pp. 43–83. [Repr. in **B-III** and republished in ***SATMI*, XIIIe**.]

49 'Early Islamic astronomy', an essay review of Fuat Sezgin, *Geschichte des arabischen Schrifttums*, VI: Astronomie, Leiden: E.J. Brill, 1978, in *Journal for the History of Astronomy* 12 (1981), pp. 55–9.

50 A review of Kenneth Brecher & Michael Feirtag, eds., *Astronomy of the Ancients*, Cambridge, Mass., 1979, in *Technology and Culture* 22 (1981), pp. 300–301.

1982

51 'On the astronomical orientation of the Kaaba' (with Gerald S. Hawkins), *Journal for the History of Astronomy* 13 (1982), pp. 102–9. [Repr. in **C-XII**.]

52 'Some astronomical observations from thirteenth-century Egypt' (with Owen Gingerich), *Journal for the History of Astronomy* 13 (1982), pp. 121–8. [Repr. in **A-VII**.]

53 'Astronomical alignments in medieval Islamic religious architecture', *Annals of the New York Academy of Sciences* 385 (1982), pp. 303–12. [Repr. in **C-XIII**.]

54 'Faces of the Kaaba', *The Sciences* (The New York Academy of Sciences) 22:5 (May/June, 1982), pp. 16–20, and 22:6 (September, 1982), p. 2.

55 'Willy Hartner, Ibn Yūnus and the meridian degree', *Centaurus* 26 (1982), pp. 218–19.

56 'Indian astronomy in fourteenth-century Fez: The versified *Zīj* of al-Qusanṭīnī' (with E.S. Kennedy), *Journal for the History of Arabic Science* 6 (1982), pp. 3–45. [Repr. in **A-VIII**.]

57 A review of Emilie Savage-Smith & M.B. Smith, *Islamic Geomancy and a Thirteenth-Century Divinatory Device*, Malibu, Ca.: Undena, 1980, in *Archaeoastronomy – The Bulletin of the Center for Archaeoastronomy* (College Park, Md.) 5 (1982), pp. 42–3. [Repr. in **B-XXII**.]

1983

58 ***Mathematical Astronomy in Medieval Yemen – A Bio-Bibliographical Survey*, (Publications of the American Research Center in Egypt), Malibu, Ca.: Undena, 1983, xiii+98 pp. and 10 pls.**

Reviews:
Marina Tolmacheva in *Bulletin of the Middle East Studies Association*, ca. 1984-85, pp. 104–5.
J.N. Mattock in *Bulletin of the School of Oriental and African Studies* (University of London), 48:2 (1985), p. 423.

59 **E.S. Kennedy, Colleagues and Former Students, *Studies in the Islamic Exact Sciences*, Beirut: American University of Beirut, 1983 (co-editor with Mary Helen Kennedy).**

60 'A report on the Azhar Manuscript Library', *Newsletter of the American Research Center in Egypt*, no. 122 (Summer, 1983), pp. 41–50.

61 'The astronomy of the Mamluks', *ISIS* 74 (1983), pp. 531–55. [Repr. in **A-III** and ***Islam and Science***, III, pp. 317–341.]

62 'al-Khwārizmī and new trends in mathematical astronomy in the ninth century', *Occasional Papers on the Near East* (New York University, Hagop Kevorkian Center for Near Eastern Studies) 2 (1983).

63 'Nasṭūlus the astrolabist once again' (with Paul Kunitzsch), *Archives Internationales d'Histoire des Sciences* 33 (1983), pp. 342–3. [Repr. in **B-V**; see also no. 27.]

64 'Mathematical astronomy in medieval Yemen', in R.B. Serjeant & Ronald Lewcock, eds., *Sanʿā': An Arabian Islamic City*, London: World of Islam Festival Trust, 1983, pp. 34–5.

65 'Al-Bazdawī on the qibla in early Islamic Transoxania', *Journal for the History of Arabic Science* 7 (1983/1986), pp. 3–38. [Repr. in **E-IX**.]

66 A review of Heinrich Suter, *Die Mathematiker und Astronomen der Araber und ihre Werke*, Amsterdam: Oriental Press, 1982 reprint of the 1900 original text, in *Journal for the History of Astronomy* 14 (1983), pp. 62–3.

1984

67 'The astronomy of the Mamluks: A brief overview', *Muqarnas* 2 (1984), pp. 73–84.

68 'Architecture and astronomy: The ventilators of medieval Cairo and their secrets', *Journal of the American Oriental Society* 104 (1984), pp. 97–133. [A revised version is in ***SATMI*, VIIb**.]

1985

69* 'Five minor works of al-Khwārizmī' [in Russian], *Proceedings of the International Conference on Khorezmi, Tashkent and Urgench, 1983,* Tashkent, 1985, pp. 91–5.

70 'The sacred direction in Islam: A Study of the Interaction of Religion and Science in the Middle Ages', *Interdisciplinary Science Reviews* 10:4 (1985), pp. 315–28.

71 'Osmanische astronomische Handschriften und Instrumente', in *Türkische Kunst und Kultur aus osmanischer Zeit*, 2 vols., Recklinghausen: Aurel Bongers, 1985, II, pp. 373–8. [Repr. in **B-XIV**.]

72 'Astronomy for landlubbers and navigators: The case of the Islamic Middle Ages', *Revista da Universidade de Coimbra* 32 (1985), pp. 211–23.

73 'The medieval Yemeni astrolabe in the Metropolitan Museum of Art in New York', *Zeitschrift für Geschichte der arabisch-islamischen Wissenschaften* 2 (1985), pp. 99–122, and 4 (1987/88), pp. 268–9 (corrections). [Repr. in **B-II** and republished in ***SATMI*, XIVa**.]

74 A review of Paul Kunitzsch, *Glossar der arabischen Fachausdrücke in der mittelalterlichen europäischen Fachliteratur*, Göttingen, 1983, in *ISIS* 76 (1985), p. 435.

75 A review of Ahmad Saeed Khan, *A Bibliography of the Works of Abu'l-Raihan al-Biruni*, New Delhi, 1982, in *Ghanita-Bhāratī* 7 (1985), pp. 43–4.

76 A review of Ali Abdallah Al-Daffa & John S. Stroyls, *Studies in the Exact Sciences in Medieval Islam*, New York, N.Y., 1984, in *Bulletin of the Middle East Association of North America* 19 (1985), pp. 243–5. [Reviewed for Islamicists – see also no. 85.]

1986

77 ***A Catalogue of the Scientific Manuscripts in the Egyptian National Library* [in Arabic], vol. 2: Descriptive catalogue arranged chronologically according to subjects – Indexes of authors and titles, Cairo: General Egyptian Book Organization, 1986, ix + 1299 pp.** [See no. 47.]

78 ***A Survey of the Scientific Manuscripts in the Egyptian National Library*, (Publications of the American Research Center in Egypt), Winona Lake, Ind.: Eisenbrauns, 1986, 331 pp.** [Based on nos. 47 and 77, and arranged as a supplement to the standard bio-bibliographical literature. See no. 242 for an index.]

79 **[A] *Islamic Mathematical Astronomy*, London: Variorum, 1986.** Contents:

I Some reflections on the history of Islamic astronomy;
II On the astronomical tables of the Islamic Middle Ages (no. 14);
III The astronomy of the Mamluks (no. 61);
IV Mathematical astronomy in medieval Yemen (no. 35);
V A double-argument table for the lunar equation attributed to Ibn Yūnus (no. 7);
VI Ibn al-Majdī's tables for calculating ephemerides (no. 46);
VII Some astronomical observations from thirteenth-century Egypt (no. 52);
VIII Indian astronomy in fourteenth-century Fez: The versified *Zīj* of al-Qusanṭīnī (no. 56);
IX Ibn Yūnus' *Very Useful Tables* for reckoning time by the sun (no. 4);
X Astronomical timekeeping in fourteenth-century Syria (no. 24);
XI al-Khalīlī's auxiliary tables for solving problems of spherical astronomy (no. 3);
XII Astronomical timekeeping in Ottoman Turkey (no. 36);
XIII al-Khalīlī's *qibla* table (no. 13);
XIV On medieval Islamic multiplication tables (no. 8);
XV Supplementary notes on medieval Islamic multiplication tables (no. 32);
XVI A handlist of the Arabic and Persian astronomical manuscripts in the Maharaja Mansingh II Library in Jaipur (no. 45);
XVII Islamic mathematics and astronomy. An essay review of the chapters on mathematics and astronomy in S.H. Nasr, *Islamic Science: An Illustrated Study* (no. 30);
XVIII Islamic mathematics. A review of A.A. Daffa, *The Muslim Contribution to Mathematics* (no. 41);

Addenda and corrigenda; indexes.
[See no. 132 for the 2nd edn.]

Reviews:
Sonja Brentjes in *Historia Mathematica* 16 (1989), p. 295.
Jan Hogendijk in *Mathematical Reviews* (1989), no. 89e:01053.
F. Jamil Ragep in *Nuncius – Annali di Storia della Scienza* 6 (1991), pp. 211–13.

80 ***From Deferent to Equant: Studies in the History of Science in the Ancient and Medieval Near East in Honor of E.S. Kennedy* (co-editor with George Saliba), *Annals of the New York Academy of Sciences* (500), 1986, xv + 771 pp.**

81 'Some early Islamic tables for determining lunar crescent visibility', in D.A. King & George Saliba, eds., *From Deferent to Equant*: *Studies in the History of Science in the Ancient and Medieval Near East in Honor of E.S. Kennedy*, *Annals of the New York Academy of Sciences* 500 (1986), pp. 185–225. [Repr. in **C-II**.]

82 'Some Ottoman schemes of sacred geography', *Proceedings of the II. International Symposium on the History of Turkish and Islamic Science and Technology, Istanbul, 1986*, 2 vols., Istanbul: Istanbul Technical University, 1986, I, pp. 45–57.

83 'The earliest Islamic mathematical methods and tables for finding the direction of Mecca', *Zeitschrift für Geschichte der arabisch-islamischen Wissenschaften* 3 (1986), pp. 82–149. with corrections listed *ibid.* 4 (1987/88), p. 270. [Repr. in **C-XIV**.]

84 A review of Galina P. Matvievskaya & Boris A. Rosenfeld, *Mathematicians and Astronomers of the Islamic Middle Ages (VIII-XVII Centuries) and their Works* [in Russian], in *Historia Mathematica* 13 (1986), pp. 306–8.

85 A review of Ali Abdallah Al-Daffa & John S. Stroyls, *Studies in the Exact Sciences in Medieval Islam*, New York, N.Y., 1984, in *Historia Mathematica* 13 (1986), pp. 303–6. [Reviewed for historians of mathematics – see also no. 76.]

86 A review of Sharon Gibbs & George Saliba, *Planispheric Astrolabes from the National Museum of American History*, Washington, D.C., 1984, in *ISIS* 77 (1986), pp. 711–13.

1987

87 **[B]** ***Islamic Astronomical Instruments*, London: Variorum, 1987.** [See no. 163 for a reprint.] Contents:

I Astronomical instrumentation in the medieval Near East;
II The medieval Yemeni astrolabe in the Metropolitan Museum of Art in New York (no. 73);
III The origin of the astrolabe according to the medieval Islamic sources (no. 48);
IV A note on the astrolabist Nasṭūlus/Basṭūlus (no. 27);
V Nasṭūlus the astrolabist once again (no. 63);
VI The ᶜAbd al-A'imma astrolabe forgeries (no. 2);
VII On the early history of the universal astrolabe in Islamic astronomy and the origin of the term *shakkāzīya* in medieval scientific Arabic (no. 33);
VIII The astrolabe of ᶜAlī al-Wadāᶜī;
IX The astronomical instruments of Ibn al-Sarrāj: A brief survey;
X An analog computer for solving problems of spherical astronomy: The *shakkāzīya* quadrant of Jamāl al-Dīn al-Māridīnī (no. 10);
XI New light on the *Zīj al-ṣafā'iḥ* of Abū Jaᶜfar al-Khāzin (no. 43);
XII Ibn al-Shāṭir's *Ṣandūq al-yawāqīt*: An astronomical compendium (no. 21);
XIII An Islamic astronomical instrument: A review of W. Brice, C. Imber & R. Lorch, *The Dā'ire-yi Muᶜaddel of Seydī ᶜAlī Re'īs* (no. 42);
XIV Osmanische astronomische Handschriften und Instrumente (no. 71);
XV Three sundials from Andalusia (no. 26);
XVI Le cadran solaire de la mosquée d'Ibn Ṭūlūn au Caire (no. 28);
XVII The sundial on the West Wall of the Madrasa of Sultan Qāytbāy in Jerusalem (no. 38);
XVIII A fourteenth-century Tunisian sundial for regulating the times of Muslim prayer (no. 20);
XIX Ibn Yūnus and the pendulum: A history of errors (extract of no. 34);
XX Medieval mechanical devices: A review of D.R. Hill, *The Book of Knowledge of Ingenious Mechanical Devices* (no. 17);
XXI On Arabic water-clocks: A review of D.R. Hill, *On the Construction of Water-Clocks* (no. 23);

XXII Islamic geomancy – A review of E. Savage-Smith & M.B. Smith, *Islamic Geomancy and a Thirteenth-Century Divinatory Device* (no. 57);
Addenda and corrigenda; indexes.

Reviews:
E.S. Kennedy in *Annals of Science* 45 (1988), pp. 544–5.
Sharon Gibbs Thibodeau in *ISIS* 81 (1990), pp. 101–2.

88 'The astrolabe of ᶜAlī al-Wadāᶜī' (previously unpublished), in **B-VIII**.

89 'The astronomical instruments of Ibn al-Sarrāj' (previously unpublished), in **B-IX**. [See ***SATMI*, XIVb-5.1** for a more detailed description.]

90 'Science in medieval Syria', in Harvey Weiss, ed., *Ebla to Damascus: Art and Archaeology of Ancient Syria*, Washington, D.C.: Smithsonian Institution Traveling Exhibition Service, 1985, pp. 497–507.

91 'Universal solutions in Islamic astronomy', in J. Lennart Berggren & Bernard R. Goldstein, eds., *From Ancient Omens to Statistical Mechanics: Essays on the Exact Sciences Presented to Asger Aaboe, Acta Historica Scientiarum Naturalium et Medicinalium* (Copenhagen) 39 (1987), pp. 121–32. [Repr. in **C-VI**; a revised version is in ***SATMI*, VIa**.]

92* 'Astronomie im mittelalterlichen Yemen', in Werner Daum, ed., *Jemen*, Innsbruck: Pinguin & Frankfurt am Main: Umschau, 1987, pp. 276–81 and 297–302. [See no. 106 for the original.]

93 'Makka. iv. As centre of the world' [sacred geography], *The Encyclopaedia of Islam*, new edition, vol. VI, fascs. 101–2, Leiden: E.J. Brill, 1987, pp. 180–87. [Repr. in **C-X**.]

94 A review of Edward S. Kennedy, David Pingree & Fuad Haddad, *The Book of Reasons behind Astronomical Tables (Kitāb fī ᶜilal al-zījāt) by ᶜAlī ibn Sulaymān al-Hāshimī*, in *Journal for the History of Astronomy* 18 (1987), pp. 284–6.

95 A review of Živa Vesel, *Les encyclopédies persanes : Essai de typologie et de classification des sciences*, Paris, 1986, in *Bulletin of the Middle East Studies Association of North America* 21 (1987), pp. 115–16.

1988

96 'Universal solutions to problems of spherical astronomy from Mamluk Egypt and Syria', in Farhad Kazemi & Robert B. McChesney, eds., *A Way Prepared: Essays on Islamic Culture in Honor of Richard Bayly Winder*, New York: New York University Press, 1988, pp. 153–84, repr in **C-VII**. [A revised version is in ***SATMI*, VIb**.]

97 'A medieval account of algebra before al-Khwārizmī', *al-Masāq: Studia Arabo-Islamica Mediterranea* 1 (1988), pp. 25–32. [See now no. 258.]

98 'Ibn Yūnus on lunar crescent visibility', *Journal for the History of Astronomy* 19 (1988), pp. 155–68. [Repr. in **C-III**.]

99 A review of Charles Pellat, *Cinq calendriers égyptiens*, Cairo, 1986, in *Journal of the American Research Center in Egypt* 25 (1988), pp. 252–3.

1989

100 'Some Arabic copies of Vettius Valens' table for calculating the duration of life', in Gerhard Endress, ed., *Symposium Graeco-Arabicum II*, Amsterdam: B.R. Grüner, 1989, pp. 25–8. [An introduction to the materials presented in no. 224.]

101 'al-Marrākushī', *The Encyclopaedia of Islam*, new edition, vol. VI, fascs. 107–8, Leiden: E.J. Brill, 1989, p. 598.

102 'Maṭlaᶜ' [rising-points], *The Encyclopaedia of Islam*, new edition, vol. VI, fascs. 111–12, Leiden: E.J. Brill, 1989, pp. 839–40. [Repr. in **C-XI**.]

103 'Maṭāliᶜ' [ascensions], *The Encyclopaedia of Islam*, new edition, vol. VI, fascs. 111–12, Leiden: E.J. Brill, 1989, pp. 792–4.

104 A review of Anton H. Heinen, *Islamic Cosmology: A Study of as-Suyūṭī's al-Hay'a al-sanīya fi-l-hay'a al-sunnīya*, Beirut, 1982, in *Journal of the American Oriental Society* 109 (1989), pp. 124–7.

1990

105 'An overview of the sources for the history of astronomy in the medieval Maghrib', *Actes du 2e Colloque Maghrébin de l'Histoire des Mathématiques Arabes, Tunis, 1-3 Dec. 1988*, Tunis: Institut supérieur de l'Éducation et de la Formation continue, n.d. [*ca.* 1990], pp. 125–57. [See no. 196 for a revised, expanded version.]

106 'Astronomy in medieval Yemen', in Werner Daum, ed., *Yemen – 3000 Years of Art and Civilization in Arabia Felix*, Innsbruck: Pinguin & Frankfurt/Main: Umschau, n.d. [*ca.* 1990], pp. 300–308. [See also no. 92.]

107 'Astronomy' in M.J.L. Young, J.D. Latham & R.B. Serjeant, eds., *Religion, Learning and Science in the 'Abbasid Period*, (a volume of the *Cambridge History of Arabic Literature*), Cambridge, *etc.*: Cambridge University Press, 1990, pp. 274–89. [Submitted in 1975!]

108* 'Die Sterne weisen nach Mekka – Arabische Astronomie im Dienste des Islam', in Uwe Schultz, ed., *Scheibe, Kugel, Schwarzes Loch – Die wissenschaftliche Eroberung des Kosmos*, Munich: C.H. Beck, 1990, pp. 104–17. [Reprinted as a paperback Frankfurt am Main & Leipzig: Insel Verlag, 1996.]

109 'A survey of medieval Islamic shadow schemes for simple timereckoning', *Oriens* 32 (1990), pp. 191–249. [A new version is in ***SATMI*, III**.]

110 'Between Europe and China: aspects of the astronomical traditions of the lands of Islam', in Isaia Iannaccone & Adolfo Tamburello, eds., *Dall'Europa alla Cina: contributi per una storia dell'astronomia*, Naples: Istituto Universitario Orientale, 1990, pp. 55–66.

111 'Mayl' [declination], *The Encyclopaedia of Islam*, new edition, vol. VI, fascs. 113–14, Leiden: E.J. Brill, 1990, pp. 914–15.

112 'Mīḳāt. ii. Astronomical aspects' [time-keeping], *The Encyclopaedia of Islam*, new edition, vol. VII, fascs. 115–16, Leiden: E.J. Brill, 1990, pp. 27–32. [Repr. in **C-V**.]

113 A review of Emilie Savage-Smith, *Islamicate Celestial Globes: Their History, Construction and Use*, Washington, D.C., 1985, in *ISIS* 81 (1990), pp. 762–4.

1991

114 'Lunar crescent visibility predictions in medieval Islamic ephemerides', in S. Seikaly, R. Baalbaki, P. Dodd, eds., *Quest for Understanding – Arabic and Islamic Studies in Memory of Malcolm H. Kerr*, Beirut: American University of Beirut, 1991, pp. 233–51. [Repr. in **C-IV**.]

115 'Medieval astronomical instruments: A catalogue in preparation', *Bulletin of the Scientific Instrument Society* 31 (Dec., 1991), pp. 3–7.

116* 'Strumentazione astronomica nel mondo medievale islamico', in Gerard L'E. Turner, ed., *Gli strumenti*, Turin: Giulio Einaudi, 1991, pp. 154–89 and 581–5. [See no. 125 for a summary and ***SATMI*, X** for a revised version of the original English text.]

117 'Science in the service of religion: The case of Islam', *Impact of Science on Society* (UNESCO), no. 159 (1991), pp. 245–62. [Repr. in **C-I**; also published in French, Portuguese, Russian, Chinese, Japanese and Korean. See nos. 143 and 164 for German and Italian versions, respectively.]

118 '[Yemeni astrolabe, dated 1291]', in Richard Ettinghausen *et al.*, *Islamische Kunst – Meisterwerke aus dem Metropolitan Museum of Art, New York* [in German and English], Berlin: Staatliche Museen Preußischer Kulturbesitz, 1981, pp. 146–7.

119 'Mizwala' [sundial], *The Encyclopaedia of Islam*, new edition, vol. VII, fascs. 117–18, Leiden: E.J. Brill, 1991, pp. 210–11. [Repr. in **C-VIII**.]

120 A review of Mohammed Ilyas, *Astronomy of Islamic Times for the Twenty-First Century*, London & New York, 1988, in *ISIS* 82 (1991), pp. 348–9.

121 A review of David Pingree, *The Astronomical Works of Gregory Chioniades*, I: *The Zīj al-ᶜAlā'ī*, Amsterdam, 1985–86, and Alexander Jones, *An Eleventh-Century Manual of Arabo-Byzantine Astronomy*, Amsterdam, 1987, in *ISIS* 82 (1991), pp. 116–18.

1992

122 'Qibla charts, qibla maps, and related instruments' (with Richard Lorch), a chapter in J.B. Harley & David Woodward, eds., *The History of Cartography*, vol. 2, book 1: *Cartography in the Traditional Islamic and South Asian Societies*, Chicago, Ill. & London: The University of Chicago Press, 1992, pp. 189–205. [See now no. 194.]

123* 'Los cuadrantes solares andalusíes', in Juan Vernet, Julio Samsó, *et al.*, eds., *El legado científico andalusí*, Madrid: Ministerio de Cultura, 1992, pp. 89–102.

124 '[Andalusī astronomical instruments]', in Jerrilynn D. Dodds, ed., *Al-Andalus – The Art of Islamic Spain*, New York: The Metropolitan Museum of Art, 1992, pp. 376–83.

125 'Some remarks on Islamic astronomical instruments', *Scientiarum Historia* (Brussels) 18:1 (1992), pp. 5–23.

126 'Astronomical instruments between East and West' [summary], *Medium aevum quotidianum* (Krems) 27 (1992), pp. 125–30. [See no. 148 for the full version.]

127+ 'Die Astrolabiensammlung des Germanischen Nationalmuseums', in Gerhard Bott, ed., *Focus Behaim-Globus*, 2 vols., Nuremberg: Germanisches Nationalmuseum, 1992, I, pp. 101–14, and II, pp. 568–602, 640–43.

128+ 'Weltkarten zur Ermittlung der Richtung nach Mekka', *ibid.*, I, pp. 167–71, and II, 686–91. [See now no. 194.]

129 'Some remarks on Islamic scientific manuscripts and instruments and past, present and future research', in John Cooper, ed., *The Significance of Islamic Manuscripts*, London: Al-Furqan Islamic Heritage Foundation, 1992, pp. 115–44. [Arabic version on pp. 159–93 of the Arabic translation *Ahammiyyat al-makhṭūṭāt al-islāmiyya*, published simultaneously.]

130 'The ciphers of the monks and the astrolabe of Berselius reconsidered', in Sergei S. Demidov, Menso Folkerts, David A. Rowe & Christoph J. Scriba, eds., *Amphora – Festschrift für Hans Wussing zum 65. Geburtstag*, Basel, Boston, Mass. & Berlin: Birkhäuser, 1992, pp. 375–88. [See now no. 203.]

1993

131 **[C] *Astronomy in the Service of Islam*, Aldershot (U.K.): Variorum, 1993.** Contents:

I Science in the service of religion – the case of Islam (no. 117);
II Some early Islamic tables for determining lunar crescent visibility (no. 81);
III Ibn Yūnus on lunar crescent visibility (no. 98);
IV Lunar crescent visibility predictions in medieval Islamic ephemerides (no. 114);
V Mīḳāt: astronomical timekeeping (no. 112);
VI Universal solutions in Islamic astronomy (no. 91);
VII Universal solutions to problems of spherical astronomy from Mamluk Egypt and Syria (no. 96);
VIII Mizwala: sundials (no. 119);
IX Ḳibla: sacred direction (no. 39);
X Makka: as the centre of the world (no. 93);
XI Maṭlaᶜ: astronomical rising-points (no. 102);
XII On the orientation of the Kaaba (no. 51);
XIII Astronomical alignments in medieval Islamic religious architecture (no. 53);
XIV The earliest Islamic mathematical methods and tables for finding the direction of Mecca (no. 83);

Addenda; indexes.

Reviews:
Anonymous in *Newsletter of the Organisation of the Islamic Conference Research Centre for Islamic History, Art, and Culture* (Istanbul) 35 (Dec., 1994), p. 32.
George Saliba in *ISIS* 86 (1995), pp. 97–8.
Paul Kunitzsch in *Zeitschrift der Deutschen Morgenländischen Gesellschaft* 146 (1996), pp. 597–8.
Raymond Mercier in *Journal for the History of Astronomy* 27 (1996), pp. 275–6.

132 ***Islamic Mathematical Astronomy*, 2nd revised edition, Aldershot (U.K.): Variorum, 1993.** [See no. 79 for the 1st edn.]

133 '1992 – A good year for medieval astronomical instruments', *Bulletin of the Scientific Instrument Society* No. 36 (March, 1993), pp. 17–18.

134* 'L'astronomie en Syrie à l'époque islamique', in Sophie Cluzan, Eric Delpont & Jeanne Mouliérac, eds., *Syrie, Mémoire et Civilisation*, Paris: Flammarion (Institut du Monde Arabe), 1993, pp. 386–95, and ['Instruments astronomiques syriens'], pp. 432–43 and 480, and pp. 485–7 (bibliography (confused)). [An expanded English version of the descriptions of instruments is in ***SATMI*, XIVb**.]

135 'Rewriting history through instruments: The secrets of a medieval astrolabe from Picardy', in Robert G. Anderson, James A. Bennett & Will F. Ryan, eds., *Making Instruments Count – Essays on Historical Scientific Instruments presented to Gerard L'Estrange Turner*, Aldershot (U.K.): Variorum, 1993, pp. 42–62. [Repr. in **D-III**; see now no. 203.]

136 'Some medieval astronomical instruments and their secrets', in Renato Mazzolini, ed., *Non-Verbal Sources in Science before 1900*, Florence: Leo S. Olschki, 1993, pp. 29–52.

137+ 'Über historische Modelle des Universums in drei und in zwei Dimensionen – die Armillarsphäre und das Astrolab' and descriptions of one early printed work and one manuscript as well as of various instruments, in Uwe Müller, ed., *450 Jahre Copernicus 'De revolutionibus' – Astronomische und mathematische Bücher aus Schweinfurter Bibliotheken*, Schweinfurt: Stadtarchiv (Veröffentlichung Nr. 9), 1993, pp. 123–37, then 167–9 (no. 20), 351–3 (no. 169), and 361–81 (nos. 177–81).

138+ 'Die Geschichte der Naturwissenschaften: ein wahrhaft interdisziplinäres Fach – Von Astrolabien bis zu Zahlensystemen, mit Exkursen in die Architektur, die Kunst, die Religion und die Volkskunde'. Text of a lecture delivered on the occasion of the celebration of the 50th anniversary of the Institut für Geschichte der Naturwissenschaften, Frankfurt am Main, November, 1993, photocopied.

1994

139+ 'Vergessene Schätze des Mittelalters – In Frankfurt erscheint der erste Katalog mittelalterlicher astronomischer Instrumente', *Forschung Frankfurt* (Johann Wolfgang Goethe-Universität, Frankfurt am Main), 11. Jahrgang (1993), Nr. 4, pp. 1–13. [Published in 1994.]

140 'Mathematics applied to aspects of religious ritual in Islam', in Ivor Grattan-Guinness, ed., *Companion Encyclopaedia of the History and Philosophy of the Mathematical Sciences*, 2 vols., London: Routledge & Kegan Paul, 1994, I, pp. 80–84.

141 'The astrolabe dedicated to Cardinal Bessarion by Regiomontanus in 1462' (with Gerard L'E. Turner), in Gianfranco Fiaccadori, ed., *Bessarione e l'Umanesimo*, Naples: Vivarium, 1994, pp. 340–67. [An early version of the next paper (prepared for the exhibition at the Biblioteca Nazionale Marciana, Venice, during 27.4–31.5.1994), with many printing errors. See the next entry.]

142 'The astrolabe presented by Regiomontanus to Cardinal Bessarion in 1462' (with Gerard L'E. Turner), *Nuncius*: *Annali di Storia della Scienza* (Florence) 9:1 (1994), pp. 165–206. [Repr. in Gerard L'E. Turner, *Renaissance Astrolabes and Their Makers*, Aldershot (U.K.): Variorum, 2003, IV, and in **D-XI**.]

143+ 'Astronomie im Dienste des Islam', in Anton von Gotstedter, ed., *Ad radices – Festband zum fünfzigjährigen Bestehen des Instituts für Geschichte der Naturwissenschaften Frankfurt am Main*, Stuttgart: Franz Steiner, 1994, pp. 99–124. [See no. 117 for the original English text.]

144+ 'Ein vergessenes Zahlensystem des mittelalterlichen Mönchtums', in Anton von Gotstedter, ed., *Ad radices – Festband zum fünfzigjährigen Bestehen des Instituts für Geschichte der Naturwissenschaften Frankfurt am Main*, Stuttgart: Franz Steiner, 1994, pp. 405–20. [See now no. 203.]

145* 'Le plus vieil astrolabe d'Europe', *Qantara – Cultures en mouvement* (Institut du Monde Arabe, Paris), No. 11 (April–June 1994), p. 51.

146 'Poor judgement at Nuremberg', *Bulletin of the Scientific Instrument Society*, No. 41 (June, 1994), p. 1. [Comments on the fact that the richest collection of historical scientific instruments in Germany was put back into storage after the 1992–93 exhibition 'Focus Behaim Globus'. The collection was later displayed in an appropriate fashion.]

147 'Folk astronomy in the service of religion: The case of Islam', in Clive L.N. Ruggles & Nicholas J. Saunders, eds., *Astronomies and Cultures*, Niwot, Co.: University Press of Colorado, 1993 [published 1994], pp. 124–38, with a summary in Clive L.N. Ruggles, ed., *Archaeoastronomy in the 1990s*, Loughborough (U.K.): Group D Publications, Ltd., 1993, p. 346.

148 'Astronomical instruments between East and West', in Harry Kühnel, ed., *Kommunikation zwischen Orient und Okzident* (*Sitzungsberichte der Österreichischen Akademie der Wissenschaften, Phil.-Hist. Klasse*, vol. 619, Veröffentlichungen des Instituts für Realienkunde des Mittelalters und der frühen Neuzeit, vol. 16), Vienna: Österreichische Akademie der Wissenschaften, 1994, pp. 143–98. [Repr. in **D-I**.]

149 'Some illustrations in Islamic scientific manuscripts and their secrets', in George N. Atiyeh, ed., *The Book in the Islamic World: The Written Word and Communication in the Middle East*, Albany, N.Y.: State University of New York, & Washington, D.C.: The Library of Congress, 1994, pp. 149–77. [Repr. in **E-III**.]

150 'Rub^{c}' [quadrant], *The Encyclopaedia of Islam*, new edition, vol. VIII, fascs. 139–40, Leiden: E.J. Brill, 1994, pp. 574–5.

151 'Ru'yat al-hilāl' [lunar crescent visibility], *The Encyclopaedia of Islam*, new edition, vol. VIII, fascs. 141–2, Leiden: E.J. Brill, 1994, pp. 649–50.

1995

152 'Applications of folk astronomy and mathematical astronomy to aspects of Muslim ritual', *The Arabist (Budapest Studies in Arabic)*, 13–14 (1995) (Alexander Fodor, ed., *Proceedings of the XIVth UEAI Congress, Budapest, 1988*, Part I), pp. 251–74.

153++ 'Himmel über Tanger', in Hans Joachim Tischleder, ed., *Tanger – Frankfurt – Zum Beispiel*, Frankfurt am Main: Deutsch-Marokkanische Kulturinitiative, 1995, pp. 168–77.

154 'Making instruments talk – Some medieval astronomical instruments and their secrets', *Bulletin of the Scientific Instrument Society*, No. 44 (March, 1995), pp. 5–12.

155 'Sāḳ' [leg in mathematics and astronomy], in *The Encyclopaedia of Islam*, new edition, vol. VIII, fascs. 143–4, Leiden: E.J. Brill, 1995, p. 872.

156 'The orientation of medieval Islamic religious architecture and cities', *Journal for the History of Astronomy* 26 (1995), pp. 253–74. [A new version is in ***SATMI*, VIIa**.]

157 'Samt' [direction, world-maps centred on Mecca], *The Encyclopaedia of Islam*, new edition, vol. VIII, fascs. 145–6, Leiden: E.J. Brill, 1995. [See now no. 194.]

158 'Donald Routledge Hill (1922–1994)', *Arabic Science and Philosophy* 5 (1995), pp. 297–302. [Obituary notice.]

159+ 'Aspekte angewandter Wissenschaften in Moscheen und Klöstern', *Berichte zur Wissenschaftsgeschichte (Organ der Gesellschaft für Wissenschaftsgeschichte)* 18 (1995), pp. 85–95 and 137–49. [The English original is in ***SATMI*, VIII**.]

160 'A forgotten Cistercian system of numerical notation', *Cîteaux – Commentarii Cistercienses* 46:3–4 (1995), pp. 183–217. [See now no. 203.]

161 'Early Islamic astronomical instruments in Kuwaiti collections', in Arlene Fullerton & Géza Fehérvári, eds., *Kuwait: Art and Architecture – A Collection of Essays*, Kuwait (no publisher stated), 1995, pp. 76–96.

162 'Shafaḳ' [twilight], in *The Encyclopaedia of Islam*, new edition, vol. IX, fascs. 149–50, Leiden: E.J. Brill, 1995, pp. 179–80.

163 ***Islamic Astronomical Instruments*, London: Variorum, 1987, reprinted Aldershot: Variorum, 1995.** [See no. 87.]

164* 'La scienza al servizio della religione: il caso dell'Islām', in Clelia Sarnelli Cerqua, Ornella Marra & Pier Giovanni Pelfer, eds., *La civiltà islamica e le scienze, Atti del Simposio Internazionale, Firenze, Palazzo Panciatichi, 23 Novembre 1991*, Florence: CUEN, 1995, pp. 129–50. [A translation of no. 117.]

165 A review of Tzvi Langermann, *Ibn al-Haytham's On the Configuration of the World*, New York & London, 1990, in *Journal of History of Astronomy* 26 (1995), pp. 84–5.

166 A review of Raymond d'Hollander, *L'astrolabe – les astrolabes du Musée Paul Dupuy*, Toulouse: Le Musée Paul Dupuy & L'association française de topographie, 1993, in *Annals of Science* 52 (1995), pp. 531–3.

167 A review of Daniel Martin Varisco, *Medieval Agriculture and Islamic Science – The Almanac of a Yemeni Sultan*, Seattle WA & London: University of Washington Press, 1994, in *Yemeni Update*, no. 36 (Winter/Spring 1995), pp. 36 and 45.

1996

168 'Astronomy in Islamic society: Qibla, gnomonics and timekeeping', in Rushdi Rashed, ed., in collaboration with Régis Morelon, *Encyclopaedia of the History of Arabic Science*, 3 vols., London & New York, N.Y.: Routledge, 1996, I, pp. 128–84. [Repr. in ***Islam and Science***, IV, pp. 391–450; see no. 179 for the French translation.]

169 'On the role of the muezzin and the *muwaqqit* in medieval Islamic society', in F. Jamil Ragep & Sally P. Ragep, with Steven J. Livesey, eds., *Tradition, Transmission, Transformation: Proceedings of Two Conferences on Premodern Science Held at the University of Oklahoma*, Leiden, New York, N.Y., & Cologne: E.J. Brill, 1996, pp. 285–346. [A new version is in ***SATMI*, V**.]

170 'Islamic astronomy', in Christopher Walker, ed., *Astronomy before the Telescope*, London: British Museum Press, 1996, pp. 143–74. [Repr. in **E-I**.]

171 'Shakkāziyya' [universal stereographic projections], in *The Encyclopaedia of Islam*, new edition, vol. IX, fascs. 151–2, Leiden: E.J. Brill, 1996, pp. 251–3.

172 'The neglected astrolabe', in Menso Folkerts, ed., *Mathematische Probleme im Mittelalter – Der lateinische und arabische Sprachbereich*, (Wolfenbütteler Mittelalter-Studien, Band 10), Wiesbaden: Otto Harrassowitz, for the Herzog August-Bibliothek, Wolfenbüttel, 1996, pp. 45–55.

173 'The medieval Catalan astrolabe of the Society of Antiquaries, London' (with Kurt Maier), in Josep Casulleras & Julio Samsó, eds., *From Baghdad to Barcelona. Studies in the Islamic Exact Sciences in Honour of Prof. Juan Vernet*, (Anuari de Filologia (Universitat de Barcelona) XXX (1996) B-2), 2 vols., Barcelona: Instituto 'Millás Vallicrosa' de Historia de la Ciencia Árabe, 1996, II, pp. 673–718. [Repr. in **D-IV**.]

174 'The earliest European astrolabe in the light of other early astrolabes', in Wesley Stevens, Guy Beaujouan & Anthony J. Turner, eds., *The Oldest Latin Astrolabe*, (*Physis – Rivista di storia della scienza* (Rome), Nuova serie 32:2–3 (1995) [published in September, 1996], pp. 189–450), pp. 359–404 (The remarks on pp. 384–5 beginning 'An astrolabe from ca. 1300 … ' should be marked 'Added in proof'.) [Repr. in **D-II**.]

1997

175 'Der Frankfurter Katalog mittelalterlicher astronomischer Instrumente', in Gerhard Endress & Remke Kruk, eds., *The Ancient Tradition in Christian and Islamic Hellenism – Studies on the Transmission of Greek Philosophy and Sciences dedicated to H.J. Drossaart Lulofs on his ninetieth birthday*, [contains the Proceedings of the Third Symposium Graeco-Arabicum held at the University of Leiden on March 26–8, 1991, and additional contributions], Leiden: Research School CNWS, School of Asian, African, and Amerindian Studies, 1997, pp. 145–64.

176 Articles in Helaine Selin, ed., *Encyclopaedia of the History of Science, Technology and Medicine in Non-Western Cultures*, Dordrecht: Kluwer Academic Publishers, 1997, as follows: 'Astronomical Instruments in the Islamic World' (pp. 86–8); 'Astronomy in the Islamic World' (pp. 125–34); 'Ibn al-Shāṭir' (pp. 412–14); 'Ibn Yūnus' (pp. 438–40); 'al-Khalīlī' (pp. 477–8); 'Maps and Mapmaking: Islamic World Maps Centered on Mecca' (pp. 577–8 and frontispiece); 'al-Māridīnī, Jamāl al-Dīn and Badr al-Dīn' (pp. 601–2); and 'Religion and Science in Islam, I: Technical and Practical Aspects' (pp. 857–61).

177 'Two Iranian world maps for finding the direction and distance to Mecca', *Imago Mundi – The International Journal for the History of Cartography* 49 (1997), pp. 62–82 and colour plate 1. [See now no. 194.]

178+ 'Astrolabe picard et numérotation cistercienne', *Musée des arts et métiers (Paris) – La revue*, Décembre 1997, pp. 47–55. [See now no. 203.]

179* 'Astronomie et société musulmane : qibla, gnomonique, mīqāt', in Rushdi Rashed, ed., in collaboration with Régis Morelon, *Histoire des sciences arabes*, 3 vols., Paris: Éditions du Seuil, 1997, I, pp. 173–215. [See no. 168 for the English original.]

1998

180 'Mamluk astronomy and the institution of the *muwaqqit*', in Thomas Phillipp & Ulrich Haarmann, eds., *The Mamluks in Egyptian Politics and Society*, Cambridge, *etc.*: Cambridge University Press, 1998, pp. 153–62. [Repr. in **E-V**.]

181 'Astrolabe' in Robert Bud & Deborah Warner, eds., *Instruments of Science: A Historical Encyclopedia*, New York, N.Y.: Garland Publishing, 1998, pp. 32–4.

182* 'Les instruments scientifiques en terre d'Islam', in Sophie Makariou, ed., *L'apparence des cieux – Astronomie et astrologie en terre d'Islam*, Paris: Éditions de la Réunion des musées nationaux, 1998, pp. 74–95. [Catalogue of an exhibition held at the Musée du Louvre during 18.6.–21.9.1998.]

183 'Taʿdīl ... ' [three short articles on planetary equations, interpolation and the equation of time] in *The Encyclopaedia of Islam*, new edition, vol. X, fascs. 165–6, Leiden: E.J. Brill, 1998, p. 55.

184 'Taḳī al-Dīn' in *The Encyclopaedia of Islam*, new edition, vol. X, fascs. 165–6, Leiden: E.J. Brill, 1998, pp. 132–3.

185 'Ṭāsa' [magnetic compass] in *The Encyclopaedia of Islam*, new edition, vol. X, fascs. 167–8, Leiden: E.J. Brill, 1998, pp. 312–13.

186 **Editor: Donald R. Hill, *Studies in Medieval Islamic Technology – From Philo to al-Jazari – From Alexandria to Diyar Bakr*, Aldershot: Variorum, 1998.**

187 'Two Iranian world-maps for finding the direction and distance to Mecca', in Ž. Vesel, H. Beikbaghban and B. Thierry de Crussol des Epesse, eds., *La science dans le monde iranien à l'époque islamique, Actes du colloque tenu à l'Université des Sciences Humaines de Strasbourg 6-8 juin 1995*, (Bibliothèque iranienne 50), Tehran: Institut Français de Recherche en Iran, 1998, pp. 3–24. [See now no. 194.]

1999

188 'Islamische Weltkarten mit Mekka als Mittelpunkt – Die Wiederentdeckung einer bemerkenswerten Tradition mittelalterlicher Kartographie', in Peter Eisenhardt, Frank Linhard and Kaisar Petanides, eds., *Der Weg der Wahrheit – Aufsätze zur Einheit der Wissenschaftsgeschichte – Festgabe zum 60. Geburtstag von Walter G. Saltzer*, Hildesheim: Olms, 1999, pp. 93–107. [See now no. 194.]

189 'The Toledo astrolabe', *Christie's Magazine* (London), April 1999, pp. 62–3. [See no. 215 for a detailed study.]

190 'The monumental Syrian astrolabe in the Maritime Museum, Istanbul', *Aydın Sayılı Özel Sayısı*, I–III, a special issue of *Erdem* (Ankara: Atatürk Kültür Merkezi), in three parts (9:25–27), Ankara: Türk Tarih Kurumu Basımevi, 1996–97, II, pp. 729–35 and 10 plates. [A contribution to a memorial volume for Professor Dr. Aydin Sayılı, first seen in 1999.] [Republished in ***SATMI*, XIVc**.]

191 'Wine-gauging at Damme: The evidence of a late medieval manuscript' (co-author with Ad Meskens, Germain Bonte, Jacques de Groote and Mieke de Jonghe), *Histoire et mesure* (Paris: C.R.H.-C.N.R.S.) 14 (1999), pp. 51–77. [See also no. 203.]

192 'Bringing astronomical instruments back to earth: The geographical data on medieval astrolabes (to *ca.* 1100)', in Arjo Vanderjagt & Lodi Nauta, eds., *Between Demonstration and Imagination: Essays in the History of Science and Philosophy Presented to John D. North*, Leiden: E.J. Brill, 1999, pp. 3–53. [A new version is in ***SATMI*, XVII**.]

193 'Time and space in Islam', in Kristen Lippencott, ed., *The Story of Time*, London: Merrell Holberton Publishers, in association with the National Maritime Museum, Greenwich, 1999, pp. 56–9.

194 ***World-Maps for finding the direction and distance to Mecca: Innovation and tradition in Islamic science*, Leiden: Brill, and London: Al-Furqan Islamic Heritage Foundation, 1999, xxix + 638 pp.** [See no. 177 for a summary, and ***SATMI*, VIIc** for new information on the mathematics behind the grids on these maps.] Contents:

I First orientations ...
  1 Aspects of Islamic science;
  2 The determination of the sacred direction in Islam;
  3 The main sources of Safavid mathematical geography;
II New discoveries – Two Mecca-centred world-maps from Safavid Islam ...
  4 The instruments on which the maps are engraved;
  5 The geographical data on the maps;
  6 The cartographic grids;
  7 The makers of the Safavid instruments;
  8 Traces of European influence on the instruments;
  9 Further reflections on Mecca-centred world-maps;
  10 Epilogue // Appendixes ...
    A Reconstruction of the Timurid geographical table;
    B The geographical information on various Safavid instruments;
    C Localities and coordinates featured on the Safavid world-maps;
    D The geographical table of al-Khāzinī;
    E Miscellaneous Iranian sources;
    F Miscellaneous Egyptian, Syrian and Ottoman Turkish sources;
    G Indexes of place-names.

Reviews:

J. Lennart Berggren in *Journal of the American Oriental Society* 121:3 (2001), pp. 512–14.

Charles Burnett in *Annals of Science* 59 (2002), pp. 328–9.

Emilia Calvo in ***Suhayl*** *– International Journal for the History of the Exact and Natural Sciences in Islamic Civilisation* (Barcelona) 1 (2000), pp. 363–6.

Benno van Dalen in *Journal of the Royal Asiatic Society* (London), Series 3, 12 (2002), pp. 371–3.

Elly Dekker, 'Cartographic Grids from Iran: An Early Version of the Retro-Azimuthal Orthographic Projection?', *The Cartographical Journal* 37 (2000), pp. 109–16. [Dekker cannot accept that the ellipse segments representing the latitude curves on the grid are approximated by arcs of circles or that the inspiration for the grids is Islamic.]

Owen J. Gingerich in *History of Science* 38 (2000), pp. 245–7. [Misunderstands the structure of the maps and the grids and proposes a construction for the grids that is invalid.]

Jan Hogendijk in *Historia Mathematica* 30 (2003), pp. 85–7. [Mentions newly-discovered materials on the mathematics underlying the grids from early Islamic sources.]

Muzaffar Iqbal in *Islamic & Science* 1:1 (2003), pp. 135–42. [A Muslim perspective on a Western 'positivist' operation.]

John D. North in *Bibliotheca Orientalis* (Leiden), 57 (2000), cols. 747–50. [Raises problems partly solved by the materials presented in ***SATMI*, XIIa** and **XIIb** (universal horary quadrant and universal horary dial) and **VIIc** (the mathematics underlying the grids on the Mecca-centred world-maps).]

F. Jamil Ragep in *Journal for the History of Astronomy* 32:2 (2001), pp. 171–2. [Unfortunately mentions the procedure for constructing the grids proposed by Owen Gingerich without realizing that it is invalid.]

Roser Puig in *ISIS* 92 (2001), pp. 360–62.

George Saliba in *Mathematical Reviews* (2001) (accessible on the Internet under MathSciNet, No. 2001h:01008, pp. 1–5).

Emilie Savage-Smith in *Bulletin of the Scientific Instrument Society* 66 (2000), pp. 32–5. [Much confused.]

Roberto Tottoli in *Quaderni di Studi Arabi* 18 (2000), pp. 239–40.

195 'Aspects of Fatimid astronomy: From hard-core mathematical astronomy to architectural orientations in Cairo', in Marianne Barrucand, ed., *L'Égypte Fatimide: son art et son histoire – Actes du colloque organisé à Paris les 28, 29 et 30 mai 1998*, Paris: Presses de l'Université de Paris-Sorbonne, 1999, pp. 497–517. [Repr. in **E-IV**.]

196 'On the history of astronomy in the medieval Maghrib', in *Études philosophiques et sociologiques dédiées à Jamal ed-Dine Alaoui*, Université Sidi Mohamed Ben Abdallah, Publications de la Faculté des Lettres et des Sciences Humaines Dhar El Mahraz - Fès, N° Spécial 14 (Département de Philosophie, Sociologie et Psychologie), Fez, 1998 [published 1999], pp. 27–61. [A revised and updated version of no. 105 – see now no. 239 and **E-VI**.]

2000

197 'Cataloguing medieval Islamic astronomical instruments', an essay review of Francis Maddison and Emilie Savage-Smith, *Science, Tools and Magic*, vol. XII (in 2 parts) of *The Nasser D. Khalili Collection of Islamic Art*, general editor Julian Raby, The Nour Foundation, London, in association with Azimuth Editions and Oxford University Press, 1997, in *Bibliotheca Orientalis* (Leiden) 57 (2000), cols. 247–58.

198 'The culmination of Islamic mathematical geography: World-maps for finding the direction and distance to Makkah', *Newsletter / Hadīth al-Dār*, Dar al-Athar al-Islamiyyah, Kuwait, 7 (1997) [published 2000], pp. 19–21 (English / Arabic). [See no. 194.]

199 'The star-names on three 14th-century astrolabes from Spain, France and Italy', in Menso Folkerts and Richard P. Lorch, eds., *Sic itur ad astra. Studien zur Geschichte der Mathematik und Naturwissenschaften. Festschrift für den Arabisten Paul Kunitzsch zum 70. Geburtstag*, Wiesbaden: Otto Harrassowitz, 2000, pp. 307–33. [Repr. in **D-VII**.]

200 'Islamic world-maps centred on Mecca: The rediscovery of a remarkable tradition of medieval cartography', in Ekmeleddin İhsanoğlu and Feza Günergun, eds., *Science in Islamic Civilisation – Proceedings of the International Symposia 'Science Institutions in Islamic Civilisation' and 'Science and Technology in the Turkish and Islamic World'*, Istanbul: Research Centre for Islamic History, Art and Culture (IRCICA), 2000, pp. 111–21. [See no. 194.]

201 'Too many cooks ... – A newly-rediscovered account of the first Islamic geodetic measurements', ***Suhayl*** – *International Journal for the History of the Exact and Natural Sciences in Islamic Civilisation* (Barcelona) 1 (2000), pp. 207–41. [Repr. in ***Islam and Science***, IV, pp. 451–85, and **E-X**.]

202 'Mathematical astronomy in Islamic civilisation', in Helaine Selin, ed., *Astronomy across Cultures: The History of Non-Western Astronomy*, Dordrecht, Boston & London: Kluwer Academic Publishers, 2000, pp. 585–613.

2001

203 ***The Ciphers of the Monks – A Forgotten Number Notation of the Middle Ages*, (Boethius – Texte und Abhandlungen zur Geschichte der Mathematik und der Naturwissenschaften, ed. Menso Folkerts, Band 44), Stuttgart: Franz Steiner Verlag, 2001, 506 pp.** [See nos. 160, 144 and 178 for summaries in English, German and French.] Contents:

I Introduction; the main varieties; the cultural context;
II The English ciphers; John of Basingstoke; the Acropolis shorthand; the Tironian notes;
III The horizontal ciphers of the Cistercians;
IV The ciphers on the astrolabe of Berselius, made in 14th-century Picardy;
V The French vertical ciphers in the manuscript sources;
VI The fate of the monastic ciphers in the Renaissance and thereafter;
VII The ciphers falling between the cracks of modern scholarship and emerging therefrom;

Apps:
A Bibliographical notes;
B The survival of the Roman numerals in medieval Europe;
C Ancient Greek and medieval alphanumerical notations;
D The introduction of the 'Hindu-Arabic' numerals in Europe;
E Sundry numeral notations and symbols in medieval and later sources;
F Aspects of medieval astronomy;
G The principle and use of the astrolabe;
H On medieval European astronomical instruments;
J The quatrefoil on medieval astrolabe retes;
K Astronomical instrumentation in Northern France in the 14th century;
L The Picard astrolabe with monastic ciphers;
M The Virgin of Berselius;
N Non-historical reflections on the ciphers.

Reviews:
Julio Samsó in ***Suhayl*** *– International Journal for the History of the Exact and Natural Sciences in Islamic Civilisation* 2 (2001) pp. 409–11.
Martin Hellmann, 'Neue kurzschriftgeschichtliche Erkenntnisse im Zusammenhang mit den "griechischen" und "chaldäischen" Zahlzeichen', published in *Archiv für Stenographie, Textverarbeitung, Bürotechnik* (Forschungs- und Ausbildungsstätte für Kurzschrift und Textverarbeitung, Bayreuth), 2002, with a longer version at: www.forschungsstaette.de/Rezensionen/King_2001.htm. [Contains important reflections on the connection between the ciphers as they appear in England in the 13th century and the 'Acropolis' shorthand notation from 4th-century-B.C. Athens.]
Eberhard Knobloch in *Mathematical Reviews* 2002, (accessible on the Internet under MathSciNet, No. 2002k:01013). [Announcement only.]
Jens Høyrup in *Centaurus* 42 (2000), pp. 246–7.
Catherine Eagleton in *British Journal for the History of Science*, 38 (2005), pp. 138–9.
Alain Boureau in *Histoire et mesure* 18 (2006), with several errors of fact and interpretation.

204 'Astronomical handbooks and tables from the Islamic world (750–1900): An interim report' (co-author with Julio Samsó, with a contribution from Bernard R. Goldstein), ***Suhayl*** *– International Journal for the History of the Exact and Natural Sciences in Islamic Civilisation* (Barcelona) 2 (2001), pp. 9–105. [A shorter illustrated version is in no. 208.]

205 'The astrolabe depicted in the intarsia of the *Studiolo* of Archduke Federico in Urbino', in *La scienza del Ducato di Urbino – The Science of the Dukedom of Urbino*, Flavio Vetrano, ed., Urbino: Accademia Raffaello, 2001, pp. 101–39. [The original version is now in **D-X**.]

206 A review of Ekmeleddin İhsanoğlu, ed., *Osmanlı astronomi literatürü tarihi (History of Astronomy Literature during the Ottoman Period)*, 2 vols., and *Osmanlı matematik literatürü tarihi (History of Mathematics Literature during the Ottoman Period)*, 2 vols., (IRCICA Studies and Sources on the History of Science, nos. 7–8), Istanbul: Islâm Tarih, Sanat ve Kültür Araştırma Merkezi (IRCICA), 1996 and 1999, in *ISIS* 92:2 (2001), pp. 357–9.

2002

207 A website featuring the table of contents of the Frankfurt catalogue of medieval instruments, listing these chronologically according to region:
www.uni-frankfurt.de/fb13/ign/instrument-catalogue.html

208 Article 'Zīdj' [= astronomical handbooks and tables] in *The Encyclopedia of Islam*, Leiden: Brill, 2002, vol. XI, fasc. 187, pp. 496–508. [Illustrated; a longer version is in no. 204.]

209* 'Weltkarten zum Auffinden der Richtung und Entfernung nach Mekka', *Spektrum Iran* (Kulturabteilung der Botschaft der Islamischen Republik Iran, Berlin) 15:1 (2002), pp. 5–10. [See no. 194.]

210 'A *vetustissimus* Arabic text on the *quadrans vetus*', *Journal for the History of Astronomy* 33 (2002), pp. 237–255. [Repr. in **D-VIII**; see also ***SATMI*, XIIa**.]

211 'Medieval monastic ciphers in Renaissance printed texts', in *Verfasser und Herausgeber mathematischer Texte der frühen Neuzeit*, Rainer Gebhardt, ed., Annaberg-Buchholz: Adam-Ries-Bund (Schriften, Band 14), 2002, pp. 51–62.

212 'A recently discovered sixteenth-century Spanish astrolabe' (by Roberto Moreno, with David A. King and Koenraad Van Cleempoel), *Annals of Science* 59 (2002), pp. 331–62. [In the published version, all three contributors are listed as co-authors.]

213 'Notes on Yemeni astronomy in the Rasulid period', an essay review of Daniel Martin Varisco and G. Rex Smith, eds., *The Manuscript of al-Malik al-Afḍal al-ᶜAbbās b. ᶜAlī b. Dā'ūd b. Yūsuf b. ᶜUmar b. ᶜAlī ibn Rasūl (d. 778/1377) – A Medieval Arabic Anthology from the Yemen*, [London?]: E.J.W. Gibb Memorial Trust, 1998, in *Yemen Update* 44 (2002), accessible at: www.aiys.org/webdate/kngrev.html.

2003

214 'The Renaissance of astronomy in Baghdad in the ninth and tenth centuries – A list of publications, mainly from the past 50 years', website at:
www.uni-frankfurt.de/fb13/ign/astronomy_in_baghdad.bibliography.html

215 'An astrolabe from 14th-century Christian Spain with inscriptions in Latin, Hebrew and Arabic – A unique testimonial to an intercultural encounter', ***Suhayl*** – *International Journal for the History of the Exact and Natural Sciences in Islamic Civilisation* (Barcelona) 3 (2002/03), pp. 9–156. [A new version is in ***SATMI*, XV**.]

216+ 'Ein vergessenes Zahlensystem des mittelalterlichen Mönchtums', in *Jahresbericht des Physikalischen Vereins Frankfurt am Main – 1994, 170. Vereinsjahr*, Frankfurt, 2003 [!], pp. 47–62. [Repr. from no. 144.]

217* 'L'astronomia al servizio dell'Islam / Astronomie im Dienste des Islam', in *Nel segno di Aldebaran / Im Zeichen von Aldebaran – L'Islam e la Scienza / Islam und Wissenschaft*, Bolzano, 2003, pp. 12–19. [Catalogue of an exhibition in the Centro Culturale Trevi, Bolzano, during 8.–28.03.2003.]

218* 'Zur Geschichte des Astrolabiums in der Welt des Islam', in *Das Astrolabium: Beobachtungs- und Rechengerät freisichtiger Himmelskunde*, Hermann Mucke, ed., Vienna: Österreichischer Astronomischer Verein (Internationale Fachabende), 2003, pp. 79–88.

219 A review of Gerald L'E. Turner, *Elizabethan Instrument Makers: The Origins of the London Trade in Precision Instrument Making*, Oxford: Oxford University Press, 2000, in *Journal of the History of Collections* (Oxford) 15:1 (2003), pp. 147–50.

220 'A remarkable Italian astrolabe from *ca.* 1300 – Witness to an ingenious Islamic tradition of non-standard astrolabes', in *MUSA MUSAEI: Studies on Scientific Instruments and Collections in Honour of Mara Miniati*, Marco Beretta, Paolo Galluzzi and Carlo Triarico, eds., Florence: Leo S. Olschki, 2003, pp. 29–52. [Repr. in **D-V**; a new version is in ***SATMI*, XIIId**.]

221 '14th-century England or 9th-century Baghdad? New insights on the origins of the elusive astronomical instrument called the *Navicula de Venetiis*', in *Astronomy and Astrology from the Babylonians to Kepler – Essays Presented to Bernard R. Goldstein on the Occasion of his 65th Birthday*, Peter Barker, Alan C. Bowen, José Chabás, Gad Freudenthal and Y. Tzvi Langermann, eds., 2 pts., *Centaurus* 45 (2003) and 46 (2004), I, pp. 204–26. [Repr. in **D-IX**; see also ***SATMI*, XIIb**.]

222 'Astronomy in the service of Islam', in *Cosmology through Time: Ancient and Modern Cosmologies in the Mediterranean Area – Conference Proceedings*, Astronomical Observatory of Rome, Monteporzio Catone, June 17–20, 2001, Sergio Colafrancesco and Giuliana Giobbi, eds., Milan: Mimesis, 2003, pp. 143–52.

223 'The cult of St. Wilgefortis in Flanders, Holland, England and France', in *Am Kreuz – Eine Frau: Anfänge – Abhängigkeiten – Aktualisierungen*, Sigrid Glockzin-Bever and Martin Kraatz, eds., in *Ästhetik – Theologie – Liturgik* (Münster: LIT Verlag), 26 (2003), pp. 55–97. [A revised version with many more illustrations is available: see X2 at the end of this list.]

2004

224 'A Hellenistic astrological table deemed worthy of being penned in gold ink: The Arabic tradition of Vettius Valens' auxiliary function for finding the length of life', in *Studies in the History of the Exact Sciences in Honour of David Pingree*, Charles Burnett, Jan P. Hogendijk, Kim Plofker and Michio Yano, eds., Leiden & Boston: Brill, 2004, pp. 666–714. [A new version is in **E-VII**.]

225* 'Astrolabis de la Catalunya medieval', in *La ciència en la història dels Països Catalans*, I: Dels àrabs al Renaixement, Barcelona: Institut d'Estudis Catalans, 2004, pp. 161–204.

226 'Reflections on some new studies on applied science in Islamic societies (8th–19th centuries)', *Islam & Science* 2:1 (2004), pp. 43–56. [Available on the Internet at www.highbeam.com (enter title at their site); repr. in ***Islam and Science***, III, pp. 3–16.]

227 'Towards a history from Antiquity to the Renaissance of sundials and other instruments for reckoning time by the sun and stars', an essay review of Hester Higton, ed., *Sundials at Greenwich – A Catalogue of the Sundials, Nocturnals, and Horary Quadrants in the National Maritime Museum*, Oxford: Oxford University Press, 2002, and *eadem*, *Sundials – An Illustrated History of Portable Dials*, London: Philip Wilson, 2001, *Annals of Science* 61:3 (2004), pp. 377–389. [Available at: http://www.tandf.co.uk/journals/articlecollections/annalsofscience/documents/scientific_instruments.pdf.]

228 'From inscriptions to context: Some Islamic astronomical instruments and their secrets', in *Text & Context in Islamic Societies*, Irene A. Bierman, ed., Reading (U.K.): Ithaca, 2004, pp. 87–130. [Repr. in **E-II**.]

229 'Islamic astronomical instruments and some newly-discovered examples of transmission to Europe', in *Mediterranean. Splendour of the Medieval Mediterranean. 13th-15th Centuries*, Elisenda Guedea, ed., Barcelona: Institut Europeu de la Mediterrània (IeMed) & Lunwerg Editores, 2004, pp. 400–423 and 606–7 (bibliography).

230 **[*SATMI*-1] *In Synchrony with the Heavens – Studies in Astronomical Timekeeping and Instrumentation in Islamic Civilization*, vol. 1: *The Call of the Muezzin. Studies I-IX*, (Islamic Philosophy, Theology and Science – Texts and Studies, vol. LV:1), Leiden & Boston: Brill, 2004, lvii + 930 pp.** Contents:

I A survey of tables for timekeeping by the sun and stars (previously unpublished);
II A survey of tables for regulating the times of prayer (previously unpublished);
III A survey of arithmetical shadow-schemes for time-reckoning (see no. 109);
IV On the times of prayer in Islam (previously unpublished);
V On the role of the muezzin and the muwaqqit in medieval Islamic societies (no. 169);
VIa Universal solutions in Islamic astronomy (see no. 91);
VIb Universal solutions from Mamluk Syria and Egypt (see no. 96);
VIIa On the orientation of medieval Islamic architecture and cities (see no. 156);
VIIb Architecture and astronomy: The ventilators of medieval Cairo and their secrets (see no. 68);
VIIc Safavid world-maps centred on Mecca (supplement to no. 194);
VIII Aspects of practical astronomy in mosques and monasteries (see no. 159);
IX When the night sky over Qandahar was lit only by stars ... ... (previously unpublished).

Reviews:
Imad-ad-Dean Ahmad, in *Archeoastronomy: The Journal of Astronomy in Culture* 19 (2005), p. 103.
Benno van Dalen in *Abstracta Iranica* 29 (2006).
Charles Burnett in *The Medieval Review* (2005).
Muzaffar Iqbal in *Islam & Science* 4:1 (2006), pp. 79–83.
George Saliba, an essay review 'Islamic astronomy at its best' in *Journal for the History of Astronomy* 37 (2006), pp. 233–8.
Mercè Comes in ***Suhayl*** – *International Journal for the History of the Exact and Natural Sciences in Islamic Civilisation* 8 (2008), pp. 278–80.

Julio Samsó in *Journal of Semitic Studies* 52 (2007), pp. 415–17.
Giorgio Strano in *Nuncius – Journal of the History of Science* 23:2 (2008), pp. 388–90 (also no. 231).

2005

231 **[*SATMI-2*] *In Synchrony with the Heavens – Studies in Astronomical Timekeeping and Instrumentation in Islamic Civilization*, vol. 2: *Instruments of Mass Calculation. Studies X-XVIII*, (Islamic Philosophy, Theology and Science – Texts and Studies, vol. LV:2), Leiden & Boston: Brill, 2005, lxxvi + 1066 pp.** Contents:

X Astronomical instrumentation in the Islamic world (previously unpublished);
XI An approximate formula for timekeeping (750–1900) (previously unpublished);
XIIa On the universal horary quadrant for timekeeping by the sun (previously unpublished);
XIIb On universal horary dials for timekeeping by the sun and stars (previously unpublished);
XIIIa The neglected astrolabe – A supplement to the standard literature on the favourite astronomical instrument of the Middle Ages (previously unpublished);
XIIIb The oldest astrolabe in the world, from 8th-century Baghdad (previously unpublished);
XIIIc Astrolabes from late-9th- and 10th-century Baghdad (previously unpublished);
XIIId A medieval Italian testimonial to a forgotten Islamic tradition of non-standard astrolabes (no. 220);
XIIIe The origin of the astrolabe according to medieval Islamic sources (see no. 48);
XIVa An astrolabe made by the Yemeni Sultan al-Ashraf (see no. 73);
XIVb Some astronomical instruments from medieval Syria (see no. 134);
XIVc A monumental astrolabe from 13th-century Damascus (see no. 190);
XIVd An astrolabe for the Sultan Ulugh Beg (previously unpublished);
XIVe Two astrolabes for the Ottoman Sultan Bāyazīd II (see no. 236);
XIVf Brief remarks on astronomical instruments from Muslim India (previously unpublished);
XIVg A universal astrolabe from 17th-century Lahore (previously unpublished);
XV An astrolabe from medieval Spain with inscriptions in Hebrew, Arabic and Latin (see no. 215);
XVI The geographical data on early medieval Islamic instruments (see no. 192);
XVII The quatrefoil as decoration on astrolabe retes (previously unpublished);
XVIII A checklist of Islamic astronomical instruments to *ca.* 1500, ordered chronologically by region (previously unpublished).

Reviews:
Benno van Dalen in *Abstracta Iranica* 29 (2006).
George Saliba, an essay review 'Islamic astronomy at its best' in *Journal for the History of Astronomy* 37 (2006), pp. 233–8.
Muzaffar Iqbal in *Islam & Science* 4:1 (2006), pp. 79–83.
Mercè Comes in ***Suhayl*** *– International Journal for the History of the Exact and Natural Sciences in Islamic Civilisation* 8 (2008), pp. 278–80.
Julio Samsó in *Journal of Semitic Studies* 52 (2007), pp. 415–17, also available at doi:10.1093/jss/fgm027
Giorgio Strano in *Nuncius – Journal of the History of Science* 23:2 (2008), pp. 388–90 (also no. 230).

232 'The sacred geography of Islam', in *Mathematics and the Divine – A Historical Study*, T. Koetsier and L. Bergmans, eds., Dordrecht: Elsevier, 2005, pp. 161–78. [Repr. in **E-VIII**.]

233a* 'Astronomie und Mathematik als Gottesdienst: Das Beispiel Islam', in Jochen Brüning and Eberhard Knobloch, eds., *Die mathematischen Wurzeln der Kultur – Mathematische Innovationen und ihre kulturellen Folgen*, Munich: Wilhelm Fink Verlag, 2005, pp. 91–123. [Translation not seen by the author before publication.]

233b* Article 'Kibla' in *Religion in Geschichte und Gegenwart*, 4. Auflage, Tübingen: RGG4, 2005.

2006

234 'An introduction to Ioannes Regiomontanus' acrostic, Cardinal Basileios 'Ioannes' Bessarion's agenda, and Piero Della Francesca's enigma' (based on two remarkable discoveries by Berthold Holzschuh), *Preprints of the Institute for the History of Science, Johann Wolfgang Goethe University, Frankfurt am Main*, 4th series, no. 2 (2006), *ca.* 50 pp. – available at: web.uni-frankfurt.de/fb13/ign/Code.htm.

235 'Astrolabes and angels, epigrams and enigmas:~ Regiomontanus, Cardinal Bessarion and Piero della Francesca *conveniunt in unum*' (based on two remarkable discoveries by Berthold Holzschuh), *Preprints of the Institute for the History of Science, Johann Wolfgang Goethe University, Frankfurt am Main*, 4th series, no. 1 (2006), *ca.* 300 pp. [See now no. 238.]

236 'Two astrolabes for the Ottoman Sultan Bayezit II', in *Essays in Honour of Ekmeleddin İhsanoğlu*, 2 vols., Istanbul: Research Centre for Islamic History, Art and Culture (IRCICA), 2006, I, pp. 439–59. [A new version is in ***SATMI***, **XIVe**.]

2007

237 Article 'Astrolabes, quadrants and computing devices', in *The Encyclopedia of Islam*, 3rd edn., Leiden: Brill, 2007.

238 ***Astrolabes and Angels, Epigrams and Enigmas – From Regiomontanus' Acrostic for Cardinal Bessarion to Piero della Francesca's* Flagellation of Christ *– An essay inspired by two remarkable discoveries by Berthold Holzschuh* (*Boethius – Texte und Abhandlungen zur Geschichte der Mathematik und der Naturwissenschaften*, Band 56, Menso Folkerts, ed.), Stuttgart: Franz Steiner, 2007, pp. xi+348 pp. with CD-ROM.** Contents:

I Cardinal Bessarion and his two astrolabes;
II Bessarion's new astrolabe decoded or Regiomontanus' agenda revealed;
III Piero's painting and Regiomontanus' acrostic;
IV A new hypothesis to explain Piero's painting;
V The universality of the painting.

Reviews:
Marco Böhlandt in *ISIS* 100 (2009), pp. 903–4.
Uta Lindgren in *Sudhoffs Archiv* 93:1 (2009), pp. 119–20.
Arianna Borrelli in *NTM – Berichte zur Wissesnschaftsgeschichte* 32:1 (2009), pp. 109–10.
Emmanuel Poulle in *Archives internationales d'histoire des sciences* 59 (2009), pp. 366–9.
Peter Schreiner in *Byzantinische Zeitschrift* 102 (2009), pp. 246–7.
Michael H. Shank in *Journal for the History of Astronomy* 42 (2011), pp. 391–403.

239 'On the history of astronomy in the medieval Maghrib', in *Études d'histoire des sciences arabes*, Mohammed Abattouy, ed., Casablanca: Fondation du Roi Abdul Aziz Al Saoud pour les Etudes islamiques et les Sciences humaines, 2007, pp. 175–218. [A new version of nos. 105 and 196; repr. in **E-VI**.]

240 'A world-map in the tradition of al-Bīrūnī (*ca.* 1040) and al-Khāzinī (*ca.* 1120) presented by Sirāj al-Dīn al-Sajāwandī (1210)', *Mélanges offerts à Hossam Elkhadem par ses amis et ses élèves*, Frank Daelemans, Jean-Marie Duvosquel, Robert Halleux & David Juste, eds., *Archives et bibliothèques de Belgique // Archief- en bibliotheekwezen in België* 83, Brussels, 2007, pp. 131–60. [Repr. in **E-XI**.]

241 Articles 'Ibn Yūnus', 'Ibn al-Shātir', 'al-Khalīlī', 'Abu 'l-ᶜUqūl', and 'Hasan Husayn and Muhammad Husayn', in Thomas A. Hockey, ed., *Biographical Encyclopedia of Astronomers*, 2 vols., Dordrecht: Kluwer Academic Publishers, 2007.

242 Benno van Dalen, with an introduction by DAK, 'An index of authors for the *Survey of Cairo Scientific Manuscripts*', ***Suhayl*** *– International Journal for the History of the Exact and Natural Sciences in Islamic Civilisation* 7 (2007), pp. 9–46. [See no. 78 for the original publication.]

243 'Henry C. King (1915–2005)', an obituary in *Journal for the History of Astronomy* 38:1 (2007), pp. 526–7.

244 'Asger Aaboe (1922–2007)', an obituary in *ISIS* 98 (2007), pp. 796–8.

2008

245 'Mathematical geography in 15th-century Egypt – An episode in the decline of Islamic science', *Islamic Thought in the Middle Ages – Studies in Text, Transmission and Translation, in Honour of Hans Daiber*, Anna Akasoy & Wim Raven, eds. (Islamic Philosophy, Theology and Science, Texts and Studies, Hans Daiber, ed., vol. LXXV), Leiden & Boston: Brill, 2008, pp. 319–44. [A new version is in **E-XII**.]

246 'An astrolabe from Einbeck datable *ca.* 1330', in *Mathematics Celestial and Terrestrial – Festschrift für Menso Folkerts zum 65. Geburtstag*, Joseph Dauben, Stefan Kirschner, Andreas Kühne, Paul Kunitzsch & Richard P. Lorch, eds., *Acta Historica Leopoldina*, Nummer 54, Halle (Saale): Deutsche Akademie der Naturforscher Leopoldina, 2008, pp. 161–78. [Repr. in **D-VI**.]

247 'An instrument of mass calculation made by Nasṭūlus in Baghdad *ca.* 900', ***Suhayl*** *– International Journal for the History of the Exact and Natural Sciences in Islamic Civilisation* 8 (2008), pp. 93–119, repr. in ***Islam and Science***, III, pp. 243–69.

248 'Islamic astronomical instruments and some examples of transmission to Europe', in Emilia Calvo, Mercè Comes, Roser Puig and Mònica Rius, eds., *A Shared Legacy – Islamic Science East and West – Homage to Professor J. Millàs Vallicrosa*, Barcelona: Universitat de Barcelona, Publicacions i edicions, 2008, pp. 321–61.

2009

249 Article 'Badīᶜ al-Asṭurlābī', in *Encyclopaedia of Islam*, 3rd edn, Leiden: Brill, 2009.

250 'The geometry of Piero's *Flagellation of Christ* and the geometry of the epigram on the astrolabe of Regiomontanus that inspired it', in Rocco Sinisgalli, ed., *L'arte della matematica nella prospettiva*, Foligno (PG): C.B. Cartei & Bianchi Editore, 2009, pp. 189–91 and 407–12 (illustrations).

2010

251 'Edward Stewart Kennedy, (1912–2009)', an obituary in *Journal for the History of Astronomy* 41 (2010), pp. 117–19.

252 'Astronomical instruments', in *Handbook of Medieval Studies*, Albrecht Classen, ed., Berlin and New York: De Gruyter, 2010, pp. 126–30.

253 Contributions to biographical notes on E.S. Kennedy and a complete bibliography of his works (the last with Benno van Dalen), in ***Suhayl*** *– International Journal for the History of the Exact and Natural Sciences in Islamic Civilisation* 9 (2009/10), pp. 185–214.

254 'An illustration of the Caliph al-Ḥākim together with his astronomer/astrologer Ibn Yūnus', in *Ismaili and Fatimid Studies in Honor of Paul E. Walker*, Bruce D. Craig, ed., Chicago: University of Chicago, Center for Middle East Studies, 2010, pp. 151–9.

2011

255 **[D]** ***Astrolabes from Medieval Europe*****, Farnham: Ashgate-Variorum, 2011.**

Contents:

I Astronomical instruments between East and West (no. 148);
II The earliest European astrolabe in the light of other early astrolabes (no. 174);
III Rewriting history through instruments: The secrets of a medieval astrolabe from Picardy (no. 135);
IV The medieval Catalan astrolabe of the Society of Antiquaries, London (no. 173);
V A remarkable Italian astrolabe from *ca.* 1300 – Witness to an ingenious Islamic tradition of non-standard astrolabes (no. 220);
VI An astrolabe from Einbeck datable *ca.* 1330 (no. 246);
VII The star-names on three 14th-century astrolabes from Spain, France and Italy (no. 199);
VIII A *vetustissimus* Arabic text on the *quadrans vetus* (no. 210);
IX 14th-century England or 9th-century Baghdad? New insights on the origins of the elusive astronomical instrument called the *Navicula de Venetiis* (no. 221);
X The astrolabe depicted in the intarsia of the *Studiolo* of Archduke Federico in Urbino (no. 205);
XI The astrolabe presented by Regiomontanus to Cardinal Bessarion in 1462 (no. 142);
XII An ordered list of European astrolabes up to *ca.* 1500 (previously unpublished).

Reviews:

Elly Dekker, 'King on Astrolabes', *Journal for History of Astronomy* 43:2 (2012), pp. 249–50.

Patrick Gauthier Dalché in *Francia-Recensio* 3 (2012), 2 pp.

256 A review of Julio Samsó, *Astronomy and Astrology in al-Andalus and the Maghrib*, Aldershot: Ashgate-Variorum, 2007, and *idem*, *Astrometeorología y astrología medievales*, Barcelona, 2008, in *Journal of Islamic Studies* (Oxford), 22:3 (2011), pp. 473–5, also available at doi: 10.1093/jis/etr068.

257 A review of Catherine Eagleton, *Monks, Manuscripts and Sundials – The Navicula in Medieval England*, Leiden & Boston: Brill, 2010, in *Speculum – A Journal of Medieval Studies*, October, 2011. [Illustration removed by reviews editor!]

258 'The invention of algebra in Zabid: Between legend and fact', in *Islamic Philosophy, Science, Culture, and Religion: Studies in Honor of Dimitri Gutas*, David Reisman and Felicitas Opwis, eds., Leiden: Brill, 2011, pp. 223–31.

259 'From a heavenly Arabic poem to an enigmatic Judaeo-Arabic astrolabe' (with Mohamed Abu Zayed and Petra Schmidl), ***Suhayl*** *– International Journal for the History of the Exact and Natural Sciences in Islamic Civilisation* 10 (2011), pp. 1–58.

2012

260 **[E] *Islamic Astronomy and Geography*, Farnham: Ashgate-Variorum, 2012.**

Contents:

I Islamic astronomy (no. 170);
II From inscriptions to context: Some Islamic astronomical instruments and their secrets (no. 228);
III Some illustrations in Islamic scientific manuscripts and their secrets (no. 149);
IV Aspects of Fatimid astronomy: From hard-core mathematical astronomy to architectural orientations in Cairo (no. 195);
V Mamluk astronomy and the institution of the *muwaqqit* (no. 180);
VI On the history of astronomy in the medieval Maghrib (see no. 239);
VII A Hellenistic astrological table deemed worthy of being penned in gold ink: The Arabic tradition of Vettius Valens' auxiliary function for finding the length of life (no. 224, newly formatted);
VIII The sacred geography of Islam (no. 232);
IX Al-Bazdawī and the qibla in early Islamic Transoxania (no. 65);
X Too many cooks ... – A new account of the earliest Muslim geodetic measurements (no. 201);
XI A world-map in the tradition of al-Bīrūnī (*ca.* 1040) and al-Khāzinī (*ca.* 1120) presented by Sirāj al-Dīn al-Sajāwandī (1210) (no. 240);
XII Mathematical geography in 15th-century Egypt – An episode in the decline of Islamic science (no. 245, newly formatted).

261 'An Ottoman astrolabe full of surprises', to appear in 2012 in a *Festschrift* for J. Lennart Berggren, edited by Nathan Sidoli and Glen Van Brummelen.

WORKS NOT PUBLISHED YET

X1 'Two newly-rediscovered Abbasid astrolabes', to appear in ***Suhayl*** *– International Journal for the History of the Exact and Natural Sciences in Islamic Civilisation.*

X2 'The woman on the cross: the bearded virgin St. Wilgefortis'. [A revised version of no. 223 with more illustrations; a French translation by Jeremy Nicklin is available.]

X3 'Astronomy in the service of Islam', to appear in *Handbook of Archaeoastronomy and Ethnoastronomy*, Clive Ruggles, ed., published by Springer Science and Business Media.

X4 A review of Gianni Ferrari, *Le meridiane dell'antico Islam*, 2nd edn., Modena, 2012, for *Journal for the History of Astronomy*.

X5 'The Two Traditions of Sacred Geography in Medieval Islamic Texts and their Influence on the Orientation of Islamic Religious Architecture', to appear in *The World of Islamic Art – Papers in Memory of Ernst J. Grube*, Doris Abouseif and Dalu Jones, eds.

I

Islamic Astronomy

The nature of Islamic astronomy

From the ninth century to the fifteenth, Muslim scholars excelled in every branch of scientific knowledge. In particular their contributions to astronomy and mathematics are impressive. There are an estimated 10,000 Islamic astronomical manuscripts and close to 1,000 Islamic astronomical instruments preserved in libraries and museums in the Near East, Europe and North America, but it is clear that even if all of them were properly catalogued and indexed – and we are still very far from this state of affairs – the picture that we could reconstruct of Islamic astronomy, especially for the eighth, ninth and tenth centuries, would be quite deficient. Most of the available manuscripts and instruments date from the later period of Islamic astronomy, that is, from the fifteenth to the nineteenth century, and although some of these are based or modelled on earlier works many of the early works are extant in unique copies and others have been lost almost without trace, that is, we sometimes know only of their titles. The thirteenth-century Syrian scientific biographer Ibn al-Qifṭī relates that the eleventh-century Egyptian astronomer Ibn al-Sanbadī heard that the manuscripts in the library in Cairo were being catalogued and so he went to have a look at the works relating to his field. He found 6,500 manuscripts relating to astronomy, mathematics and philosophy. Not one of these survives amongst the 2,500 scientific manuscripts preserved in Cairo today.

The surviving manuscripts thus constitute but a small fraction of those that were actually copied; nevertheless they preserve for us a substantial part of the Islamic scientific heritage, certainly enough of it for us to judge its level of sophistication. Only in the past few decades has the scope of the activity and achievements of Muslim scientists become apparent, and the days are long past when they were regarded merely as transmitters of superior ancient knowledge to ignorant but eager Europeans. Islamic astronomy is to be viewed on its own terms. The fact that only a small part of the available material, mainly Greek and Indian material in Arabic garb, was indeed transmitted to Europe is to be viewed as an accident of Islamic history. There is no need to apologise for using the expression 'Islamic astronomy'. Within a few decades of the death of the

I

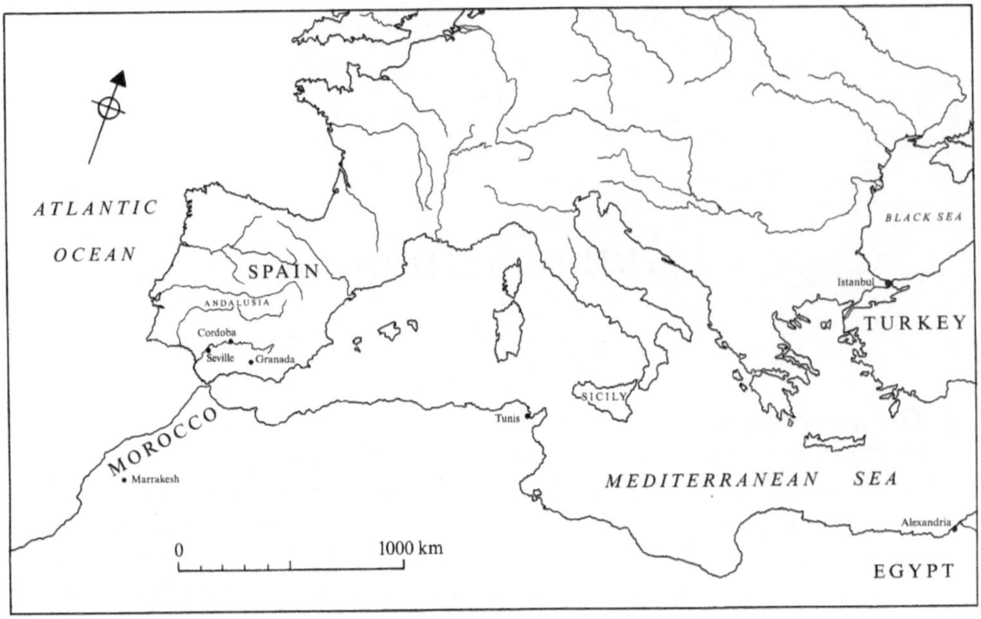

47 *(above and opposite)* The Islamic world. [Inserted by the original publisher.]

Prophet Muḥammad in 632 the Muslims had established a commonwealth stretching from Spain to Central Asia and India. They brought with them their own folk astronomy, which was then mingled with local traditions, and they discovered the mathematical traditions of the Indians, Persians and Greeks, which they mastered and adapted to their needs. Early Islamic astronomy was thus a pot-pourri of pre-Islamic Arabian star-lore and Indian, Persian and Hellenistic astronomy, but by the tenth century Islamic astronomy had acquired very distinctive characteristics of its own. A. I. Sabra labels this process 'appropriation and naturalisation'.

We should point out at the outset that astronomy flourished in Islamic society on two different levels: folk astronomy, devoid of theory and based solely on what one can see in the sky, and mathematical astronomy, involving systematic observations and mathematical calculations and predictions. Folk astronomy was favoured by the scholars of the sacred law (*fuqahā'*), not least because of various religious obligations that demanded a basic knowledge of the subject; these legal scholars generally had no time (or need) for mathematical astronomy. That discipline was fostered by a select group of scholars, most of whose activities and pronouncements were, except in the case of astrological predictions, of little interest to society at large.

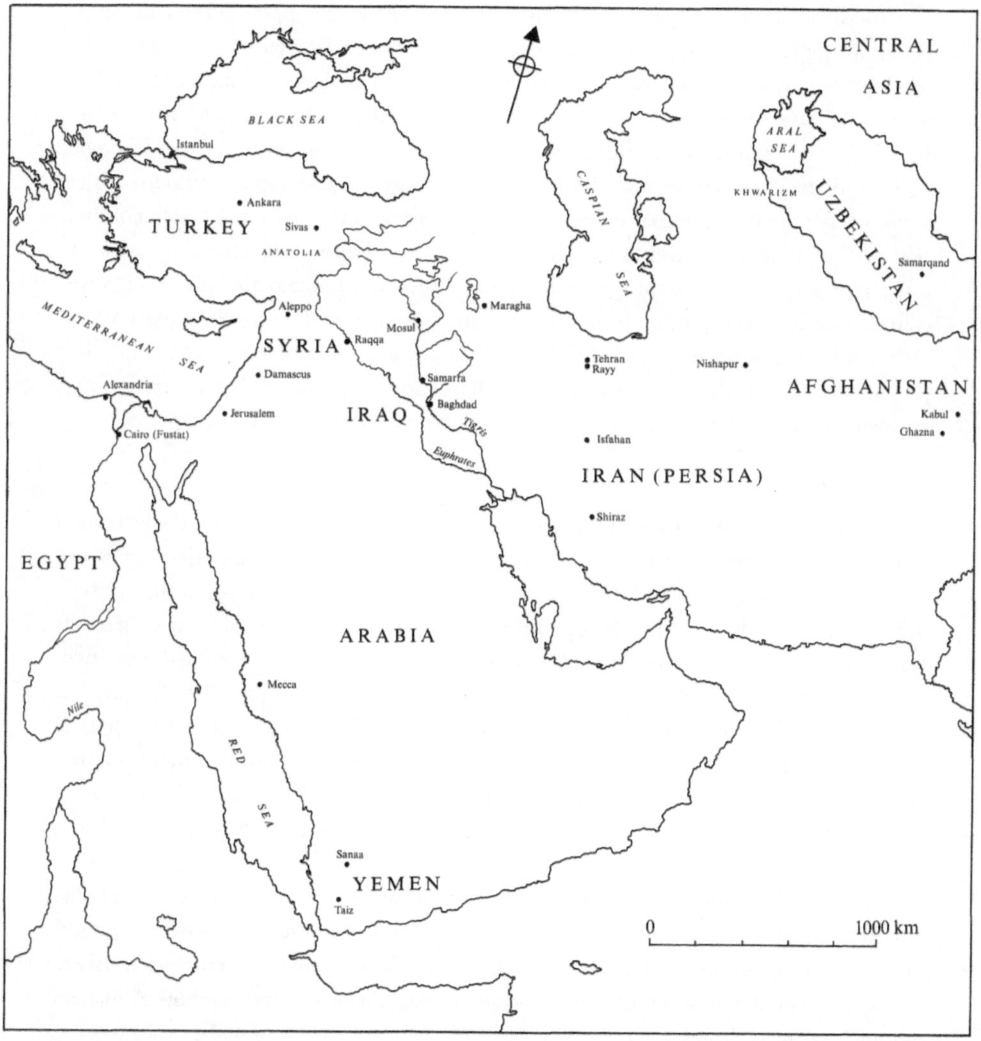
CENTRAL ASIA
BLACK SEA
Istanbul
Ankara
TURKEY
Sivas
ANATOLIA
CASPIAN SEA
ARAL SEA
KHWARIZM
UZBEKISTAN
Samarqand
Maragha
Aleppo
Mosul
MEDITERRANEAN SEA
SYRIA
Raqqa
Damascus
Samarra
Tehran
Rayy
Nishapur
AFGHANISTAN
Alexandria
Baghdad
Jerusalem
IRAQ
Tigris
Kabul
Ghazna
Cairo (Fustat)
Isfahan
Euphrates
IRAN (PERSIA)
Shiraz
EGYPT
ARABIA
Mecca
Nile
RED SEA
Sanaa
YEMEN
Taiz
0
1000 km

As we shall see, the astronomers also played their part in applying their discipline to certain aspects of Islamic religious practice. It was not Islam that encouraged the development of astronomy but the richness of Islamic society, a multiracial, highly-literate, tolerant society with one predominant cultural language, Arabic. But neither did Islam, the religion, stand in the way of scientific progress. The Prophet had said: 'Seek knowledge, even as far as China'. To be sure, over-zealous orthodox rulers occasionally pursued, killed or otherwise attacked 'scientists' or destroyed or burnt their libraries, but these were exceptions. The scholars of the religious law, who saw themselves as the representatives of Islam, generally ignored the pronouncements of the scientists, even on matters relating to religious practice. Astronomy was the most important of the Islamic sciences, as we can judge by the volume of the associated textual tradition, but a discussion of it in the broader context of the various branches of knowledge, which has been attempted several times elsewhere, is beyond the scope of this chapter.

Arab star-lore

The Arabs of the Arabian peninsula before Islam possessed a simple yet developed astronomical folklore of a practical nature. This involved a knowledge of the risings and settings of the stars, associated in particular with the cosmical settings of groups of stars and simultaneous heliacal risings of others, which marked the beginning of periods called *nawʾ*, plural *anwāʾ*. These *anwāʾ* eventually became associated with the twenty-eight lunar mansions (Fig. 48), a concept apparently of Indian origin. A knowledge of the passage of the sun through the twelve signs of the zodiac, associated meteorological and agricultural phenomena, the phases of the moon, as well as simple time-reckoning using shadows by day and the lunar mansions by night, formed the basis of later Islamic folk astronomy, which flourished separately from mathematical astronomy in Islamic society.

More than twenty compilations on the pre-Islamic Arabian knowledge of celestial and meteorological phenomena as found in the earliest Arabic sources of folklore, poetry and literature, are known to have been compiled during the first four centuries of Islam. The best known is that of Ibn Qutayba, written in Baghdad about the year 860. Almanacs enumerating agricultural, meteorological and astronomical events of significance to local farmers were also compiled: several examples of these survive from the medieval Islamic period, one such being for Córdoba from the year 961. The Yemen possessed a particularly rich tradition of folk astronomy, and numerous almanacs were compiled there.

Since the sun, moon and stars are mentioned in the Quran, an extensive literature dealing with what may well be labelled Islamic folk cosmology arose. This was inevitably unrelated to the more 'scientific' Islamic tradition based first on Indian sources and then predominantly on Greek ones. Since it is also stated in the Quran that man should use these celestial bodies to guide him, the scholars of the religious law occupied themselves with folk astronomy. We shall mention below various treatises dealing

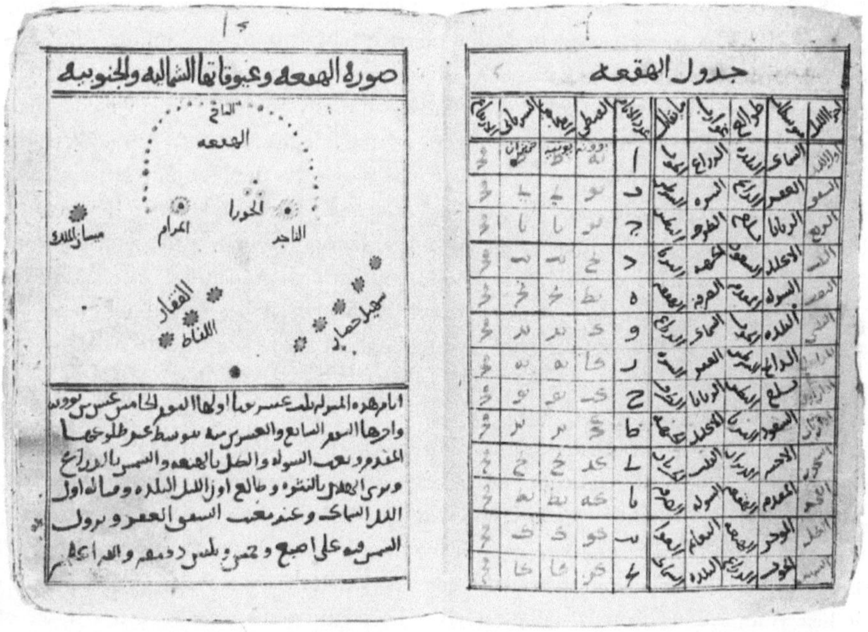

48 An illustration of the stars of one of the lunar mansions (*al-haqʿa*) in an Egyptian treatise on folk astronomy. The table on the right identifies the mansions culminating, rising and setting, and the mansions opposite them (at 180°) at different times of the night when the sun is in that mansion; it also gives the date in the Coptic, Western and Syrian calendars and the midday solar altitude when the sun is in each of the 13 degrees of the mansion ($360 \div 28 \approx 13$). The associated text repeats some of this information and adds the midday shadow length. (Chester Beatty Library, Dublin, MS 4538)

with simple time-keeping and the determination of the direction of Mecca by non-mathematical means.

Persian and Indian sources

The earliest astronomical texts in Arabic seem to have been written in Sind and Afghanistan, areas conquered by the Muslims already in the seventh century. Our knowledge of these early works is based entirely on citations from them in later works. They consisted of text and tables and were labelled *zīj* after a Persian word meaning 'cord' or 'thread' and by extension 'the warp of a fabric', which the tables vaguely resemble. The Sasanian *Shahriyārān Zīj* in the version of Yazdigird III (see p. 135) was translated from Pahlavi into Arabic as the *Shāh Zīj*, and the astronomers of the Caliph al-Manṣūr chose an auspicious moment to found his new capital Baghdad using probably an earlier Pahlavi version of this *zīj*. The various horoscopes computed by Māshā'allāh (Baghdad, *c.* 800) in his astrological world history are based on it.

Significant for the subsequent influence of Indian astronomy in the Islamic tradition was the arrival of an embassy sent to the court of al-Manṣūr from Sind *c.* 772. This embassy included an Indian well versed in astronomy and bearing a Sanskrit astronomical text apparently entitled the *Mahāsiddhānta* and based partly on the *Brāhmasphuṭasiddhānta* (see pp. 136–7). The Caliph ordered al-Fazārī to translate this text into Arabic with the help of the Indian. The resulting *Zīj al-Sindhind al-kabīr* was the basis of a series of *zījes* by such astronomers as al-Fazārī, Ya'qūb ibn Ṭāriq, al-Khwārizmī, Ḥabash, Ibn Amājūr, al-Nayrīzī, and Ibn al-Ādamī, all prepared in Iraq before the end of the tenth century. The *Sindhind* tradition (see pp. 133 and 137) flourished in Andalusia, mainly through the influence there of the *Zīj* of al-Khwārizmī (see below). As a result, the influence of Indian astronomy is attested from Morocco to England in the late Middle Ages.

Greek sources

Ptolemy's *Almagest* was translated at least five times in the late eighth and ninth centuries. The first was a translation into Syriac and the others were into Arabic, the first two under the Caliph al-Ma'mūn in the middle of the first half of the ninth century, and the other two (the second an improvement of the first) towards the end of that century. All of these were still available in the twelfth century, when they were used by Ibn al-Ṣalāḥ for his critique of Ptolemy's star catalogue. The translations gave rise to a series of commentaries on the whole text or parts of it, many of them critical and one, by Ibn al-Haytham (*c.* 1025), actually entitled 'Doubts about Ptolemy' (*al-Shukūk*). The most commonly used version of the *Almagest* in the later period was the recension of the late ninth-century version by the polymath Naṣīr al-Dīn al-Ṭūsī in the mid thirteenth century. Various other works by Ptolemy, notably the *Planetary Hypotheses* and the *Planisphaerium*, and other Greek works, including the short treatises by Autolycus, Aristarchus, Hypsicles and Theodosius, and works on the construction known as the analemma for reducing problems in three dimensions to a plane, were also translated into Arabic; most of these too were later edited by al-Ṭūsī. In this way Greek planetary models, uranometry and mathematical methods came to the attention of the Muslims. Their redactions of the *Almagest* not only contained reformulations and paraphrases of its contents but they also 'corrected, completed, criticized and brought [the contents] up to date both theoretically and practically' (Saliba); most of this material has not been studied in modern times.

Developments in astronomy

Theoretical astronomy

The geometrical structure of the universe conceived by Muslim astronomers of the early Islamic period (*c.* 800–1050) is more or less that expounded in Ptolemy's *Almagest*, with the system of eight spheres being regarded essentially as mathematical models. However, already in Ptolemy's *Planetary Hypotheses* these models are taken as representing physical

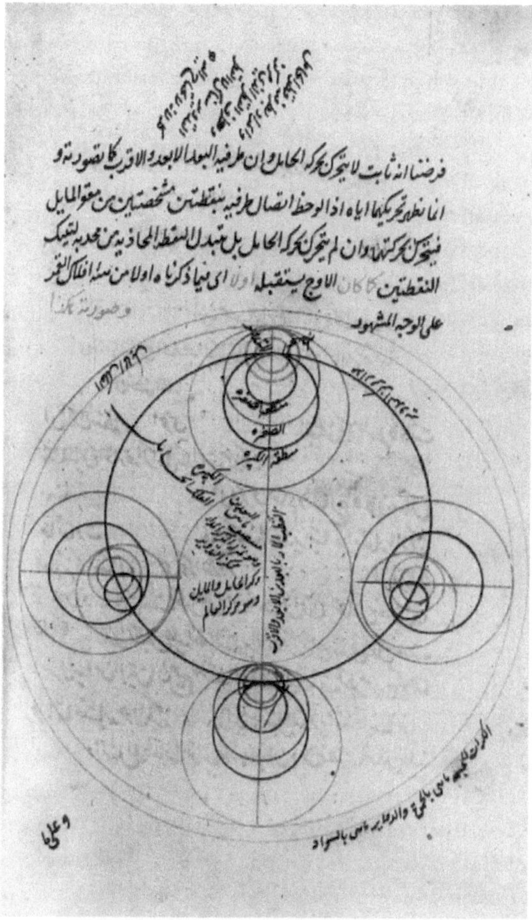

49 A non-Ptolemaic planetary model for the moon, found in a copy of the treatise on planetary astronomy (*al-Tuḥfa al-shāhiyya*) written by Quṭb al-Dīn al-Shīrazī in Sivas (Anatolia) in AD 1285. Only since the 1950s have these models been investigated by modern scholars; the discovery that a series of Muslim astronomers concerned themselves with such models from the eleventh to the sixteenth centuries and developed models without the problems inherent in the Ptolemaic ones has promoted considerable interest in medieval Islamic planetary theory. (Egyptian National Library, Cairo, MS K3758)

reality; this text also became available in Arabic. Several early Muslim scholars wrote on the sizes and relative distances of the planets, and one who proposed a physical model for the universe was Ibn al-Haytham (fl. *c.* 1025). In order to separate the two motions of the eighth sphere, the motion of the fixed stars due to the precession of the equinoxes and the motion of the fixed stars due to the apparent daily rotation, he proposed a ninth sphere to impress the apparent daily rotation on the others.

Of considerable historical interest are various Arabic treatises on the notion of the trepidation of the equinoxes. This theory, developed from Greek sources, found followers who believed that it corresponded better to the observed phenomena than a simple theory of uniform precession. The mathematical models proposed were complex and have only recently been studied properly (notably those of Pseudo-Thābit (date unknown) and Ibn al-Zarqāllu (Andalusia, *c.* 1070), who seems to have relied on his predecessor Ṣā'id al-Andalusī). The theory of trepidation continued to occupy certain Muslim scholars (in the late period mainly in the Maghrib), as it did European scholars well into the Renaissance. The history of this notion has yet to be written.

Other significant Islamic modifications to Ptolemaic planetary models, devised to overcome the philosophical objections to the notion of an equant and the problem of the variation in lunar distance inherent in Ptolemy's lunar model, belong to the later period of Islamic astronomy (Fig. 49). There were two main schools, one of which reached its fullest expression in Maragha in North-west Iran in the thirteenth century (notably with al-Ṭūsī and his colleagues) and Damascus in the fourteenth (with Ibn al-Shāṭir), and the other of which developed in Andalusia in the late twelfth century (notably with al-Biṭrūjī). The latter tradition was doomed from the outset by a slavish adherence to (false) Aristotelian tenets and mathematical incompetence. The former was based on sophisticated modifications to Ptolemy's models, partly inspired by new observations; Ptolemy himself would have been impressed by it, as have been modern investigators, for the tradition has been rediscovered and studied only in the latter half of this century. In the 1950s E. S. Kennedy discovered that the solar, lunar and planetary models proposed by Ibn al-Shāṭir in his book *The Final Quest Concerning the Rectification of Principles* (*Nihāyat al-su'l*) were different from those of Ptolemy, indeed that they were mathematically identical to those of Copernicus some 150 years later. In this work Ibn al-Shāṭir 'laid down the details of what he considered to be a true theoretical formulation of a set of planetary models describing planetary motions, and actually intended as alternatives to the Ptolemaic models' (Saliba). He maintained the geocentric system, whereas Copernicus proposed a hypothesis, which he was unable to prove, that the sun was at the centre of things. Nevertheless this important discovery raised the interesting question whether Copernicus might have known of the works of the Damascene astronomer (see p. 202). Since the 1950s we have progressed to a new stage of inquiry: we now know that there was a succession of Muslim astronomers from the eleventh century to the sixteenth who concerned themselves with models different from those of Ptolemy, all designed to overcome what were seen as flaws in them. The question we may now ask is: was Copernicus influenced by any of these Muslim works? The answer is unsatisfactory, namely, that he must have been; definitive proof is, however, still lacking.

Mathematical astronomy – the tradition of the zījes

The Islamic *zījes* constitute an important category of astronomical literature for the historian of science, by virtue of the diversity of the topics dealt with, and the information that can be obtained from the tables. In 1956 E.S. Kennedy published a survey of about 125 Islamic *zījes*. We now know of close to 200, and material is available for a revised version of the *zīj* survey. To be sure, many of these works are lost, and many of the extant ones are derived from other *zījes* by modification, borrowing, or outright plagiarism. Nevertheless, there are enough *zījes* available in manuscript form to reconstruct a reasonably accurate picture of the Islamic activity in this field.

Most *zījes* consist of several hundred pages of text and tables; the treatment of the material presented may vary considerably from one *zīj* to another. The following aspects of mathematical astronomy are handled in a typical *zīj*:

1) chronology
2) trigonometry
3) spherical astronomy
4) solar, lunar and planetary mean motions
5) solar, lunar and planetary equations
6) lunar and planetary latitudes
7) planetary stations
8) parallax
9) solar and lunar eclipses
10) lunar and planetary visibility
11) mathematical geography (lists of cities with geographical co-ordinates), determination of the direction of Mecca
12) uranometry (tables of fixed stars with co-ordinates)
13) mathematical astrology.

As noted above, already in the eighth century in India and Afghanistan there were compiled a number of Arabic *zījes*. These earliest examples, based on Indian and Sasanian works, are lost, as are the earliest examples compiled at Baghdad in the eighth century. With the *zījes* compiled in Baghdad and Damascus in the early ninth century under the patronage of the Caliph al-Ma'mūn we are on somewhat firmer ground. These follow either the tradition of the *Almagest* and *Handy Tables* or the Indian tradition. Manuscripts exist of the *Mumtaḥan Zīj* of Yaḥyā ibn Abī Manṣūr and the Damascus *Zīj* of Ḥabash, each of which was based on essentially Ptolemaic theory rather than Indian. The *Zīj* of al-Khwārizmī, based mainly on the Persian and Indian traditions, has survived only in a Latin translation of an Andalusian recension. The *Ṣābi' Zīj* of al-Battānī of Raqqa *c.* 910; the *Ḥākimī Zīj* of Ibn Yūnus, compiled in Cairo at the end of the tenth century; the *zīj* called *al-Qānūn al-Mas'ūdī* by al-Bīrūnī, compiled in Ghazna about 1025; the *Zīj* of Ibn Isḥāq, compiled in Tunis *c.* 1195; the *Īlkhānī Zīj* of Naṣīr al-Dīn al-Ṭūsī, prepared in Maragha in the mid thirteenth century; and the *Sulṭānī Zīj* of Ulugh Beg from early fifteenth-century Samarqand: these are amongst the most important later works of this genre, and also the most influential.

The only *zījes* from the early period of Islamic astronomy that have been published with translation and commentary are those of al-Khwārizmī (in the much modified later recension) and al-Battānī. The Arabic text of the *Zīj* of al-Bīrūnī has been published and a Russian translation and commentary are available. The observation accounts in the introduction of the *Ḥākimī Zīj* of Ibn Yūnus and the texts (but not the tables) of the *Zīj*es of Ibn al-Bannā' (Marrakesh, *c.* 1300) and of Ulugh Beg have been published and translated. Also a Byzantine translation of one of the *zīj*es of al-Fahhād (Iran, *c.* 1150; see p. 108) has been published. No other *zīj*es have received such attention.

Although the *zīj*es are amongst the most important sources for our knowledge of Islamic mathematical *astronomy*, it is important to observe that they generally contain extensive tables and explanatory text relating to mathematical *astrology* as well. Islamic astrological texts form an independent corpus of literature, mainly untouched by modern scholarship. Often highly sophisticated mathematical procedures are involved. It should also be pointed out that in spite of the fact that astrology was anathema to Muslim orthodoxy, it has always been (and still is) widely practised in Islamic society.

We shall now consider various aspects of the *zīj*es and the related literature.

NUMERICAL NOTATION AND BASIC MATHEMATICAL AUXILIARY TABLES All early Islamic astronomical tables have entries written in Arabic alphanumerical notation and expressed sexagesimally, that is, to base 60. A number written in letters equivalent to '23 30 17 seconds' (Ulugh Beg's value for the obliquity) stands for $23 + \frac{30}{60} + \frac{17}{3600}$ degrees, that is, 23°30'17". In sexagesimal arithmetic, more so than in decimal arithmetic, it is useful to have a multiplication table at hand, and such tables, with 3,600 or even 216,000 entries, were available.

Already in the early ninth century Muslim astronomers had restyled the cumbersome Indian sine function using the Greek base 60 (which the Greeks had used for their even more cumbersome chord function). Likewise the Indian shadow functions, unknown in Greek astronomy, were adopted with different bases (12, 6, $6\frac{1}{2}$ and 7, and also 60 and occasionally 1). Most *zīj*es contain tables of the sine and (co)tangent function for each whole, or half, or quarter degree of arc. Entries are generally given to three sexagesimal digits, corresponding roughly to five decimal digits. But certain Muslim scholars compiled more extensive sets of trigonometric tables that were not included in *zīj*es. Already in the early tenth century al-Samarqandī prepared a set of tables of the tangent function with entries to three sexagesimal digits for each minute of arc. Later in the same century Ibn Yūnus tabulated the sine function to five sexagesimal digits, equivalent to about nine decimal digits, for each minute of arc, also giving the differences for each second. He also tabulated the tangent function for each minute of arc, and the solar declination for each minute of solar longitude. His trigonometric tables were not sufficiently accurate to warrant this number of significant figures, and indeed over four centuries were to elapse before the compilation in Samarqand of the magnificent trigonometric tables in the *Sulṭānī Zīj* of Ulugh Beg, which display the values of the sine and tangent to five sexagesimal digits for each minute of argument and are generally accurate in the last digit.

Planetary tables and ephemerides Given the Ptolemaic models and tables of the mean motion and equations of the sun, moon and planets such as were available to Muslim astronomers in the *Almagest* and *Handy Tables*, or the corresponding tables based on Indian models that exemplify the *Sindhind* tradition, Muslim astronomers from the ninth to the sixteenth century sought to improve the numerical parameters on which these tables were based. Most of the leading Muslim astronomers of the early period made solar observations and computed new solar equation tables. Ibn Yūnus is the only astronomer from the first four centuries of Islam known to have compiled a new set of lunar equation tables. The majority of Islamic planetary equation tables are Ptolemaic, and where exceptions do occur, such as in the tables of Ibn al-Aʿlam and Ibn Yūnus for Mercury, we find that they are based on a Sasanian parameter rather than on any new observations. Pending a new edition of Kennedy's survey of Islamic *zīj*es to include all available parameters as well as bio-bibliographical data and a reclassification of the 200-odd known *zīj*es into clearly defined families, it is better not to say more at this time.

Ptolemy used the same data as Hipparchus for his determination of the solar apogee and hence obtained the same result. The Muslims thus inherited the notion that the solar apogee is fixed with respect to the fixed stars (although the planetary apogees move with the motion of precession), and it is to their credit that their earliest observations established that the solar apogee had moved about 15° since the time of Hipparchus. Most early Muslim astronomers accepted the *Mumtaḥan* value of 1° in $66\frac{2}{3}$ Persian years (actually a parameter attested in earlier Persian sources) for both precession and the motion of the apogees. Ibn Yūnus possessed all the necessary data that could be used to demonstrate that the motion of the solar apogee is not the same as the motion due to precession, but he chose to use the same value for both, 1° in $70\frac{1}{4}$ Persian years, which happens to be remarkably close to the actual rate of precession. Al-Bīrūnī (Central Asia, *c.* 1025) seems to have been the first to distinguish the proper motion of the solar apogee from the motion of precession (this discovery is sometimes erroneously attributed to al-Battānī). It was Ibn al-Zarqāllu (Andalusia, *c.* 1070) who was the first to assign a numerical value to both motions, although he also subscribed to the theory of trepidation.

All Islamic *zīj*es contained tables of mean motions and equations for computing solar, lunar and planetary positions for a given time (Fig. 50). Some of the equation tables are arranged in a form more convenient for the user (so that one simply has to enter the mean motions, and calculations are avoided). Auxiliary tables were sometimes available for generating ephemerides without the tedious computation of daily positions from mean-motion and equation tables. From the ninth to the nineteenth century Muslim astronomers compiled ephemerides displaying solar, lunar and planetary positions of each day of the year, as well as information on the new moons and astrological predictions resulting from the position of the moon relative to the planets. Al-Bīrūnī described in detail how to compile ephemerides in his astronomical and astrological

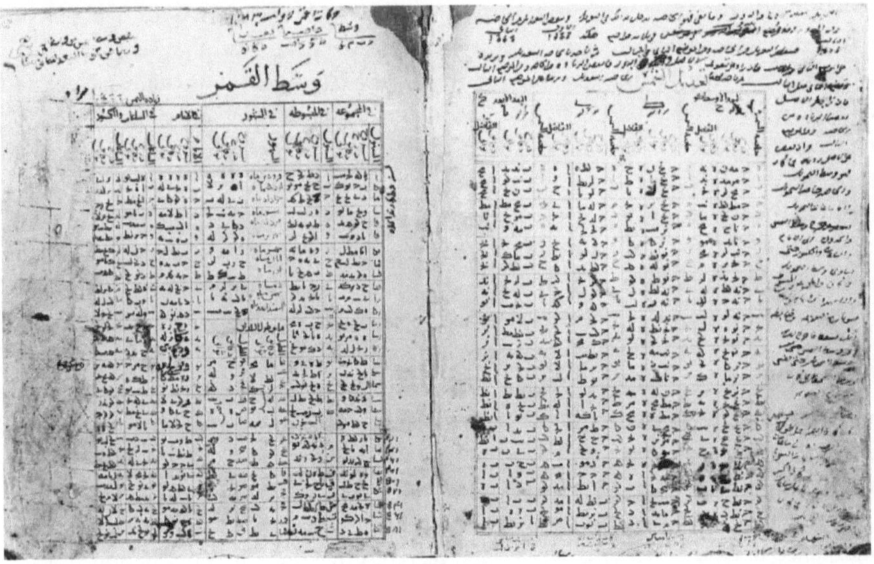

50 Astronomical tables typical of the type found in Islamic *zījes*. These serve the solar equation (right) and the lunar mean motion (left). They are found in a Yemeni manuscript of the *Zīj* of the Persian astronomer Kūshyār ibn Labbān, compiled *c.* AD 1000 and copied here *c.* AD 1250. The manuscript has various marginalia giving modifications to the tables for the longitude of Sanaa. The Yemen was an important centre of astronomy in the Middle Ages. (Egyptian National Library, Cairo, MS DM 400)

handbook *Instruction in the Art of Astrology* (*Tafhīm*). Manuscripts of ephemerides had a high rate of attrition since the tables could be dispensed with at the end of the year: the earliest complete extant examples are from fourteenth-century Yemen, discovered in Cairo in the 1970s and still unpublished; on the other hand, literally hundreds of ephemerides survive from the late Ottoman period.

STELLAR CO-ORDINATES AND URANOGRAPHY Most *zījes* contain lists of stellar co-ordinates in either the ecliptic or the equatorial system, or occasionally in both systems. A survey of the stellar co-ordinates in Islamic *zījes*, which has not yet been conducted, would be a valuable contribution to the history of Islamic astronomy, and could help determine the extent to which original observations were made by Muslim astronomers. An impressive amount of research on Arabic star names and their later influence in Europe has been conducted in the last few years by P. Kunitzsch.

In his *Book of Constellation Figures* (*Ṣuwar al-kawākib*, Col. Pl. IX) the tenth-century Shiraz astronomer al-Ṣūfī presented lists of stellar co-ordinates as well as illustrations of the constellation figures from the Hellenistic tradition and also information on the lunar mansions following the Arab tradition. Later Islamic works on uranography are mostly restricted to Persian and Turkish translations of al-Ṣūfī, although some astrological works also contain illustrations of the constellations that have recently attracted the attention of historians of Islamic art.

Spherical astronomy and spherical trigonometry Most *zīj*es contain in their introductory text the solutions of the standard problems of spherical astronomy, such as, to give only one example, the determination of time from solar and stellar altitude. Rarely is any explanation given of how the formulae outlined in words in the text were derived. There were two main traditions. In the first, the problems relating to the celestial sphere are reduced to geometric or trigonometric problems on a plane. The construction known as the analemma was a singularly powerful tool for solutions of this kind. In the second, the problems are solved by applications of rules of spherical trigonometry. Both techniques are ultimately of Greek origin, and Muslim scholars made substantial contributions to each.

There is some confusion about these contributions in the modern literature. It has been assumed by modern writers that when a medieval writer used a medieval formula that is mathematically equivalent to the modern formula derived by a specific rule of spherical trigonometry, the medieval scholar must have known the equivalent of the modern rule of spherical trigonometry. In fact, however, the medieval formula may have been derived without using spherical trigonometry at all. The first known Islamic treatise dealing with spherical trigonometry independently from astronomy is by the eleventh-century Andalusian Ibn Muʿādh. The contributions to spherical astronomy by such scholars as Thābit ibn Qurra, al-Nayrīzī, Abu l-Wafāʾ al-Būzajānī, al-Khujandī, Kūshyār ibn Labbān, al-Sijzī, Abū Naṣr, are outlined in the recently rediscovered *Keys to Astronomy* (*Maqālīd*) of al-Bīrūnī, also from the eleventh century.

Already in the work of Ḥabash in the mid ninth century we find a Muslim astronomer at ease with both spherical trigonometric methods and analemma constructions for solving problems of spherical astronomy. In the *zīj*es of scholars of the calibre of Ibn Yūnus and al-Bīrūnī we find various methods for solving all of the standard problems of medieval spherical astronomy. The auxiliary trigonometric tables compiled by such scholars as Ḥabash, Abū Naṣr (*c.* 1000) and al-Khalīlī (*c.* 1360) for solving all of the problems of spherical astronomy for any latitude are a remarkable testimony to their mastery of the subject.

Applications of astronomy to aspects of religious practice

The lunar calendar

The Muslim calendar is lunar and the civil months begin with the first sighting of the lunar crescent. The precise determination of the beginnings and ends of the months is particularly important for Ramaḍān, the sacred month of fasting, and various other religious festivals. The legal scholars were content to rely on direct sighting of the crescent or on alternating 29 and 30-day months to regulate the calendar. But the subject of lunar crescent visibility was generally treated in *zījes* (Fig. 51), and a wide variety of methods and tables were devised to facilitate the solution of this problem. Al-Khwārizmī, for example, compiled a table of the minimum ecliptic elongation of the sun and moon for each zodiacal sign, computed for the latitude of Baghdad, and another early Muslim astronomer, perhaps from Andalusia, compiled a similar table for each of the climates, no doubt inspired by Ptolemy's planetary visibility tables. A few early Islamic astronomers, notably Thābit ibn Qurra and Ibn Yūnus, postulated conditions that appear to be considerably more sophisticated, although they are not yet fully understood. Others like al-Battānī and Ibn al-Zarqāllu, merely played around with numbers, deriving complicated procedures from a basic notion that visibility occurs when the ecliptic elongation is 12° or 1 day's relative motion of the sun and moon after conjunction. Unfortunately no Muslim astronomer has left us any observational data on crescent visibility, and with a few possible exceptions yet to be identified none of the various sets of conditions proposed by the Muslim astronomers appears to be based on such data.

Nowadays there is often confusion about the beginning of Ramaḍān, such as could never have occurred in medieval times. This confusion results from the fact that the crescent may be seen in some locations and not in others, and no less from the reluctance of the religious scholars, who have the final say in announcing the new month, to listen to the astronomers.

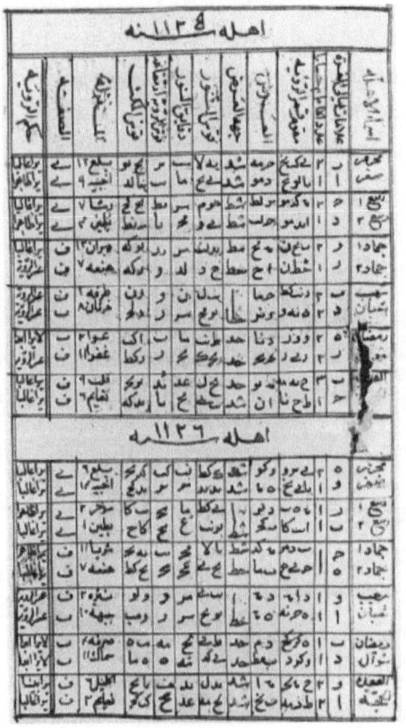

51 Extracts from an Egyptian set of tables displaying calculations for the visibility of the lunar crescent. These display for each month of the lunar years 1125 and 1126 Hijra (= AD 1713–14), for sunset on the first day of each month in the civil calendar, the lunar longitude and latitude, angular separation of the sun and moon, and difference in their setting times, followed by a prediction: '[the crescent] will be seen clearly', 'probably' (literally, mostly), 'with difficulty', 'not at all'. In the last case, the month would officially begin the next evening. (Egyptian National Library, Cairo, MS DṢ 155)

The times of prayer

In Islam the times of prayer are astronomically determined. The standard definitions of the five prayer-times, still in use today, are briefly as follows: The Muslim day begins at sunset, and the interval during which the first prayer (*maghrib*) is to be performed lasts from sunset to nightfall. The interval for the second prayer (*'ishā'*) begins at nightfall and lasts until daybreak. The third prayer (*fajr*) is performed in the interval between daybreak and sunrise. The permitted time for the fourth prayer (*ẓuhr*) begins when the sun has crossed the meridian and ends when the interval for the fifth prayer (*'aṣr*) begins, namely, when the shadow of an object equals its midday shadow increased by the length of the object. The interval for the fifth prayer may last until the shadow increases again by the length of the object or until sunset. In medieval Andalusian practice the fourth prayer begins when the shadow has increased beyond its midday minimum by one quarter of the length of the object. Isolated instances of the performance of another prayer (*ḍuḥā*) in the morning, at a time after sunrise equal to the time remaining till sunset after the *'aṣr*, are attested in medieval sources. The definitions of the daytime prayers in terms of shadow increases relate the times to the seasonal hours (one-twelfth divisions of the length of daylight), the connection being provided by an approximate Indian formula for time-keeping known to the Muslims in the eighth century. This formula associates the end of the ninth seasonal hour (mid-afternoon) with a shadow increase of 1 gnomon-length, and the end of the tenth hour with a shadow increase of 2 gnomon-lengths.

Clearly the times of the prayers as defined above can be regulated without difficulty by observation, assuming a clear sky. A genre of literature dealing with time-keeping from a non-mathematical point of view was compiled by specialists in folk astronomy, some of whom were also noted legal scholars. These works consist mainly of discussions of Quranic verses and Prophetic statements on these two subjects, embellished with descriptions of simple procedures for time-keeping by shadows and by the lunar mansions. Most surviving examples are of thirteenth-century or later Yemeni and Hejazi provenance, but the trend was set already in the ninth century. Since they involved the religious traditions, they formed part of the corpus of Islamic legal literature, and they were subject to different interpretations by different legal schools. In the first few centuries of Islam, as far as we can tell from the scant evidence available, the times of prayer were regulated by the muezzins themselves, and these were selected as much because of their fine voices as for their ability in folk astronomy.

This notwithstanding, it was found convenient to have tables at hand displaying the length of the prayer-times for each day of the year or each degree of solar longitude. The earliest such tables we have are a set for Baghdad displaying the shadow at the *ẓuhr* and the *'aṣr* for each 6° of solar longitude, and another, found only in a thirteenth-century Iraqi *zīj*, but probably also dating from the ninth century, displaying the time from sunset to daybreak (based on an approximate Indian formula for time-keeping) and the midday shadow, and the solar altitudes at midday and the beginning and end of the *'aṣr* for each day of the Syrian calendar. About the middle of the tenth century 'Alī ibn Amājūr compiled two tables displaying the time of day as a function of solar meridian altitude and instantaneous altitude. The first, based on an accurate formula, was computed specifically for Baghdad, and the second, based on an approximate Indian formula, is universal and works quite well for all latitudes. In both cases values are given in seasonal day-hours and minutes for each degree of both arguments.

These early Islamic tables for time-keeping, of which so few examples survive, began a tradition that reached its zenith in thirteenth-century Cairo and fourteenth-century Damascus (Fig. 52). Most of the corpora of tables for time-keeping compiled for these two centres and others such as Jerusalem (Fig. 53), Alexandria, Maragha, Tunis and Taiz, belong to the later period of Islamic astronomy. Some remarkable tables were produced: one, compiled in Egypt in the late thirteenth century, has three arguments (solar or stellar altitude and meridian altitude and half arc of visibility) and displays the time of day or night for any terrestrial latitude; it has over 400,000 entries. It has only recently become apparent that European astronomers in later centuries compiled tables similar in their conception; few of these have been studied yet. Indeed the history of such tables extends from the ninth to the nineteenth century in the Islamic world and from the fourteenth century onwards in Europe and eventually also North America.

It was apparently in Egypt in the thirteenth century that the office of the *muwaqqit* or mosque astronomer responsible mainly for the times of prayer was developed. Most

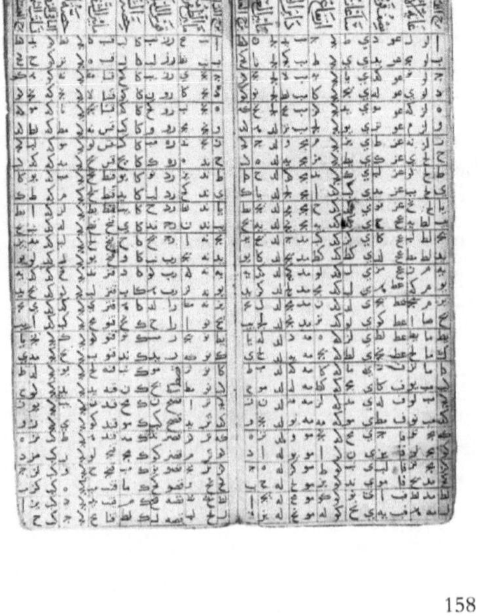

52 An extract from the prayer-tables for the latitude of Damascus, computed in the mid fourteenth century by the *muwaqqit* al-Khalīlī. Twelve functions relating to time-keeping are tabulated across the page for each degree of solar longitude (corresponding roughly to each day of the year); this double page serves the sign of Aquarius. Similar tables have been found for localities between Morocco and Central Asia, Crete and the Yemen. Such tables enabled the *muwaqqit*s to inform the muezzins when the time for each of the five prayers had arrived so that they could summon the faithful to prayer. (Bibliothèque Nationale, Paris, MS ar. 2558)

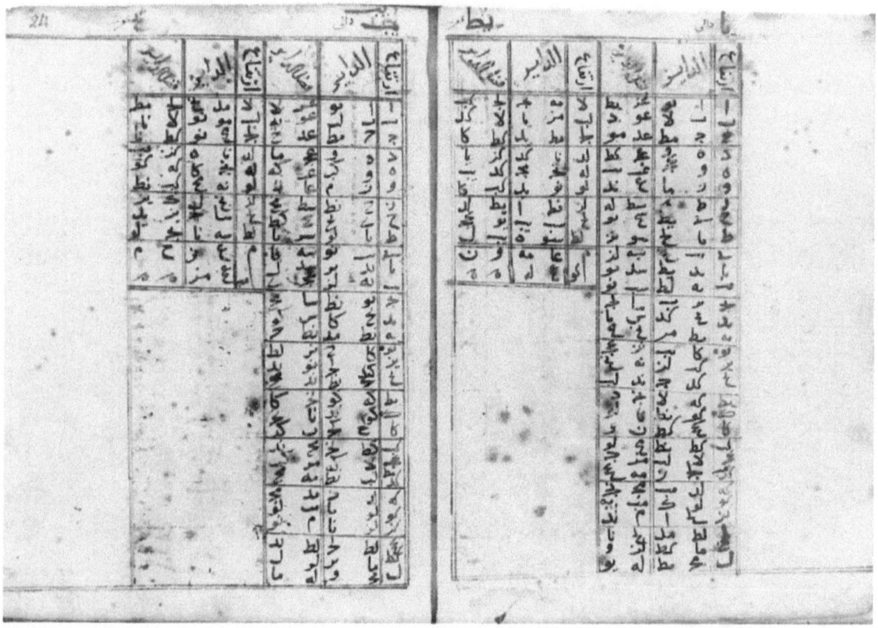

53 An extract from the tables for time-keeping for the sun, compiled for the latitude of Jerusalem by the fourteenth-century *muwaqqit* al-Karakī. For each degree of solar longitude (here Aquarius 11° and 12°, also serving Scorpio 19° and 18°) and for each degree of solar altitude up to the maximum, the time since sunrise and hour-angle are tabulated in equatorial degrees and minutes. The first tables of this kind were prepared in Baghdad in the ninth century and numerous examples for other locations were compiled during the next millennium. (Leipzig University Library, MS 808)

of the Egyptian and Syrian astronomers of consequence from the fourteenth and fifteenth centuries were *muwaqqit*s. The office was particularly important in the Ottoman Empire and in many mosques from Anatolia to the Balkans one can still see the buildings attached to mosques where the *muwaqqit*s kept their books and instruments. There does not appear to have been a similar office in the Muslim East, but we do have evidence that there was a team of *muwaqqit*s in Granada around 1300.

The tables that modern Muslims use to regulate their prayers, published in newspapers, pocket diaries and wall calendars, have a history of over a millennium, only recently documented. The muezzin's call to prayer at five specific times is one of the most distinctive features of modern Islamic life.

The sacred direction

The Quranic injunction that prayer and other ritual acts should be performed facing the Kaaba in Mecca caused no problems for the first generations of Muslims who found

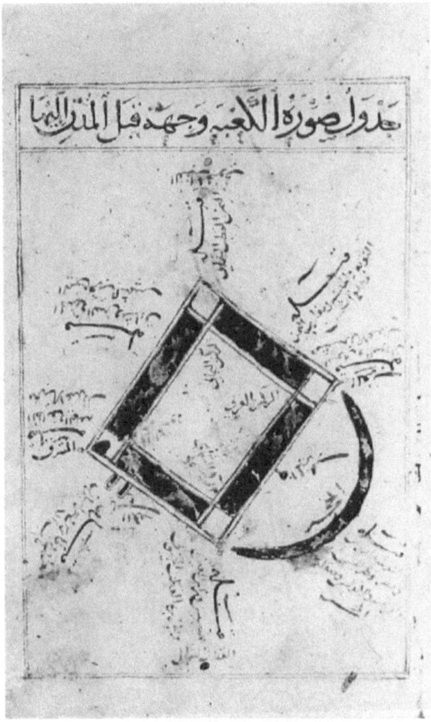

54 An illustration of the Kaaba, with each of the corners and four walls associated with regions of the Islamic commonwealth, and the *qibla* in each region defined in terms of an astronomical horizon phenomenon. From a copy of the cosmography of the fifteenth-century Syrian Ibn-al-Wardī. (Bibliothèque Nationale, Paris, MS ar. 2186)

themselves in localities as far apart as Andalusia and Sind. They had not the means to determine the *qibla*, or direction of Mecca, in the localities they settled, so they simply used the direction of the pilgrim road to Mecca or the cardinal directions; as a result, the earliest mosques and the *miḥrāb*s in their *qibla* walls face Mecca only roughly. But the Muslims knew that the Kaaba was aligned in certain astronomical directions (major axis towards the rising of Canopus, minor axis solstitially aligned), and for them the Kaaba was at the centre of their world (Fig. 54). They devised a simple expedient to face the segment of the perimeter corresponding to their location: one should stand in the same direction as one would be standing in if one were actually in front of that segment of the perimeter of the Kaaba. Different schemes of sacred geography were developed by the specialists in folk astronomy and the legal scholars, in which the world was divided into sectors about the Kaaba and the *qibla* in each sector was defined in terms of the rising or setting of the sun or a certain fixed star. This kind of information about the *qibla* was proposed by the legal scholars, and is reflected in the orientations of medieval mosques, only a minority of which are aligned in directions that could have been proposed by the astronomers.

By the early ninth century Ptolemy's list of geographical co-ordinates was available to Muslim scholars and the co-ordinates of Mecca and Baghdad had been investigated by a team commissioned by the Caliph al-Ma'mūn. Further, exact geometrical and trigonometric procedures had been devised for determining the *qibla* from geographical co-ordinates. (The problem of the determination of the *qibla* is easily transformed into a problem of spherical astronomy by considering the zeniths of the localities involved.) al-Bīrūnī's treatise on mathematical geography (*Taḥdīd*), the ultimate goal of which was the determination of the *qibla* at Ghazna, is the most important work of its kind from the medieval period. Since these exact procedures for determining the *qibla* were rather complicated, approximate methods were also derived. Already in the ninth century a table was compiled based on one such method and it displays the direction of Mecca as a function of terrestrial latitude and longitude. Over the centuries several such tables were compiled. We return later to some of the various instruments used for finding the *qibla*.

Only in the eighteenth century did it become possible for the first time to measure longitude differences correctly, and only then could it become obvious that most medieval longitude co-ordinates were incorrect and that even the *qibla*s derived by a correct mathematical procedure but based on these coordinates were also off by a few degrees. Some medieval mosques that are still in use have two *miḥrāb*s indicating the *qibla*, the second being based on modern calculations. Nowadays some Muslims use pocket-compasses provided with lists of *qibla*s for the major cities of the world.

Observation programmes and regional schools of astronomy

Al-Ma'mūn's circle

In the early ninth century the Abbasid Caliph al-Ma'mūn patronised observations first in Baghdad and then in Damascus, gathering the best available astronomers to conduct observations of the sun and moon. Some of the results were incorporated into a *zīj* called *al-Mumtaḥan*, 'tested', although the details of the activities at the two observation posts are somewhat confusing. The *Mumtaḥan Zīj* was apparently compiled in Baghdad by Yaḥyā ibn Abī Manṣūr, but upon his death, according to Ḥabash, the Caliph ordered his colleague Khālid al-Marwarrūdhī to prepare some new instruments and conduct a one-year programme of solar and lunar observations in Damascus in order to compile a new *zīj*. According to Ḥabash this was done, but no such *zīj* is otherwise known to have been prepared before Ḥabash's own *Damascus Zīj*.

These observations, like later ones, were mainly directed towards determining the local latitude and current value of the obliquity, and towards deriving improved parameters for the Ptolemaic planetary models and more accurate star positions. The armillary sphere, the meridian quadrant and the parallactic ruler were known to the Muslims from the *Almagest*, and they added new scales and other modifications, often building larger instruments even when smaller ones would have sufficed. Our knowledge of the instruments used by al-Ma'mūn's astronomers is meagre. An armillary sphere used by Yaḥyā in Baghdad was said to display markings for each 10' of arc, but even contemporary astronomers were not impressed by the precision of the results obtained using it. A mural quadrant made of marble with a radius of about 5 m was used in Damascus, as well as a vertical gnomon made of iron standing about 5 m high. Al-Ma'mūn also patronised measurements of the longitude difference between Baghdad and Mecca (by simultaneous observations of a lunar eclipse) in order to establish the *qibla* at Baghdad properly, as well as measurements of the length of one degree of terrestrial latitude. Most of what is known about the Baghdad and Damascus observatories is provided by Ibn Yūnus and al-Bīrūnī (see p. 163), and the available manuscripts of the *Zīj*es of Yaḥyā and Ḥabash await detailed study.

Other observational programmes

Besides the officially-sponsored observations conducted in Baghdad and Damascus in the early ninth century, there are numerous instances of other series of observations conducted in different parts of the Muslim world.

The two brothers called Banū Mūsā made observations in their own house in Baghdad and also in nearby Samarra about 30 years after the *Mumtaḥan* observations. They also arranged for simultaneous observations of a lunar eclipse in Samarra and Nishapur in order to determine the difference in longitude between the two cities. In view of their proficiency in mathematics, it is most unfortunate that neither of the two *zīj*es compiled by them has survived.

Al-Battānī carried out observations during the period 887 to 918 in Raqqa in North Syria. He appears to have financed his observational activity himself, and although we have no description of the site where he made his observations, the instruments mentioned in the *zīj* based on his observations include an armillary sphere and mural quadrant, as well as a parallactic ruler, an astrolabe, a gnomon and a horizontal sundial.

The observational activities of the Baghdad family known as the Banū Amājūr were almost contemporary with those of al-Battānī in Raqqa. Father and two sons, and also a freed family slave, all made observations and each compiled a *zīj*, none of which survives. In the accounts of their eclipse observations recorded by Ibn Yūnus it appears that the place where they conducted their observations had some kind of a balcony fitted with slits for observation, but the details are obscure. A particularly interesting account of a solar eclipse in the year 928 that they observed by reflection in water includes a remark that the altitude of the sun was measured on an instrument marked for each third of a degree.

A large mural quadrant was erected at Rayy (near modern Tehran) about the year 950 but we have information only on its use to establish the local latitude and obliquity of the ecliptic. In Shiraz not long thereafter al-Ṣūfī used an armillary sphere with diameter about 5 m to derive the same parameters and to 'observe' equinoxes and solstices. Al-Ṣūfī is best known for his work on the fixed stars, but it seems that this was based more on 'observation' with the naked eye than on 'measurement', looking at the heavens with precision instruments and making estimates of positions. Another contemporary astronomer who conducted observations on which we have no information other than the main parameters of his *zīj* was Ibn al-Aʿlam. The observations of both al-Ṣūfī and Ibn alʿAlam were patronised by the Buwayhid ruler ʿAḍud al-Dawla.

In the late tenth century the distinguished mathematician and astronomer Abu l-Wafāʾ al-Būzajānī made observations in Baghdad. Most of these appear to have been directed towards the determination of the solar parameters, and the obliquity of the ecliptic and the latitude of Baghdad, although Abu l-Wafāʾ also collaborated with al-Bīrūnī in Khwarizm (modern Khiva in Uzbekistan) on the simultaneous observation of a lunar eclipse in the year 997. We have no information on the nature of the site where Abu l-Wafāʾ made his observations, other than its location in a specific quarter of Baghdad.

Contemporaneous with the activity of Abu l-Wafāʾ was the establishment in 998 of an observatory in the garden of the Baghdad residence of the Buwayhid ruler Sharaf al-Dawla. The organisation of a building and programme of observations was entrusted to Abū Sahl al-Qūhī, a mathematician of considerable standing. We know from contemporary historical records that a special building was erected for the observations, which in turn were witnessed by 'judges, scientists and scholars of note, astronomers, and engineers'. In view of the favourable conditions under which this observatory was established, and the competence of its director, it is somewhat surprising that the two recorded 'observations' that were 'witnessed' by so many dignitaries were the entry of the sun into Cancer and Libra in the year 988. Al-Bīrūnī describes the main instrument which was constructed as a hemisphere of radius 12.5 m on which the solar image was projected through an aperture at the centre of the hemisphere. Activity at the observatory stopped in 989 with the death of Sharaf al-Dawla, so that the institution lasted not much more than a year.

In 994 Abū Maḥmūd al-Khujandī made a measurement of the obliquity using a meridian sextant of about 20 m radius. This instrument was erected in Rayy but al-Khujandī confessed to al-Bīrūnī that it was so large that the centre of the sextant had become displaced from its intended position.

The Egyptian astronomer Ibn Yūnus made a series of observations of eclipses, conjunctions and occultations, as well as equinoctial and solstitial observations. We are extremely fortunate to have not only his reports of these observations but also his citations of earlier observations of the same kind made by individuals such as Ḥabash and the Banū Amājūr. Ibn Yūnus's purpose in making these observations and recording them in the introduction to his *Zīj* is somewhat obscured by the fact that he does not list those observations or present those calculations with which he derived his new solar, lunar and planetary parameters. Neither does he mention any locations for his observations other than his grandfather's house in Fustat and a nearby mosque in al-

Qarāfa. The popular association of Ibn Yūnus with an observatory on the Muqaṭṭam Hills outside Cairo is, as A. Sayılı has shown, a myth. Nevertheless, Ibn Yūnus mentions at least one instrument, probably a meridian ring, that was provided by the Fatimid Caliphs al-ʿAzīz and al-Ḥākim. In a later medieval Egyptian source Ibn Yūnus is reported to have received 100 dinars a day from al-Ḥākim, and it may be that such extremely high payments were made to Ibn Yūnus when he was making satisfactory astrological predictions for the Caliph. Al-Ḥākim made an abortive attempt to found an observatory in Cairo, but this was after the death of Ibn Yūnus in 1009. At some time during his reign there was an armillary sphere in Cairo with nine rings, each large enough that a man could ride through them on horseback.

The observations of al-Bīrūnī were conducted between 990 and *c.* 1025 in several localities between Khwarizm (modern Khiva) and Kabul. His recorded observations include determinations of equinoxes and solstices, eclipses, and determinations of the obliquity and local latitude.

The corpus of tables known as the *Toledan Tables* was compiled in the eleventh century, based on observations directed by Ṣāʿid al-Andalusī and continued by Ibn al-Zarqāllu. Only the mean motion tables in this corpus of tables are original; most of the remainder were lifted from the *Zīj*es of al-Khwārizmī and al-Battānī.

In the thirteenth century there was a substantial observational programme at Maragha. The results are impressive only in so far as theoretical astronomy is concerned (see p. 150). Otherwise the trigonometric and planetary tables in the major production of the Maragha astronomers were modified or lifted *in toto* from earlier sources. This is not a happy outcome for a generously endowed observatory fitted with the latest observational instruments, known to us only from texts. In the early fifteenth century the scene had moved to Samarqand in Central Asia: there a group of astronomers directed by the astronomer-prince Ulugh Beg did impressive work. Only the 40-m meridian sextant survives from the observatory. These men produced a set of tables which, however, it would be foolish to judge before they have been properly studied. The same is true for the short-lived observatory in Istanbul under the direction of Taqi l-Dīn (1577; see frontispiece).

Regional schools of astronomy

After the tenth century there developed regional schools of astronomy in the Islamic world, with different interests and concentrations. They also had different authorities (for example, in the furthest East al-Bīrūnī and al-Ṭūsī, and in Egypt Ibn Yūnus). The main regions were Iraq; Iran and Central Asia; Muslim Spain; Egypt and Syria; the Yemen; the Maghrib; and later also the Ottoman lands. Only recently have the complex tradition of Muslim Spain (tenth–fourteenth centuries), the colourful tradition of Mamluk Egypt and Syria (thirteenth–early sixteenth centuries), the distinctive tradition of Rasulid Yemen (thirteenth–sixteenth centuries), and the staid tradition of the Maghrib (twelfth–nineteenth centuries) been studied. The traditions of Ottoman Turkey and Mogul India are currently being researched.

Astronomical instruments

As noted in the previous section most Islamic observational instruments are lost and known to us only through texts. The state of documentation of the other, smaller Islamic astronomical instruments that do survive leaves much to be desired. Many of the most important instruments are still unpublished, and much that has been written on instruments is on a very amateur level. For these reasons a project is currently underway in Frankfurt to catalogue all Islamic instruments (and European ones) to *c.* 1550 as well as various historically significant later Islamic pieces.

Also the most important writings on instruments have not yet received the attention they deserve. For example, a hemispherical observational instrument for a fixed latitude was devised by the tenth-century astronomer al-Khujandī, the leading instrument-maker of the early period, and this was modified in the twelfth century to serve all latitudes. There are no surviving examples and the available manuscripts have yet to be studied. An important work on instruments was compiled in Cairo *c.* 1280 by Abū ʿAlī al-Marrākushī; this has yet to be subjected to a detailed analysis. The author simply collected all of the treatises on instruments known to him and incorporated them into his book. An exciting find of the 1980s was a treatise by the fourteenth-century Aleppo astronomer Ibn al-Sarrāj, the leading instrument-maker of the later Islamic period. In this the author described every kind of instrument known to him as well as those invented by himself. This treatise is currently being studied.

Armillary spheres and globes

In the eighth century al-Fazārī wrote a treatise on the armillary sphere, called in Arabic *dhāt al-ḥalaq*, which means 'the instrument with the rings'. No early Islamic armillary spheres survive, but several other treatises on it were compiled over the centuries. The earliest treatise in Arabic dealing with the celestial globe, called *dhāt al-kursī*, 'the instrument with the stand [literally, the 'throne' on which it sits]' or simply *al-kura*, 'sphere', was written by Qusṭā ibn Lūqā in the ninth century. This treatise by Qusṭā, who was one of the most important translators of Greek works into Arabic, remained popular for a millennium. Of the various surviving celestial globes, which number over 100, none predates the eleventh century. A globe made in Mosul in 1274–5 is preserved in the British Museum (Col. Pl. X).

The spherical astrolabe, unlike the armillary sphere and the celestial globe, appears to be an Islamic development. Various treatises on it were written from the tenth to the sixteenth century, and only one complete instrument, from the fourteenth century, survives. Ḥabash wrote on the spherical astrolabe, the armillary sphere, and the celestial globe, as well as on various kinds of planispheric astrolabes.

Astrolabes

Al-Fazārī also wrote on the use of the astrolabe. The tenth-century bibliographer Ibn al-Nadīm states that al-Fazārī was the first Muslim to make such an instrument; he also informs us that at that time the construction of astrolabes was centred on Harran and spread from there. Several early astronomers, including Ḥabash, al-Khwārizmī and al-Farghānī, wrote on the astrolabe, and introduced features not found on earlier Greek instruments, such as the shadow squares and trigonometric grids on the backs and the azimuth curves on the plates for different latitudes, as well as the universal plate of horizons. Also extensive tables were compiled in the ninth century to facilitate the construction of astrolabes.

Another important development to the astrolabe occurred in Andalusia in the eleventh century, when Ibn al-Zarqāllu devised the single universal plate (*ṣafīḥa*) called *shakkāziyya* and the related plate called *zarqālliyya* with two sets of *shakkāziyya* markings for both equatorial and ecliptic co-ordinate systems. The latter was fitted with an alidade bearing a movable perpendicular straight-edge (transversal). Several treatises on these two instruments exist in both Western and Eastern traditions of later Islamic astronomy; the Europeans knew of them as the *saphea*. Ibn al-Zarqāllu's contemporary, ʿAlī ibn Khalaf, wrote a treatise on a universal astrolabe that did not need plates for different latitudes. This treatise exists only in Old Spanish in the *Libros del Saber*, and was apparently

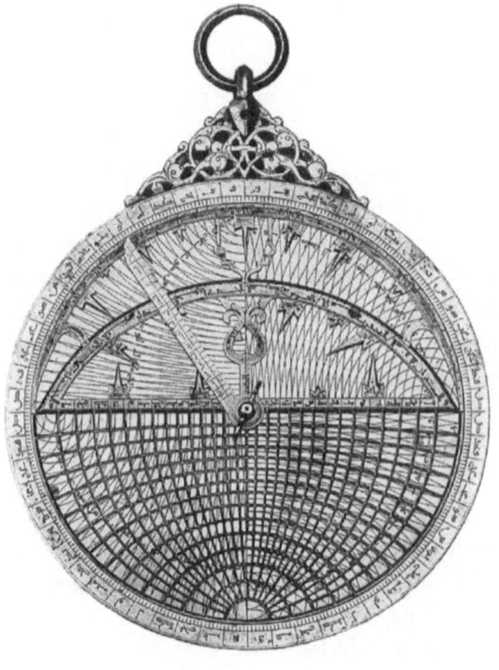

55 The front of the universal astrolabe of Ibn al-Sarrāj, dated AD 1329. This remarkable instrument not only represents the culmination of Islamic astrolabe-making, but has no equal in sophistication amongst instruments from the European Renaissance. Whereas the standard astrolabe requires a different plate for each latitude, that of Ibn al-Sarrāj has plates that serve all latitudes; indeed, the various components can be used in five different ways to solve all the problems of spherical astronomy for any latitude. IC no. 140 (Benaki Museum, Athens)

not known in the Islamic world outside Andalusia. The instrument was further developed in Syria in the early fourteenth century: Ibn al-Sarrāj devised in Aleppo a remarkable astrolabe that can be used universally in five different ways (Fig. 55).

The astrolabes made by Muslim craftsmen show a remarkable variety within each of several clearly defined regional schools. We may mention the simple, functional astrolabes of the early Baghdad school; the splendid astrolabe of al-Khujandī of the late tenth century, which started a tradition of zoomorphic ornamentation that continued in the Islamic East and in Europe for several centuries; the very different astrolabes of the Andalusian school in the eleventh century and the progressive schools of Iran in the thirteenth and fourteenth centuries; and the remarkable instruments from Mamluk (thirteenth and fourteenth-century) Egypt and Syria. The British Museum possesses a spectacular Mamluk astrolabe by ʿAbd al-Karīm al-Miṣrī. After about 1500 the construction of astrolabes continued in the Maghrib, in Iran and in India until the end of the nineteenth century. Many of these, especially those from the Islamic East, were objects of the finest workmanship. Such is the splendid astrolabe made in Isfahan in 1712 for Shāh Ḥusayn, also now in the British Museum (Col. Pl. XI).

Quadrants

Another category of observational and computational devices to which Muslim astronomers made notable contributions was the quadrant, of which we can distinguish three main varieties. First, the sine quadrant with an orthogonal grid. This instrument, in a simpler form, was described already by al-Khwārizmī and was widely used throughout the Islamic period. Some Islamic astrolabes display such a trigonometric grid on the back. The grid can be used together with a thread and movable marker (or the alidade of an astrolabe) to solve all of the standard problems of spherical astronomy for any latitude. Second, the horary quadrant with fixed or movable cursor. This instrument is described already in an anonymous ninth-century Iraqi source and was likewise commonly used for centuries (albeit usually without the cursor, which is not essential to the function of the device). A set of arcs of circles inscribed on the quadrant display graphically the solar altitude at the seasonal hours (approximately, according to an Indian formula). Other Islamic quadrants from the ninth century onwards had markings for the equinoctial hours. The instrument can be aligned towards the sun so that the time can be determined from the observed altitude using the grid. Again, this kind of markings was often marked on the back of astrolabes. Third, the astrolabic quadrant displaying one half of the altitude and azimuth circles on an astrolabe plate for a fixed latitude, and a fixed ecliptic. The effect of the daily rotation is achieved by a thread and bead attached at the centre of the instrument rather than by the movable astrolabe rete. The quadrant with astrolabic markings on one side and a trigonometric grid (Fig. 56) on the other generally replaced the astrolabe all over the Islamic world (with the notable exceptions of Iran, India and the Yemen) in the later period of Islamic astronomy.

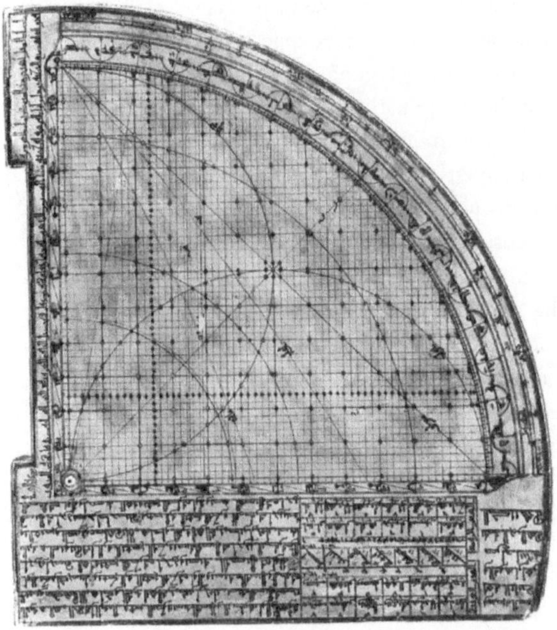

56 A trigonometric grid for solving all the various problems of spherical astronomy without calculation. The grid, a quarter-circle of sexagesimal orthogonal markings like modern graph-paper and with radius 60 units, was developed from simpler varieties first used in Baghdad in the early ninth century. One medieval European quadrant with similar markings has been preserved. This grid, found on an astrolabic quadrant made in Damascus *c.* AD 1800, is remarkable for all the additional markings, lines, arcs of circles and curves, which facilitate the solution of specific problems relating to the astronomically defined times of prayer. (Private collection)

Sundials

We learn from Islamic tradition that the pious Umayyad Caliph ʿUmar ibn ʿAbd al-ʿAzīz (Damascus, *c.* 718) used a sundial, probably a Graeco-Roman one, to regulate the times of the daytime prayers in terms of the seasonal hours. The earliest sundials described in the Arabic astronomical sources are planar, usually horizontal, but also vertical and polar. The mathematical theory for computing the shadow for the seasonal hours at different times of the year and the corresponding azimuths was available from Indian sources, which seem to have inspired the Islamic tradition more than any of the available Greek works. Already the treatise on sundial construction by al-Khwārizmī contains extensive tables displaying the polar co-ordinates of the intersections of the hour lines with the solstitial shadow traces on horizontal sundials for twelve different latitudes. The treatise on sundial theory by Thābit ibn Qurra contains all the necessary mathematical theory for constructing sundials in any plane; likewise impressive from a theoretical point of view is the treatise on gnomonics by his grandson Ibrāhīm.

The earliest surviving Islamic sundial, apparently made in Córdoba about the year 1000 by the Andalusian astronomer Ibn al-Ṣaffār, displays the shadow traces of the equinoxes and solstices, and the lines for the seasonal hours as well as for the times of the two daytime prayers. There is a world of difference between this simple, carelessly constructed piece and the magnificent sundial made in the late fourteenth century by Ibn al-Shāṭir, so devised that it can be used to measure time with respect to any of the

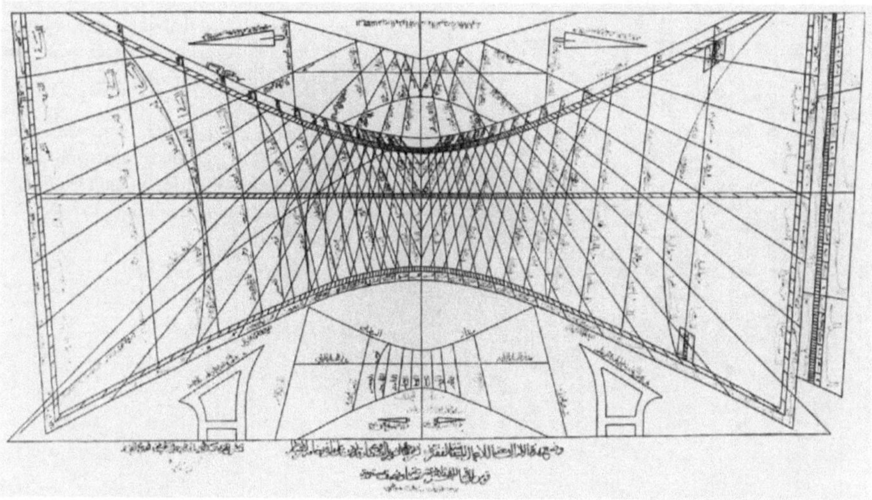

57 The sundial made by Ibn al-Shāṭir for the main minaret of the Umayyad Mosque in Damascus in the year AD 1371–2 was broken by accident in the nineteenth century. The *muwaqqit* who broke it whilst trying to realign it was sufficiently well versed in gnomonics to make this copy, which is still *in situ* on the minaret. Several pieces of the original were discovered in excavations of the drainage system of the mosque in 1958 and are now displayed in the garden of the Archaeological Museum, Damascus. (Alain Brieux, Paris)

five daily prayers (Fig. 57). In the late period of Islamic astronomy a sundial was to be found in most of the major mosques.

Miscellaneous

Several multipurpose instruments were devised by Muslim astronomers. Notable examples are the rule (*mīzān*) of al-Fazārī, fitted with a variety of non-uniform scales for various astronomical functions, and the compendium of Ibn al-Shāṭir, comprising a magnetic compass and *qibla*-indicator, a universal polar sundial, and an equatorial sundial. Of particular interest is a circular brass instrument made in Isfahan *c.* 1700 which is engraved with a world map with Mecca at the centre, and a cartographical grid so devised that the *qibla* for some 150 cities between Spain and China can be read off the outer scale and the distance from Mecca can be read off the non-uniform scale on the diametrical rule (Fig. 58). Muslim interest in projections preserving direction and distance can be traced back several centuries to al-Bīrūnī and Ḥabash.

There are several Islamic treatises on eclipse computers and planetary equatoria (see p. 184) for determining the positions of the planets for a given date. With these the standard problems of planetary astronomy dealt with in *zījes* are resolved mechanically, without calculation. Treatises on eclipse computers are known from the early tenth

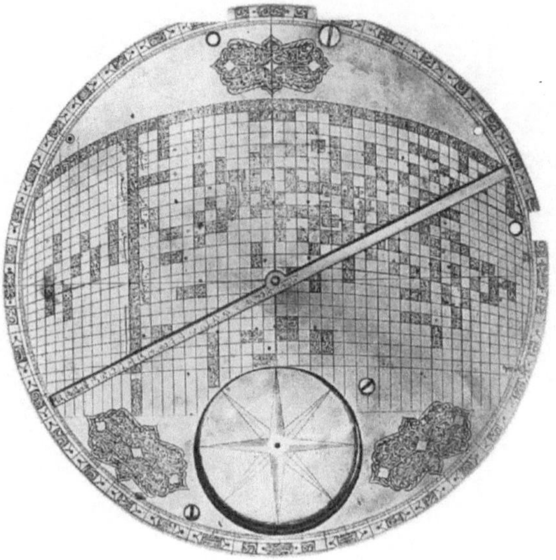

58 A cartographical grid with Mecca at the centre, so devised that one can read the direction and distance to Mecca for any locality in the Islamic commonwealth. The instrument, made in Isfahan *c.* AD 1700 but inspired by an earlier tradition, was originally fitted with a magnetic compass and an hour dial adjustable for any latitude for finding the time of day, all now missing. (Private collection)

century, and al-Bīrūnī in the early eleventh describes such an instrument in detail. A newly discovered manuscript (not yet available for research) contains a treatise by the tenth-century Iranian astronomer Abū Ja'far al-Khāzin called *Zīj al-Ṣafā'iḥ*, the *Zīj of Plates*, describing an equatorium. The sole known example of this instrument, made in the twelfth century, is, alas, incomplete: it is in the form of an astrolabe with tables engraved on the mater and additional markings for the foundation of an equatorium. Otherwise the only known early Islamic treatises on planetary equatoria are from eleventh-century Andalusia. The most interesting aspect of the equatorium described by Ibn al-Zarqāllu is the ellipse drawn on the plate for the centre of the deferent of Mercury; it seems that he was the first to notice this characteristic of Mercury's deferent. Al-Kāshī, the leading astronomer of early fifteenth-century Samarqand, has left us a description of a planetary equatorium with which not only ecliptic longitudes but also latitudes could be determined and eclipses calculated.

Transmission to Europe

The Europeans learned of Islamic astronomy through Spain, a region where, because of political problems and the difficulty of communications, the most up-to-date writings were not always available. This explains, for example, how it came to pass that the Europeans came across two major works of Muslim astronomers from the East, al-Khwārizmī and al-Battānī, at a time when these works were no longer widely used in the Islamic East. It also explains why so few Eastern Islamic works became known in Europe. None of the Eastern Islamic developments to Ptolemy's planetary theory were known in Andalusia or in medieval Europe. Al-Biṭrūjī's unhappy attempt to develop planetary models confused Europeans for several centuries; he must be worth reading, they naïvely thought, because he was trying to reconcile Ptolemy with Aristotle. As far as astronomical time-keeping was concerned, this does not seem to have been of much concern to the Muslims in Spain; hence nothing of consequence was transmitted.

On the other hand, some early Eastern Islamic contributions, later forgotten in the Islamic East, were transmitted to Spain and thence to Europe; they have been considered European developments because evidence to the contrary has seemed to be lacking. A good example is the horary quadrant with movable cursor (the so-called *quadrans vetus*), which was invented in Baghdad in the ninth century and (at least in the version with the cursor) virtually forgotten in the Islamic East thereafter; it came to be the favourite quadrant in medieval Europe. What, if any, astronomical knowledge was transmitted through Islamic Sicily remains a mystery, and nothing of consequence is known to have been learned about the subject by the Crusaders.

In the European Renaissance there was no access to the latest Islamic works. So the Europeans contented themselves with new editions of the ancient Greek works, with occasional, almost nostalgic, references to Albategnius (al-Battānī), Azarquiel (Ibn al-Zarqāllu), Alpetragius (al-Biṭrūjī) and the like. A few technical terms derived from the Arabic, such as alidade, azimuth, almucantar, nadir, saphea, and zenith, and a few star-names such as Aldebaran, Algol, Altair and Vega, survived. When the Europeans did come to learn of some of the major Islamic works and to try to come to terms with them it was as orientalists and historians of astronomy, for by this time the Islamic materials other than observation accounts were of historical rather than scientific interest. Thanks to orientalists such as the Sédillots in Paris, works that had been completely unknown to Europeans and mainly forgotten by Muslims were published, translated and analysed. Islamic astronomy was highly respected by such scholars and others, like the historian of astronomy J.-B. Delambre, who, innocent of Arabic, took the trouble to read what his colleagues had written about the subject. But Islamic astronomy, indeed Islamic science in general, received a blow beneath the belt from P. Duhem, a physicist and philosopher ignorant of Arabic, who simply ignored what scholars such as the Sédillots had written. His thesis, that the Arabs were incapable of scientific thought and that whatever merits their science may have had were due to the intellectually superior Greeks, still has many followers, but only amongst those totally ignorant of the research of the past 150 years.

In the period after *c.* 1500 Islamic astronomy declined. All of the problems had been solved, some many times over. Much of the innovative activity had led into a cul-de-sac, from which it would not emerge until modern times, thanks to investigations of manuscripts and instruments. Not that interest in astronomy died down. From Morocco to India the same old texts were copied and studied, recopied and restudied, usually different texts in each of the main regions. But there was no new input of any consequence. Astronomy continued to be used as the handmaiden of astrology, and for the regulation of the calendar and the prayer-times. Where there appears to have been some innovation – such as, for example, in the remarkable device made in Isfahan *c.* 1700 that correctly displays the direction and distance of Mecca for any locality – one can be confident of the existence of an earlier tradition. True some European '*zījes*', notably those of Cassini and Lalande, were translated into Turkish and their tables adapted for the longitudes of Istanbul and later Damascus. But the old traditions died hard, and Muslim astronomers for several centuries spent more time copying old treatises and tables than compiling new ones.

Conclusion

The reader will have seen that during the millennium beginning *c.* 750 and especially in the period up to *c.* 1050, although also in the period up to *c.* 1500, Muslim astronomers did first-rate work, most of which was not known in medieval Europe at all. Those few Islamic works from the early period that were transmitted, notably the *Zīj*es of al-Khwārizmī and al-Battānī (especially through the *Toledan Tables*) and the banal summary of the *Almagest* by al-Farghānī, convey only an impression of classical astronomy in Arabic garb. But they were in no way representative of contemporary Islamic astronomy in the East, and whilst the Europeans laboured for centuries to come to terms with them, Muslim astronomers were making substantial contributions to their subject that have only been revealed by modern scholarship.

Anyone who leafs through the pages of the two volumes of Fuat Sezgin's *Geschichte des arabischen Schrifttums* listing the manuscript sources for Islamic astronomy and astrology up to *c.* AD 1050, or my survey of the 2,500 scientific manuscripts in the Egyptian National Library, which serves mainly the period after that, will observe the wealth of material relating to this subject that remains untouched by modern scholarship. Very few Islamic astronomical works have been published or have received the attention that they merit. Three out of close to 200 Islamic *zīj*es have been published in the optimum way (text, translation and commentary). We have no published edition of the Arabic versions of the *Almagest* (except for the star catalogue), or of any Arabic recensions or commentaries. Many of the published Arabic scientific texts were printed in Hyderabad, most with no critical apparatus. In view of the lack of a published corpus of texts and the improbability that such a corpus will materialise, there is an obvious need for reproduction of manuscripts of particular importance. Otherwise the historian of Islamic astronomy will be forced to continue to rely mainly on microfilms of manuscripts, which some libraries are unable or unwilling to supply. Likewise most of the historically important Islamic astronomical instruments are still unpublished, although the catalogue currently in preparation in Frankfurt promises to make them better known.

In 1845 L. A. Sédillot, whose privilege it was to have access to the rich collection of Arabic and Persian scientific manuscripts in the Bibliothèque Nationale in Paris, wrote: 'Each day brings some new discovery and illustrates the extreme importance of a thorough study of the manuscripts of the East.' Sédillot, like his father before him but like few historians of Islamic science since, also realised the importance of Islamic astronomical instruments. Given the vast amount of manuscripts and instruments now available in libraries and museums elsewhere in Europe, the United States and the Near East, and the rather small number of people currently working in this field, Sédillot's statement is no less true now than it was a century and a half ago.

Bibliography

Dictionary of Scientific Biography, 16 vols. New York: Charles Scribner's Sons, 1970–80. (See especially the articles (in chronological order) al-**F**azārī, **Y**a'qūb ibn Ṭāriq, **M**āshā'allāh, **Y**aḥyā ibn Abī Manṣūr, **A**bū Ma'shar, al-**Kh**wārizmī, **Ḥ**abash, Ibn **Q**utayba, Banū **M**ūsā, **Th**ābit ibn Qurra, al-**B**attānī, **A**bu l-Wafā' al-Būzajānī, Ibn **Y**ūnus, **I**bn al-Haytham, al-**B**īrūnī, al-**Z**arqālī, al-**B**iṭrūjī (supplement), al-**Ṭ**ūsī, **I**bn al-Shāṭir, al-**Kh**alīlī (supplement), al-**K**āshī, and **U**lugh Beg.)

Encyclopaedia of Islam, new edn, 8 vols to date, Leiden: E. J. Brill, 1960 to present. (See especially the articles on individual astronomers: al-**F**azārī, **Ḥ**abash, al-**B**attānī, **I**bn Yūnus, **I**bn al-Haytham, al-**B**īrūnī, al-**B**iṭrūdjī, al-**M**arrākushī, and on specific topics: '**I**lm al-hay'a [=astronomy]', '**N**udjūm, aḥkām al- [=astrology]', '**M**inṭaḳa [=ecliptic]' and '**M**ayl [=declination]', '**A**nwā' [=calendrical divisions]', '**Ḳ**utb [=celestial pole]', '**L**ayl wa-nahār [=night and day]', '**M**adjarra [=Milky Way]', '**M**anāzil [=lunar mansions]', '**N**udjūm [=stars]', '**H**ilāl [=lunar crescent]' and '**R**u'yat al-hilāl [=lunar crescent visibility]', '**M**īḳāt [=astronomical time-keeping]', '**Ḳ**ibla [=sacred direction]', '**M**aṭāli' [= astronomical risings]', and '**M**akka IV: As centre of the world [=sacred geography]'; '**A**sṭurlāb [= astrolabe]', '**R**ub' [=quadrant]', '**S**amt [=direction and Mecca-centred world-maps]', '**S**hakkāziyya [= universal projection]' and '**M**izwala [=sundial]'.)

Berggren, J.L. 1986. *Episodes in the Mathematics of Medieval Islam*. Berlin, Heidelberg and New York: Springer-Verlag.

Dalen, B. van 1993. *Ancient and Mediaeval Astronomical Tables – Mathematical Structure and Parameter Values*. Utrecht University.

Goldstein, B.R. 1985. *Theory and Observation in Ancient and Medieval Astronomy*. London: Variorum.

Gunther, R.T. 1932. *The Astrolabes of the World*, 2 vols. Oxford University Press. Reprinted 1976 in one vol. London: The Holland Press (long outdated and full of errors).

Heinen, A. 1982. *Islamic Cosmology* . . . Beirut and Wiesbaden: Franz Steiner.

Kennedy, E.S. 1956. A survey of Islamic astronomical tables. *Transactions of the American Philosophical Society*, new ser., 46. Reprinted, n.d. (*c.* 1990).

Kennedy, E.S. *et al*, 1983. *Studies in the Islamic Exact Sciences*. Beirut: American University of Beirut Press.

Kennedy, E.S. and Ghanem, I. 1976. *The Life and Work of Ibn al-Shāṭir: An Arab Astronomer of the Fourteenth Century*. Aleppo: Institute for the History of Arabic Science.

Kennedy, E.S. and Kennedy, M.H. 1987. *Geographical Coordinates of Localities from Islamic Sources*. Frankfurt am Main: Institut für Geschichte der Arabisch-Islamischen Wissenschaften.

King, D.A. 1983. *Mathematical Astronomy in Medieval Yemen – A Bio-Bibliographical Survey*. Publications of the American Research Center in Egypt. Malibu, Ca.: Undena Publications.

King, D.A. 1986a. *A Survey of the Scientific Manuscripts in the Egyptian National Library*. Publications of the American Research Center in Egypt. Winona Lake, Ind.: Eisenbrauns.

King, D.A. 1986b. *Islamic Mathematical Astronomy*.

London: Variorum. 2nd rev. edn 1993. Aldershot: Variorum.

King, D.A. 1987. *Islamic Astronomical Instruments.* London: Variorum. Repr. 1995. Aldershot: Variorum.

King, D.A. 1991. Medieval astronomical instruments – a catalogue in preparation. *Bulletin of the Scientific Instrument Society* 31, 3–7.

King, D.A. 1992a. Some remarks on Islamic astronomical instruments. *Scientiarum Historia* 18, 5–23.

King, D.A. 1992b. Some remarks on Islamic scientific manuscripts and instruments and past, present and future research. In *The Significance of Islamic Manuscripts*, ed. John Cooper, pp. 115–44. London: Al-Furqan Islamic Heritage Foundation.

King, D.A. 1993. *Astronomy in the Service of Islam.* Aldershot: Variorum.

King, D.A. and Saliba, G. (eds) 1986. *From Deferent to Equant: Studies in the History of Science in the Ancient and Medieval Near East in Honor of E. S. Kennedy*, Annals of the New York Academy of Sciences, 500.

Kunitzsch, P. 1989. *The Arabs and the Stars.* Northampton: Variorum.

Kunitzsch P. and Smart, T. 1986. *Short Guide to Modern Star Names and Their Derivations.* Wiesbaden: Otto Harrassowitz.

Lorch, R.P. 1995. *Arabic Mathematical Sciences – Instruments, Texts, Transmission.* Aldershot: Variorum.

Mayer, L.A. 1956. *Islamic Astrolabists and their Works.* Geneva: A. Kundig.

Pingree, D. 1973. The Greek influences on Early Islamic mathematical astronomy. *Journal of the American Oriental Society* 93, 32–43.

Ragep, F. Jamil 1993. *Naṣīr al-Dīn al-Ṭūsī's Memoir on Astronomy (al-Tadhkira fī ʿilm al-hayʾa)*, 2 vols. Berlin, Heidelberg and New York, Springer-Verlag.

Sabra, A.I. 1987. The appropriation and subsequent naturalisation of Greek science in medieval Islam. *History of Science* 25, 223–43.

Saliba, G. 1991. The astronomical tradition of Maragha: a historical survey and prospects for future research. *Arabic Science and Philosophy* 1, 67–99.

Saliba, G. 1994. *A History of Arabic Astronomy: Planetary Theories during the Golden Age of Islam.* New York University Press.

Samsó, J. 1992. *Las ciencias de los antiguos en al-Andalus.* Madrid: Mapfre.

Samsó, J. 1994. *Islamic Astronomy and Medieval Spain.* Aldershot: Variorum.

Savage-Smith, E. 1985. *Islamicate Celestial Globes – Their History, Construction, and Use.* Washington, D.C.: Smithsonian Institution Press.

Sayılı, A. 1960. *The Observatory in Islam.* Publications of the Turkish Historical Society, ser. VII, no. 38.

Sezgin, F. 1974, 1978, 1979. *Geschichte des arabischen Schrifttums*, V: Mathematik, VI: Astronomie, VII: Astrologie, Meteorologie und Verwandtes. Leiden: E. J. Brill.

Storey, C.A. 1958. *Persian Literature: A Bio-Bibliographical Survey*, vol. II, pt 1. London: Luzac and Co.

Suter, H. 1900. Die Mathematiker und Astronomen der Araber und ihre Werke. *Abhandlungen zur Geschichte der mathematischen Wissenschaften* 10 (1900), and 14 (1902), 157–85. Both reprinted 1982. Amsterdam: The Oriental Press.

Toomer, G.J. 1968. A Survey of the Toledan Tables. *Osiris* 15, 5–174.

Van Brummelen, G. 1991. The numerical structure of al-Khalīlī's auxiliary tables. *Physis*, new ser., 28, 667–97

Varisco, D.M. 1993. *Medieval Agriculture and Islamic Science – The Almanac of a Yemeni Sultan.* Seattle, Wa.: University of Washington Press.

Vernet, J. 1978. *La cultura hispanoárabe en Oriente y Occidente.* Barcelona: Ariel. German language edn 1984: *Die spanisch-arabische Kultur in Orient und Okzident.* Zurich and Munich. French language edn 1985: *Ce que la culture doit aux Arabes d'Espagne.* Paris.

Wright, R.R. 1934. *The Book of Instruction in the Elements of Astrology by . . . al-Bīrūnī.* London: Luzac & Co.

II

From Inscriptions to Context: Some Islamic Astronomical Instruments and their Secrets

Introductory remarks

Close to one thousand "medieval" scientific instruments survive from the Islamic world (750–1900) and the European Middle Ages (950–1500). Muslim craftsmen, some of whom were professional astronomers, have left us some particularly remarkable instruments attesting to their initiative and their competence, and their European counterparts were highly indebted to their Muslim predecessors. The inscriptions and/or decoration on these instruments are important keys to understanding their historical context, and to establishing their authenticity where this is in doubt. Figure 1 shows the astronomers of the Istanbul Observatory in the late sixteenth century with their instruments and books: it is the task of the historian of Islamic astronomy to explain what these astronomers were doing, how they were doing it, and why. Alas, this chapter cannot include the dozens of illustrations of instruments and manuscripts which I used in my lecture at the Levi Della Vida Conference.

Our concern here is mainly with Islamic instruments, though we shall also consider some European ones inspired by Islamic precursors. Instrumentation constituted a significant aspect of Islamic astronomy,[1] in particular astronomical timekeeping[2] in which instruments are certainly as important as astronomical tables.[3] Also, these objets d'art constitute a substantial corpus of Islamic metalwork, hitherto much neglected for their potential contribution to the history of art.[4] A small subset of instruments serves all latitudes and provides "universal" solutions to problems of spherical astronomy, a theme pursued with singular enthusiasm by Muslim astronomers over several centuries.[5] Furthermore, the individual instruments reflect the predilections of the regional schools of astronomy in which they were developed or in which they were constructed.[6] And last but not least, the story of the way in which some of these have come down to us, for the rare cases where this is known, or how they survived the vicissitudes of the twentieth century, is often of considerable interest.

A. General remarks on instruments

1. The sources for the history of Islamic astronomical instrumentation and the state of the art

Our knowledge of astronomical instrumentation in the Islamic world between the eighth and nineteenth centuries is derived essentially from two main sources:

1. the instruments which survive in various museums and private collections around the world; and
2. the treatises on the construction and use of instruments which are preserved in manuscript form in libraries mainly in Europe and the Muslim world.

The surviving Islamic instruments are legion: over 600 astrolabes, over 150 globes, and a few dozen each of quadrants and sundials, although most postdate the creative period of Islamic science that lasted from the eighth to the fifteenth century. Dozens of treatises survive, and not a few of these were compiled during the early period of Islamic science. Some describe instruments of which there are no surviving examples but which are far more interesting than the standard astrolabe, quadrant or sundial. Many more manuscripts have yet to be unearthed in the various uncataloged collections of Arabic, Persian and Turkish scientific manuscripts, particularly those in libraries in Turkey, Iran and India.[7]

In the nineteenth century, Orientalists began to discover Islamic instruments and produced some of the best studies of such objects that have appeared to this day.[8] Our knowledge of Muslim achievements in astronomical instrumentation took a monumental step forward with the publication of Robert Gunther's *Astrolabes of the World* in 1932, based heavily on the instruments now housed in the Museum of the History of Science in Oxford, which is still the largest and most important collection in the world. But Gunther was not an Arabist and had problems with the inscriptions on the Islamic astrolabes he surveyed, with several unfortunate consequences. Alas, his work has never been corrected or updated. Furthermore, he started a regrettable trend in some circles, namely that it was considered acceptable to work on Islamic instruments without a proper knowledge of Arabic. Numerous nineteenth- and early twentieth-century writings on Islamic instruments scattered in learned journals are now available in the form of reprints in twelve substantial volumes with a total of close to 5,000 pages.[9] Catalogs of individual collections of Islamic instruments are few and far between.[10] A very useful survey of Muslim instrument-makers was published by Leo A. Mayer in 1956.[11] There is currently some scholarly and public interest in this subject, much of which goes back to the insights and aspirations of Derek de Solla Price who, in about 1970, set about taking stock of all known Islamic and European astrolabes (about three times the 300-odd pieces known to Gunther).[12] The first survey of Islamic celestial globes, by Emilie Savage-Smith, was published in 1985;[13] and the long-awaited repertory of Islamic astrolabists and their works by Alain Brieux and Francis Maddison is shortly to appear.[14] I am attempting to compile a catalog of Islamic (and also European) astrolabes, quadrants and sundials up to around 1500, but this is another long-term project.[15]

2. *Basic astronomical notions*

To understand the function of these instruments it is convenient to consider the heavens as fixed on a sphere about the observer which rotates relative to his local horizon (see Figure 2). The celestial sphere appears to rotate about a celestial axis, once in a day-and-night. The altitude of the northern celestial pole above the horizon is the same as the latitude of the locality on the terrestrial sphere. The celestial meridian is the great-circle passing through the celestial pole and the zenith, the point directly above the observer; the meridian, the great-circle through the zenith and the celestial pole, defines the north and south points on the local horizon. The east and west points are defined by the perpendicular great-circle, called the prime vertical. The celestial equator is the great-circle perpendicular to the celestial axis, and it passes through the east and west points of the horizon. Celestial bodies appear to traverse the sky on small-circles parallel to the celestial equator. Their altitude above the horizon and their direction around the horizon, called azimuth, can be measured directly by observation.

The sun appears to move against the background of the fixed stars during the course of a year: its path, called the ecliptic, is inclined to the celestial equator at an angle of about 23½°, called the obliquity. This angle decreases perceptibly over the centuries. The moon and planets appear to move along the ecliptic (in the direction opposite to that of the apparent daily rotation of the celestial sphere), but they, unlike the sun, also wander from side to side of it. To find the positions of the sun on the ecliptic or of the moon and planets with respect to it, the ancients developed mathematical models, either arithmetical or geometrical. The so-called fixed stars also move parallel to the ecliptic perceptibly over the centuries with a motion known as precession. It was necessary to determine their positions for a specific epoch and update star catalogs from time to time. Celestial coordinates were measured either with respect to the ecliptic (longitude and latitude) or with respect to the celestial equator (right ascension and declination).

Knowing the position of the celestial body in a celestial coordinate system and being able to measure its position relative to the horizon enables one to determine the time, in fractional units of the day-and-night interval, defined either as 24 hours or 360°. In the Middle Ages, as in Antiquity, time was also measured in seasonal hours, twelfth divisions of the length of day or night; but these seasonal hours have the disadvantage that they vary from one latitude to another and change in length throughout the year (see below). One can also determine the instantaneous intersections of the ecliptic with the horizon and the meridian, divide the ecliptic further into so-called "houses," and ascertain which planets are in which house; such configurations were thought by some to have some influence on affairs on earth.

3. A classification of medieval Islamic instruments

It is convenient to distinguish between two main categories of Islamic instruments, namely observational and non-observational instruments. Those instruments used by Muslim astronomers for observations followed closely in the tradition of the devices described by Ptolemy of Alexandria (fl. c. AD 140): the armillary sphere – a physical representation of specific astronomically significant circles on the celestial sphere, such as the horizon, the meridian, the celestial equator and the ecliptic; the mural quadrant, for measuring the meridian altitudes of celestial bodies; and the parallactic ruler, for measuring the zenith distance of a celestial body.[16]

Instruments whose primary function is not observational are mainly for solving problems of spherical astronomy, the mathematics of configurations on the celestial sphere about the observer. The main problems are related to timekeeping, using either the risings and settings of the sun and stars over the local horizon or their culminations across the local meridian and an observed altitude.[17] Such instruments include:

- *the celestial sphere*, a model of the universe in which the sun and stars are represented on a sphere that can rotate about the celestial axis, so that risings and settings can be simulated over any horizon;
- *the "analogue computer" known as the astrolabe* (see Figure 3), for representing – in two dimensions rather than three – the positions of the sun and the fixed stars with respect to the local horizon, the passage of time being measured by the amount of rotation between two such positions;[18]
- *mathematical grids* like the sine quadrant for obtaining numerical solutions to problems of trigonometry without calculation;
- *sundials* and other devices for measuring the time of day by means of shadows; and
- *the equatorium*, used for determining planetary positions according to geometric models of the Ptolemaic kind for the sun, moon and five naked-eye planets.

With the exception of the trigonometric grids and numerous instruments in the form of quadrants, these instruments were known to the Muslims from Greek sources. However, what the Muslim astronomers did with them constitutes a singularly colorful chapter in the history of astronomy.

4. Treatises on instruments

Treatises on the construction and use of various instruments abound, but only a few have been published.[19] Some deal with varieties of instruments of which there are no surviving examples.[20] The two major Islamic treatises on the various aspects of instrumentation were:

- The *Kitāb al-Mabādiʾ wa-'l-ghāyāt fī ʿilm al-mīqāt*, meaning something like "An A to Z on astronomical timekeeping," by the Maghribi astronomer Abū ʿAlī al-Marrākushī (Cairo, c. 1280), dealing with all of the standard instruments of his day, and including a detailed survey of spherical astronomy with numerous tables, as well as a series of texts culled from earlier sources on the use of different instruments. The first half of this treatise was published in French translation and the second half summarized by the Sédillots *père et fils* in Paris in the mid-nineteenth century, but the work merits a new study.[21]
- An anonymous untitled treatise that can be attributed to Najm al-Dīn al-Miṣrī (Cairo, c. 1325), in which he describes all of the unusual

instruments known to him and all those invented by himself. This treatise, discovered in 1982 in a manuscript of the Chester Beatty Library in Dublin, is full of surprises, and an edition with translation and commentary has now been published.[22]

Another important work, dealing mainly with the construction of the astrolabe, is *Al-Istīʿāb fī wujūh ṣināʿat al-asṭurlāb* by al-Bīrūnī (Ghazna, early eleventh century), alas as yet not properly published. Al-Bīrūnī relied partially on his teacher al-Sijzī, who was familiar with developments in Baghdad in the ninth and tenth centuries. See below on other works of consequence on instrumentation by such personalities as Ḥabash al-Ḥāsib (Baghdad, ninth century) and Abū Jaʿfar al-Khāzin (Baghdad, tenth century).

5. Schools of astronomical instrumentation

The most important contributions to instrumentation were made by individuals working alone. As the leading lights in this field we may mention Ḥāmid ibn ʿAlī al-Wāsiṭī of Baghdad and Ḥāmid ibn Khiḍr al-Khujandī of Rayy (also of Baghdad?) from the tenth century; ʿAlī ibn Khalaf and Ibn al-Zarqāllu from the eleventh, both active in Toledo; and Ibn al-Sarrāj from the fourteenth, working in Aleppo.

There were schools of instrument-makers, often several generations of the same family, functioning in the following major centers:

- *Baghdad in the ninth and tenth centuries.* Thanks to Ibn al-Nadīm, we have the names of the most important of the men (and one woman) involved,[23] and, happily, a few of their instruments, notably those of Nasṭūlus and al-Wāsiṭī (see Section B1 below). Also, the splendid astrolabe of al-Khujandī (see Section B2) was made for use in Baghdad. In the twelfth century, Baghdad again became important for instrumentation, notably with Hibatallāh (see Section B10) and his followers.
- *Various centers in al-Andalus in the eleventh century* (we have a dozen instruments),[24] and especially in Cordova and Toledo for universal astrolabes and plates (for these we have texts but no instruments).[25]
- *Isfahan from the eleventh to the thirteenth century.* We have several instruments, especially those of Ḥāmid ibn Maḥmūd al-Iṣfahānā and his son Muḥammad,[26] and Muḥammad ibn Abī Bakr al-Ibarī.[27]

- *Marrakesh and Seville in the early thirteenth century*: see the numerous instruments of Abū Bakr ibn Yūsuf[28] and al-Khamāʾirī,[29] respectively, and their many imitators in later centuries.
- *Damascus and Cairo in the thirteenth century*, whence the spectacular, monumental instruments of ʿAbd al-Raḥmān ibn Sinān al-Baʿlabakkī[30] and ʿAbd al-Karīm al-Miṣrī.[31]
- *Delhi in the fourteenth century*, where large and richly decorated instruments were made, none alas extant, and known only from a textual source.[32]
- *Northern Iran and Central Asia in the fourteenth and fifteenth centuries.* We have several instruments by the al-Kirmānī family, of which the most important are those of Muḥammad ibn Jaʿfar al-Kirmānī known as Jalāl (see Section B5).
- *Granada in the fourteenth century*, represented by the Ibn Bāṣo father-and-son team; a treatise by the father and several instruments by the son survive.[33]
- *Damascus in the fourteenth century*: several unusual instruments, notably by Ibn al-Shāṭir (see Sections B12–13) and also al-Mizzī,[34] and numerous treatises.
- *Lahore from the sixteenth to the eighteenth century*, beginning with Mawlānā Maqṣūd Harawī,[35] then mainly one family stemming from Allāh-dād (or Ilāh-dād),[36] and generally more adventurous than their colleagues in Isfahan in that they incorporated numerous features known to us mainly from a much earlier textual tradition.
- *Isfahan (and also Meshed) from the seventeenth to the nineteenth century.*[37] Several prolific astrolabists, all competent craftsmen with an eye for beauty but not particularly innovative from a scientific point of view (although see Section B14), including Muḥammad Muqīm Yazdī, Muḥammad Zamān Mashhadī,[38] Muḥammad Mahdī, Muḥammad Khalīl, ʿAbd al-ʿAlī,[39] Muḥammad Ṭāhir, Muḥammad Amīn and ʿAbd al-Aʾimma.[40]
- *Istanbul from the sixteenth to the nineteenth century*: numerous late quadrants available, surprisingly few astrolabes.[41]
- *Marrakesh and Meknes from the seventeenth (?) to the nineteenth century*, especially the prolific al-Baṭṭūṭī around 1700.[42]

With this background, we now turn to some instruments of singular historical significance, most of which have been studied only in the past twenty years.

B. Some examples of Islamic instruments

1. The oldest surviving Islamic astrolabes

In the late tenth century Ibn al-Nadīm stated that the Muslims came in contact with the astrolabe in Ḥarrān.[43] He added that the first Muslim to make an astrolabe was the eighth-century astronomer al-Fazārī.[44] Alas, we do not have an astrolabe made by al-Fazārī, but there is one from the eighth century: until April 2003, it was preserved in the Archaeological Museum in Baghdad; it bears a signature of an Ottoman named Aḥmad ibn Kamāl. The information given in a description from the 1950s together with the very poor illustrations[45] suffice to establish that this is the oldest known Islamic astrolabe, partly reworked by an Ottoman craftsman. Some of the original *kūfī* inscriptions are still visible. Also, the Ottoman *naskhī* star-names occasionally refer to the wrong stars: during the period of close to a thousand years between the time this piece was made and the time it was fixed up, the star-positions had changed so much that the Ottoman mistook some stars on the rete for the ones that would have been in their positions on a rete of his own time. The simple design of this rete, not found on any other Islamic astrolabe, is identical with that of the sole surviving Byzantine astrolabe, which is dated 1062 and is considerably larger.[46] Even the choice of stars is identical.[47] Furthermore, this is the only Eastern Islamic astrolabe to have plates serving each of the seven geographical climates of Antiquity.[48] It clearly predates all of the other early Islamic astrolabes that survive (see below), and may even predate the main activity of the astronomers of al-Maʾmūn in the early ninth century, by which time the Muslims were in full control of the construction and use of the standard astrolabe.[49] In any case, it is an object of considerable historical importance.

Ibn al-Nadīm also presented a list of the Muslim instrument-makers known to him from the eighth century to his own time.[50] A few astrolabes from the hands of some of these craftsmen survive, and they bear witness to the simple design modified from the standard Byzantine design[51] and to the changes and additional markings that were introduced on the instruments, mainly in the ninth century.[52] These modifications include a plate of horizons for all latitudes, several kinds of scales showing shadow-lengths corresponding to a range of solar altitudes, various kinds of trigonometric grids for facilitating operations with sines and cosines and two kinds of markings in the form of a quadrant for finding the time from the solar altitude, either accurately for a specific latitude or approximately for all latitudes.[53]

2. An astrolabe that survived the Iraqi invasion of Kuwait

A spectacular astrolabe, the most sophisticated piece surviving from the early Islamic period, bears the signature of the distinguished astronomer Ḥāmid ibn al-Khiḍr al-Khujandī and is dated AH 374 (984/5).[54] In 1990 the owner fled Kuwait with this piece in his pocket; all of his other astronomical instruments were stolen by Iraqi troops and have not yet been recovered. Not only do we find on this instrument all of the innovations mentioned in ninth- and tenth-century treatises on the astrolabe, such as, for example, an horary quadrant for the latitude of Baghdad and plates for the astrological doctrine known as "casting the rays", but in addition it is an object of exquisite beauty: see Figure 4. There are some zoomorphic designs on the rete[55] as well as a quatrefoil, generally considered a specifically medieval European motif.[56] Al-Khujandī clearly made other instruments, and these established a tradition that was highly influential in later instrumentation, if mainly in Iran, though not only, for at least one spectacular fourteenth-century French astrolabe bears strong resemblance in its design and decoration to that of al-Khujandī.[57]

3. An astrolabe made by a sultan

An astrolabe in the Metropolitan Museum of Art in New York bears the signature of the Rasulid Yemeni Sultan al-Ashraf and is dated AH 690 (1291). In the 1950s it was declared a fake: no Yemeni sultan could have made an astrolabe, at best the piece must have been made in Egypt. But more recently about a hundred Yemeni astronomical manuscripts have been surveyed. One of these was a copy of a book by al-Ashraf himself, dealing with the construction of the astrolabe, the sundial and the magnetic compass (see Section B13); this is extant in a precious holograph in Cairo with a copy thereof in Tehran.[58] There are two *ijāza*s at the end by two of al-Ashraf's teachers approving various astrolabes that he made, one of which can be identified as the instrument now in New York. The astrolabe is competently made but is of the standard variety. What is more important is that its historical context has been firmly established.[59]

4. Astrolabes made for tourists

One of the unusual features of that Yemeni piece is the use of both *abjad* and Hindu-Arabic numerals on the outer scale of the front rim.[60] In 1989 the Deutsches Schiffahrtsmuseum in Bremerhaven sent me photographs of an astrolabe they had been offered by a German sea captain who had acquired it in Bombay. I suggested they tell the sea captain to take it back to India and get his money back, which he apparently did, for six months later a Lufthansa pilot brought precisely the same instrument to me in Frankfurt; he too had picked it up in Bombay. It is a not unattractive modern fake, and on the outer scale it has both *abjad* and Hindu-Arabic numerals. In fact, it was made in New Delhi in the late 1980s, possibly by someone who had surely seen an illustration of al-Ashraf's astrolabe, although the rete design is Maghribi and other features are Iranian.[61]

Another attractive fake astrolabe was on offer for $75,000 in Fez a few years ago, and was purchased, for what price I do not know, by someone who then hawked it to Christie's and Sotheby's in London. Christie's sent me photos, and I told them it was a fake. Several weeks later, on the very day when I was at Sotheby's checking a few dozen supposedly fake Islamic astrolabes for a possible gem,[62] the same instrument was brought in for assessment. The bearer, already *maẓlūm* after his visit to Christie's, left Sotheby's *maẓlūm khāliṣ*. The piece bears an inscription with a dedication to Ibn Baṭṭūṭa, conveniently dated so that he could have taken it with him on his major voyage. But most fakes, like this one, are easily recognizable, and there are plenty more genuine instruments with more urgent demands on our attention.

5. An astrolabe made for a prince

An elegant astrolabe preserved in Copenhagen is dated AH 830 (1426/7) and bears the signature of one Jalāl al-Kirmānī, already known by several other instruments. An inscription mentioning the name of a ruler has been deliberately obliterated, but we can tell who he was, for there are special markings on the instrument for finding the qibla at Samarqand and Herat (see Figure 5).[63] The ruler who commuted between those two cities at that time was the prince Ulugh Beg,[64] then governor of Khurasan and Mazandarān, and a later astronomer mentions Jalāl al-Dīn al-Asṭurlābī as the instrument-maker at the observatory in Samarqand.[65] Who obliterated his patron's name? Ulugh Beg, by then sultan of Transoxania and Khurasan, was murdered by his own son (who also had astronomical interests), so we know of at least one person who didn't like him. But the astrolabe does not show a hefty dent, and neither does

Ulugh Beg's skull; indeed, his mortal remains attest to the fact that he died from a knife wound.

6. The most sophisticated astrolabe ever made

A spectacular universal astrolabe signed by Ibn al-Sarrāj in AH 729 (1328/9), now preserved in the Benaki Museum in Athens, reveals how little we know about the development of instrument-making and the transmission of scientific ideas within the medieval Muslim world (see Figure 6). Gunther was impressed by this piece, remarking that "if this instrument, which we have not examined, be all of one period, it is an important and early example."[66] There is a manuscript in the Egyptian National Library of a treatise by Ibn al-Sarrāj on the universal astrolabe, without any indication that virtually the same instrument was described by the eleventh-century Andalusi astronomer ʿAlī ibn Khalaf.[67] But the Benaki universal astrolabe is far more developed than the "standard" one described in either treatise: it contains a set of plates that serve other universal functions and a universal grid with a nonuniform diametral scale on the back. The mystery of how these were to be used was solved by the discovery of manuscripts now in Princeton and Dublin of a treatise by the mid-fifteenth-century Egyptian astronomer al-Wafāʾī, who once owned the Benaki instrument (his name is engraved before those of three other later owners on the outer rim). In the treatise, al-Wafāʾī stated that Ibn al-Sarrāj had not provided a treatise on its use: this he would do himself. Al-Wafāʾī described the five different ways in which the various components can be used universally. This quintuply-universal astrolabe is the most sophisticated astrolabe ever made, and even astronomers in sixteenth-century England, where the straightforward universal astrolabe again became popular,[68] would have had severe problems trying to figure out how to use it.[69]

7. Europeans make their first astrolabe

A European astrolabe acquired by Marcel Destombes[70] in 1961 from a dealer in Toulouse (significantly just across the border from Spain) was described by him in a groundbreaking article as "Carolingian" and dated to tenth-century Catalonia.[71] One of the plates serves the latitude of 41°30′, associated with "ROMA ET FRANCIA" (see Figure 7). The piece is clearly the oldest surviving European astrolabe, and hails from the milieu in which eager Europeans first came in contact with Islamic science.

However, its features and inscriptions raised a host of problems which could not be solved at the time, and the instrument was therefore generally considered to be a fake. Destombes was so shattered by the reaction of academics in Paris that he bequeathed the piece, along with the rest of his collection, to the Institut du Monde Arabe in Paris, rather than any national institution.[72] A special session of the Nineteenth International Congress of the History of Science held at Saragossa in 1993 was devoted to the instrument, for, if it is genuine, it is of unparalleled importance; many of the speakers nevertheless remained unconvinced.[73] The piece has, however, been analyzed in light of several dozen early European astrolabes, and, at least in the opinion of this writer, is indeed from the milieu claimed by Destombes.[74] The term *Francia* is to be interpreted as derived from the concept *bilād al-ifranj*, appropriate for Catalonia and points north in the tenth century.[75] The star-pointers, which are unlabelled, are also inaccurately situated, especially when viewed in light of contemporaneous or slightly later instruments and star-tables.[76] The distinctive original inscriptions correspond to those attested in other tenth-century Catalonian inscriptions,[77] and the later (thirteenth-century?) Latin inscriptions contain orthographic curiosities (an R in RISES [*sic*] for the P in PISCES) also found on another late-thirteenth-century Catalan astrolabe preserved by the Society of Antiquaries in London.[78] The "kitchen Latin" inscriptions on the calendrical scale (for example, MAIAS for MAIUS) have their parallels in other late provincial Latin inscriptions. The fact that the style of the instrument bears no relation to any known Andalusi astrolabe (we have one from the tenth century and several from the eleventh) or to any ʿAbbāsid astrolabe (we have one from the eighth or early ninth and several from the tenth century), as well as the inclusion of Rome on the plates (featured on several early Andalusi astrolabes) and a calendrical scale on the back (rarely featured on ʿAbbāsid astrolabes), I take as evidence for the existence of a hypothetical Roman tradition in astrolabe-making, hitherto unknown.[79] Destombes and his astrolabe have been revindicated, but alas, it is a fact that once an artifact is dubbed a fake, the piece remains forever tarnished.

8. An astrolabe made by a Jew, engraved by a Christian and finished by a Muslim

An astrolabe with three layers of inscriptions in cursive Hebrew, Latin and Arabic appeared out of the blue at an auction in France in 1998. The piece was made in Christian Spain, probably in Toledo in the mid-fourteenth century (see Figure 8). It was started by a Jew, partially completed by a Christian well versed in the Latin of the scriptoria, and then – after a very evident rupture in the engraving – finished by a *mudéjar* who left us his name, simply Masʿūd. Perhaps the Jew and the Christian died suddenly in the Black Death. Masʿūd's inscriptions, some in silver-inlayed cartouches, reveal traces of Spanish Arabic and also bear witness to a certain amount of dissimulation, as if he were a prisoner of the Christians.[80] He certainly wanted out, seeking a *hijra* to the Dār al-Islām, and he also wanted to take his astrolabe with him, for he equipped it with a special set of markings for the latitudes of Algiers and Mecca. It is not clear why an otherwise sensible man would have wanted to go to a place like Algiers, which in the Middle Ages was singularly devoid of intellectual activity. (In the thirteenth century the traveler al-ʿAbdarī stated that trying to find a scholar there was like looking for a horse full of camel's eggs.) But Masʿūd apparently did go, and it may be that we have identified his son: there is an Ibn Masʿūd of Andalusian origin (alias Ibn *F-r-m-j-h*), born in Tlemcen at the right time, and with an interest in astronomical instruments. A treatise by him on the astronomical quadrant survives in the Biblioteca de El Escorial.[81] Masʿūd's astrolabe turns out to be the only surviving European instrument from the central lands of Christian Spain from before c. 1500 (altogether four astrolabes are known from medieval Catalonia, the first being that of Destombes). It has been possible to reconstruct the essentials of how this piece came to be what it is. On no other medieval instrument are the lives and fates of three individuals involved in its construction so poignantly indicated, albeit dimly; it is a perfect fruit of *convivencia* between Jews, Christians and Muslims, even though al-Andalus was no utopia and the interfaith participation may not always have been voluntary.[82]

9. Two genuine astrolabes mutated by a forger and a competent craftsman, respectively

In 1960 the Institute for the History of Science at Frankfurt University purchased two Islamic astrolabes which are still in its possession. A rather exorbitant price was paid for one, not least because it was large and imposing but also because it seemed to be associated with the celebrated thirteenth-century polymath Naṣīr al-Dīn al-Ṭūsī. The other was acquired for a pittance, not least because it was small. Alas, the inscription providing a dubious connection with Naṣīr al-Dīn turned out to be a deliberate forgery; the piece was in fact seventeenth-century Indo-Iranian.[83] The other piece was ostensibly from seventeenth-century Isfahan, signed and dated on the back by the well-known astrolabist Muḥammad Khalīl. However, the body of the instrument (the mater) had been reworked and an original inscription deliberately obliterated, and some of the plates are several centuries older than others. Close inspection reveals part of an original inscription with a date, AH 588 (1182/3), and a rather noncommittal name, Muḥammad ibn Aḥmad ibn ʿAlī ibn Muḥammad. Muḥammad Khalīl had simply provided a new rete and a couple of plates, including one for Isfahan. So for a modest sum the Institute acquired one of the thirty oldest Islamic astrolabes, albeit substantially modified.[84]

10. An instrument that survived World War II

A most unusual-looking astrolabic instrument which just before World War II was in a private collection in Munich was until recently known only from an incomplete set of photographs. After the War it proved impossible to trace the owner, and the instrument was presumed lost. The photos revealed that the piece was made by (Badīʿ al-Zamān) Hibatallāh ibn Ḥusayn al-Baghdādī (al-Asṭurlābī) in AH 514 (1120/1), who announced in an inscription that this was the elusive *Zīj al-Ṣafāʾiḥ*, "astronomical handbook on the plates (of an astrolabe),"[85] invented by the tenth-century scholar Abū Jaʿfar al-Khāzin. The mater is indeed covered with astronomical tables, a considerable feat of engraving, and there was an astrolabic plate with standard markings for latitude 33°, serving Baghdad, but there was no rete. A study of what was visible of the instrument in the photos caused us to lament its loss and the fact that we do not have a copy of al-Khāzin's treatise on it, which is known from citations by such leading scholars as Abū Naṣr ibn ʿIrāq and al-Bīrūnī.[86] In 1997 an incomplete photocopy of a manuscript of this treatise in a library in India became available. In the same year, during a visit to the Museum für Islamische Kunst in Berlin to have a look at the seventeenth-century astrolabe of Muḥammad Zamān,[87] I was informed that the lost instrument from Munich had been located in the vaults of the Museum. The treatise is worth more than one doctoral dissertation and the instrument has more components than were shown in the old photos. Research on both will have to wait until the right doctoral students show up and the matter of ownership of the instrument is clarified.

11. A sundial telling all about the times of prayer

Most medieval Islamic sundials[88] bear markings for the times of the *ẓuhr* and *ʿaṣr* prayers.[89] The definitions for these prayer-times which became standard are not specifically mentioned either in the Qurʾān or the *ḥadīth*; they appear for the first time in the *fiqh* books of the eighth century. The origin of the curious definitions of the *ẓuhr* in western Islamic practice and of the *ʿaṣr*, both in terms of the increase of the shadow over its midday minimum, had before the 1970s never been explained.

Now, a marble sundial made in Tunis in AH 746 (1345/6) bears markings for these two prayers and also for the *ḍuḥā* prayer during the morning (see Figure 9).[90] The markings for the *ḍuḥā* and the *ʿaṣr* are symmetrical with respect to the meridian, a feature that led to an investigation of the reason why they were defined in this way. The answer is found in medieval texts on folk astronomy,[91] several of which note a simple approximate arithmetical formula apparently of Indian origin, which became known to the Muslims in the eighth century, linking the time of day in seasonal hours to shadow increases.[92] The aim of the new definitions was to link the beginning of the *ḍuḥā* with the moment when three seasonal hours of daylight had passed (i.e., mid-morning), and the beginning of the *ʿaṣr* with the moment when nine hours had passed (i.e., mid-afternoon).[93] It was the inscription *ḍuḥā* on the Tunisian sundial that led to the realization that the definitions derive from the association of the daylight prayers with the seasonal hours, with the associated historical implications.[94] But this turned out to be nothing new: the whole matter, including the prayer-times of other religions, is discussed by al-Bīrūnī (Ghazna, c. 1025) in his treatise *On Shadows*.[95]

12. A sundial displaying time relative to the times of the five prayers

Most medieval Islamic sundials were modest in size and in potential. However, the astronomer Ibn al-Shāṭir, chief *muwaqqit* of the Umayyad Mosque in Damascus, constructed in AH 773 (1371/2) a horizontal sundial in marble, some 2 m x 1 m in size. This was erected on a platform on the southern side of the main minaret of the Mosque, and it was undoubtedly the most splendid sundial of the Middle Ages. It could be used to measure time after sunrise in the morning and time before sunset in the afternoon, as well as time before and after midday. This means that it measured time relative to the *ẓuhr* and *maghrib* prayers, and the *ʿaṣr* curve enabled regulation of time relative to that prayer also. There are even curves for measuring time with respect to daybreak and nightfall. So with this instrument, which actually comprises three different sundials sharing the same gnomon, one can measure time with respect to midday, sunrise or sunset, and with respect to the times of prayer. The original was broken in the nineteenth century by the leading *muwaqqit* at the Mosque, Muḥammad ibn Muṣṭafā al-Ṭanṭāwī, while he was trying to improve its alignment, and the pieces, recovered during excavations in 1958, are in the garden of the Archaeological Museum in Damascus; al-Ṭanṭāwī's beautifully executed replica is still in situ on the balcony of the main minaret of the Umayyad Mosque (see Figure 11).[96]

13. Magnetic compasses and compendia

Two new sources enable us to document the history of the compass in Islamic civilization more precisely than was hitherto possible.[97] These devices were of use in determinations of the qibla, or sacred direction in Islam.[98]

First, the Rasulid Yemeni Sultan al-Ashraf (see Section B3 above), in his compendium of astronomical instruments compiled in around 1295, described in detail a compass-bowl (*ṭāsa*) with a graduated rim, in which a compass-needle would float on water or some other fluid (see Figure 10). My suspicion that al-Ashraf was drawing on Iranian or Central Asian sources for the idea of this compass is strengthened by the existence of an early sixteenth-century Syrian instrument of precisely this kind which is clearly in a tradition of much earlier Iranian instruments. This is a ceramic bowl preserved in the Archaeological Museum in Damascus that could have been filled with water and would have had a floating magnetic needle to establish the cardinal directions. It was made in Damascus by one Thābit, otherwise unknown to us, and dedicated to Sultan Selīm I and thus datable to c. 1516–20 (see Figure 12).[99] This remarkable object is fully within the tradition of Syrian blue-and-white ceramic tiles, which began in the fifteenth century. However, the geographical data on it is clearly Iranian in origin but so badly corrupted that it is barely intelligible.[100] The Damascus bowl is precisely the kind of instrument described by the Yemeni Sultan al-Ashraf. There is, I believe, sufficient evidence to conclude that such qibla-bowls were in circulation in Central Asia or in Iran in the thirteenth century, if not before.

Second, the Egyptian astronomer Ibn Sāmʿūn, writing in about 1300, describes a "dry" compass sometimes fitted with a small horizontal sundial. His description is followed by a list of qiblas of numerous localities.[101]

Ibn al-Shāṭir (see Section B12) devised a compendium, that is, a multifunctional instrument, in the form of a box with a hinged lid and various movable parts. Some parts of this "jewel box" (*ṣandūq al-yawāqīt*) made by Ibn al-Shāṭir himself survive in Aleppo, and a single plate from another one made by the Cairo astronomer al-Wafāʾī (see Section B6) is preserved in Istanbul. A treatise preserved in Berlin explains the way in which the ensemble is to be used.[102] The box contained a magnetic compass for aligning it in the cardinal directions and a universal polar sundial on a plate that could be attached to the lid and erected in a plane perpendicular to the celestial equator for any of a series of latitudes. Also, sights could be attached to the lid to measure the hour-angle at any of these latitudes, and a set of horizons engraved on the lid enabled the user to compute the instantaneous configuration of the ecliptic relative to the appropriate horizon. In addition, the plate bore markings indicating the qiblas or local directions of Mecca for a series of localities.

Another kind of compendium described by al-Wafāʾī consists of a circular base fitted with a magnetic compass and a universal equatorial sundial.[103] His instrument was popular in Ottoman Turkey for several centuries, and several examples survive. The fact that the magnetic compass does not necessarily point north–south was first noted in the Islamic world by al-Wafāʾī.

14. World-maps for finding the qibla and distance to Mecca

In 1989 a remarkable device surfaced at Sotheby's in London: a circular brass instrument with a diameter of 22.5 cm, which, on the basis of the elegant decoration and Persian inscriptions, is datable to around 1675 (see Figure 13). On the circular plate is engraved a world-map with the positions of some 150 localities from al-Andalus to China, on a complex mathematical grid which preserves direction and distance to Mecca at the center. The meridians are represented by a series of vertical parallels that get closer to each other the further one moves away from the central meridian of Mecca, and the parallels of latitudes by arcs of circles which approximate arcs of ellipses. The inscriptions indicate that if one lays the diametrical rule over a given city, the qibla can be read from the outer scale and the distance to Mecca from the nonuniform diametrical scale. Maps with the property of preserving direction and distance to a non-polar center, in this case Mecca, were first devised in Europe in the early decades of the twentieth century,[104] so the context of the map presented something of a problem since it was clearly a genuine Safavid piece.

Before the piece could be properly assessed, a second specimen showed up at an instrument-dealer's in Paris. This was signed by a certain Muḥammad Ḥusayn, who is known from textual sources as one of the leading instrument-makers of late seventeenth-century Isfahan. Unfortunately, no other pieces bearing his signature have been preserved (and also the piece looks as if it was made in around 1800). It was immediately apparent that the two world-maps were examples of a series, in fact, careful copies of a more detailed world-map.

It seemed reasonable to suspect that there might be some European influence behind the grid (the instruments on which the maps are engraved were originally fitted with European-type sundials and magnetic compasses). But my investigation of contemporaneous European cartography and of the activities of the Europeans resident in or visiting Iran in the seventeenth century produced no evidence of such European initiative. Also, seventeenth-century Safavid works on the qibla and even on astronomy in general proved to be disappointing, both in themselves and for my purposes. Now, the geographical data (longitudes and latitudes) are from a Timurid source, a geographical table with 274 entries apparently compiled in Kish near Samarqand in the mid-fifteenth century. The anonymous compiler calculated the qibla and Mecca-distances of all of the localities mainly accurately to within a few seconds, so the question arose as to whether he might also have devised a graphic solution to the problem he so ably solved by calculation. But only the table is preserved for us, in an eighteenth-century recension in which the qiblas and distances are rounded to minutes.

There were two earlier Muslim scholars who, alone in the history of Islamic science, were not only concerned with the qibla problem but also with mappings preserving direction and distance to the center. The first was Ḥabash (in c. 825) and the second al-Bīrūnī (in c. 1025). Ḥabash appears to have been the first to solve the problem of the determination of the qibla by two distinct accurate methods, one of which involves the mathematical formula relating the qibla and the distance to Mecca which underlies the Safavid grids.[105] Ḥabash also designed an astrolabe based on a related principle preserving direction and distance to the center, in this case the celestial pole.[106] There is no concrete evidence that Ḥabash first conceived a Mecca-centered grid preserving direction and distance to the center based on a formula that he derived. However, an anonymous ʿAbbāsid treatise, possibly an early work of Ḥabash, derives the qibla approximately by cartographical methods that could be applied to, or were derived by consideration of, maps.[107] Likewise, al-Bīrūnī, who was familiar with some of the works of Ḥabash, wrote on the mapping underlying Ḥabash's astrolabe and on others which are also usually regarded as European developments. Only two of a dozen titles by al-Bīrūnī dealing with mathematical geography and qibla determinations have been preserved for us, and in these two there is no mention of a Mecca-centered grid. However, failing a viable alternative hypothesis, I considered that it is possible that a work of Ḥabash or al-Bīrūnī (we are talking about a few sentences of text) which had lain dormant for several centuries might have come to the attention of some Timurid (or even Safavid) astronomer who thought it would be a good idea to make such a world-map. In any case, these two Safavid world-maps, which constitute the only properly drawn Islamic world-maps of any kind to have survived the vicissitudes of time, have led to a detailed investigation of the available evidence.[108] The world-maps remain something of an enigma, and I am hopeful that further investigations of Arabic and Persian scientific manuscripts, on the one hand, and/or materials relating to the transmission of mathematical geography and cartography between Europe and Iran in both directions on the other, may cast more light on this fascinating problem.[109]

In the Fall of 2001, I learned of a third Safavid world-map in a private collection in the United States. This piece, currently on display in the Arthur M. Sackler Museum at Harvard University, is signed by Ḥasan Ḥusayn, an individual whose existence I had hypothesized.[110] Alas, this new piece tells us little about the inspiration for the grids. However, in December 2001, my colleague Dr. Jan Hogendijk of Utrecht informed me that he had found two textual references to the solution of the qibla problem by means of conic sections, one in a treatise from tenth-century Baghdad and the other in a treatise from eleventh-century Isfahan. The first author was known to al-Bīrūnī and the second author knew of al-Bīrūnī. Since the latitude curves on the Safavid grids are ellipses approximated by circular arcs, it seems that here we have come a major step closer to the "missing link" in the story of these world-maps.[111] Hogendijk's forthcoming publication on these new sources is awaited with anticipation.

15. One of the most refined instruments of medieval Europe

From the Muslim world no device is known with which one can find for any latitude the time of day from the solar altitude without calculation. However, the ingenious European universal dial for timekeeping by the sun known as the *navicula de Venetiis*, "little ship of the Venetians," of which a few survive, mainly from fourteenth-century England,[112] has long been thought to be based on an Islamic invention.[113] I have recently assessed the scant evidence that the main mathematical components of the *navicula*, not necessarily organized in the form of a ship,[114] have their origin in the mind of Ḥabash.[115]

In a unique manuscript in Oxford there is a treatise by Ḥabash on an instrument for finding the time of night by the stars.[116] In the form of an astrolabe with throne, the instrument bears mathematical data relating to stellar coordinates, inserted graphically. With the aid of trigonometric grids and various nonuniform scales on the alidade, this data can be manipulated to yield for any latitude the time of night and the azimuth of the star from its altitude, without any calculation. Alas, the operations are not described in the treatise, which deals solely with the construction and is not complete, and have so far defied satisfactory explanation. In addition, the tenth-century encyclopedist Abū ʿAbdallāh al-Khwārizmī mentions a *lawḥ al-sāʿāt* as one of the instruments for timekeeping known to him, and an Ottoman manuscript contains an illustration of a rectangular device simply labeled "*mustaṭīl*" which appears to be some centuries older than the illustration and not necessarily of European inspiration. It is clear that the inventor of the main markings on the *navicula* knew very well how to use the instrument to yield an accurate result.[117] My main reason for suspecting Ḥabash to be behind the clever mathematical device which underlies the *navicula* when used properly is the thought that one would only devise a universal instrument for the stars if one had already devised one for the sun. But on the standard medieval *navicula* there were two additional sets of markings: a shadow square and a universal horary quadrant for determining the time of day approximately for any latitude; these we know for certain were invented in ninth-century Baghdad.[118]

Concluding remarks

We have dealt mainly with some rather exceptional instruments; the reader should bear in mind that many Islamic astrolabes and quadrants are more run-of-the-mill pieces. There is still a great deal of work to be done on instruments and groups of instruments of particular historical importance, and for some aspects – notably metallurgical analyses and construction techniques – future investigators will have to start almost from scratch. Likewise, much can still be written on the decoration and changing design of astrolabe retes. There is also more work to be done on textual sources relating to instrumentation. Both instruments and texts should be considered in light of contemporaneous activity in astronomy. The imminent publication of the *Répertoire* of all known Islamic astrolabists and their works by Francis Maddison and the late Alain Brieux[119] is calculated to arouse new interest in these valuable historical sources. In order to put these in their historical perspective, interested readers can now consult a provisional table of contents of the Frankfurt catalog – still a long way from being complete – on the internet.[120]

NOTES

The following abbreviations are used throughout:

AIOS: Fuat Sezgin et al., eds., *Arabische Instrumente in orientalistischen Studien*, 6 vols. (Frankfurt: Institut für Geschichte der Arabisch-Islamischen Wissenschaften an der Johann Wolfgang Goethe-Universität, 1990–91).

Destombes, *Contributions*: Günther Schilder, Peter van der Krogt and Steven de Clercq, eds., *Marcel Destombes (1905–1983): Contributions sélectionnées à l'histoire de la cartographie et des instruments scientifiques* (Utrecht: HES Publishers, 1987).

DSB: *Dictionary of Scientific Biography*, 16 vols. (New York: Charles Scribner's Sons, 1970–80).

EI[2]: *The Encyclopaedia of Islam, New Edition* (Leiden: E. J. Brill, 1960–).

Frankfurt IGN Festband: Anton von Gotstedter, ed., *Ad radices: Festband zum fünfzigjährigen Bestehen des Instituts für Geschichte der Naturwissenschaften, Frankfurt am Main* (Stuttgart: Franz Steiner, 1994).

Gunther, *Astrolabes*: Robert T. Gunther, *The Astrolabes of the World*, 2 vols. (Oxford: Oxford University Press, 1932; reprinted in 1 vol., London: Holland Press, 1976).

Kennedy, *Studies*: Edward S. Kennedy, *Astronomy and Astrology in the Medieval Islamic World* (Aldershot: Ashgate, 1998).

Kennedy et al., *Studies*: E. S. Kennedy, colleagues and former students, *Studies in the Islamic Exact Sciences* (Beirut: American University of Beirut, 1983).

Kennedy Festschrift: David A. King and George Saliba, eds., *From Deferent to Equant: A Volume of Studies in the History of Science in the Ancient and Medieval Near East in Honor of E. S. Kennedy* (New York: New York Academy of Sciences, 1987).

Khalili Catalogue: Francis Maddison and Emilie Savage-Smith, *Body and Spirit, Mapping the Universe*, pt. 1 of *Science, Tools and Magic*, vol. 12 of *The Nasser D. Khalili Collection of Islamic Art* (London: Nour Foundation in association with Azimuth Editions and Oxford University Press, 1997).

King, *SATMI*: David A. King, *Studies in Astronomical Timekeeping in Medieval Islam*, 12 parts (in press with E. J. Brill).

King, *Studies A*: David A. King, *Islamic Mathematical Astronomy* (London: Variorum, 1986; 2nd rev. edn., Aldershot: Variorum, 1993).

King, *Studies B*: David A. King, *Islamic Astronomical Instruments* (London: Variorum, 1987; reprint, Aldershot: Variorum, 1995).

King, *Studies C*: David A. King, *Astronomy in the Service of Islam* (Aldershot: Variorum, 1993).

King, *World-Maps*: David A. King, *World-Maps for Finding the Direction and Distance to Mecca: Innovation and Tradition in Islamic Science* (Leiden: Brill, and London: Al-Furqān Islamic Heritage Foundation, 1999).

Kuwait Catalogue: Arlene Fullerton and Géza Fehérvári, eds., *Kuwait: Arts and Architecture; A Collection of Essays* (Kuwait: n.p., 1995).

Linton Catalogue: *Scientific Instruments and Rare Books from the Collection of Leonard Linton, Point Lookout, N. Y.* (Paris: Alain Brieux, 1980).

Lorch, *Studies*: Richard Lorch, *Arabic Mathematical Sciences: Instruments, Texts, Transmission* (Aldershot: Variorum, 1995).

Madrid MAN 1992 Exhibition Catalogue: Juan Vernet and Julio Samsó, eds., *El legado científico andalusí* (Madrid: Ministerio de Cultura, 1992).

Paris IMA 1993–94 Exhibition Catalogue: Sophie Cluzan, Eric Delpont and Jeanne Mouliérac, eds., *Syrie: Mémoire et civilisation* (Paris: Institut du Monde Arabe, 1993).

Vernet Festschrift: Josep Casulleras and Julio Samsó, eds., *From Baghdad to Barcelona: Studies in the Islamic Exact Sciences in Honour of Prof. Juan Vernet*, 2 vols. (Barcelona: Instituto Millás Vallicrosa de Historia de la Ciencia Árabe, 1996).

1 For an overview see David King, "Islamic Astronomy," in *Astronomy before the Telescope*, ed. Christopher Walker (London: British Museum Press, 1996; reprint, 1999), 143–74. Much of the information in that article is based on essays collected in Kennedy et al., *Studies*; Kennedy, *Studies*; Lorch, *Studies*; and King, *Studies A–C*.

Useful collections of articles on Islamic science are in the *Kennedy Festschrift* and the *Vernet Festschrift*. The pioneering work by Aydın Sayılı, *The Observatory in Islam and its Place in the General History of the Observatory* (Ankara: Türk Tarih Kurumu Basımevi, 1960; reprint, New York: Arno Press, 1981), is still extremely useful. We shall also cite various articles in the new edition of the *Encyclopaedia of Islam* (*EI*^{2}). Several Muslim astronomers are well treated in the *Dictionary of Scientific Biography* (*DSB*). The standard bio-bibliographical sources of Heinrich Suter, Carl Brockelmann and Fuat Sezgin are well known.

2 See King, *SATMI*, I–XII. For an introduction to the topic, see various papers reprinted as King, *Studies A*, IX–XII, and King, "On the role of the Muezzin and the *Muwaqqit* in medieval Islamic society," in *Tradition, Transmission, Transformation: Proceedings of two Conferences on Premodern Science held at the University of Oklahoma*, ed. F. J. Ragep and S. P. Ragep (Leiden: E. J. Brill, 1996), 285–346, to be reprinted as King, *SATMI*, V. For an overview see the articles in *EI*^{2}, s.v. "Mīḳāt" (regulation of the prayers and astronomical timekeeping) by A. J. Wensinck (legal considerations) and D. A. King (scientific aspects), of which the latter is reprinted in King, *Studies C*, V.

3 An overview of Islamic instrumentation is in King, "Strumentazione astronomica nel mondo medievale Islamico," in *Gli strumenti*, ed. Gerard L'Estrange Turner, 2nd edn. (Turin: Giulio Einaudi, 1991), 154–89 and 581–5 (bibliography), of which the English original appears as *SATMI*, VIII. For a summary see my paper, "Some Remarks on Islamic Astronomical Instruments," *Scientiarum Historia* (Brussels) 18, 1 (1992): 5–23.

4 It is worth noting that the two volumes by Leo A. Mayer entitled *Islamic Astrolabists and their Works* (Geneva: Albert Kundig, 1956) and *Islamic Metalworkers and their Works* (Geneva: Albert Kundig, 1959) are of about the same size and scope.

5 See the articles reprinted in King, *Studies C*, VI and VII, and in *SATMI*, VI (universal solutions). On some universal instruments, see Sections B6, B13 and B14, and also n. 25 below.

6 There is, alas, no overview of Islamic astronomy in the formative period. The regional schools appear in the "post-Sezgin" period, that is, in the eleventh and twelfth centuries, with their own specific interests and their own authorities. They have been surveyed in the following studies:

- *Iran and Central Asia*: E. S. Kennedy, "The exact sciences in Iran under the Saljuqs and Mongols," *The Cambridge History of Iran*, vol. 5, *The Saljuq and Mongol Periods*, ed. J. A. Boyle (Cambridge: Cambridge University Press, 1968), 659–79; idem, "The exact sciences in Timurid Iran," ibid., vol. 6, *The Timurid and Safavid Periods*, ed. Peter Jackson and Laurence Lockhart (Cambridge: Cambridge University Press, 1986), 568–80; and for the Safavid period see H. J. J. Winter, "Persian science in Safavid Times," ibid., 581–609; and now King, *World-Maps*, 128–34.
- *Egypt, Syria and the Yemen*: King, "Aspects of Fatimid astronomy," in *L'Égypte fatimide, son art et son histoire: Actes du colloque organisé à Paris les 28, 29 et 30 mai 1998*, ed. Marianne Barrucand (Paris: Presses de l'Université de Paris-Sorbonne, 1999), 497–517; idem, "The Astronomy of the Mamluks," *Isis* 74 (1983): 531–55, reprinted in King, *Studies A*, III; idem, "L'Astronomie en Syrie à l'époque islamique," in *Paris IMA 1993–94*

Exhibition Catalogue, 386–95, 432–43 and 480; and idem, *Mathematical Astronomy in Medieval Yemen: A Bio-Bibliographical Survey* (Malibu: Undena, 1983).

- *The Maghrib*: King, "On the history of astronomy in the medieval Maghrib," in *Études philosophiques et sociologiques dédiées à Jamal ed-Dine Alaoui* (Fez, 1998), 27–61; and Julio Samsó, "An outline of the history of Maghribā *Zīj*es from the end of the thirteenth century," *Journal for the History of Astronomy* 29 (1998): 93–102.
- *Al-Andalus*: See Julio Samsó, *Las ciencias de los antiguos en al-Andalus* (Madrid: Mapfre, 1992); idem, *Islamic Astronomy and Medieval Spain* (Aldershot: Variorum, 1994); and numerous other publications of the Barcelona School.
- *Ottoman Turkey*: No reliable survey is available but see Ekmeleddin İhsanoğlu, ed., *Osmanlı Astronomi Literatürü Tarihi (History of Astronomy Literature during the Ottoman Period)*, 2 vols., and *Osmanlı Matematik Literatürü Tarihi (History of Mathematics Literature during the Ottoman Period)*, 2 vols. (Istanbul: İslâm Tarih, Sanat ve Kültür Araştırma Merkezi, 1997 and 1999).
- *India*: S. M. Razaullah Ansari, "On the transmission of Arabic-Islamic astronomy to medieval India," *Archives internationales d'histoire des sciences* 45 (1995): 273–97.

See Section A5 of this chapter on regional schools of instrumentation.

7 On the delights and rewards of working on such sources, and some of the problems and frustrations, see King, "Some remarks on Islamic scientific manuscripts and instruments, and past, present, and future research," in *The Significance of Islamic Manuscripts*, ed. John Cooper (London: Al-Furqān Islamic Heritage Foundation, 1992), 115–43.

8 The best studies from the nineteenth century are those by Sédillot *fils* (see n. 21 below), Sarrus (n. 28), Woepcke (n. 24), Morley (n. 39) and Dorn (n. 34). The best from the twentieth century, still in the same tradition, are those by Josef Frank and Max Meyerhof (n. 36) and Gisela Helmecke (n. 38).

9 In addition to *AIOS*, six more volumes were published by the Institut für Geschichte der Arabisch-Islamischen Wissenschaften as part of a monumental reprint series in 105 volumes to date, namely *Islamic Mathematics and Astronomy*, vols. 91–6 (1998); these have not been cited below.

10 We shall have occasion to cite the *Linton Catalogue*; King, "Early Islamic Astronomical Instruments in Kuwaiti Collections," in the *Kuwait Catalogue*, 76–96; and the *Khalili Catalogue*, on which see my review essay, "Cataloguing Medieval Islamic Astronomical Instruments," *Bibliotheca Orientalis* (Leiden) 57 (2000), cols. 247–58. Also various exhibition catalogs could be mentioned, including *Madrid MAN 1992 Exhibition Catalogue* and the *Paris IMA 1993–94 Exhibition Catalogue*.

11 See n. 4 above.

12 Sharon Gibbs, Janice A. Henderson and Derek J. de Solla Price, *A Computerized Checklist of Astrolabes* (New Haven: Yale University, Department of History of Science and Medicine, 1973). See also Price, "An International Checklist of Astrolabes," *Archives internationales d'histoire des sciences* 8 (1955): 243–63 and 363–81.

13 Emilie Savage-Smith, *Islamicate Celestial Globes: Their History, Construction, and Use* (Washington, DC: Smithsonian Institution Press, 1985).

14 Alain Brieux and Francis R. Maddison, *Répertoire des facteurs d'astrolabes et de leurs oeuvres*, vol. 1 (Islam), in press (Paris: CNRS).

15 See King, "Medieval astronomical instruments: a catalogue in preparation," *Bulletin of the Scientific Instrument Society* 31 (1991): 3–7. For some of the earliest findings, see idem, "Astronomical instruments between East and West," in *Kommunikation zwischen Orient und Okzident: Alltag und Sachkultur*, ed. Helmut Hundsbichler (Vienna: Österreichische Akademie der Wissenschaften, 1994), 143–98. Funding has been provided by the Deutsche Forschungsgemeinschaft during 1992–5 (for Islamic instruments) and during 1999–2002 (for European instruments). For the table of contents, see n. 120 below.

16 On Islamic observational instruments, see Sayılı, *The Observatory in Islam* (cited in n. 1 above). Studies of instruments at individual observatories beyond those discussed by Sayılı include the following: Frans Bruin, "The Fakhrī sextant in Rayy," *Al-Bīrūnī Newsletter* (American University of Beirut, Physics Department) 19 (1969); Aydın Sayılı, "Al-Khāzinī's treatise on astronomical instruments," *Ankara Üniversitesi Dil ve Tarih-Coğrafya Fakültesi Dergisi* 14 (1956): 18–19; Frans Bruin, "A treatise on small instruments by . . . al-Khāzinī," *Al-Bīrūnī Newsletter* 30 (1970); Hugo J. Seemann, "Die Instrumente der Sternwarte zu Maragha nach den Mitteilungen von al-ʿUrḍī," *Sitzungsberichte der Physikalisch-Medizinischen Sozietät zu Erlangen* 60 (1928): 15–126, reprinted in *AIOS*, 6:17–129; Sevim Tekeli, "'The article on the quality of the observations' of al-ʿUrḍī," *Araştırma* (Ankara Üniversitesi Dil ve Tarih-Coğrafya Fakültesi, Felsefe Araştırmaları Enstitüsü) 8 (1970), 57–169; Frans Bruin, "The observational instruments of al-ʿUrḍī," *Al-Bīrūnī Newsletter* 10 (1968); idem, "The construction of instruments . . . by . . . al-ʿAmilī," ibid., 15 (1968); E. S. Kennedy, "Al-Kāshī's treatise on astronomical observational instruments," *Journal of Near Eastern Studies* 20 (1961): 98–108, reprinted in Kennedy et al., *Studies*, 394–404; Sevim Tekeli, "The astronomical instruments of *Zīj-i-Shāhinshāhīya*," *Araştırma* 1 (1963): 88–122; idem, "The comparison of the instruments of Taqī al Dīn and Tycho Brahe," *Ankara Üniversitesi Dil ve Tarih-Coğrafya Fakültesi Dergisi* 16 (1958): 377–86; Frans Bruin, "The instruments of Taqī al-Dīn," *Al-Bīrūnī Newsletter* 29 (1970); and S. M. Raza Ansari and S. A. Khan Ghori, "Two treatises on astronomical instruments by ʿAbd al-Munʿim al-ʿAmilī and Qāsim ʿAlī al-Qāyinī," in *History of Oriental Astronomy: Proceedings of an International Astronomical Colloquium No. 91, New Delhi, India, 13–16 November 1985*, ed. G. Swaru, A. K. Bag and K. Shukla (Cambridge: Cambridge University Press, 1987), 215–25.

17 These problems are essentially problems of conversion of coordinates between different systems: the horizon-based (with coordinates altitude and azimuth for a specific terrestrial latitude), the equatorial (declination and hour-angle), and the ecliptic (celestial latitude and longitude). Thus, for example, if one measures the altitude of the sun on a day when the solar declination and the local latitude are known, one can derive the hour-angle and the azimuth.

18 The best general study of different varieties of astrolabe is *The Planispheric Astrolabe* (Greenwich: National Maritime Museum, 1976). The popular article by John D. North, "The astrolabe," *Scientific American* 230 (1974): 96–106, also provides an excellent introduction to the most common type. On the astrolabe in Islam, see Willy Hartner, "The principle and use of the astrolabe" and "Asṭurlāb" (the latter reprinted from *EI*[2]), in *Willy Hartner, Oriens-Occidens: Ausgewählte Schriften zur Wissenschafts- und Kulturgeschichte; Festschrift zum 60. Geburtstag*, ed. Günther Kerstein et al. (Hildesheim: Georg Olms, 1968), 287–311 and 312–18. New findings on the various components and their origins are presented in King, "The neglected astrolabe," in *Mathematische Probleme im Mittelalter: Der lateinische und arabische Sprachbereich*, ed. Menso Folkerts (Wiesbaden: Harrassowitz, 1996), 45–55.

19 For these the reader is referred to King, *SATMI*, VIII, although a few newly published treatises will be mentioned below.

20 See, for example, E. S. Kennedy, *The Planetary Equatorium of Jamshīd Ghiyāth al-Dīn al-Kāshī (d. 1429)* (Princeton: Princeton University Press, 1960); and Richard Lorch, "The astronomical instruments of Jābir ibn Aflaḥ and the Torquetum," *Centaurus* (Copenhagen) 20 (1976): 11–34, reprinted in idem, *Studies*, XVI; idem, "Al-Khāzinī's 'sphere that rotates by itself'," *Journal for the History of Arabic Science* 4 (1980): 287–329, reprinted in idem, *Studies*, XI; and idem, "Naṣr ibn ʿAbdallāh's instrument for finding the Qibla," *Journal for the History of Arabic Science* 6 (1982): 123–31.

21 See Jean-Jacques Sédillot (*père*), *Traité des instruments astronomiques des Arabes composé au treizième siècle par Aboul Hhassan Ali de Maroc*, 2 vols. (Paris: Imprimerie Royale, 1834–5; reprinted in 1 vol., Frankfurt: Institut für Geschichte der Arabisch-Islamischen Wissenschaften [IGAIW], 1985), for a translation of the first half of the work; Louis-Amélie Sédillot (*fils*), "Mémoire sur les instruments astronomiques des Arabes," *Mémoires de l'Académie Royale des inscriptions et belles-lettres de l'Institut de France* 1 (1844): 1–229 (reprint, Frankfurt: IGAIW, 1989), for a summary of the second half; and Fuat Sezgin et al., *Comprehensive Collection of Principles and Objectives in the Science of Timekeeping: Jāmiʿ al-mabādiʾ wa-'l-ghāyāt* (facsimile of MS Topkapı Ahmet III 3343), 2 vols. (Frankfurt: IGAIW, 1984). The originality of the author is underestimated in my article "al-Marrākushī" in *EI*[2].

22 See François Charette, *Mathematical Instrumentation in Fourteenth-Century Syria and Egypt: The Illustrated Treatise of Najm al-Dīn al-Miṣrī* (Leiden: E.J. Brill, 2003). On Najm al-Dīn al-Miṣrī see idem, "A Monumental Medieval Table for Solving the Problems of Spherical Astronomy for All Latitudes,"

Archives internationales d'histoire des sciences 48 (1998): 11–64. A second, incomplete copy of his treatise on instrumentation surfaced recently: see *Christie's (London), Islamic Art and Manuscripts 11.4.2000 Auction Catalogue*, lot 22. This manuscript was also used by Charette for his edition.

23 See nn. 43 and 50 below.

24 See, for example, Franz Woepcke, "Über ein in der Königlichen Bibliothek zu Berlin befindliches arabisches Astrolabium," *Abhandlungen der Königlichen Akademie der Wissenschaften zu Berlin* (1855), 1–31 (separatum), reprinted in *AIOS*, 2:1–36, and also in Fuat Sezgin, ed., *Franz Woepcke: Études sur les mathématiques arabo-islamiques* (Frankfurt: IGAIW, 1986), 2:131–65, and summarized in Gunther, *Astrolabes*, 1:251–2, no. 116.

25 On the universal astrolabe and plate see the early studies of J. Millás Vallicrosa, including "Estudios sobre Azarquiel: El tratado de la azafea," *Archeion* 14 (1932): 393–419; *Don Profeit Tibbon: Tractat de l'assafea d'Azarquiel* (Barcelona, 1933); *Estudios sobre Azarquiel* (Madrid and Granada, 1943–50); and "Un ejemplar de azafea árabe de Azarquiel," *Al-Andalus* 9 (1944): 11–119. Important new studies are Roser Puig Aguilar, *Ibn al-Naqqāš al-Zarqālluh (Azarquiel): Al-Šakkāzīya* (Barcelona, 1985); idem, ."Concerning the *ṣafīḥa shakkāzīya*," *Zeitschrift für Geschichte der arabisch-islamischen Wissenschaften* 2 (1985): 123–39; idem, *Los tratados de construcción y uso de la azafea de Azarquiel* (Madrid, 1987); idem, "Al-Zarqālluh's graphical method for finding lunar distances," *Centaurus* 32 (1989): 294–309; and idem, "Instrumentos universales en al-Andalus," chapter in *Madrid MAN 1992 Exhibition Catalogue*, 67–73. See also King, *Studies B*, VII and IX, and my article "Shakkāziyya" (universal projections) in *EI²*.

26 For examples, see Gunther, *Astrolabes*, 1:117, no. 4, and *Linton Catalogue*, 83–5, no. 161.

27 See Gunther, *Astrolabes*, 1:118–20, no. 5, and *Khalili Catalogue*, 210–11, no. 122.

28 Frédéric Sarrus, "Description d'un astrolabe construit à Maroc en l'an 1208," *Mémoires de la Société du Muséum d'Histoire Naturelle de Strasbourg* 4, 2–3 (1850): 1–32 and 6 pls. (separatum dated 1852), reprinted in *AIOS*, 1:193–230.

29 See, for example, Gunther, *Astrolabes*, 1:269–77, nos. 127–30.

30 Published in King, "The monumental Syrian astrolabe in the Maritime Museum, Istanbul," in *Erdem* (Ankara: Atatürk Kültür Merkezi), special issue in memory of Aydın Sayılı, vol. 9, nos. 25–7 (1996–97), 26:729–35 and 10 pls.

31 See Gunther, *Astrolabes*, 1:233–7, nos. 103 and 104. These cry out for a detailed comparative study. New insights by J. Michael Rogers on the not unproblematic inscriptions are in Jay A. Levenson, ed., *Circa 1492: Art in the Age of Exploration* (New Haven: Yale University Press, 1991), 215–16.

32 See the important new study by Sreeramula R. Sarma, "Sulṭān, Sūri and the astrolabe," *Indian Journal for the History of Science* 35 (2000): 129–47.

33 See *Linton Catalogue*, 87–9, no. 162, for one instrument, and Emilia Calvo, "Ibn Bāṣo's universal plate and its influence on European astronomy," *Scientiarum Historia* 18, 1 (1992): 61–9, and idem, "Ibn Bāṣo's astrolabe [plate]

in the Maghrib and East," in the *Vernet Festschrift*, 2:755–67, for the most remarkable feature of these instruments.

34 Three quadrants by al-Mizzī are well served in William H. Morley, "Description of an Arabic quadrant," *Journal of the Royal Asiatic Society* 17 (1860): 322–30 and 2 pls., reprinted in *AIOS*, 1:326–36; Bernhard Dorn, "Drei in der Kaiserlichen Öffentlichen Bibliothek befindliche astronomische Instrumente mit arabischen Inschriften," *Mémoires de l'Académie Impériale des Sciences de St. Pétersbourg*, 7th ser., 9:1 (1865), reprinted in *AIOS*, 1:345–98; and Géza Fehérvári, "An eighth/fourteenth-century quadrant of the astrolabist al-Mizzī," *Bulletin of the School of Oriental and African Studies* 36 (1973): 115–17 and 2 pls.

35 There is a splendid monumental astrolabe in the Khalili Collection in London which is datable to c. 1550 and is perhaps to be attributed to Harawī. See *Khalili Catalogue*, 219, 230, 232–3 and no. 133, and King, *World-Maps*, xxix. This instrument and those of Allāh-dād show influence from Samarqand, but there was already an independent tradition in Muslim India (see n. 32 above).

36 A model publication of one of the family's productions is Josef Frank and Max Meyerhof, "Ein Astrolab aus dem indischen Mogulreiche," *Heidelberger Akten der von-Portheim-Stiftung* 13 (Heidelberg: Carl Winter, 1925), reprinted in *AIOS*, 4:307–56. See also Sreeramula R. Sarma, "The Lahore family of astrolabists and their ouvrage," *Studies in History of Medicine and Science* (Hamdard University, Hamdard Nagar, New Delhi) 13 (1994): 205–24.

37 For new light on this school, see King, *World-Maps*, 128–34 and 262–74.

38 See the splendid study by Gisela Helmecke, "Das Berliner Astrolab des Muḥammad Zamān al-Mašhadī," *Forschungen und Berichte der Staatlichen Museen zu Berlin* 25 (1985): 129–42 and pls. 31–2.

39 See the splendid study by William H. Morley, *Description of a Planispheric Astrolabe, Constructed for Sháh Sultán Husain Safawí* (London: Williams and Norgate, 1856), reprinted as an introduction to Gunther, *Astrolabes*, and in *AIOS*, 1:249–325.

40 Several fake astrolabes "signed" by him have ended up in major collections: see Owen Gingerich, David King and George Saliba, "The ʿAbd al-A'imma astrolabe forgeries," *Journal for the History of Astronomy* 3 (1972): 188–98, reprinted in King, *Studies B*, VI.

41 An important astrolabe made by al-Aḥmar al-Nujūmī al-Rūmī for the treasury of the Ottoman Sultan Bayezit II (1481–1512), dated AH 911 (1505/6), was offered at Sotheby's of London in October 1998, but aroused no interest and was not sold.

42 See *Khalili Catalogue*, 202–5 and 261–3, nos. 152–3, for two examples.

43 See n. 50 below for a new translation of Ibn al-Nadīm's remarks based on *Al-Fihrist*, ed. G. Flügel, 2 vols. (Leipzig, 1871; reprinted in 1 vol., Beirut: Khayyat's, 1964), 284–5. (The translation by Bayard Dodge in *The Fihrist of al-Nadīm: A Tenth-Century Survey of Muslim Culture* [New York: Columbia University Press, 1970], 2:670–2, is unreliable.) On this and other sources see

also King, "The origin of the astrolabe according to the medieval Islamic sources," *Journal for the History of Arabic Science* 5 (1981): 43–83, reprinted in King, *Studies B*, III.

44 See the article by David Pingree in the *DSB*, s.v. "al-Fazārī."

45 B. Fransīs and N. al-Naqshbandī, "Al-Asṭurlābāt fī Dār al-āthār al-ʿarabiyya fī Baghdād," *Sumer* (Baghdad) 13 (1957): 9–33 and 3 pls., esp. 12–13 and pls. 2–3. In April 2003 I prepared a detailed study of this, the oldest known astrolabe.

46 O. M. Dalton, "The Byzantine astrolabe at Brescia," *Proceedings of the British Academy* (London) 12 (1926): 133–46 and 3 pls., summarized in Gunther, *Astrolabes*, 1:104–8, no. 2.

47 See Burkhard Stautz, "Die früheste bekannte Formgebung der Astrolabien," in *Frankfurt IGN Festband*, 315–28, esp. 320–22; and idem, *Untersuchungen von mathematisch-astronomischen Darstellungen auf mittelalterlichen Instrumenten islamischer und europäischer Herkunft* (Bassum: Geschichte der Naturwissenschaften und der Technik, 1997), 39–43.

48 On the climates see King, "Bringing astronomical instruments back to earth: the geographical data on medieval astrolabes (to ca. 1100)," in *Between Demonstration and Imagination: Essays in the History of Science and Philosophy Presented to John D. North*, ed. Lodi Nauta and Arjo Vanderjagt (Leiden: Brill, 1999), 3–53, esp. 6–9.

49 The two treatises by al-Khwārizmī (c. 825) on the construction and use of the astrolabe, the earliest extant treatises in Arabic, have now been prepared with text, translation and commentary by François Charette and Petra Schmidl (as yet unpublished).

50 Ibn al-Nadīm's text (cited in n. 43 above) translates as follows (the names of makers from whom we have surviving instruments are italicized):

> Discussion of the Instruments and their Makers (*ṣunnāʿ*). In ancient times astrolabes were plane. The first person to make (*ʿamala*) them was Ptolemy. It is said that they were made before his time, but this is not known with exactitude. The first person to make a plane astrolabe (*sattaḥa al-asṭurlāb*) was Abiyūn al-Baṭrīq. Instruments were made (*kānat tuʿmalu*) in the city of Ḥarrān. From there they spread (*tashattatat ? wa ẓaharat*) [and] {text: *wa lākinnahā*, lit. but} they increased [in number], and in the ʿAbbāsid realm work became plentiful for the instrument-makers (*ṣunnāʿ*), from the days of al-Maʾmūn to this our own time.
>
> When al-Maʾmūn wished to make astronomical observations, he selected [Khālid] {text: Ibn Khalaf} al-Marwarrūdhī, who made [for the Caliph] an armillary sphere, the same one which is still in the possession of one of the scholars of our city. Al-Marwarrūdhī [also] made astrolabes.

The names of the makers are [Khālid] {text: Ibn Khalaf} al-Marwarrūdhī; al-Fazārī, who has already been mentioned; ʿAlī ibn ʿĪsā, the apprentice (*ghulām*) of al-Marwarrūdhī; *Khafīf, the apprentice of ʿAlī ibn ʿĪsā*, who was clever and superior; *Aḥmad ibn Khalaf, apprentice of ʿAlī ibn ʿĪsā*; Muḥammad ibn Khalaf, also an apprentice of ʿAlī; Aḥmad ibn Isḥāq al-Ḥarrānī; al-Rabīʿ ibn al-Farrās al-Ḥarrānī; [*Nasṭūlus*] {Flügel confused}, apprentice of Khafīf; ʿAlī ibn Aḥmad the engineer/geometer (*al-muhandis*), apprentice of Khafīf; *Muḥammad ibn Shaddād al-Baladī*; ʿAlī ibn ṣurad [al-] Ḥarrānī; Shujāʿ ibn . . .(?), an apprentice of [Nasṭūlus], who was with Sayf al-Dawla; Ibn Salm, an apprentice of [Nasṭūlus]; al-ʿIjlī al-Asṭurlābī, an apprentice of [Nasṭūlus]; al-ʿIjliyya, his daughter, who was with Sayf al-Dawla, [and who was] a pupil (*tilmīdha*) of [Nasṭūlus].

Some of the apprentices of Aḥmad and Muḥammad, the two sons of Khalaf, were Jābir ibn Sinān al-Ḥarrānī; Jābir ibn Qurra al-Ḥarrānī; Sinān ibn Jābir al-Ḥarrānī; Farrās ibn al-Ḥasan al-Ḥarrānī; [and] *Abu al-Rabīʿ Ḥāmid ibn ʿAlī*, an apprentice of ʿAlā ibn Aḥmad *al-muhandis*. Some of the apprentices of Ḥāmid ibn ʿAlī were Ibn Najiyya, whose name was . . .(?) . . .(?); [and] al-Būqī, whose name was al-Ḥusayn, but he changed it to ʿAbd al-Ṣamad.

[Other] early instrument-makers were ʿAlī ibn Yaʿqūb al-Raṣṣāṣ and ʿAlī ibn Saʿīd al-Uqlīdisī, [as well as] Aḥmad ibn ʿAlī ibn ʿĪsā, near[er] to our own time.

51 For example:

- An undated astrolabe by Aḥmad ibn Khalaf, preserved in the Bibliothèque Nationale de France: see Gunther, *Astrolabes*, 1:230, no. 99.
- An undated astrolabe and a solitary rete, the former signed by "Khafīf, the apprentice of ʿAlī ibn ʿĪsā," preserved in the Museum of the History of Science, Oxford. Both pieces bear later inscriptions in Armenian. Described in Francis R. Maddison, *A Supplement to a Catalogue of Scientific Instruments in the Collection of J. A. Billmeir, Esq., C.B.E.* (London: Frank Partridge & Sons, 1957), 16–18, nos. 155–6. Yet another instrument by the same maker is illustrated in a document preserved in the Uffizi: see George Saliba, "A sixteenth-century drawing of an astrolabe made by Khafīf Ghulām ʿAlī b. ʿĪsā (c. 850 A.D.)," *Nuncius: Annali di Storia della Scienza* (Florence) 6 (1991): 109–19.
- An astrolabe by Nasṭūlus, dated AH 315 (927/8), preserved in the Dār al-Athār al-Islāmiyya in Kuwait: see *Kuwait Catalogue*, 76, 78, 79, 81 and 83.
- Another unsigned and undated piece from the same milieu is the so-called "Astrolabe of Pope Sylvester II" preserved in the Museo di Storia della Scienza in Florence. See Gunther, *Astrolabes*, 1, no. 101. This piece bears

some dubious European markings on the back and is housed in a case with a highly dubious medieval Latin inscription. It has been exhibited several times.

52 For example:

- The mater of an undated astrolabe by Muḥammad ibn Shaddād (al-Baladī), formerly in Berlin, present location unknown. Described and illustrated in Dorn, "Drei astronomische Instrumente" (see n. 34 above), 115–18 (*AIOS*, 1:461–4).
- The mater of an imposing astrolabe by Ḥāmid ibn ʿAlī al-Wāsiṭī dated AH 343 (954/5). The instrument was stolen from a museum in Palermo early in the twentieth century and has not resurfaced. Fortunately, it is described and illustrated in V. Mortillaro, "Illustrazione di un astrolabio arabo del nono secolo," *Opere*, vol. 4 (Palermo, 1848), 110–35 and 2 pls., reprinted in *AIOS*, 1:162–91.

53 See EI^2, s.v. "Rubʿ" (quadrant), and now King, "A *Vetustissimus* Arabic treatise on the *Quadrans Vetus*," *Journal for the History of Astronomy* 33 (2002): 1–19, with a more detailed version in idem, *SATMI*, IXa.

54 Gunther, *Astrolabes*, 1:245, no. 111, and *Kuwait Catalogue*, 80 and 82–9.

55 See Owen Gingerich, "Zoomorphic astrolabes and the introduction of Arabic star names into Europe," in the *Kennedy Festschrift*, 89–104, esp. 97–8, where this piece is not mentioned.

56 On the quatrefoil as decoration on astrolabes, see King, *The Ciphers of the Monks: A Forgotten Number-Notation of the Middle Ages* (Stuttgart: Franz Steiner, 2001), appendix J, 380–90.

57 Illustrated in Emmanuel Poulle, *Les Instruments astronomiques du Moyen Âge* (Paris: Alain Brieux, 1983), 15.

58 On the astronomical interests and productions of al-Ashraf, see King, *Mathematical Astronomy in Medieval Yemen* (cited in n. 6 above), 27–9, no. 8.

59 A description, together with text and translation of the *ijāza*s, is in King, "The medieval Yemeni astrolabe in the Metropolitan Museum of Art in New York City," *Zeitschrift für Geschichte der arabisch-islamischen Wissenschaften* 2 (1985): 99–122, and 4 (1987–88): 268–9 (corrections), reprinted in King, *Studies B*, II.

60 Rida A. K. Irani, "Arabic numeral forms," *Centaurus* 4 (1955): 1–12, reprinted in Kennedy et al., *Studies*, 710–21; Marcel Destombes, "Les Chiffres coufiques des instruments astronomiques arabes," *Physis* 2, 3 (1960): 197–210, reprinted in Destombes, *Contributions*, 127–40.

61 Two other productions of his have recently been described as genuine items from c. 1800: see *Khalili Catalogue*, 237–8, nos. 136–7.

62 In the Middle Ages, any instrument that did not satisfy the beholder's conception of what it should look like (such as an astrolabe with a melon-shaped ecliptic on the rete, on which see n. 106 below) would most likely have been melted down and refashioned into something useful. Nowadays any instrument which

does not correspond to our ideas of what medieval instruments should look like tends to be labeled a fake. On the other hand, an instrument with a second layer of dubious inscriptions can still have a good life if it has never been studied in great detail: an example is the so-called "Astrolabe of Pope Sylvester II," on which see n. 51 above.

63 See further King, *World-Maps*, 106–8 and 128–9.

64 See *EI²*, s.v. "Ulugh Beg" (Beatrice F. Manz). On his scientific activity see also Edward S. Kennedy, "Ulugh Beg as scientist" and "The heritage of Ulugh Beg," first published in idem, *Studies*, X–XI.

65 Sayılı, *The Observatory in Islam* (cited in n. 1 above), 267.

66 Gunther, *Astrolabes*, 1:285–6, no. 140. The piece is dedicated to a Muḥammad ibn Muḥammad al-Tanūkhī, on whom I have no further information (he is not identical with the man listed twice in Brockelmann, *GAL*, S I, 499 and 520).

67 See n. 25 above.

68 See, for example, Gunther, *Astrolabes*, 2:492–501, nos. 308–9. Recently a Spanish example (showing Flemish influence) has come to light and is described in a forthcoming paper by Roberto Moreno together with the present writer and Koenraad van Cleempoel, to appear in *Annals of Science*.

69 François Charette and I are currently preparing for publication a study of the instrument and the two treatises. See already *Paris IMA 1993–94 Exhibition Catalogue*, 434–5, no. 330.

70 Destombes was a real specialist in instruments, but he is better known for his contributions to the history of cartography: see Destombes, *Contributions*.

71 Marcel Destombes, "Un astrolabe carolingien et l'origine de nos chiffres arabes," *Archives internationales d'histoire des sciences* 58–9 (1962): 1–45, reprinted in idem, *Contributions*, 153–200.

72 It is omitted from the catalog of the collection – see Jeanne Mouliérac, "La Collection Marcel Destombes," in *Astrolabica No. 5: Études 1987–1989*, ed. Anthony J. Turner (Paris: Institut du Monde Arabe and Société Internationale de l'Astrolabe, 1989), 77–126 – because even in the 1980s it was still considered dubious.

73 The proceedings were published in *Physis* (Florence: Leo S. Olschki), special issue on "The oldest Latin astrolabe," ed. Wesley M. Stevens, Guy Beaujouan and Anthony J. Turner, vol. 32, nos. 2–3 (1995): 185–450.

74 King, "The earliest known European astrolabe in the light of other early astrolabes," op. cit., 359–404.

75 Julio Samsó, "*Roma et Francia* (= *Ifranja*) in M. Destombes' Carolingian astrolabe," op. cit., 239–52.

76 Paul Kunitzsch and Elly Dekker, "The stars on the rete of the so-called 'Carolingian astrolabe'," in the *Vernet Festschrift*, 2:655–72.

77 Anscari M. Mundó, "Analyse paléographique de l'astrolabe 'carolingien'," *Physis* 32, 2–3: 303–21 and 9 figs.

78 David King and Kurt Maier, "The Medieval Catalan astrolabe of the Society of Antiquaries, London," in the *Vernet Festschrift*, 2:673–718, esp. 690.

79 See King, "Earliest known European astrolabe," 384–5.

80 There is a flavor of dissimulation in Masʿūd's rhymed inscription of ownership, as if he did not want to mention the name of God. It reads, *Ṣāḥibuhu al-faqīr Masʿūd al-wāthiq bi-'l-malik al-maʿbūd*, which translates, "Owned by the needy Masʿūd, who trusts in the King who is to be worshiped [i.e., God]." Neither did he engrave the name of Mecca on the plate for the latitude of the holiest city in Islam (although he did label the one for Algiers), nor did he label the special markings for the times of Muslim prayer on the plates for Algiers, Mecca and Jerusalem. This could be taken as a hint that he was a prisoner of the Christians, although the veiled inscriptions of Muslim prisoners and slaves under Christian domination were in general less subtle – see Peter Sjord van Koningsveld, "Muslim slaves and captives in western Europe during the late Middle Ages," *Islamic and Christian-Muslim Relations* 6 (1995): 5–23.

81 See Maravillas Aguiar Aguilar, "*Risālat Kayfīyat al-sayb fī l-ʿamal bi-l-jayb* de ʿIzz al-Dīn b. Masʿūd: Un tratado del siglo XIV acerca de los usos del cuadrante de senos" (doctoral thesis, Universidad de La Laguna, Tenerife, 1995), and idem, "En torno a ʿAbd al-ʿAzīz b. Masʿūd," *Boletín de la Asociación Española de Orientalistas* (Madrid) 32 (1996): 187–96.

82 See King, "An astrolabe from 14th-century Christian Spain with inscriptions in Latin, Hebrew and Arabic: a unique testimonial to an intercultural encounter," *Suhayl* (Barcelona) 3 (2002–03): 1–148.

83 Petra Schmidl, "Ein Astrolab aus dem 17. Jahrhundert: prachtvoll und verfälscht," in *Frankfurt IGN Festband*, 291–314.

84 Silke Ackermann, "Mutabor: Die Umarbeitung eines mittelalterlichen Astrolabs im 17. Jahrhundert," ibid., 193–209.

85 On *zījes*, see E. S. Kennedy, "A survey of Islamic astronomical tables," *Transactions of the American Philosophical Society*, n.s., 42, 2 (1956): 123–77, and now D. A. King and Julio Samsó, "Astronomical handbooks and tables from the Islamic world (750–1900): an interim report," *Suhayl* (Barcelona) 2 (2001): 9–106, esp. section 3.3.3, with a summary forthcoming in *EI²*, s.v. "Zīdj" (astronomical handbooks and tables).

86 King, "New light on the *Zīj al-Ṣafāʾiḥ* of Abū Jaʿfar al-Khāzin," *Centaurus* 23 (1980): 105–117, reprinted in King, *Studies B*, XI.

87 On the astrolabe of Muḥammad Zamān, see Helmecke, "Das Berliner Astrolab" (cited in n. 38 above).

88 See *EI²*, s.v. "Mizwala" (sundial), reprinted in King, *Studies C*, VIII.

89 On the times of Muslim prayer, see the literature cited in n. 2 above, and also the next few notes.

90 King, "A fourteenth-century Tunisian sundial for regulating the times of Muslim prayer," in *ΠΡΙΣΜΑΤΑ: Naturwissenschaftsgeschichtliche Studien – Festschrift für Willy Hartner*, ed. Y. Maeyama and W. G. Saltzer (Wiesbaden: Franz Steiner, 1977), 187–202, reprinted in King, *Studies B*, XVIII.

91 On this subject see *EI²*, s.v. "Anwā'" (division of the year) and "Layl and Nahār" (division of day and night) by Charles Pellat. See also Daniel

M. Varisco, *Medieval Folk Astronomy and Agriculture in Arabia and the Yemen* (Aldershot: Ashgate, 1997), and idem, *Medieval Agriculture and Islamic Science: The Almanac of a Yemeni Sultan* (Seattle: University of Washington Press, 1993).

92 Seasonal hours (*sāʿāt zamāniyya*) are one-twelfth divisions of the length of day or night. Since only at the equinoxes is day equal to night, the day hours are generally different from the night hours. Thus, for example, in summer the day hours are longer than "our" hours, and the night hours are shorter. See *E. J. Brill's First Encyclopaedia of Islam (1913–1936)*, reprint (Leiden: E. J. Brill, 1987), s.v. "Zamān" (time) by Willy Hartner, and also King, "On the times of Muslim prayer" (in *SATMI*, IV, 2). In the Middle Ages, and, if we take the medieval sources at face value, even before the advent of Islam, the Arabs had names for each of the seasonal hours. Inevitably there were several sets of names in circulation and no set gained precedence over the others. Again, if we take the medieval sources at face value, the names of the daylight prayers in Islam are related to the names of the corresponding seasonal hours (*SATMI*, IV, 8).

93 The difference in the significance of the name *ʿaṣr* in relation to the hours of the afternoon led to the distinction between the definitions of the Mālikīs, Shāfiʿīs and Ḥanbalīs on the one hand and the Ḥanafīs on the other (these correspond originally to the beginning and end of the tenth seasonal hour, respectively). In their Ottoman manifestation the times are referred to as *ʿaṣr-i evvel* and *ʿaṣr-i sānī*, respectively.

94 See already King, "A survey of Medieval Islamic shadow schemes for simple timereckoning," *Oriens* 32 (1990): 191–249, also in *SATMI*, III, and, more especially, idem, "On the times of Muslim prayer," cited in n. 92 above.

95 E. S. Kennedy, *"The Exhaustive Treatise on Shadows" by Abū al-Rayḥān Muḥammad b. Aḥmad al-Bīrūnī: Translation and Commentary*, 2 vols. (Aleppo: Institute for the History of Arabic Science, 1976). Also idem, "Al-Bīrūnī on the Muslim Times of Prayer," in *The Scholar and the Saint: Studies in Commemoration of Abu'l-Rayhan al-Bīrūnī and Jalal al-Din al-Rūmī*, ed. Peter J. Chelkowski (New York: Hagop Kevorkian Center for Near Eastern Studies, New York University Press, 1975), 83–94, reprinted in Kennedy et al., *Studies*, 299–310.

96 Louis A. Janin, "Le Cadran solaire de la Mosquée Umayyade à Damas," Centaurus 16 (1972): 285–98, reprinted in E. S. Kennedy and Imad Ghanem, eds., *The Life and Work of Ibn al-Shāṭir: An Arab Astronomer of the Fourteenth Century* (Aleppo: Institute for the History of Arabic Science, 1976), 107–121; also *Paris IMA 1993–94 Exhibition Catalogue*, 439, no. 334.

97 See Petra G. Schmidl, "Two early Arabic sources on the magnetic compass," *Journal of Arabic and Islamic Studies* (an electronic journal based at the University of Bergen), 1 (1996–97): 81–132, on the treatises of al-Ashraf and Ibn Samʿūn; and my overview in *EI²*, s.v. "Ṭāsa" (magnetic compass).

98 See my articles in *EI²*, s.v. "Kibla, 2, astronomical aspects" (sacred direction) and "Makka, 4, as the centre of the world" (sacred geography), reprinted in King, *Studies C*, IX–X. See also King, *World-Maps*.

99 See *Paris IMA 1993–94 Exhibition Catalogue*, 440–1, no. 336; and King, *World-Maps*, 110–14, 168–70 and 478–80.

100 See King, *World-Maps*, 110–14 and 168–70.

101 Ibid., 113–14 and 80–3.

102 These materials are evaluated in Louis Janin and David A. King, "Ibn al-Shāṭir's *Ṣandūq al-Yawāqīt*: an astronomical 'compendium,'" *Journal for the History of Arabic Science* 1 (1977): 187–256, reprinted in King, *Studies B*, XII.

103 Sevim Tekeli, "The 'Equatorial Armilla' of Izz al-Din b. Muhammad al-Wafa'i and the Torquetum," *Ankara Üniversitesi Dil ve Tarih-Coğrafya Fakültesi Dergisi* 18 (1960): 227–59; William Brice, Colin Imber and Richard Lorch, "The Dā'ire-yi Muʿaddel of Seydī ʿAlī Re'īs," *Seminar on Early Islamic Science* (University of Manchester) 1 (1976): 3–4; Muammer Dizer, "The *Dā'irat al-muʿaddal* in Kandilli Observatory, and some remarks on the earliest recorded Islamic values of the magnetic declination," *Journal for the History of Arabic Science* 1 (1977): 257–60; and, most recently, the *Khalili Catalogue*, 277–9.

104 See Carl Schoy, "Die Mekka- oder Qiblakarte (Gegenazimutale mittabstandstreue Projektion mit Mekka als Kartenmitte)," *Kartographische und schulgeographische Zeitschrift* 6 (1917): 184–5, and one map, reprinted in *Carl Schoy: Beiträge zur arabisch-islamischen Mathematik und Astronomie*, ed. Fuat Sezgin et al. (Frankfurt: IGAIW, 1988), 1:157–9.

105 It is important to keep in mind that the mapping underlying the grids preserves the sine of the distance to the center (measured as an arc on the terrestrial sphere), which accounts for the nonuniform gradation of the diametrical rule. This in itself is entirely within the Islamic tradition of instrumentation. As a result, the grid is different from and considerably simpler than the grid conceived by Schoy (see previous note).

106 See now E. S. Kennedy, P. Kunitzsch and R. P. Lorch, *The Melon-Shaped Astrolabe in Arabic Astronomy* (Stuttgart: Franz Steiner, 1999).

107 Analyzed in King, "The earliest Islamic mathematical methods and tables for finding the direction of Mecca," *Zeitschrift für Geschichte der arabisch-islamischen Wissenschaften* 3 (1986): 82–149, reprinted in King, *Studies C*, XIV.

108 King, *World-Maps*.

109 For reviews of King, *World-Maps*, at each end of the spectrum, see George Saliba in *Mathematical Reviews* (2001), no. 01008, and Emilie Savage-Smith in *Bulletin of the Scientific Instrument Society* 66 (2000): 32–5. In my book I showed that the localities were marked after the grid had been constructed, a practice known from antiquity; Owen Gingerich, in *History of Science* 38 (2000): 245–7, proposes a novel but absurd construction for the grids after first marking the localities. Elly Dekker, in "Cartographic grids from Iran: an early version of the retro-Azimuthal orthographic projection," *Cartographical Journal* 37, 2 (2000): 109–16, counters my claim that the latitude curves are simply arcs of circles, and prefers to see a rect-azimuthal projection (inevitably under European influence), first proposed in the literature on cartographical projections in 1968.

110 King, *World-Maps*, 260–2.
111 See now King, *SATMI*, Xc.
112 Several examples are described in Giuseppe Brusa, "Le navicelle orarie di Venezia," *Annali dell'Istituto e Museo di Storia della Scienza di Firenze* 5 (1980): 51–9.
113 In 1960 the late Derek de Solla Price thought that he had found the origin of the *navicula* in an Islamic instrument also shaped like a ship (see Price, "The Little Ship of Venice, a Middle English instrument tract," *Journal of the History of Medicine and Allied Sciences* 15 [1960]: 399–407, esp. 399–400). The latter device, actually part of an astrolabe rete serving as a horizon for a specific latitude, he found described in the treatise of al-Marrākushī (Sédillot *fils*, *Mémoire*, 182 and pl. 89, cited in Price, "Little Ship," 399, n. 2), but it is merely one of the many highly ingenious kinds of unusual astrolabe retes and associated paraphernalia devised by Muslim astronomers mainly in the tenth century (see now *Khalili Catalogue*, 230).
114 On the fifteenth-century Viennese tradition of these universal hour-dials, in which essentially the same instrument was also not in the form of a ship, but rather a musical instrument, see Ernst Zinner, *Deutsche und niederländische astronomische Instrumente des 11.–18. Jahrhunderts* (Munich: C. H. Beck, 1956; 2nd edn., 1967; reprint, 1972), 110–17.
115 King, *World-Maps*, 351–9.
116 Ibid., 354–8, but see now François Charette and Petra Schmidl, "A universal plate for timekeeping with the stars by Ḥabash al-Ḥāsib: text, translation and preliminary commentary," *Suhayl* 2 (2001): 107–59.
117 I should point out that there is as yet no description in the modern literature of the way in which the *navicula* can be used to derive the time of day accurately. All modern writers have followed a simplified *modus operandi* found in the fourteenth-century English sources for using the device which is approximate, and was clearly not intended by the inventor. See King, "14th-century England or 9th-century Baghdad? New insights on the origins of the elusive astronomical instrument called *Navicula de Venetiis*," to appear in a special issue of *Centaurus* dedicated to Bernard R. Goldstein, and a more detailed version in *SATMI*, IXb.
118 See n. 53 above.
119 See n. 14 above.
120 *www.uni-frankfurt.de/fb13/ign/instrument-catalogue.html*

1
The observatory in Istanbul directed by astronomer Taqī al-Dīn, here shown holding an astrolabe. From King, *World-Maps*, 9.

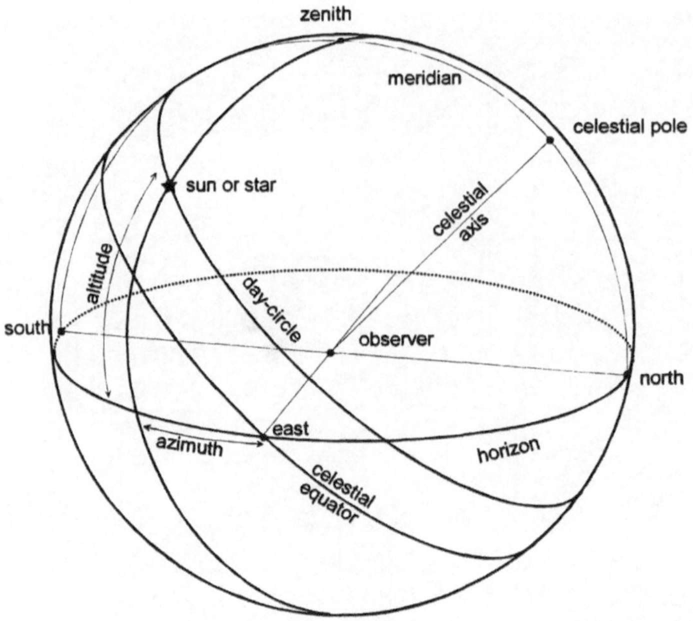

2
The celestial sphere

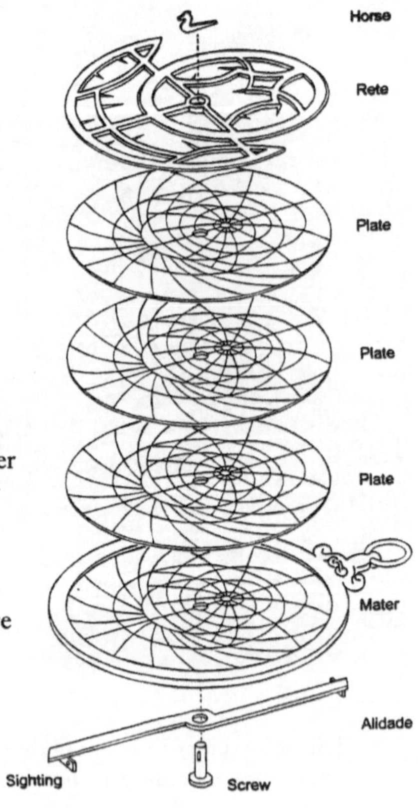

3
The components of the standard astrolabe. The "celestial" part, i.e. the rete with star-pointers and a ring for the ecliptic, can rotate over the "terrestrial" part, i.e. any one of a series of plates for a specific latitude, and thus simulate the apparent daily rotation of the celestial sphere at the latitude of the user.

4

The front of the astrolabe of al-Khujandī. Private collection. From King in *Kuwait Catalogue*, 80.

5

Throne and detail of one of the plates on the astrolabe of Jalāl al-Kirmānī. Courtesy of The David Collection, Copenhagen (inv. no. D 25/1986).

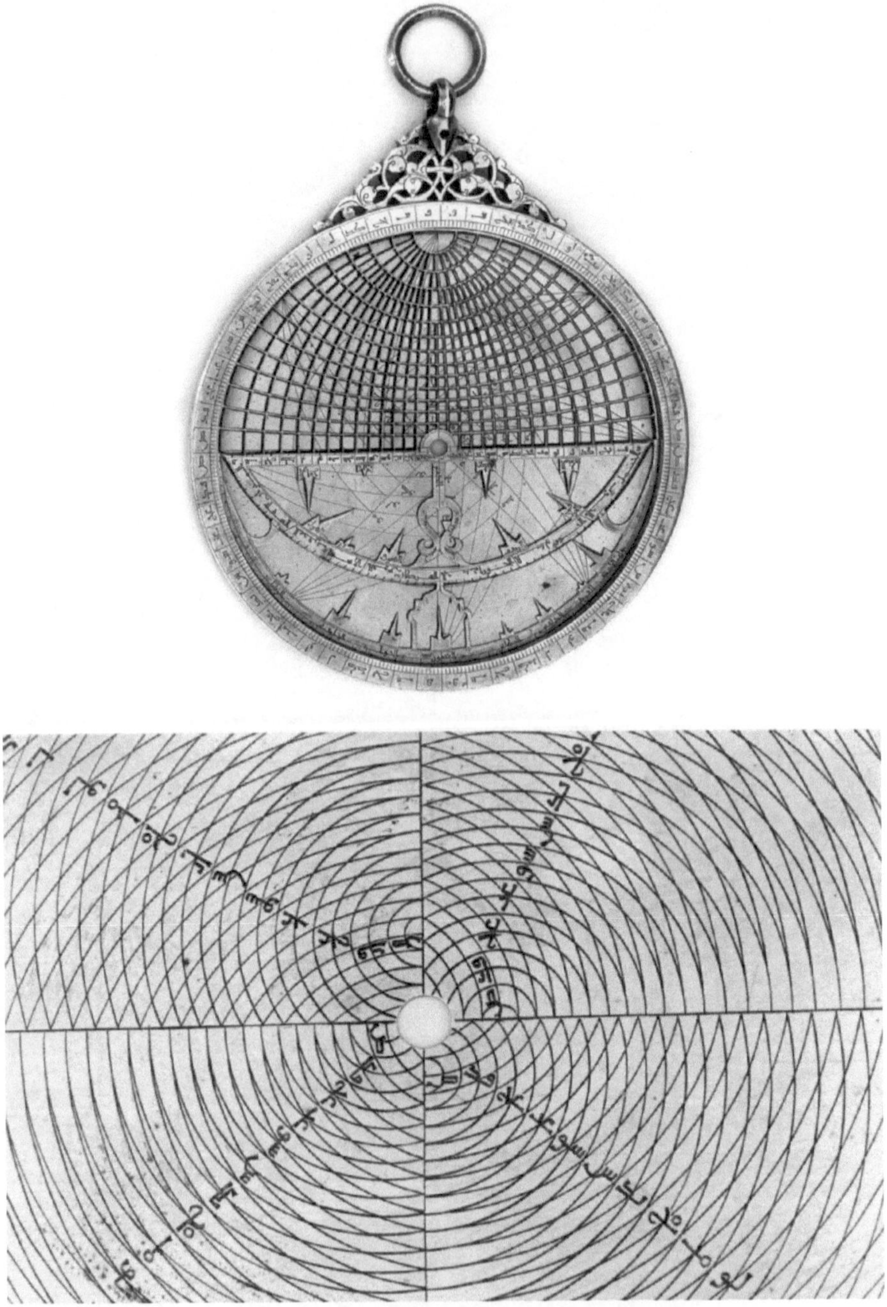

6

Front and detail of the universal astrolabe of Ibn al-Sarrāj.
Courtesy of the Benaki Museum, Athens (inv. no. 13178).

7

The front of the oldest European astrolabe (above). Photo © Institut du Monde Arabe. Detail of the plate for Rome and "Francia." Photo by David King, courtesy of IMA.

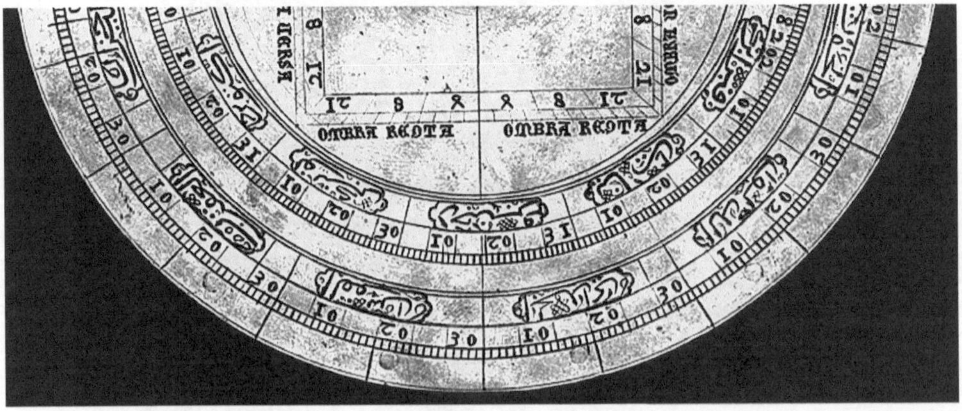

8
The front of the Toledo astrolabe and a detail of the back,
showing the Latin and later Arabic inscriptions. Courtesy of Christie's, London.

9
A fourteenth-century Tunisian sundial. Courtesy of the late Alain Brieux, from King, *World-Maps*, 110.

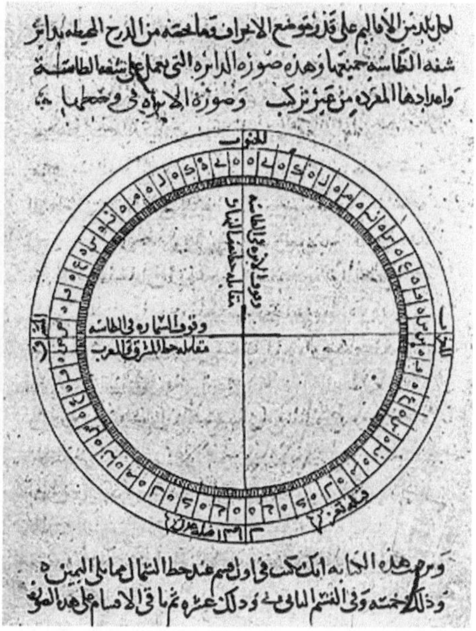

10
The compass box illustrated in the treatise of the Yemeni Sultan al-Ashraf. From King, *World-Maps*, 111.

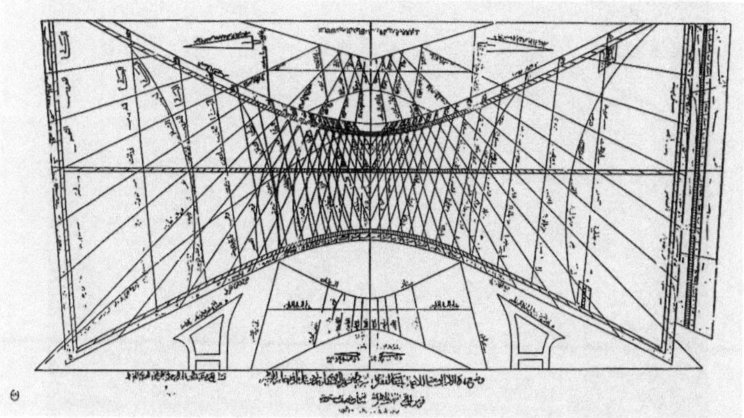

11
A replica of the sundial of Ibn al-Shāṭir.
Courtesy of the late Alain Brieux, from King, *World-Maps*, 22.

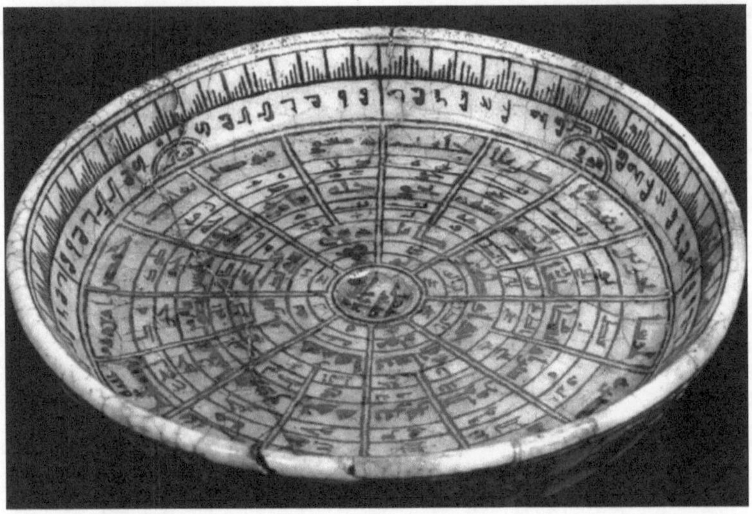

12
A ceramic qibla-bowl from early sixteenth-century Damascus.
From King, *World-Maps*, 112.

13
The first Safavid world-map. Courtesy of Christie's, London.

Some Illustrations in Islamic Scientific Manuscripts and Their Secrets

An estimated 10,000 Islamic scientific manuscripts survive in libraries around the world. Although many were copied after the creative period of Islamic science, they bear witness to a scientific tradition which knew no rival from the eighth to the fifteenth centuries. While many of these manuscripts are illustrated, the illustrations have received scant attention from historians. For example, even though some manuscripts of al-Ṣufi's *Book on Constellations* have been studied by art historians, neither a critical study of the available manuscripts nor a comparative study of the various traditions of illustration has been conducted. Moreover, there are other treatises in which constellation figures occur, and other astrological treatises in which there are illustrations that should be of interest to art historians.

Some attention has been paid to the diagrams in works on engineering (see figure 9.1), and scholars have labored over geometrical diagrams to understand the accompanying mathematical procedures (see figure 9.2). Islamic maps have also fared quite well at the hands of historians, and the state of our present knowledge of Islamic cartography is reasonably advanced. Investigations of diagrams of geometrical planetary models led, in the 1950s, to the discovery of a tradition lasting several centuries of Islamic modifications to Ptolemaic planetary theory (see figure 9.3).

My present purpose is to demonstrate the significance of various other categories of illustrations which I have worked on recently and which open up new chapters in the field of Islamic science.

SOME ILLUSTRATIONS IN TREATISES ON ASTRONOMICAL INSTRUMENTS

Before mentioning some illustrations in Islamic texts on astronomical instruments, let us look at figure 9.4, from the *Shāhinshāhnāme*, which shows the scene at the Ottoman Observatory in Istanbul. The observatory was built in 1577 and demolished in 1580, and we are lucky indeed to have this miniature.

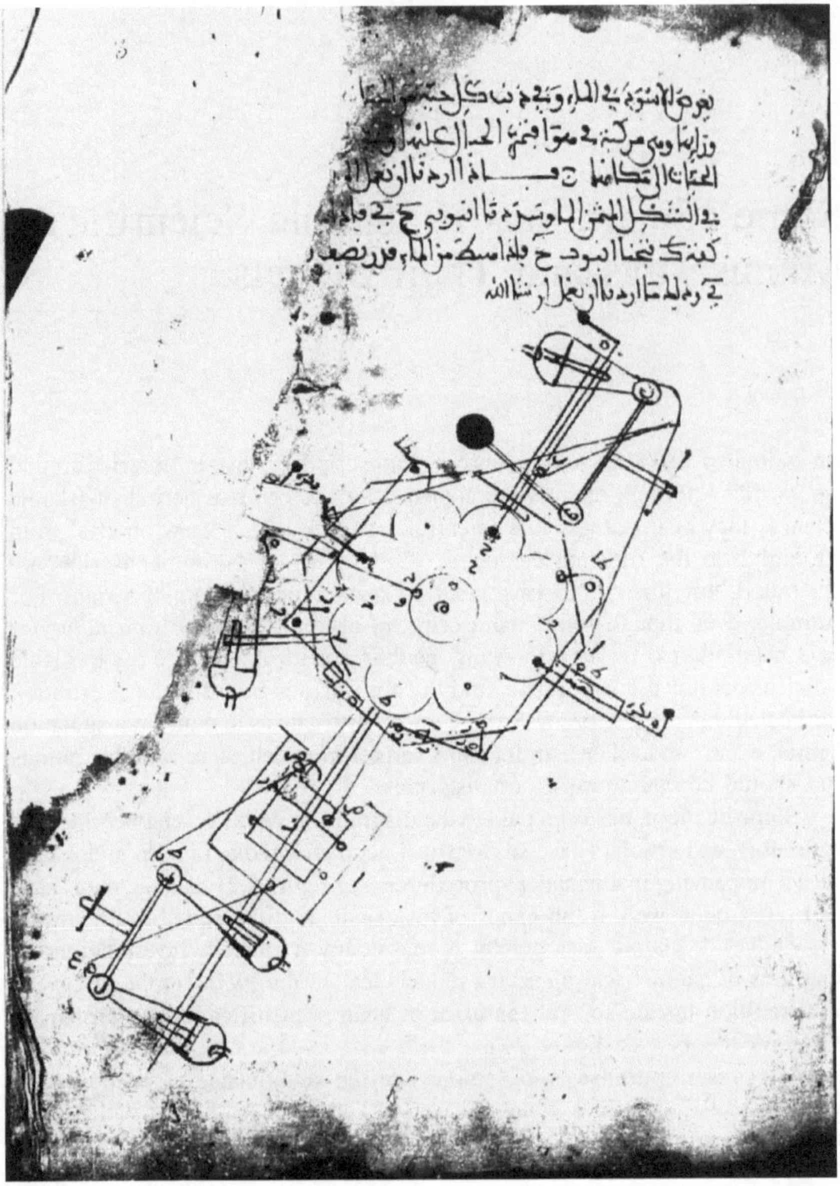

Fig. 9.1. Numerous mechanical devices are illustrated and described in this eleventh-century Andalusian treasure, discovered less than twenty years ago. (MS Florence Bibliotheca Medicea-Laurenziana Or. 152.)

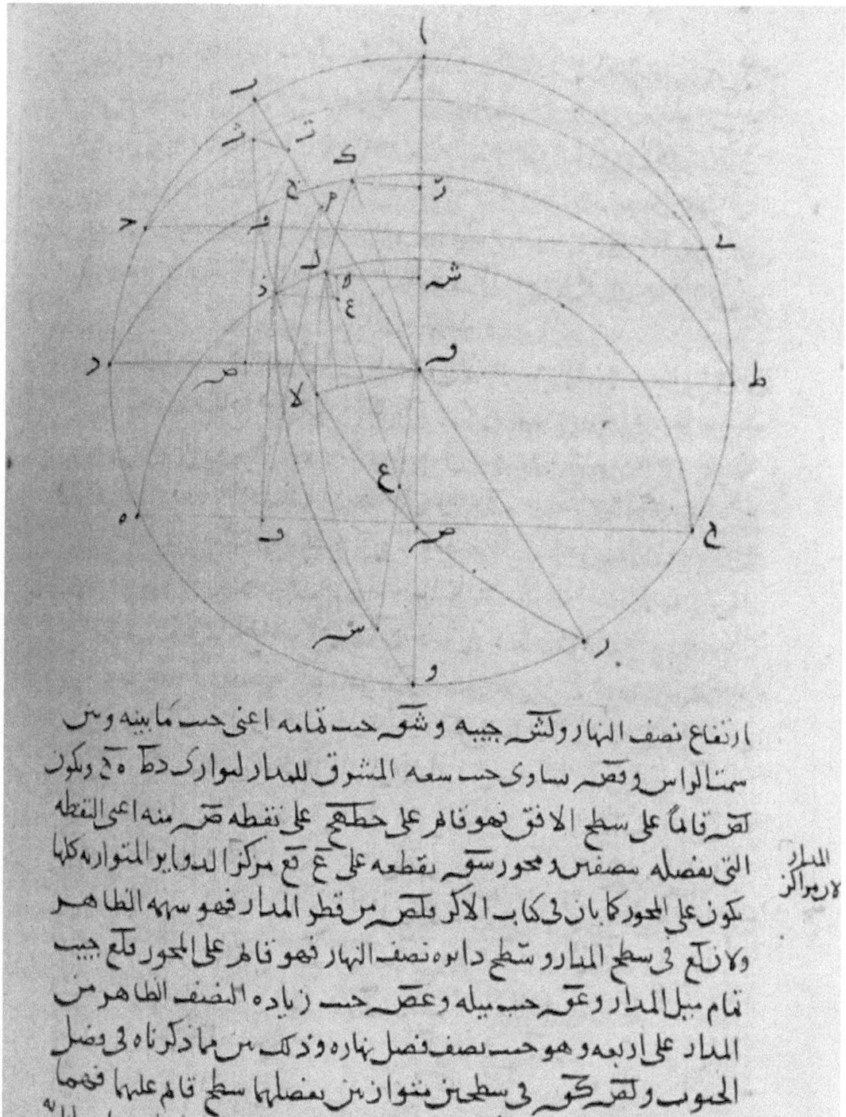
ارتفاع نصف النهار ولكن جيبه وشق حب تمامه اعني حب ما بينه وبين
سمت الراس ونقص ساوي حب سعه المشرق للمدار لموازي دط هح ونكون
لقر قائما على سطح الافق فهو قائم على خط طح على نقطه صر منه اعني النقطه
التي يفصله بنصفين ومحور سو يقطعه على ع ثم ع مركز الدواير المتوازيه كلها
يكون على المحور كما بان في كتاب الاكر فلقر من قطر المدار فهو سهمه الظاهر
ولان كع في سطح المدار وسطح دايره نصف النهار فهو قائم على المحور فكع جيب
تمام ميل المدار وعق حب ميله وعصر حب زياده النصف الظاهر من
المدار على اربعه وهو حب نصف فضل نهاره وذلك بين مما ذكرناه في فصل
الجنوب ولقر كق في سطحين متوازيين يفصلهما سطح قائم عليهما فهما
متوازيان كما بان ذلك في اصول اقليدس فزاويه صر الداخله مساويه لزاويه

Fig. 9.2. A geometrical diagram accompanying the solution of a problem in spherical astronomy. (From a manuscript in the Universiteitsbibliotheek, Leiden.)

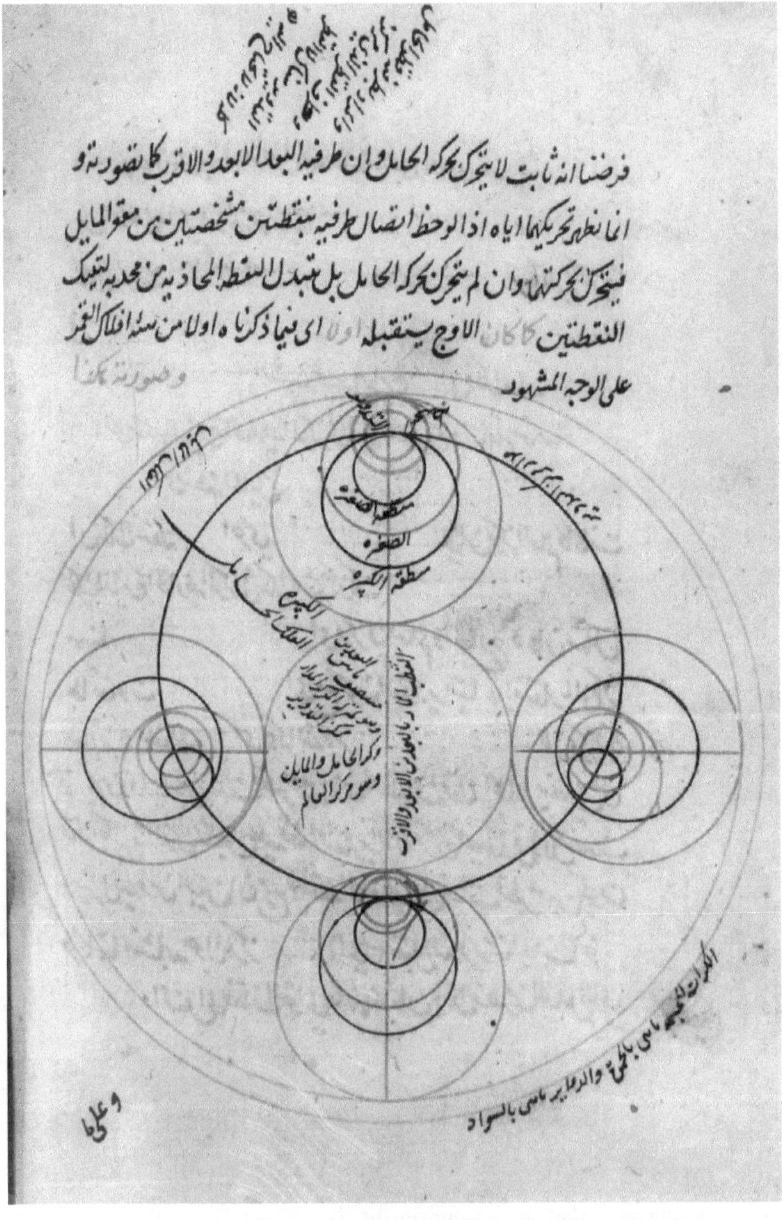

Fig. 9.3. An illustration of a non-Ptolemaic planetary model from a treatise of the thirteenth-century Maragha school. (From a manuscript in the Dār al-Kutub, Cairo.)

Fig. 9.4. The astronomers of the Istanbul Observatory with some of their instruments. This miniature is a veritable gold-mine of information which has yet to be fully exploited. (MS Istanbul University Library Yildiz 1404.)

The director, Taqī al-Dīn, is one of the two men contemplating the astrolabe. Some of the books behind him are now in the University Library in Leiden; his mark of ownership is found on several manuscripts in that library. Taqī al-Dīn himself wrote a treatise on mechanical clocks, and one such clock is featured here. Most of the instruments, notably the astrolabe and two quadrants, and various observational instruments, are of Islamic provenance; one, the terrestrial globe, is European. A terrestrial globe from the workshop of Mercator, with a dedication to Sultan Murād III dated 1579, was auctioned at Christie's of London in 1991; could this be the one illustrated here?

A project currently in progress in Frankfurt aims to catalogue all of the historically significant, surviving Islamic astronomical instruments (as well as European to ca. 1550). As many of the instruments as possible are being examined at first hand, and the relevant manuscript sources are also being exploited for information. Sometimes, manuscripts constitute the only available source of information on certain classes of instruments of which no examples have survived.

Figure 9.5 shows some of the illustrations of unusual astrolabe retia in a treatise on the astrolabe by al-Bīrūnī. These variatons on the standard astrolabe are of considerable historical interest, not least because we now know al-Bīrūnī's source for his information on them. His predecessor, al-Sijzī, actually mentions the astronomers who devised them and the patrons to whom they dedicated examples of them. Alas, all such instruments have disappeared without a trace, and it is left to the modern historian to piece together the details. Dr. Richard Lorch of Munich is currently preparing a critical analysis of all of the related textual material. Sometimes we are more fortunate in having both text and instrument, although in the following examples the instruments were known to modern scholarship before the texts.

An astrolabe in the Metropolitan Museum of Art in New York bears the signature of the Yemeni Rasulid Sultan al-Ashraf and is dated 690/1291 (see figure 9.6). A few decades ago, it was pronounced a fake: no Yemeni sultan could have made such an astrolabe. Perhaps, it was speculated, the astrolabe was made for him in Cairo, because the Yemen is a backwater, and no serious astronomy could have been practiced there. We now know of *over one hundred Yemeni astronomical manuscripts* attesting to an active tradition of astronomy in the Yemen from the tenth century to the twentieth. One of these, preserved in the Egyptian National Library, is an illustrated treatise by al-Ashraf on the construction of the astrolabe and sundial (see figure 9.7). Appended to the text, which may well be in the Sultan's own hand, is a set of *ijāzah*s by two of his teachers approving six astrolabes which he made and mentioning various features of each. One of these is clearly the New York instrument.

Another case is an astrolabe in the Benaki Museum in Athens which was made by Ibn al-Sarrāj in Aleppo in 729/1328 or 1329. It is the most sophisticated

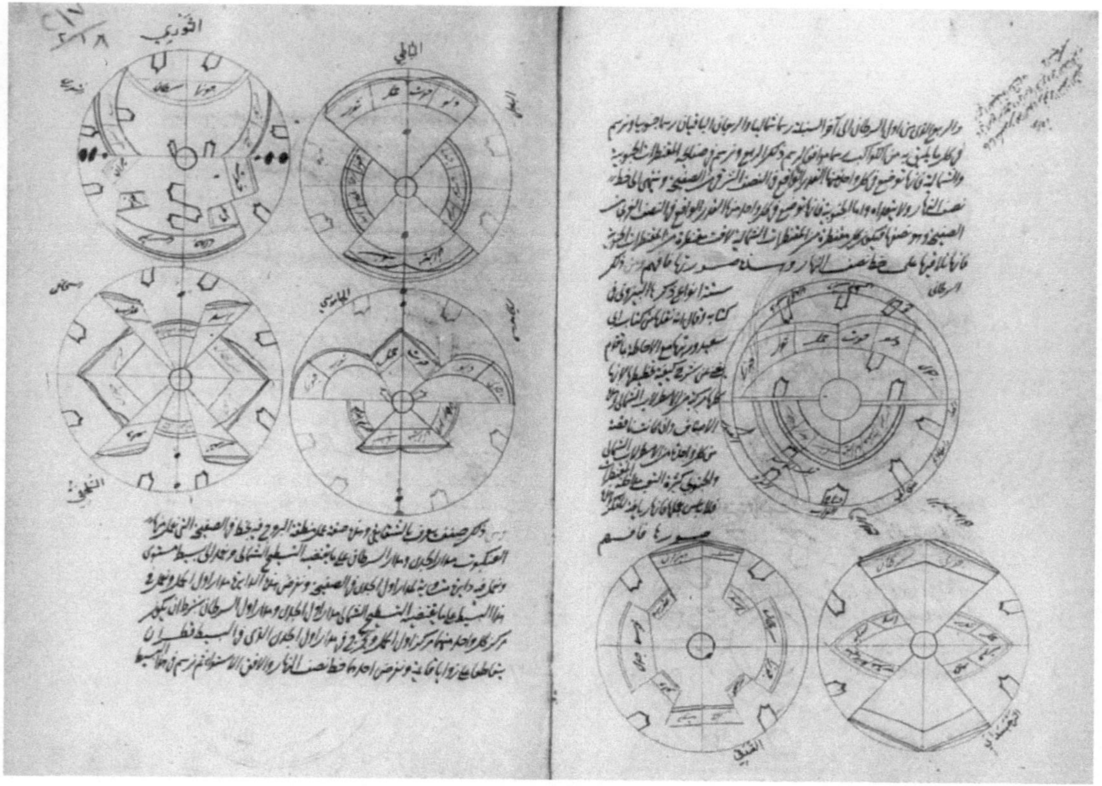

Fig. 9.5. Non-standard astrolabes illustrated in the treatise on astronomical instruments by al-Marrākushī (Cairo, ca. 1280). The text is based on al-Bīrūnī (Central Asia, ca. 1025), who abridged the discussion of them by al-Sijzī (E. Iran, ca. 1000). (MS Cairo Dār al-Kutub K 3821.)

Fig. 9.6. The back of the astrolabe of the Yemeni Sultan al-Ashraf, dated 690/1291. This is the only known Islamic astrolabe on which the back is devoted to astrological information and also one of the few on which the sun, moon, and five planets are indicated by their symbols. The very same peculiarities are featured in an illustration found in al-Ashraf's treatise on astrolabe construction (see figure 8.7). (Metropolitan Museum of Art, New York, inv. no. 91.1.535.)

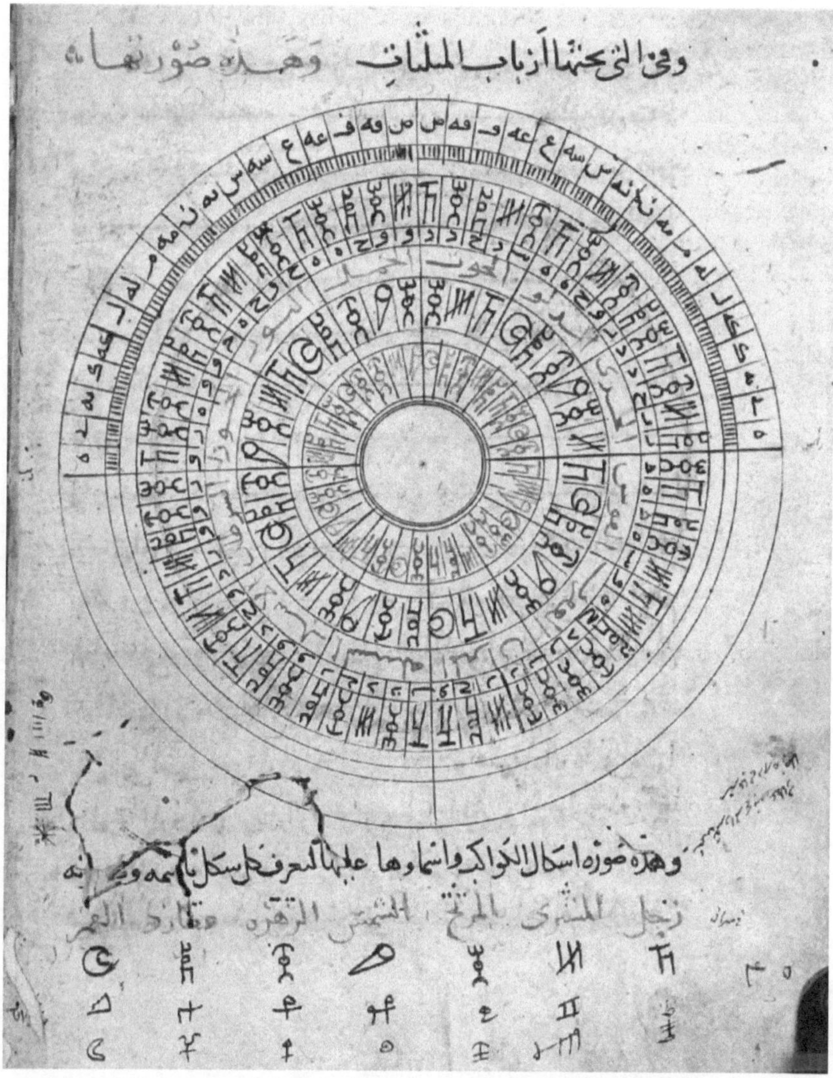

Fig. 9.7. One of numerous illustrations in al-Ashraf's treatise on astrolabe construction. If any proof were needed that the New York astrolabe was indeed due to al-Ashraf. . . but none is. The treatise also deals with sundial construction (no medieval sundials are known from the Yemen) and contains the earliest known description of a magnetic compass in any astronomical text.

astrolabe ever constructed, being universal in five different ways (see figure 9.8). (Most astrolabes have plates serving a series of latitudes; the universal astrolabe with a single plate serving all latitudes was invented in Andalusia in the eleventh century.) A manuscript in Princeton contains an account by the fifteenth-century Egyptian astronomer 'Abd al-'Azīz al-Wafā'ī of the use of Ibn al-Sarrāj's remarkable instrument. (al-Wafā'ī actually owned the astrolabe; his name is engraved on the edge of the rim.) In his treatise, al-Wafā'ī complained that Ibn al-Sarrāj had not explained its use, and so he undertook to do this himself. It is fortunate that he did, because the use of some of the parts is not at all obvious. Ibn al-Sarrāj emerges as a genius. But who was he? We knew

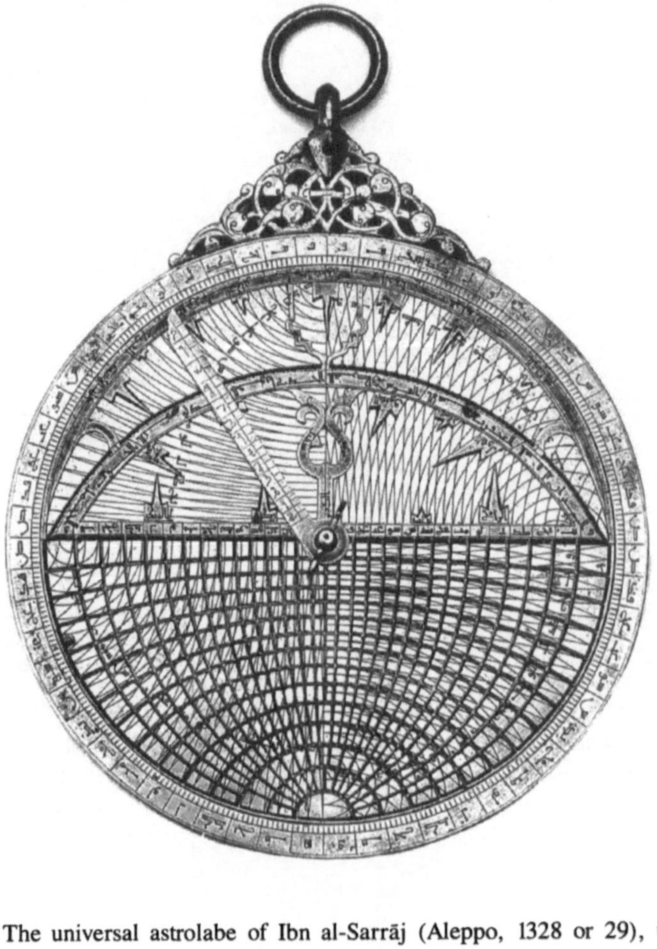

Fig. 9.8. The universal astrolabe of Ibn al-Sarrāj (Aleppo, 1328 or 29), the most sophisticated astrolabe ever made. (Benaki Museum, Athens, inv. no. 13178.)

but little about him until 1983, when a richly illustrated manuscript was discovered in the Chester Beatty Library in Dublin (miscataloged as a Persian work on the astrolabe). It turned out to be a treatise by Ibn al-Sarrāj, probably in his own hand, describing all of the different varieties of instruments known to him, and all those invented by him (see figure 9.9). Although this treatise was clearly written before he developed his quintuply-universal masterpiece, it provides ample evidence of his background and has yet to be exploited for the historical information that it contains.

QIBLAH MAPS

Another genre of illustration serves to display the sacred direction (*qiblah*) of localities in the Muslim world. The map in figure 9.10 (l.h.s), found in an eighteenth-century Egyptian treatise, is a crude cartographic attempt to demonstrate the *qiblah* in various localities, which are roughly displayed on an orthogonal latitude and longitude grid. Such maps were prepared already in the ninth century, and later others were with a more sophisticated projection of the parallels of latitude and meridians such that the former are straight lines and the latter ellipses. However, most Muslim cartographers, and surely all scientists, knew that one could not merely read the *qiblah* off a map by simply joining a locality to Mecca and measuring the direction of the line-segment. The scientists knew that the problem had to be solved on a sphere or on a mathematically acceptable two-dimensional representation thereof. Although very few such maps with orthogonal grids specifically designed to find the *qiblah* are known, in 1989 there appeared at Sotheby's in London a cartographic representation of a much more sophisticated kind, on which the meridians are no longer equally spaced, and the parallels of latitude are no longer parallel (see figure 9.11). The combination is so devised as to enable the user to read off the *qiblah* on the circular scale and to read off the distance from Mecca on the diametral rule. The instrument was clearly made in Isfahan around 1710. It is not too much to hope that somewhere in the vast manuscript sources available for further study of Islamic science, a treatise on the construction of such an instrument might one day be found. In April 1994 it was discovered that the Isfahan world-map represents a cartographic tradition that goes back at least to al-Bīrūnī (ca. 1025), who authored a dozen works on mathematical cartography, only two of which are known to have survived. A series of Islamic geographical tables from the fifteenth, fourteenth, and twelfth centuries give, in addition to the usual longitudes and latitudes, here based on al-Bīrūnī's distinctive values, the *qiblah*-values for 250 localities from Spain to China. The *qiblah*-values are given to the nearest ten minutes, and it has been shown that they can only have been derived from a world-map centered on Mecca, such as is found on the Isfahan instrument. A preliminary notice of this remarkable cartographic tradition will appear in the article "Samt" in *The Encyclopedia of Islam*, new edition.

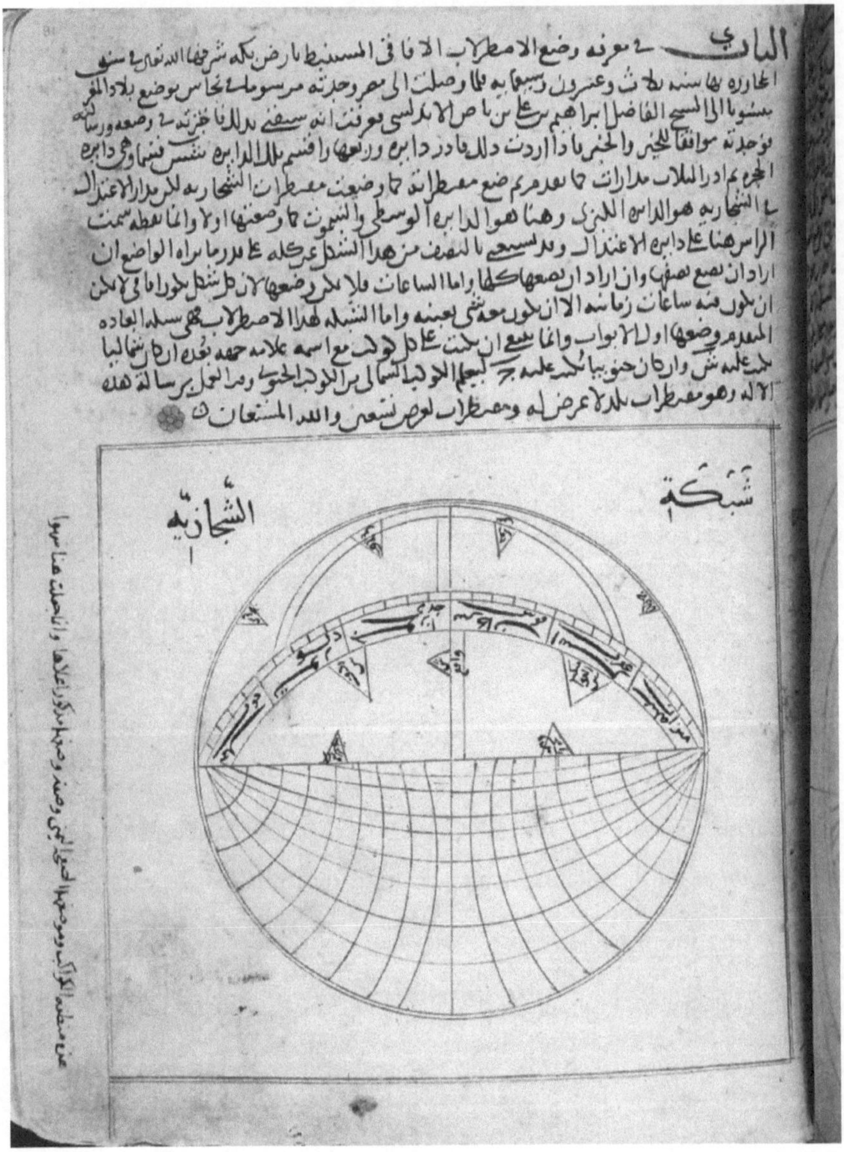

Fig. 9.9. An illustrtion of the rete of a universal astrolabe in Ibn al-Sarrāj's treatise on astronomical instruments. (MS Dublin Chester Beatty Library 102). By courtesy of the Chester Beatty Library.

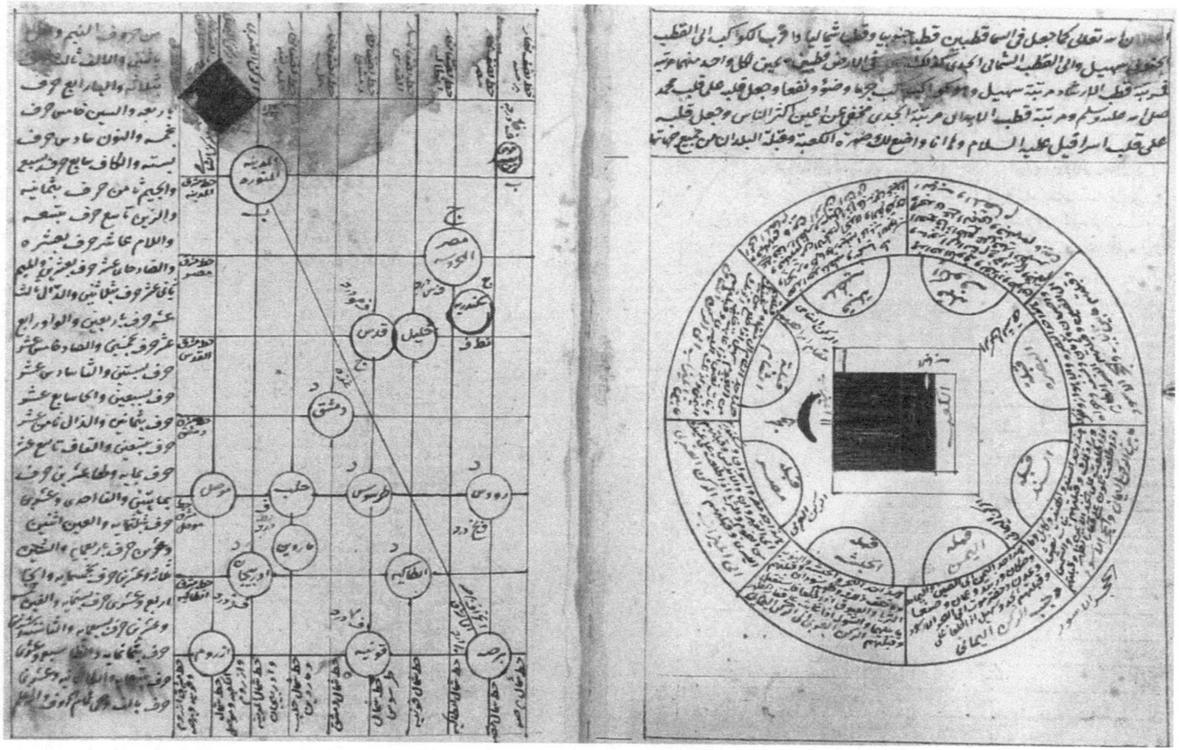

Fig. 9.10. On the left, the Ka'bah is depicted—appropriately inclined to the meridian—on a crude cartographic grid from which one can determine the *qiblah* very approximately by joining one's city to the Ka'bah (here Bursa is shown connected to the Ka'bah [via Medina]). On the right is a diagram of sacred geography showing various areas around the Ka'bah and indicating the astronomical horizon phenomena which define their *qiblahs*. (MS CAiro Dār al-Kutub Tal'at *majāmi'* 811.)

Fig. 9.11. A *qiblah*-indicator from Isfahan, ca. 1710, unique of its genre. The *qiblah*s and distances from Mecca of 150 localities marked on the cartographic grid—which can be read from the circumferential scale when the diametral rule, center at Mecca, is laid on the marker for the locality in question—are given accurately for all practical purposes. (Private collection, photograph from the Museum of the History of Science, Oxford.)

DIAGRAMS REPRESENTING THE SACRED GEOGRAPHY OF ISLAM

In various Islamic sources, by no means restricted to scientific texts but including also treatises on the sacred law of Islam and encyclopedias, there are found illustrations of the divisions of the world around the Ka'bah. The diagrams of this kind with twelve divisions of the world around the Ka'bah in the treatises of Yāqūt and al-Qazwīnī are well known (see figure 9.12). Their purpose is to identify regions of the world which have the same *qiblah*. But these diagrams are secondary, that is, they were based on more elaborate prototypes, in which the way to find the *qiblah* in each sector was described in detail. The instructions were not to use the *qiblah*s found by the astronomers using mathematical

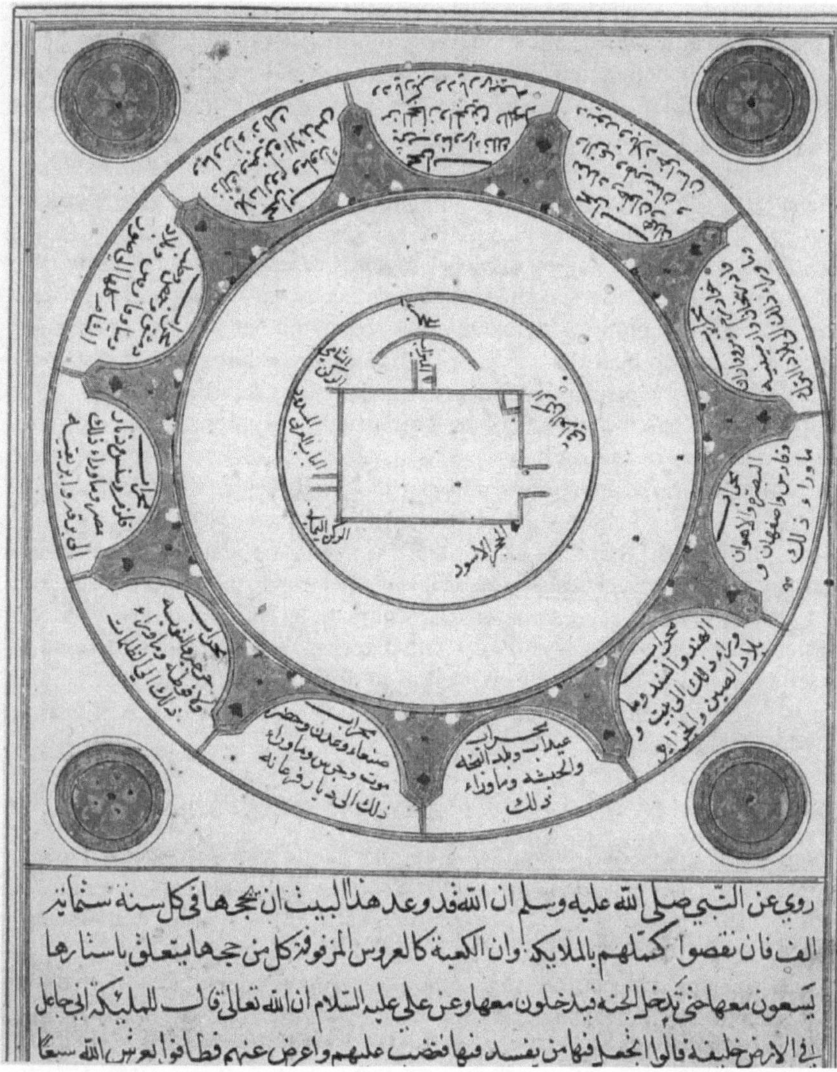

Fig. 9.12. An eleven-sector scheme of sacred geography from a manuscript of al-Qazwīnī's *Āthār al-bilād* (Iraq and Syria, ca. 1250). The information on the *qiblah* in each sector has been suppressed. Also one of the sectors in the original scheme has been ommitted by mistake (MS Dublin Chester Beatty 4163). By courtesy of the Chester Beatty Library.

formulae applied to geographical data, but rather to face the risings and settings of particular fixed stars or of the sun at various times of the year. We now know that schemes for finding the *qiblah* in different regions by such procedures were advocated from the tenth century onward, and that over the centuries some twenty different schemes of what I call "sacred geography" were formulated. Many of these were illustrated in the manuscripts (see figures 9.10 [r.h.s.] and 9.13–9.15); others were recorded only in words; and some were used on astronomical instruments (see figure 9.16). These popular methods for finding the *qiblah* were advocated by the legal scholars of Islam, who had little time for the opinions of the scientists. Their authority lay in the fact that such procedures had been used by the *ṣaḥābah* for laying out some of the earliest mosques. But why had the *ṣaḥābah* used such procedures?

The earliest Muslims knew that when they were standing in front of the walls of the Ka'bah they were facing significant astronomical directions. Also, the four corners of the Ka'bah were associated with different regions of the world, and intermediate regions with particular segments of the perimeter of the edifice. Thus, when Muslims in those regions wanted to face the appropriate part of the Ka'bah, they should face those very same directions. (This means of facing a distant object is quite reasonable if one bears in mind that the direction of a distant locality is to some extent arbitrary, let alone the direction of a particular part of a distant edifice.) The diagrams were intended to serve as *aides-memoires* to the faithful, as well as to provide pictorial representations of the entire world and the relation of its regions to the physical focus of Islamic worship, the Ka'bah.

A DIAGRAM FOR LAYING OUT THE FOUNDATIONS OF A VENTILATOR

Another remarkable diagram in Ibn al-Sarrāj's treatises on instruments purports to display the orientation of the ventilators that were a distinctive feature of the skyline of medieval Cairo (see figures 9.17 and 9.18). Why would such a diagram feature in a treatise on astronomical instruments? Since the 1970s, we have known that in the corpus of astronomical tables that was used in medieval Cairo for timekeeping and regulating the times of prayer, one table enabled the user to find by means of the sun the orientation of the ventilator (*bādahanj*), namely, winter sunrise (ca. 27° S. of E.). Inspection of the sole surviving ventilator of any consequence in modern Cairo reveals that it is aligned with the foundations of the house (the Musāfirkhāne) on whose roof it stands. Indeed, the house and the whole of the medieval city of Cairo are aligned in this direction, the major thoroughfare of the city being in the perpendicular direction.

The fact is that the entire medieval city of Cairo was oriented in the *qiblah* of the *ṣaḥābah*, towards winter sunrise. Now, the city was built alongside the old Pharaonic/Roman Canal from the Nile to the Red Sea, which at that point flowed perpendicular to that direction, so either the orientation of Cairo is

Fig. 9.13. An eight-division scheme of sacred geography from a manuscript of the treatise on cosmography by Pseudo-Ibn al-Wardī (Aleppo, ca. 1420). There are several different traditions of sacred geography in the numerous manuscripts of his treatise; the illustration in the published text shows two of these but are hopelessly corrupt. (MS Topkapi Sarayi Müzesi Kütüphanesi A 3025.)

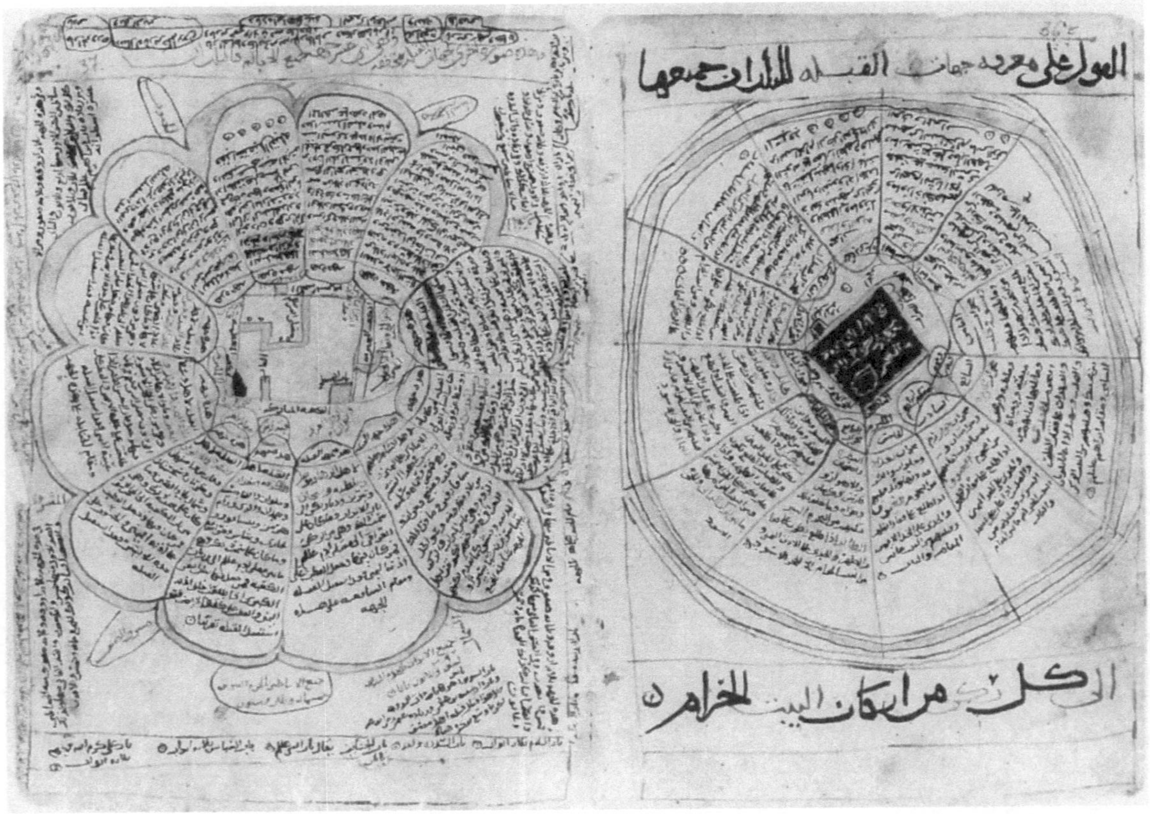

Fig. 9.14. Two different twelve-sector schemes of sacred geography with associated astronomical horizon phenomena for finding the *qiblah*, as found in a manuscript of a thirteenth-century Yemeni treatise on folk astronomy. (MS Milan Biblioteca Ambrosiana Supp. 73.)

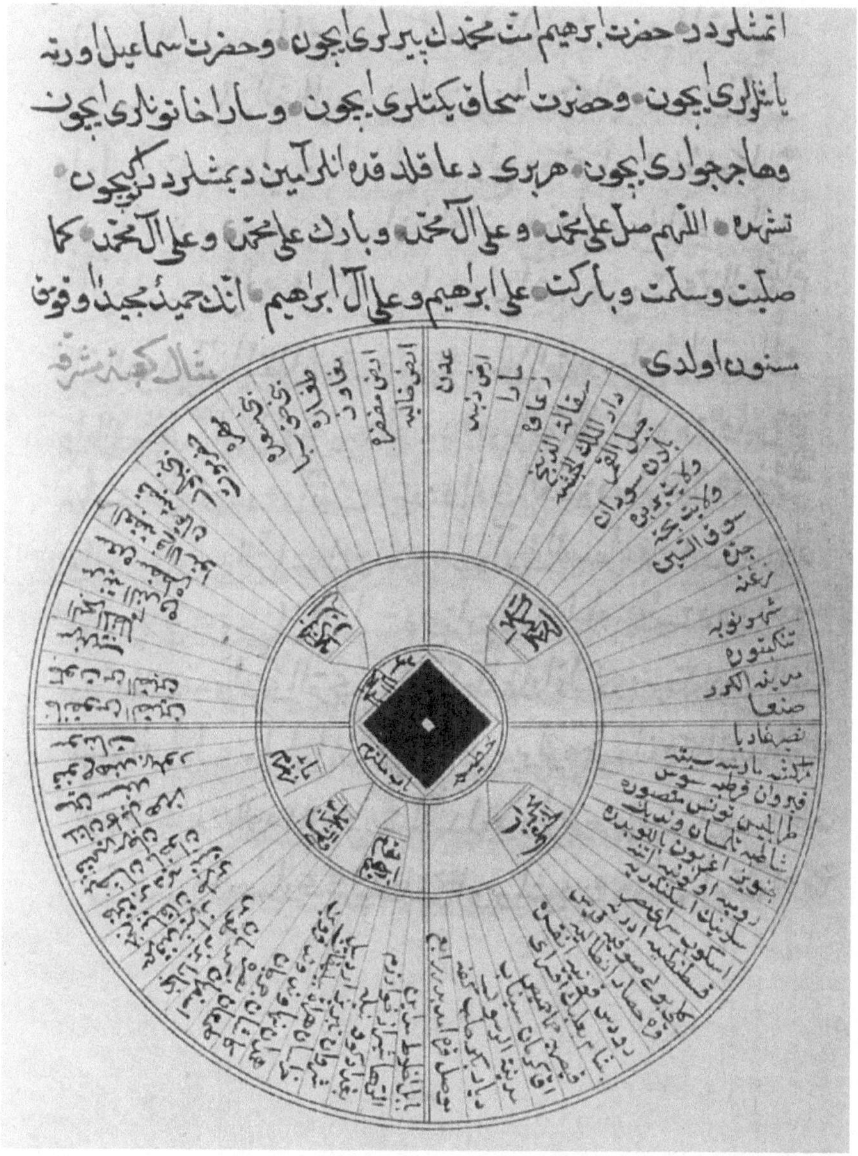

Fig. 9.15. An Ottoman scheme of sacred geography with 72 divisions of the world around the Kaʿbah, suitable for marking on a horizontal *qiblah*-indicator (see figure 9.17). (MS Topkapi Sarayı Müzesi Kütüphanesi B 179.)

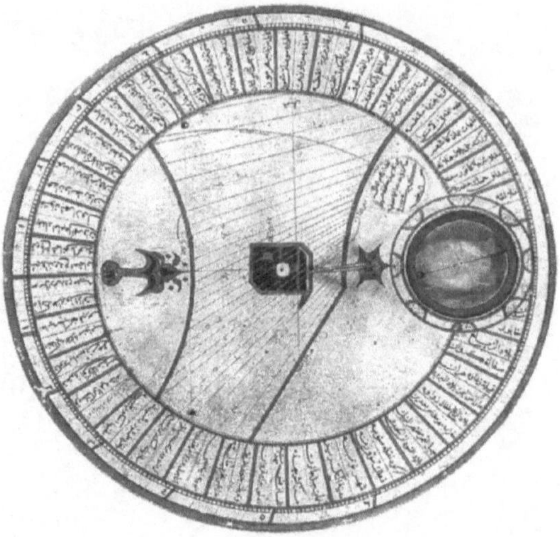

Fig. 9.16. An Ottoman sundial cum *qiblah*-indicator on which a scheme of sacred geography has been included on the circular base. (Berlin Deutsche Staatsbibliothek, Sprenger 2048.)

Fig. 9.17. The imposing ventilator on the Musāfirkhāne in Cairo, one of the few surviving examples. In medieval times, most buildings in Cairo were fitted with such ventilators, all astronomically oriented (Photo Abū Max.)

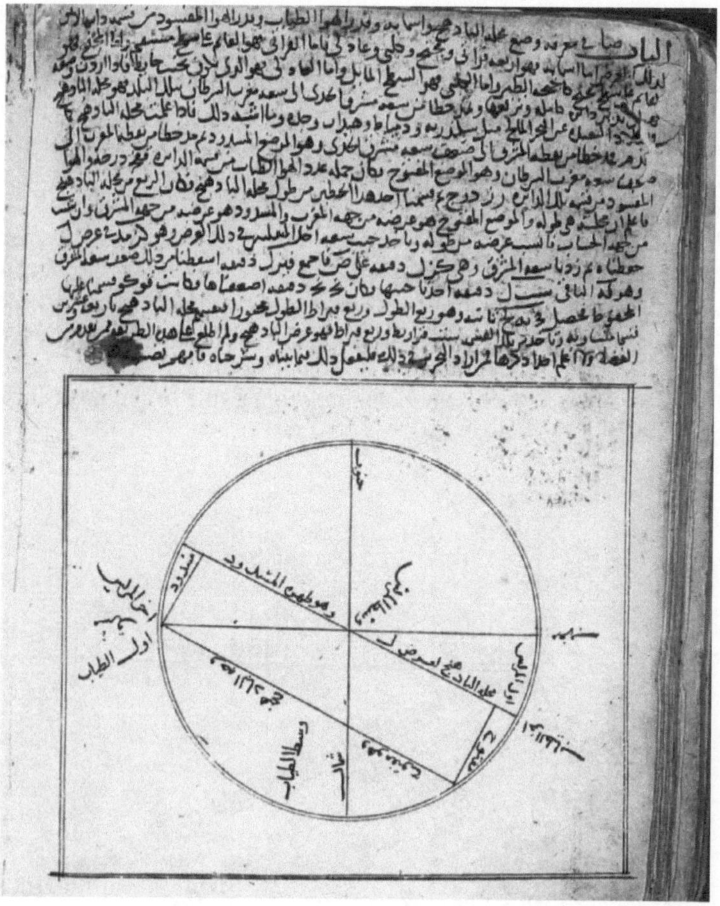

Fig. 9.18. A diagram showing the orientation of ventilators in medieval Cairo, found in the treatise on astronomical instruments of Ibn al-Sarrāj (Aleppo, ca. 1325) (MS Dublin Chester Beatty Library 102). By courtesy of the Chester Beatty Library.

fortuitous, or—and there is no evidence whatsoever for this—the site was chosen because the rectangular street pattern would be *qiblah*-oriented. A few years after the city was built, the astronomer Ibn Yūnus computed the *qiblah* in Cairo to be 37° S. of E., 10° to the south of the *qiblah* of the *ṣaḥābah*, and the Azhar and al-Ḥākim Mosques were erected 10° askew to the city-plan. Many of the religious edifices in Cairo constructed by the Mamluks have their outside walls oriented with the city plan and their inside walls and *miḥrābs* aligned with the *qiblah* of the astronomers. The City of the Dead outside Cairo is aligned with its main axis perpendicular to the *qiblah* of the astronomers; architecture in al-Qarāfah is oriented due south, in keeping with the *qiblah* favored by the Shāfiʿī school. The different *qiblahs* accepted in medieval Cairo are shown in figure 9.19. Thus, a diagram of a ventilator in an astronomical text led to the discovery of the basic orientations underlying one of the most important Islamic cities.

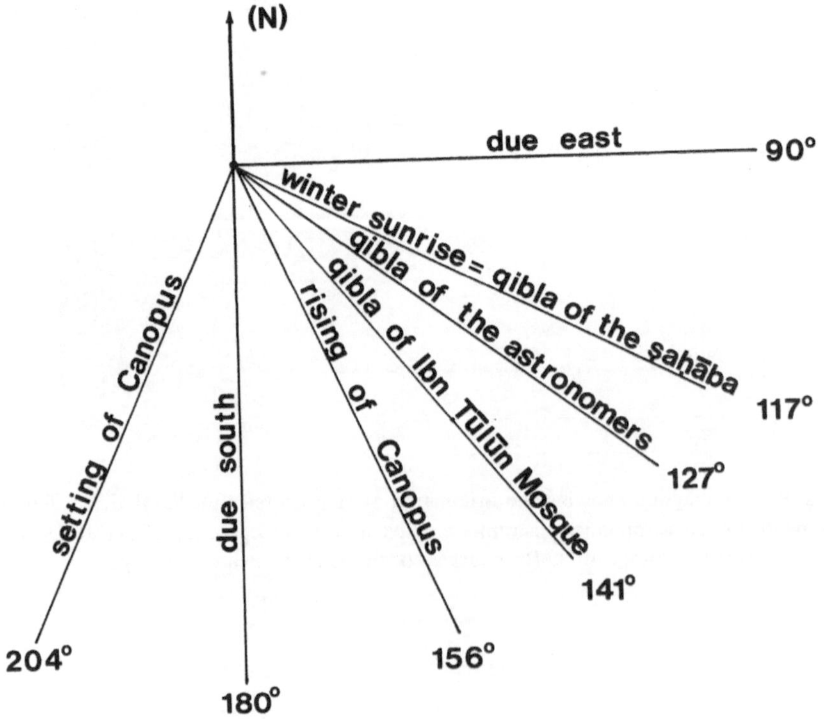

Fig. 9.19. The different *qiblahs* accepted in medieval Cairo. In the case of the southerly *qiblahs*, any direction between the rising and setting of Canopus would have been considered acceptable.

AN ILLUSTRATION OF THE SIGNIFICANCE OF THE KAʿBAH

A series of Arabic texts, mostly not illustrated and all recently investigated for the first time, discuss the orientation of the Kaʿbah itself. The base is rectangular (not square, as is often stated). See figure 9.20 [l.h.s.], with its major axis pointing toward the rising of Canopus, the brightest star in the southern sky, and its minor axis pointing toward the setting of the sun at the midwinter solstice. (At Mecca, these directions are indeed roughly perpendicular.) Thus, admonitions to face the rising of Canopus as the *qiblah* are common in Andalusia and Egypt; winter sunset in Iraq and Iran; and summer sunrise in the Sudan and Ethiopia. But there was no unity of opinion about the orientation of the Kaʿbah, let alone about the *qiblah*s in the different regions.

One notion which is *not* illustrated in any medieval Arabic text, but which is explicit in various wind-schemes described in the manuscript sources, and which underlies the very origin of the word *qiblah*, is the following (see figure 9.21). If one stands in front of (*istaqbala*) the southwest wall of the Kaʿbah (facing summer sunrise), one is facing the east (*qabūl*) wind. The north (*shamāl*) wind is on one's left (*shamāl*), blowing from Syria (*al-Shām*), and the South (*janūb*) wind on one's right (*yamīn*), blowing from the Yemen (*al-Yaman*). The west wind (*dabūr*) blows on one's back (*dabr*). This is evidence enough (and there is more) that the sacred direction (*qiblah*) of the Arabs in the pre-Islamic period was towards the east. But there are further meteorological and meteoric associations of the Kaʿbah, not the least the "Black Stone" embedded in the southeastern corner. While the diagram of the Kaʿbah shown in figure 9.20 (r.h.s.) and interpreted in figure 9.22 was taken from an eighteenth-century Egyptian manuscript, its conception was at least half a millennium older; the diagram recaptures some of these associations of the edifice. Note that the Kaʿbah is even correctly shown inclined to the meridian. The Aristotelian overtones—the elements and their qualities—are not attested elsewhere, but the geographical features and the association with the winds are. I have argued elsewhere that these pre-Islamic associations of the Kaʿbah lend credence to the Islamic belief that the Kaʿbah is a copy of a celestial counterpart, *al-bayt al-maʿmūr*. In modern terms, I suggest that this implies that it was built as an architectural representation of a pre-Islamic cosmology, featuring the sun and stars in its alignment and the winds and rains in its attributions.

CONCLUDING REMARKS

In brief, there are numerous illustrations in the Islamic scientific sources which can further our understanding of different aspects of Islamic science and even aspects of Islamic institutions. One should never rely solely on the illustrations if there is an accompanying text. But the need for facsimile editions of important manuscripts and of comparative studies of illustrated texts should be obvious.

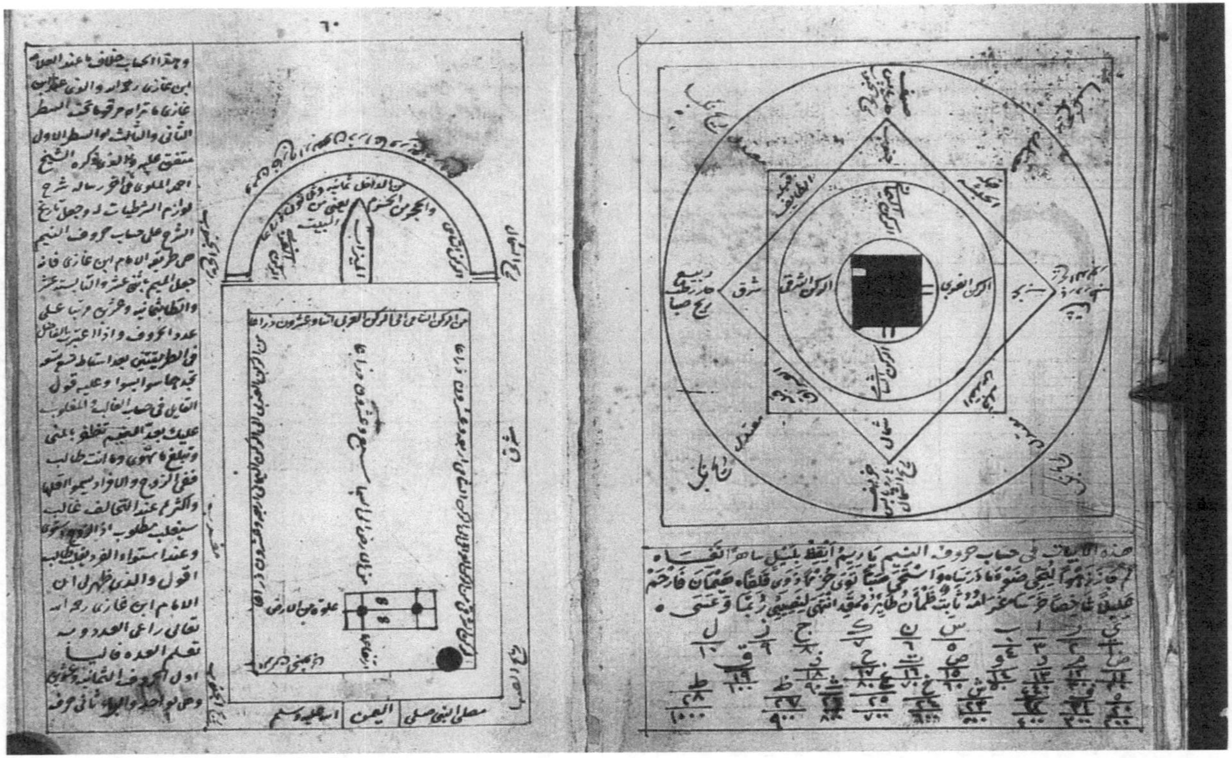

Fig. 9.20. On the left is a ground-plan of the Ka'bah (the rectangular shape is here exaggerated) and on the right a diagram which associates the edifice—incorrectly oriented—with the cardinal directions and their intermediaries, the winds, and the *qiblah*s of various localities. (MS Cairo Dār al-Kutub' Ṭal'at *majāmi'* 811.)

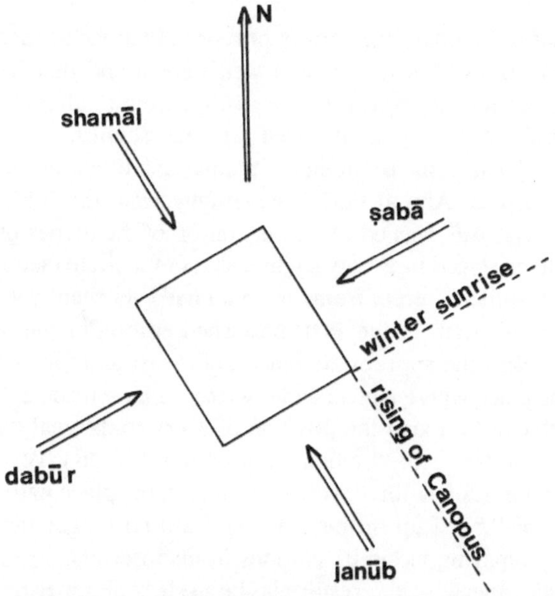

Fig. 9.21. The basic orientation of the Ka'bah and the directions of the four "cardinal" winds as recorded in various medieval texts. Actually, the minor axis is aligned with the most southerly setting point of the moon.

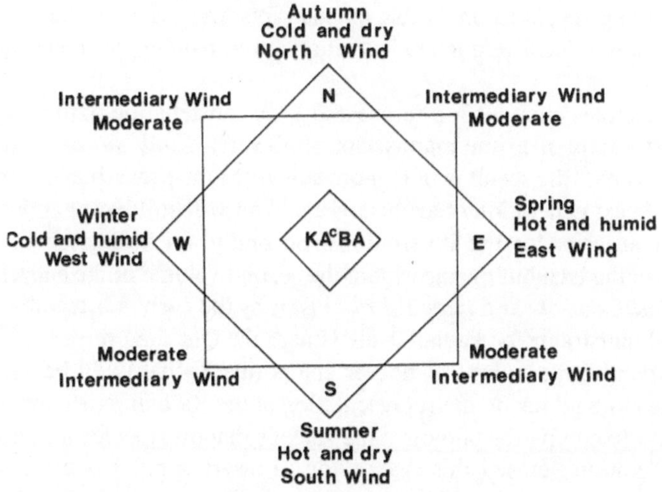

Fig. 9.22. The basic information on the Ka'bah contained in figure 9.20 (r.h.s.). The orientation of the Ka'bah has been "corrected"; in fact (see figure 9.10 [l.h.s.] and figure 9.21), its major axis is at about 30° E. of S.

One just has to compare the variety of representations of sacred geography in a few manuscripts of texts that are well known and that have already been published, but not in critical editions (for example, al-Qazwīnī, Yāqūt, and Ibn al-Wardī), to recognize the need for such studies.

In 1986, I was in the Bibliotheca Palatina Exhibition in Heidelberg, looking at a manuscript of Abū al-Fidā''s geography from the Bibliotheca Vaticana. The manuscript was opened to an illustration of the names of various regions of the world arranged in twenty-eight sectors of a circle (see figure 9.23). The diagram was quite different from any that I had ever seen, yet I knew that there was no such diagram in the 1840 published edition of that text. When I got home, I checked the appropriate place in the text and found a footnote by the editor at the place where Abū al-Fidā' wrote: "I have made a diagram (*zā'iraja*) from which can be found the position of every traditional climate and this is it. . ." The footnote read: "La figure qui se trouve ici dans l'original a été placée sous forme d'index à la fin du volume." In fact, the place names in the diagram have been swallowed up in the index and are no longer identifiable.

When preparing facsimile editions of manuscripts, strict guidelines must be followed, otherwise the result can be useless. Tampering with words that have been partially devoured by worms, removing original folio numbers, reshuffling folios without comment, and merging fragments from different copies of the same work without indication, all these contravene the most basic academic standards. Alas, they are features of the facsimile series published in recent years in Frankfurt, which has already been devastated by one reviewer who is a leading specialist on Arabic manuscripts. But the series has problems other than those which were noted by that reviewer, namely, the meddling with illustrations.

Two examples must suffice here. Firstly, the constellation figures and stars in the illustrations of a fine manuscript of al-Ṣūfī's *Kitāb ṣuwar al-kawākib* have been altered: the result is thus more suited for the proverbial coffee-table than for the next generation of art historians, who will still have to consult the original manuscript for the illustrations. Secondly, on a diagram of sacred geography in the Istanbul manuscript of the second volume of the encyclopedia entitled *Masālik al-abṣār fī mamālik al-amṣār*, by the early fourteenth-century Mamluk administrator Ibn Faḍlallāh al-ʿUmarī, the Qur'ānic verse (3: 97), *wa-man dakhalahu kāna āmin*[an], "whoever enters (the Kaʿbah) will be safe," has been written in gold ink on the representation of the Kaʿbah. Now, the modern scribe who "fixed up" the photos of the various manuscripts for the Frankfurt facsimile "edition" erased this inscription to insert another one more bold, and wrote *āmin*[an] with two *alifs* at the beginning. The illustrations in medieval manuscripts deserve more respect than this.

I hope to have shown how manuscripts in general and illustrations in particular can further our understanding of Islamic heritage. Libraries have

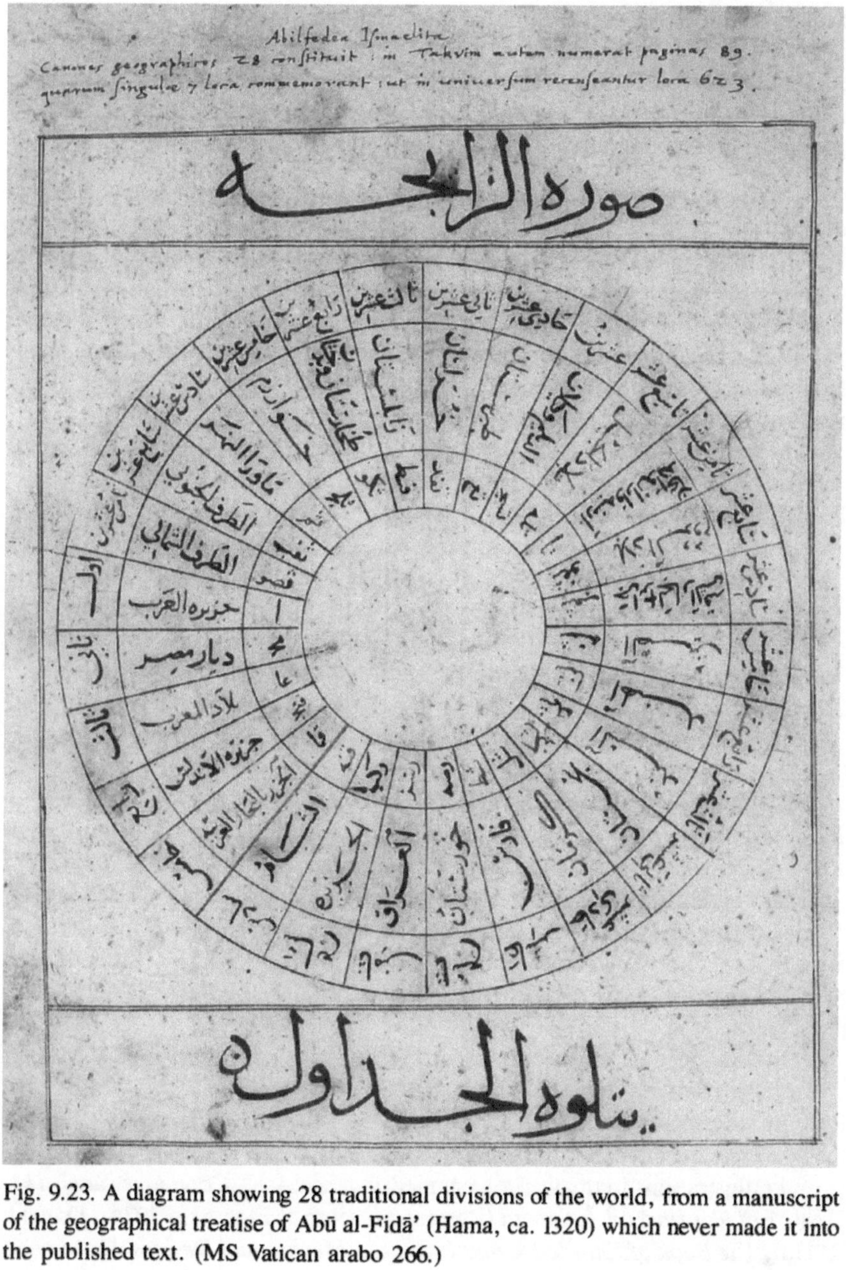

Fig. 9.23. A diagram showing 28 traditional divisions of the world, from a manuscript of the geographical treatise of Abū al-Fidā' (Hama, ca. 1320) which never made it into the published text. (MS Vatican arabo 266.)

these sources in trust and should treat them with the respect they deserve, which means cataloging them and making them available to scholars. We investigators should also treat them with respect, for they hold the key to many a mystery.

BIBLIOGRAPHICAL NOTES

Many illustrations for Islamic scientific manuscripts, some rather spectacular, are reproduced in S.H. Nasr, *Islamic Science: An Illustrated Study* (London: World of Islam Publishing Co. Ltd., 1976). Some 240 extracts from Islamic scientific manuscripts, many of them illustrated, are to be found in D.A. King, *A Survey of the Scientific Manuscripts in the Egyptian National Library* (Winona Lake, IN: Eisenbrauns, 1986). See also John E. Murdoch, *Album of Science: Antiquity and the Middle Ages* (New York: Charles Scribner's Sons, 1984), for various illustrations from Islamic scientific manuscripts.

On the activities of Taqī al-Dīn, see A. Sayılı, *The Observatory in Islam*, Publications of the Turkish Historical Society, series VII, no. 38 (Ankara: Türk Tarih Kurumu Basımevi, 1960), pp. 289–305; and A.S. Ünver, *Istanbul Rasathanesi*, Türk Tarih Kurumu Yayınlarından, V. Seri, Sayı 54 (Ankara: Türk Tarih Kurumu Basımevi, 1969).

On illustrations of constellation figures, see, for example, the works by J. Upton and E. Wellesz cited in the bibliography to the article, "al-Ṣūfī," by Paul Kunitzsch, in *The Dictionary of Scientific Biography*, 16 vols. (New York: Charles Scribner's Sons, 1970–78), 8:149–50.

On illustrations in astrological texts, see, for example, Zeren Tanındı (Akalay), "Astrological Illustrations in Islamic Manuscripts," *Proceedings of the I. International Congress on the History of Turkish-Islamic Science and Technology, Istanbul, 14–18 September 1981*, pp. 71–90; idem, "An Illustrated Astrologicai Work of the Period of Iskandar Sultan," *Akten des VII, Internationalen Kongresses für Iranische Kunst und Archäologie, München, 17–20 September 1976* (Berlin: Dietrich Relmer Verlag, 1979), pp. 418–25; and Stefano Carboni, "Two Fragments of a Jalayrid Astrological Treatise in the Keir Collection and in the Oriental Institute in Sarajevo," *Islamic Art* 2 (1987), pp. 149–86.

On non-Ptolemaic planetary theory in Islamic astronomy, see G. Saliba, "The Astronomical Tradition of Maragha: A Historical Survey and Prospects for Future Research," *Arabic Sciences and Philosophy* 1 (1991), pp. 67–99, and the works there cited.

On illustrations in treatises on engineering, see A.K. Coomaraswamy, *The Treatise of al-Jazarī on Automata* (Boston: Museum of Fine Arts, 1924); Donald R. Hill, *The Book of Knowledge and Ingenious Mechanical Devices* (Dordrecht (NL) and Boston: Dr. Reidel Publishing Company, 1974); and numerous other writings by Hill. See also Ahmad Y. al-Hasan and Donald R. Hill, *Islamic Technology: An Illustrated History* (Cambridge: Cambridge University Press;

and Paris: UNESCO, 1986). On the independent Andalusian tradition, see the chapter, "Technologia andalusi," by Donald R. Hill, in Juan Vernet, Julio Samso, et al., *El Legado Científico Andalusi* (Madrid: Ministerio de Cultura, 1992), pp. 157–72.

On Islamic cartography, see the article, "*Kharīṭa*" (Map), by S. Maqbul Ahmad, in *Encyclopedia of Islam*, new edition (Leiden: E.J. Brill, 1960–) (hereafter *EI*[2]); and various chapters in *The Historv of Cartography*, eds. J.B. Harley and David Woodward (Chicago; London: The University of Chicago Press, 1992), vol. 2, book 1: *Cartography in the Traditional Islamic and South Asian Societies*.

On al-Ashraf and his astrolabe, see D.A. King, "The Medieval Yemeni Astrolabe in the Metropolitan Museum of Art in New York," *Zeitschrift für Geschichte der arabisch-islamischen Wissenschaften* 2 (1985), pp. 99–122, repr. in *idem*, *Islamic Astronomical Instruments* (London: Variorum Reprints, 1987), 2; and on Ibn al-Sarrāj and his astrolabe, see ibid., 9. Numerous illustrations of other instruments and extracts from treatises on instrumentation are found in idem, "Strumentazione astronomica nel mondo medieval e islamico," a chapter in *Gli strumenti*, ed. Gerard Turner (Milan: Electa, 1990), pp. 154–89 and 580–85, the first of a series of richly illustrated volumes on the history of sciences, Paolo Galluzzi, general editor. An abridgement of the English original is published as "Some Remarks on Islamic Astronomical Instruments," *Scientiarum Historia* 18 (1992), pp. 5–23. On the Frankfurt project to catalogue medieval instruments, see my "Medieval Astronomical Instruments: A Catalogue in Preparation," *Bulletin of the Scientific Instrument Society* 31 (December 1991), pp. 3–7; and also no. 36 (March 1993), pp. 17–18.

On Islamic sacred geography, see my study in *Islamic Art*; a survey is in the article, "Makka: As Centre of the World," in *EI*[2]. See also D.A. King and G.S. Hawkins, "On the Orientation of the Kaʿba," *Journal for History of Astronomy* 13 (1982), pp. 102–9, repr. in King, *Astronomy in the Service of Islam* (Aldershot, UK: Variorum Reprints, 1993); idem, "Architecture and Astronomy: The Ventilators of Medieval Cairo and Their Secrets," *Journal of the American Oriental Society* 104 (1984), pp. 97–133; and idem, Richard Lorch, "Qibla Charts, Qibla Maps, and Related Instruments," in *History of Cartography* 2:1, pp. 189–205.

On facsimile editions of manuscripts and how they should not be prepared, see J.J. Witkam, "Arabic Manuscripts in Distress: The Frankfurt Facsimile Series," *Manuscripts of the Middle East* 4 (1989), pp. 175–80. For some suggestions concerning facsimile editions of Islamic scientific manuscripts which would be of use to scholars, see my paper, "Some Remarks on Islamic Scientific Manuscripts and Instrument, and Past, Present and Future Research," in *The Significance of Arabic Manuscripts*, ed. John Cooper (London: Al-Furqān Heritage Foundation, 1992), pp. 115–43 (some of the illustrations have been shuffled and no longer correspond to the captions).

IV

Aspects of Fatimid Astronomy: From Hard-Core Mathematical Astronomy to Architectural Orientations in Cairo

To Horst Jaritz,
in memory of Felicitas

Introduction

The purpose of this paper is to outline the scientific achievements of the Fatimid astronomer Ibn Yūnus and point to the implications of his works for the later development of Islamic astronomy in general, and of the specifically Islamic science of astronomical time-keeping in particular, as well as for architectural orientations in medieval Cairo.[1]

Ibn Yūnus – some biographical notes

The most able and most influential Egyptian astronomer in Fatimid times, and indeed – with the exception of Ptolemy of Alexandria (*ca.* 125 A.D.) – in Egyptian history, was Abu 'l-Ḥasan ᶜAlī ibn ᶜAbd al-Raḥmān known as Ibn Yūnus (d. 1009). He came from a respected family, his great-grandfather Yūnus having been a companion of the famous legal scholar al-Shāfiᶜī, and his father ᶜAbd al-Raḥmān a distinguished historian and scholar of *ḥadīth* (the sayings of the Prophet Muḥammad). Besides being renowned as an astronomer and astrologer, Ibn Yūnus was widely acclaimed as a poet, and some of his poems have been preserved. Unfortunately nothing of consequence is known about his early life or education. It is clear, however, from the biography of Ibn Yūnus by his late contemporary al-Musabbiḥī, preserved in fragmentary form in the writings of later historians, that Ibn Yūnus was an eccentric: he is described as a careless and absent-minded man who dressed shabbily and had a comic appearance.

We know that as a young man Ibn Yūnus witnessed the Fatimid conquest of Egypt and the foundation of Cairo in 969. In the period from 977 to 996, which corresponds roughly to the reign of

---

1 The paper which I presented at the Paris colloquium consisted essentially of a guided tour of some 100 folios of extracts from manuscripts, astronomical instruments and plans of Cairo architecture, only a few of which can be reproduced here. This written version has been compiled mainly from the article "Ibn Yūnus" in the *DSB*, together with my articles "IY's *Very Useful Tables*" and "Orientations in Medieval Cairo", generally without citations.

Felicitas Jaritz (d. 1998) was the only person who understood what I was about when I was trying to interpret actual orientations of medieval mosques and cities in the light of information found in medieval manuscripts. I remember our collaboration in Cairo during 1972-79 with pleasure and gratitude. François Charette in Frankfurt read the penultimate version of this paper and saved me from several pitfalls, if not all.

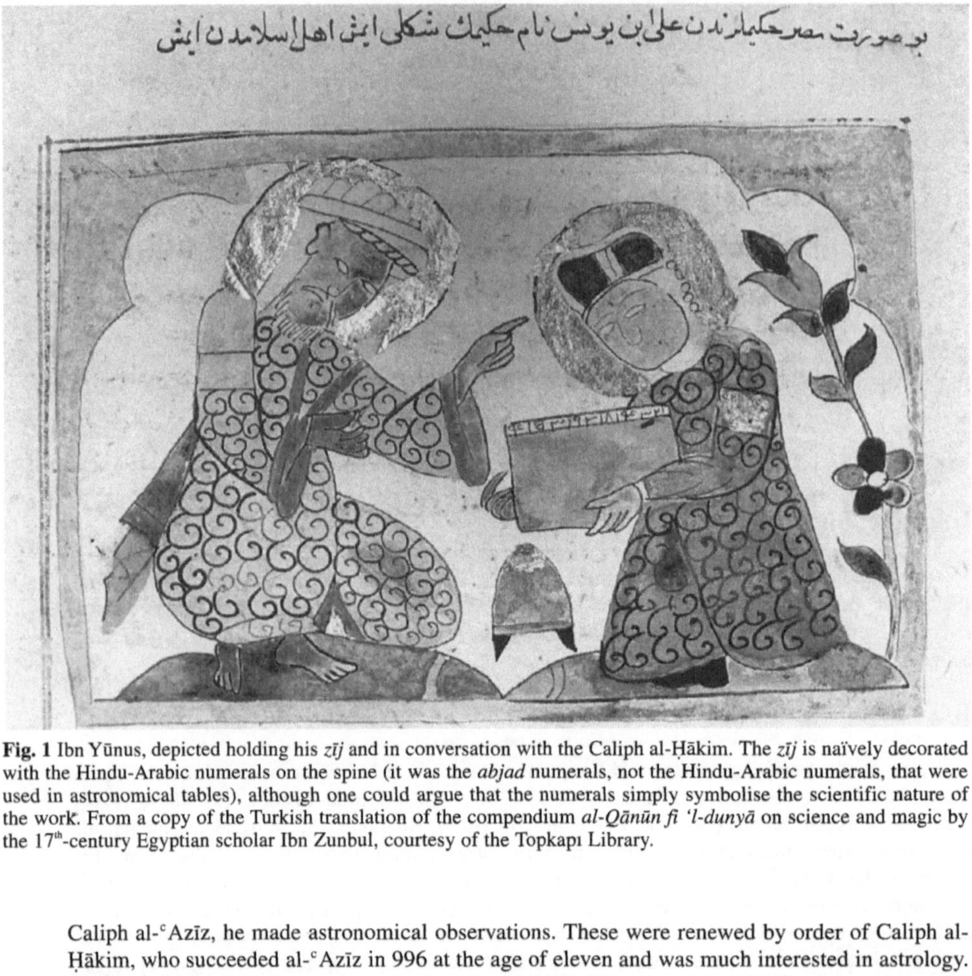

Fig. 1 Ibn Yūnus, depicted holding his *zīj* and in conversation with the Caliph al-Ḥākim. The *zīj* is naïvely decorated with the Hindu-Arabic numerals on the spine (it was the *abjad* numerals, not the Hindu-Arabic numerals, that were used in astronomical tables), although one could argue that the numerals simply symbolise the scientific nature of the work. From a copy of the Turkish translation of the compendium *al-Qānūn fī 'l-dunyā* on science and magic by the 17th-century Egyptian scholar Ibn Zunbul, courtesy of the Topkapı Library.

Caliph al-ᶜAzīz, he made astronomical observations. These were renewed by order of Caliph al-Ḥākim, who succeeded al-ᶜAzīz in 996 at the age of eleven and was much interested in astrology. Ibn Yūnus' recorded observations continued until 1003. He was a colourful character who added a distinctive Islamic flavour to contemporary astronomy – see Fig. 1.

Ibn Yūnus' major work, dedicated to al-Ḥākim

Ibn Yūnus' major work was *al-Zīj al-Ḥākimī al-kabīr*, *zīj* meaning an astronomical handbook with tables. It is a particularly fine representative of a class of astronomical handbooks, numbering perhaps 200, compiled in the Islamic world between the 8th and the 19th centuries.[2] The *Zīj* of Ibn Yūnus was dedicated to the Caliph al-Ḥākim and was aptly named *al-kabīr* ("large"): the text of the surviving first 44 of the 81 chapters of the original work is twice the length of the text of the *Zīj* by al-Battānī (Raqqa *ca.* 910) and contains more than twice as many tables as the earlier work. The only extant chapters of the *Ḥākimī Zīj* (1-44) are in two manuscripts now in Leiden and

2 See Kennedy, "*Zīj* Survey", and the forthcoming article "Zīdj" in *EI*².

Fig. 2 The title folio of the Leiden manuscript of the first quarter of the *Ḥākimī Zīj*, with biographical information on the author taken from the biographical dictionary of Ibn Khallikān recorded in the hand of Ibn Abi 'l-Fatḥ al-Ṣūfī, the leading Egyptian astronomer of the late 15th century. The folio bears a mark of ownership by the Ottoman astronomer Taqiyyu 'l-Dīn ibn Maᶜrūf, director of the Istanbul Observatory *ca.* 1577. From MS Leiden Or. 143, p. 1, courtesy of the Universiteitsbibliotheek.

Oxford, together comprising about 300 folios. A manuscript in Paris contains an anonymous abridgement of part of the *Zīj* and is a source for some additional text up to Chapter 57, as well as Chapters 77-81 dealing with mathematical astrology. The *Ḥākimī Zīj* was thus a monumental work, probably in about 500 folios, and thus one of the largest compilations of its genre – see Fig. 2.

The *Ḥākimī Zīj* deals with the standard topics of Islamic astronomy but is further distinguished from all other extant *Zīj*es by beginning with a list of observations made by Ibn Yūnus and of observations made by some of his predecessors, quoted from their works. Despite the critical attitude of Ibn Yūnus toward these earlier scholars and his careful recording of their observations and some of his own, he completely neglects to describe the observations that he used in establishing his own planetary parameters, nor does he indicate whether he used any instruments for these observations. Indeed, the *Ḥākimī Zīj* is a poor source of information about the instruments used by Ibn Yūnus. In his account of measurements of the latitude of Fusṭāṭ and of the obliquity of the ecliptic from solar meridian altitudes at the solstices, Ibn Yūnus states that he used an instrument provided by each of the Caliphs al-ᶜAzīz and al-Ḥākim. Although he describes it only by mentioning that the divisions for each minute of arc were clearly visible on its scale, the instrument was probably a large meridian ring. The only other instruments he mentions are an astrolabe and a gnomon.

Ibn Yūnus explains in the introduction to his *Zīj* that the work is intended to replace the *Mumtaḥan Zīj* of Yaḥyā ibn Abī Manṣūr, prepared for the Abbasid Caliph al-Ma'mūn in Baghdad *ca.* 830. Ibn Yūnus reports the observations of some astronomers before his own time, in which what was observed was at variance with what was calculated with the tables of the *Mumtaḥan Zīj*. When reporting his own observations, Ibn Yūnus often compares what he observed with what he had computed with the *Mumtaḥan* tables.

The observation accounts

From the introduction and Chapters 4, 5, and 6 of the *Ḥākimī Zīj*, which contain the observation accounts, it is clear that Ibn Yūnus was familiar with the *Zīj*es of Ḥabash al-Ḥāsib, al-Battānī, and al-Nayrīzī, as well as the *Mumtaḥan Zīj*. The observations made by Ḥabash quoted by Ibn Yūnus are not found in the two extant versions of Ḥabash's *Zīj*. Ibn Yūnus also records observations made by al-Māhānī, whose works are not extant. He lists the planetary parameters of the *Mumtaḥan Zīj*, and this has enabled the positive identification of at least the planetary tables in the only extant manuscript of this early work, which contains considerable spurious material. Ibn Yūnus also quotes observations made by the Banū Amājūr family in Baghdad; their five *zīj*es are not extant. Other works quoted by Ibn Yūnus, although not necessarily directly, are the *zīj*es of al-Nihāwandī, Ibn al-Ādamī, the Banū Mūsā, Abū Maᶜshar, Ibn al-Aᶜlam, ᶜAbd al-Raḥmān al-Ṣūfī, and Muḥammad al-Samarqandī; none of these works is extant, and Ibn Yūnus' references provide valuable information about them.

The observations described by Ibn Yūnus are of conjunctions of planets with each other and with Regulus (a bright star conveniently close to the ecliptic), solar and lunar eclipses, and equinoxes; he also records measurements of the obliquity of the ecliptic (Chapter 11) and of the maximum lunar latitude (Chapter 38). All of these accounts are notable for their lack of information on observational procedures. The following passage is a translation of one of Ibn Yūnus' accounts of a planetary conjunction that he had observed:

> "A conjunction of Venus and Mercury in Gemini, observed in the western sky: The two planets were in conjunction after sunset on the night whose morning was Monday, the 13th day of Jumādā II 390 Hijra era. The time was approximately eight equinoctial hours after midday on Sunday, which was the fifth day of Khardādh, 369 Yazdigird era. Mercury was north of Venus and their latitude difference was a third of a degree. According to the *Mumtaḥan Zīj* their longitude difference was four and a half degrees." [Caussin, p. 217.]

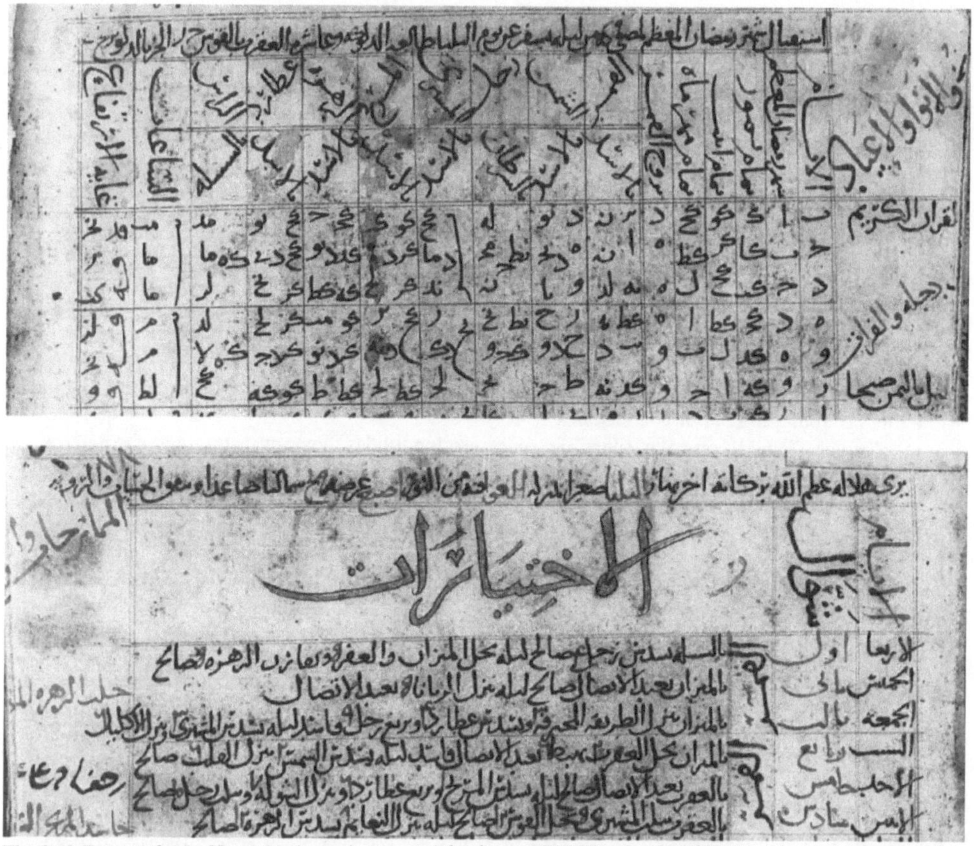

Fig. 3a-b Extracts from a Yemeni ephemeris (*taqwīm*) for the year 727 Hijra [= 1326/27], the oldest surviving complete example of this genre. The tables display for each day of the year, here for the first few days of Ramaḍān, the positions of the moon, sun and five naked-eye planets, and (on the opposite page) the astrological significance of the position of the moon relative to the planets. At the head of the table is found information on the conjunction of the sun and moon preceding the beginning of the month and on the possibility of lunar crescent visibility the next evening. The tables were computed with the planetary tables of the *Ḥākimī Zīj*, and we can anticipate that Fatimid ephemerides for Cairo contained the same kind of information, presented in the same format. From MS Cairo ENL DM 817, fols. 87v-88r, courtesy of the Egyptian National Library.

The Sunday mentioned was 19 May 1000, and computation with modern tables confirms that there was a conjunction in longitude that evening and that Mercury was indeed one-third degree north of Venus. About forty such planetary conjunctions observed by Ibn Yūnus are described in the *Zīj*.

The following passage is a translation of Ibn Yūnus' account of the lunar eclipse that occurred on 22 April 981 (Oppolzer no. 3379):

> "This lunar eclipse was in the month of Shawwāl, 370 Hijra era, on the night whose morning was Friday, the third day of Urdibihisht, 350 Yazdigird era. We gathered to observe this eclipse at al-Qarāfa in the Mosque of Ibn Naṣr al-Maghribī. We perceived first contact when the altitude of

the moon was approximately 21°. About a quarter of the lunar diameter was eclipsed, and re-emergence occurred about a quarter of an hour before sunrise." [Caussin, p. 187.]

Some of the thirty eclipses reported by Ibn Yūnus were used by Simon Newcomb (1878) in his determination of the secular acceleration of the moon. Other observations recorded in the *Ḥākimī Zīj* have been used with some abandon by R. Newton (1970), and, more recently still, but with considerably more respect for the sources, by F. R. Stephenson and various colleagues at the University of Durham.[3]

Calendrics and planetary astronomy

The first chapter of the *Ḥākimī Zīj* is the longest of the extant ones and deals with the Muslim, Coptic, Syrian and Persian calendars. There are detailed instructions for converting a date in one calendar to any of the other calendars, and extensive tables for that purpose. There are also tables for determining the dates of Lent and Easter in both the Syrian and the Coptic calendars. Such tables are found in several Islamic *zījes*.[4]

Chapters 7 and 9, on solar, lunar and planetary longitudes, contain instructions for determining true longitudes from the tables of mean motions and equations. These are the tables that would be needed to compile an ephemeris displaying true longitudes for each day of a given year[5] – see Fig. 3. No theory is described, but the theory underlying the instructions and tables is entirely Ptolemaic. The mean motions differ from those used by Ibn Yūnus' predecessors, and his values for the sun and moon were deemed sufficiently reliable by Naṣīr al-Dīn al-Ṭūsī to be used in the *Īlkhānī Zīj* some 250 years later. Ibn Yūnus' planetary tables are computed for both the Muslim and Persian calendars, and define the mean positions of the sun, moon, and planets, as well as the astrologically-significant pseudo-comet *al-Kayd*, for over 2,700 Muslim and 1,800 Persian years from the respective epochs 622 and 632.

For the year 1003, Ibn Yūnus gives the solar apogee as Gemini 26;10° and the maximum solar equation as 2;0,30°, corresponding to a double eccentricity of 2;6,10° (where the solar deferent radius is 60). No solar observations made by Ibn Yūnus are recorded in the *Zīj*. He changes the values of the lunar epicyclic radius and the eccentricity from Ptolemy's 5;15° and 10;19°, also used in the *Mumtaḥan Zīj*, to 5;1,14° and 11;7°, respectively (the latter is not used consistently), again without explanation. His planetary equation tables are identical with those of Ptolemy's *Handy Tables* and the *Mumtaḥan Zīj* for Saturn, Jupiter, and Mars. For Venus, Ibn Yūnus assumes an eccentricity exactly half that of the sun and uses an epicyclic radius of 43;42 rather than Ptolemy's 43;10. For Mercury he adopts a maximum equation of 4;2°, an Indian parameter previously used in the *Zīj* of al-Khwārizmī, rather than Ptolemy's 3;2°. Ibn Yūnus' tables of equations for the moon, Venus, and Mercury contain the same inconsistencies as al-Battānī's tables for Venus, in that some of the columns are not adjusted for the new parameters; this is a fairly common feature in Islamic *zījes*. There is evidence that Ibn Yūnus was not altogether satisfied with his determination of the planetary apogees: the *Ḥākimī Zīj* contains three different sets of values (Chapters 6, 8 and 9).

In his discussion of solar and lunar distances (Chapters 55, 56), Ibn Yūnus assumes a maximum solar parallax of 0;1,57°, instead of Ptolemy's value of 0;2,51°. Chapters 59-75, on parallax and eclipse theory and the associated tables, are not found in the surviving manuscripts of the *Zīj*.

In Chapter 38, on lunar and planetary latitudes, Ibn Yūnus states that he found the maximum lunar latitude to be 5;3°. Although he says that he measured it many times and repeatedly found this value, he does not say how the measurements were made. He did not pursue the suggestion of

3 See Stephenson & Said, "Medieval Islamic Eclipse Observations", A-C.

4 See now the article "Ta'rīkh. 2. Era chronology in astronomical handbooks", in *EI*².

5 See the article "Taḳwīm" in *EI*².

the Banū Amājūr that he quotes, namely, that the maximum lunar latitude was not constant. His planetary latitude tables are derived from those in the *Almagest*, except in the case of Venus, for which he used values originally taken from the *Handy Tables*.

Ibn Yūnus measured the position of Regulus as Leo 15;55° in the year 1003. His value for the motion of the fixed stars is 1° in 70 1/4 Persian years (of 365 days) and apparently was computed by using his own observation of Regulus and that of Hipparchus; it is the most accurate of all known Islamic values. He had information at his disposal from which he might have deduced that the motion of the planetary apogees was different from the motion of the fixed stars, but he chose to conclude that the apogees moved at the same rate as the stars (Chapter 8).

Ibn Yūnus is named as the author of an extensive set of tables, called *al-Taᶜdīl al-muḥkam* and extant in a unique copy, displaying the equations of the sun and moon; the latter are of particular interest. They are based on those in the *Ḥākimī Zīj* but are arranged so as to facilitate computation of the lunar position: the equation is tabulated as a function of the double elongation and mean anomaly, both of which can be taken from the mean motion tables; thus there is no need to find the true anomaly. The table, which accurately defines the Ptolemaic lunar equation for Ibn Yūnus' parameters, contains over 34,000 entries.[6]

The lunar crescent visibility of Ibn Yūnus is not found in the surviving manuscripts of his *Zīj* but has been recovered in later sources.[7] It imposes conditions on the apparent distance between the sun and moon and the time difference between sunset and moonset, depending on the relative positions of the sun and moon on their respective orbits. It would be interesting to know how these conditions were derived.

Trigonometry

The trigonometric functions used by Ibn Yūnus are functions of arcs rather than angles, and are computed for radius 60, as was standard in Islamic works. Chapter 10 of the *Zīj* contains a table of sines for each 0;10° of arc, computed to four significant sexagesimal digits. The values are seldom in error by more than ±2 in the fourth digit. Ibn Yūnus determined the sine of 1° to be 1;2,49,43,28 (to base 60), using a method equivalent to interpolating linearly between the values of sin x/x for x = 15/16° and 9/8°. He then improved this value by a rather dubious technique to obtain 1;2,49,43,4. The accurate value to this degree of precision is 1;2,49,43,11. Ibn Yūnus' younger contemporary al-Bīrūnī was able to calculate the chord of a unit circle subtended by an angle of 1° correctly to five significant sexagesimal digits. Although in Chapter 11 of the *Zīj* Ibn Yūnus tabulates the cotangent function to three sexagesimal digits for each ten minutes of arc, he does not take full advantage of it. Many of the methods he suggests throughout the *Zīj* require divisions of sines by cosines, and he uses the cotangent function, which he calls the shadow (*al-ẓill*), only when the argument is an altitude arc.

There are other sets of tables preserved in the manuscript sources and attributed to Ibn Yūnus that are distinct from those in the *Ḥākimī Zīj* but based on them. Thus Ibn Yūnus appears to be the author of tables of the sine and tangent functions for each minute of arc, as well as tables of solar declination for each minute of solar longitude. These sine tables display values of the sine function to five sexagesimal digits, which is roughly equivalent to nine decimal digits – see Fig. 4. The values are often in error in the fourth sexagesimal digit, however, so that it was a premature undertaking. Indeed, over four centuries were to pass before the compilation of the trigonometric tables in the *Sulṭānī Zīj* of Ulugh Beg in Samarqand, in which values are also given to five sexagesimal digits for each minute of arc, but are generally correct.

---

6 Analyzed in King, "IY's Lunar Equation Table".
7 See *idem*, "IY on Lunar Crescent Visibility".

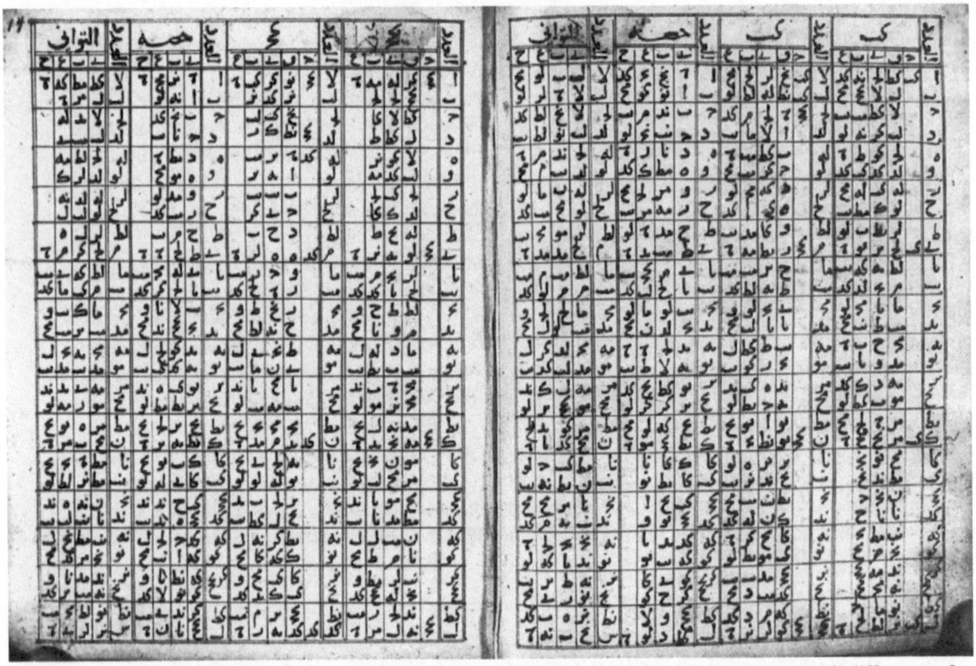

Fig. 4 An extract for arguments 22°-23° from some tables of the sine function for each minute with differences for the seconds, attributed to Ibn Yūnus and based on those in his *Zīj*. Entries are given to five sexagesimal digits and are based on those of Ibn Yūnus for each 10'; note that those entries have a zero added as the last digit. Intermediate entries are computed by linear interpolation, which raises doubts as to the authorship, because Ibn Yūnus was well aware that a second-order interpolation scheme, such as he presented in the *Ḥākimī Zīj*, would have been more appropriate. From MS Berlin DSB Ahlwardt 5752, fols. 13v-14r, courtesy of the Deutsche Staatsbibliothek.

Spherical astronomy

In spherical astronomy – the mathematics of the apparent daily rotation of the heavens – Ibn Yūnus reached a very high level of sophistication (Chapters 12-54). Although none of his formulae is explained, it seems probable that most of them were derived by means of orthogonal projections and analemma constructions, rather than by the application of the rules of spherical trigonometry that were being developed by Muslim scholars in Iraq and Persia. Altogether, there are several hundred formulae outlined in the *Ḥākimī Zīj*, many of which are trivially equivalent. These are stated in words and without recourse to any symbols, as was usual in medieval mathematical astronomy. For each method outlined, Ibn Yūnus generally gives at least one numerical example. The problems of spherical astronomy discussed in the *Ḥākimī Zīj* are more varied than those in most major Islamic *zīj*es, and the following examples are intended to illustrate the scope of the treatment.

Ibn Yūnus describes several methods for computing right and oblique ascensions (Chapters 13-14). He also computes both, the latter for each degree of the ecliptic and for each degree of terrestrial latitude from 1° to 48° (beyond which limit, according to his contemporary in Central Asia Abū Naṣr ibn ʿIrāq, "there is no one who studies this sort of thing or even thinks about it"). Ibn Yūnus discusses in great detail the determination of time and solar azimuth from solar altitude, and it will be clear from the tables mentioned below that he devoted much effort to these problems. Certain functions that he discusses in the text are also tabulated, such as the solar altitude in the prime vertical and the rising amplitude of the sun (that is, the distance of the rising sun from the east point). The problem of finding solar altitude from solar azimuth (Chapter 24) is not so simple as the inverse problem; but Ibn Yūnus solves it in several ways, including the use of an algebraic method. He also tabulates the solar altitude for certain azimuths, such as that of the qibla, the direction of Mecca (Chapter 28), and ten different azimuths (Chapter 24), to be used for finding the meridian. Several geometric solutions to the problem of determining the qibla, a favourite of the Islamic astronomers, are also outlined. One of Ibn Yūnus' solutions is equivalent to successive applications of the cosine rule and sine rule for spherical triangles, but is derived by a projection method that was also used by the late contemporary Cairene scholar Ibn al-Haytham.

Particularly elegant solutions are presented for finding the meridian from three solar observations on the same day (Chapter 23) and for finding the time between two solar observations on the same day (Chapter 33). The latter problem is solved by a direct application of the cosine rule for plane triangles, the earliest attested use of this rule. Ibn Yūnus transforms ecliptic to equatorial coordinates (Chapter 39) by a method equivalent to the cosine rule for spherical triangles but again probably derived by means of an analemma construction. His sundial theory (Chapters 26, 27, 35) is also of considerable sophistication. It deals with horizontal and vertical sundials, the latter oriented in the meridian, the prime vertical, or a general direction inclined to both. He stresses the precautions to be taken when setting up a gnomon on a marble slab to ensure that it is aligned correctly.

Mathematical Astrology

The section of the *Zīj* dealing with astrological calculations (Chapters 77 – 81), although partially extant in the anonymous abridgement of the work in the Paris manuscript, has never been studied. Ibn Yūnus was famous as an astrologer and, according to his biographers, devoted much time to making astrological predictions. An independent astrological treatise attributed to Ibn Yūnus has been shown to be spurious (see below).

Ibn Yūnus and the times of Muslim prayer

Ibn Yūnus' second major work was part of the corpus of spherical astronomical tables for timekeeping that was used in Cairo until the 19th-century. It is difficult to ascertain precisely how many tables in this corpus, which later became known – if only in a single manuscript which happened to be the first one I encountered – as the *Kitāb Ghāyat al-intifāᶜ* ("Very Useful Tables"), were actually computed by Ibn Yūnus. Some appear to have been compiled by the late 13th-century astronomer al-Maqsī. The corpus exists in numerous manuscripts, each containing different arrangements of the tables or only selected sets of tables; and in its entirety the corpus consists of about 200 pages of tables, most of which contain 180 entries – see Fig. 5. The tables are generally rather accurately computed and are all based on Ibn Yūnus' values of 30;0° for the latitude of Cairo and 23;35° for the obliquity of the ecliptic.[8]

The main tables in the corpus display the time since sunrise (*al-dā'ir*), the time remaining to midday (*faḍl al-dā'ir*), and the solar azimuth (*al-samt*) as functions of the solar altitude and solar longitude. Entries are tabulated for each degree of solar altitude and longitude, and each of the three sets contains over 10,000 entries. The remaining tables in the corpus are of spherical astronomical functions, some of which relate to the determination of the five daily prayers of Islam.

8 All of the tables are analyzed in King, "IY's *Very Useful Tables*". On the problems of the attribution see *idem*, "Mamluk Astronomy", p. 540-542.

The times of Muslim prayer are defined with reference to the apparent daily motion of the sun across the sky and vary throughout the year. The prayers must be performed within certain intervals of time, which are variously defined. The following general definitions underlie the tables in the corpus. The day is considered to begin at sunset, and the evening prayer is performed between sunset and nightfall. The permitted interval for the night prayer begins at nightfall. The interval for the morning prayer begins at daybreak and the prayer must be completed by sunrise. The period for the noon prayer begins when the sun is on the meridian, and that for the afternoon prayer begins when the shadow of any object is equal to its midday shadow plus the length of the object.[9]

Examples of functions relating to the prayer times, tabulated for each degree of solar longitude, include the following:

1. The length of morning and evening twilight (*ḥiṣṣat ṭulūᶜ al-fajr* and *ḥiṣṣat maghīb al-shafaq*), defining the permitted times for the morning and evening prayers, based on the assumption that twilight appears or disappears when the sun reaches a particular angle of depression below the horizon. (The angles suggested by Ibn Yūnus in the *Ḥākimī Zīj* are 18° for both phenomena, but in a later work he proposed 20° and 16° for morning and evening, respectively. The main twilight tables in the corpus are based on 19° and 17°.)

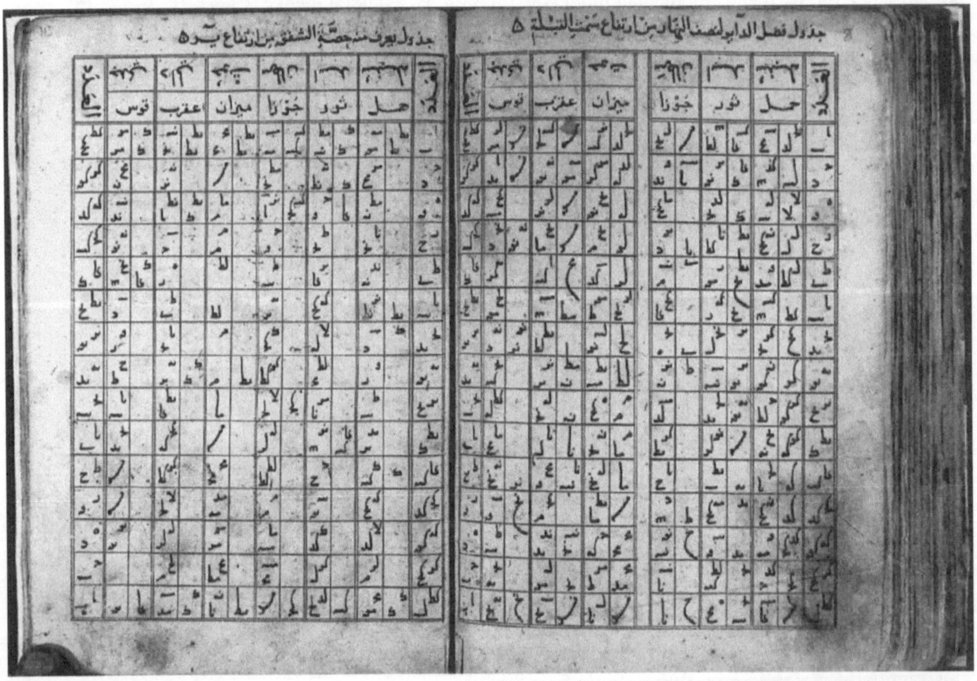

Fig. 5 An extract from the corpus of tables for astronomical timekeeping used in medieval Cairo, of which a substantial part was compiled already by Ibn Yūnus. This extract shows the time remaining until midday at the moment when the sun is in the azimuth of the qibla at Cairo (on the right), and the duration of evening twilight (on the left), based on a solar depression of 17°. In all of the tables, entries are given for each degree of solar longitude, corresponding roughly to each day of the year, and exploiting the symmetry of the tabulated functions. From MS Dublin CB 3673, courtesy of the Chester Beatty Library.

9 See the article "Mīḳāt. ii. Astronomical aspects" *in EI²*, reprinted in King, *Studies*, C-V.

2. The time from nightfall to daybreak, defining the permitted interval for the night prayer.
3. The time from sunrise to midday.
4. The time from midday to the beginning of the time for the afternoon prayer, defining the interval for the noon prayer (*al-ẓuhr*); and the time from the beginning of the afternoon prayer to sunset, defining the interval for the afternoon prayer (*al-ᶜaṣr*).
5. Corrections to the semidiurnal arc for the effect of refraction at the horizon, based on the assumption that the true horizon is about 2/3° below the visible horizon. (These corrections, which are specifically attributed to Ibn Yūnus, represent the earliest attested quantitative estimate of the effect of refraction on horizon phenomena.)
6. The altitude of the sun when it is in the azimuth of the qibla, as well as the corresponding time. (The underlying direction is 37° S of E – see below.)
7. The altitude of the sun when it is in the azimuth of the back of the ventilator (*al-bādahanj*). (The underlying direction is 27° S of E – see below.)

Virtually all later Egyptian prayer tables until the 19th century were based on those in this main corpus. In certain cases the original tables were well-disguised, the entries being written out in words for each day of the Coptic year or a given Muslim year. In the 16th century some of the tables were recomputed for an updated value of the obliquity. The impressive developments in astronomical time-keeping in 13th-century Yemen and 14th-century Syria, particularly the tables of Abu 'l-ᶜUqūl for Taiz and of al-Khalīlī for Damascus, also owe their inspiration to the main Cairo corpus.[10]

Some widespread fallacies about Ibn Yūnus

The following examples show how historical fallacies and myths can percolate through the modern literature and how attempts to suppress them are generally disregarded.

a) Ibn Yūnus and his observatory

Ibn Yūnus made observations, therefore he must have worked in a well-equipped observatory. Such was the rationale behind various claims about Ibn Yūnus' "observatory", to be found in two-dozen-odd sources between 1835 and most recent times. In his observation accounts Ibn Yūnus mentions only the roof of his grandfather's house in al-Qarāfa and the Mosque of Ibn Naṣr al-Maghribī in Fusṭāṭ. Aydın Sayılı (1960) made a serious attempt to destroy the observatory myth,[11] but it is too nice to drop out of the literature.[12]

There are two sources, however, that might cast a little more light on the situation if their reliability could be established. First, the historian Ibn Ḥammād (*fl. ca.* 1200) mentions a copper instrument, resembling an astrolabe three cubits in diameter, that a contemporary of his had seen and associated with the Ḥākimī observations. Likewise the Yemenite Sultan al-Ashraf (*fl. ca.* 1290), who was an astronomer, records that al-Ḥākim had an armillary sphere consisting of nine rings, each of which weighed 2,000 pounds and was large enough for a man to ride through on horseback. The possibility that this large instrument was that known to have been constructed in Cairo about 1125 – over a century after the death of Ibn Yūnus – cannot yet be discounted.[13]

There is evidence that al-Ḥākim had a house on the Muqaṭṭam hills overlooking Cairo, which may have contained astronomical instruments: Ibn Yūnus is known to have visited this house on one occasion to make observations of Venus. Nevertheless, al-Ḥākim's unsuccessful attempt to build an observatory in Cairo took place after Ibn Yūnus' death; and the only locations mentioned by Ibn

10 See *idem*, "Muezzins and Muwaqqits".
11 Sayılı, *The Observatory in Islam*, p.130-156, with a list of the two-dozen-odd sources on p. 130, n. 21.
12 As, for example, Nasr, *Islamic Science*, p. 112 (see my review reprinted in King, *Studies*, A-XVII, p. 217).
13 See Sayılı, *The Observatory in Islam*, p. 167-175.

Yūnus in his own accounts of his observations (see above) hardly merit the appellation 'observatory'. A note written in the 15th century on the title folio of the Leiden manuscript of the *Ḥākimī Zīj* – see Fig. 2 – states that Ibn Yūnus' observations were made in the area of Birkat al-Ḥabash in Fusṭāṭ.

b) Ibn Yūnus as the discoverer of the principle of the pendulum

Ibn Yūnus made observations and therefore must have known the principle of the pendulum for time-keeping. The association of the pendulum with the Arabs is found in several sources from the 17th century to the present, and from *ca.* 1800, thanks to Thomas Young, the Arabs became Ibn Yūnus. Even Sédillot-*fils* recorded this connection, at first sceptically and later with some conviction, despite the lack of any evidence. In some of the more recent sources he becomes confused with Kamāl al-Dīn Ibn Yūnus (d. 1242). Around 1920 Eilhard Wiedemann made an attempt to kill the myth,[14] and in 1979 I had a try.[15] But again, the association is very nice, so it will continue to be repeated.[16] At an International Symposium on Islamic Science in Aleppo in 1979 I stated in public that I had looked at all the known manuscripts associated with Ibn Yūnus and had found no mention of the pendulum: one of the participants said that he felt very sorry for me.

c) Ibn Yūnus associated with the world-map of the Fatimid Caliph al-Muᶜizz

al-Muᶜizz (*reg.* 953-975) commissioned a world-map in the year 353 H [= 964], which is described by the 15th-century historian al-Maqrīzī. Ibn Yūnus is known to have produced extensive geographical tables[17] and to have worked for al-Muᶜizz's son al-ᶜAzīz (*reg.* 975-996), to whom a substantial geographical work was dedicated in the year of his accession: therefore he must have been involved with the map. But if he was, he would have been a precocious teenager or even a child. The association of Ibn Yūnus with the map goes back to Étienne Quatremère (1811),[18] and is found in the article on maps in the *Encyclopaedia of Islam* and even in an article by E. S. Kennedy in the new encyclopaedia on Islamic science recently edited in Paris.[19]

d) Ibn Yūnus as the discoverer of the prosthaphaeresis formula of trigonometry

In his discussion of the determination of time from solar altitude (Chapter 15) Ibn Yūnus proposed two alternative procedures which are equivalent by virtue of the prosthaphaeresis formula. Therefore he must have known the formula, which is important in the history of trigonometry. J.-B. Delambre *ca.* 1820 noted the two procedures and their equivalence and Ibn Yūnus as the discoverer of the formula is mentioned in the history of trigonometry by A. von Braunmühl (1900), in the article on Ibn Yūnus in the *Encyclopaedia of Islam*, and in the history of medieval mathematics by A. P. Youschkevitsch (*ca.* 1960). But in fact he never proposed the formula and his alternative methods can be derived from scratch without any knowledge of it.[20]

---

14 Wiedemann, "Pendulum", A-B.

15 King, "IY and the Pendulum", and Charette, *Orientalisme et histoire des sciences*, p. 146-151, where the development is more clearly documented.

16 As, for example, in Nasr, *Islamic Science*, p. 101 (see my review reprinted in King, *Studies*, A-XVII, p. 217-218).

17 Kennedy & Kennedy, *Islamic Geographical Coordinates*, p. xxxvi. See also Lelewel, *Géographie du MA*, I, p. 43-62, 165-177, and Atlas, p. 3-4 and pl. II (for a reconstructed map based on the coordinates).

18 The story is related in King, *World-Maps*, Section I.1.7.

19 Article "Kharīṭa" by S. Maqbul Ahmad in *EI*², especially p. 1079a. The association with Ibn Yūnus is cemented in Shawkat, "Kharā'iṭ", and mentioned in Kennedy, "[Islamic] Mathematical Geography", p. 185.

20 An attempt to kill the myth of this association is in Daffa & Stroyls, *Studies*, section 3, albeit without any reference to primary sources (see my review in *Historia mathematica* 13 (1986), p. 303-306, esp. p. 304-305). See also King, *Ibn Yūnus*, p. 7 and 149, and *idem*, "IY's *Very Useful Tables*", p. 360.

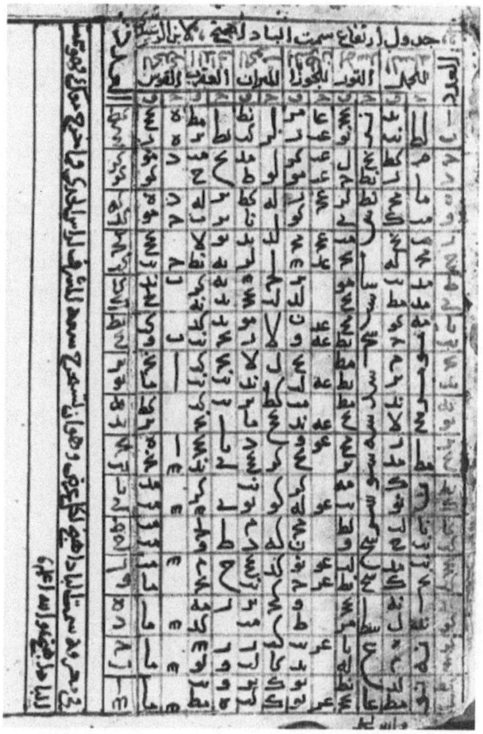

Fig. 6 A table in the corpus usually associated with Ibn Yūnus for aligning the major axis of the rectangular base of a ventilator towards winter sunrise. The function tabulated is the solar altitude in that azimuth. This particular table is here, but nowhere else, attributed to a Mamluk astronomer Ibn al-Rashīdī, otherwise known only by his set of tables for time-keeping for the latitude of Jerusalem. But it may be that he only added a correction for the effect of refraction at the horizon on the solar altitude. The theory that the visible horizon is 2/3° above the actual horizon, which accounts for the entry 0°41' at the winter solstice (see the entry at lower left), is, however, generally associated with Ibn Yūnus. From MS Cairo ENL DM 758, fol. 4r, courtesy of the Egyptian National Library.

e) Ibn Yūnus as the author of a treatise on a candle-clock

A short treatise on a candle-clock with a series of lamps one of which goes out at each seasonal hour of the night is attributed to Ibn Yūnus al-Asṭurlābī in the unique manuscript in St. Joseph's, Beirut.[21] The author is in fact Yūnus (ibn al-Ḥusayn) al-Asṭurlābī, otherwise known to us as the maker of a celestial globe dated 1144/45 now in the Musée du Louvre[22] and from citations in the treatise on mechanical devices of al-Jazarī (*ca.* 1200).[23]

f) Ibn Yūnus as the author of a treatise on the heliacal risings of Sirius

An astrological treatise on the purported significance of the situation of the moon at the time of the heliacal risings of Sirius is attributed to Ibn Yūnus in some of the available manuscripts. The attribution results from a corruption of al-Ḥakīm, the "true" author (who cites Ptolemy and Hermes), to al-Ḥākimī, then identified as Ibn Yūnus.[24]

Ibn Yūnus and the qibla at Cairo-Fustat

Ibn Yūnus was the first to compute the qibla of Miṣr (Cairo-Fusṭāṭ) according to a correct trigonometric formula applied to geographical data that is reasonably accurate by medieval stan-

21 For a translation and commentary see Kennedy & Ukashah, "IY's Candle-Clock".

22 Savage-Smith, *Islamicate Globes*, p. 217-218 (n. 2); and *Paris ML 1998 Exhibition Catalogue*, p. 70 and 80 (n. 52).

23 Hill, transl., *al-Jazarī*, p. 87, and my review reprinted in King, *Studies*, B-XX, especially p. 286, n. 2.

24 See King, *Ibn Yūnus*, p. 11, and Sezgin, *GAS*, VII, p. 173 (in both the attribution to Ibn Yūnus is accepted).

dards. His result rounds to 37° S of E. In addition he tabulated the altitude of the sun throughout the year when it has this azimuth – see already Fig. 5. His qibla was taken seriously by all later Egyptian astronomers and no less by quite a few architects, for it can be seen reflected in the city-plan of medieval Cairo.

On architectural orientations in medieval Cairo

The various qibla directions used in medieval Cairo are mentioned in several medieval historical and legal sources. The key to their interpretation turned out to be one of the tables attributed to Ibn Yūnus – but not necessarily by him – displaying the altitude of the sun when it is in the direction of the back axis of the ventilators – see Fig. 6.[25] This direction is winter sunrise, at 27° S

Fig. 7 The ventilator on the roof of the Musāfirkhāne, the most impressive of the few ventilators which survive in Cairo to this day. It is not known when such devices were first introduced, but the fact that in Mamluk times they bore a Persian name (*bādahanj*) is surely significant (the modern term *malqaf* is more recent). ʿAbd al-Laṭīf al-Baghdādī, who visited Cairo about the year 1200, commented that "one sees hardly any houses without ventilators".

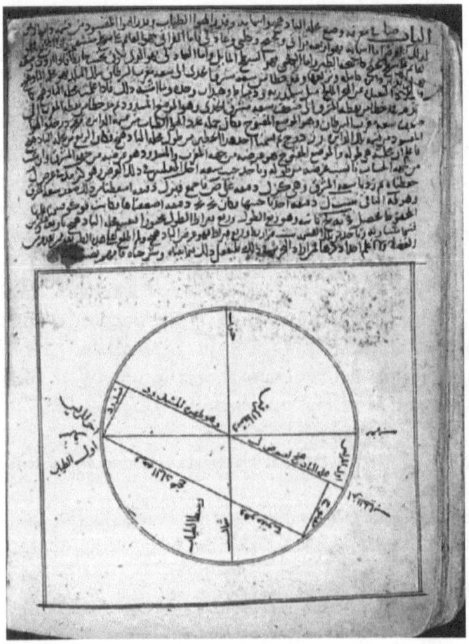

Fig. 8 Information on the orientation of ventilators as presented in an anonymous treatise on astronomical instruments probably compiled in Aleppo *ca.* 1315. The author was thought by this writer to have been the Aleppo astronomer Ibn al-Sarrāj, but closer investigation by François Charette suggests that the author could have been the Cairene astronomer Najm al-Dīn al-Miṣrī, known for his enormous set of tables for timekeeping (420,000 entries!). In any case the text contributes new information on the ventilators of medieval Cairo and their orientation, which in turn casts new light on the orientation of the entire Fatimid city of al-Qāhira. Note, for example, that the west side of the ventilator is it be open, as in the one shown in Fig. 7: this results from Egyptian folklore concerning the winds, not from any serious considerations of the optimal orientation for a wind-catcher, the actual orientation being already dictated by the city-plan. From MS Dublin CBL Persian 102, fol. 52v, courtesy of the Chester Beatty Library.

25 See King, "Orientations in Medieval Cairo".

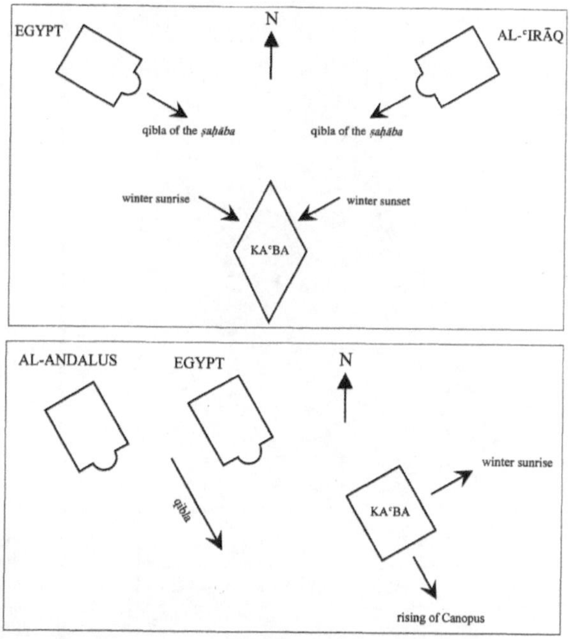

Fig. 9a-b The qibla of the *ṣaḥāba* in Fusṭāṭ was towards winter sunrise, intended to "face" the north-western wall of the Kaᶜba, assumed to be "facing" this direction. Another qibla was towards the rising of Canopus, intended to "face" the north-west wall of the Kaᶜba by being "parallel" to the axis of that edifice. Yet another, associated with the Shāfiᶜīs, was due south, corresponding to the qibla of the Prophet when he was in Medina, "facing" the northern corner of the Kaᶜba. These divergent qiblas can be better understood when it is realized firstly that the Kaᶜba itself is astronomically aligned, and secondly that there were divergent notions about its orientation, two of which are displayed here. For more information see King & Hawkins, "Orientation of the Kaᶜba".

of E. These ventilators were standard features of both secular and religious architecture in late Fatimid, Ayyubid, Mamluk, and early Ottoman Egypt (see Fig. 7), and were oriented so as to be open towards the main axis of the Fatimid city. Their structure and orientation, as well as the different varieties, are discussed in several medieval astronomical texts – see Fig. 8. The realisation that they were astronomically aligned led to the discovery that the entire city of al-Qāhira is astronomically aligned. But it is, of course, aligned with the Red Sea Canal, which is fortuitously oriented perpendicular to winter sunrise. And winter sunrise was the qibla of the *ṣaḥāba*, the Companions of the Prophet, in Egypt. The Kaᶜba itself is astronomically aligned, a fact that was known in the Middle Ages but which has only recently become known in modern times. Hence astronomical risings and settings were widely used for the qibla in different parts of the Islamic world, as was the case in Egypt, where the qibla of the *ṣaḥāba* was towards winter sunrise – see Fig. 9.[26] Thus the Fatimid city is qibla-oriented, but in the qibla of the *ṣaḥāba*. The Azhar Mosque and the Mosque of al-Ḥākim are skew to the street plan, for they are oriented in the qibla of Ibn Yūnus – see Fig. 10. The Mamluk religious architecture along the main axis of the city is aligned on the outside with the qibla of the *ṣaḥāba* and on the inside with the qibla of the astronomers. One can see the 10° difference in these orientations on the window-frames of the mosques and madrasas.

The City of the Dead is aligned with its main axis along a *wādī*. But that *wādī* is fortuitously perpendicular to the qibla of Ibn Yūnus – see Fig. 11. Thus all of the religious and mortuary architecture along the main axis is conveniently oriented in the qibla of the astronomers. A glance at modern tombs in the district (oriented at about 45° S of E, the qibla used today), some of which are visible from the minarets in the quarter, reveals the discrepancy in the medieval qibla, which results from the inaccuracy of the (very sensitive) longitude difference between Cairo and Mecca.

---

26 See the article "Makka. iv. As centre of the world" in *EI*², reprinted in King, *Studies*, C-X, and also *ibid.*, C-XII and XIII.

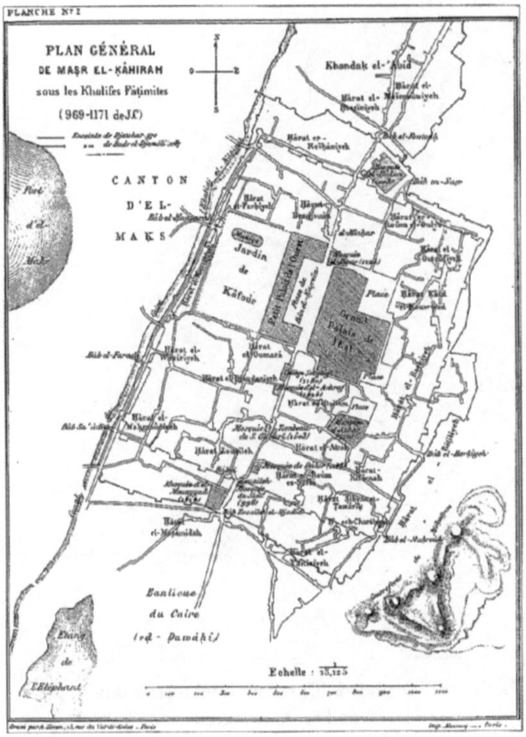

Fig. 10 The major axis of the Fatimid city is parallel to the Red Sea Canal, which is providentially perpendicular to the qibla of the *ṣaḥāba*, that is, winter sunrise, at *ca.* 27° S of E. But already the Azhar Mosque and the Mosque of al-Ḥākim are oriented in the qibla of Ibn Yūnus, some 10° further south. The divergence is noticeable already on this early reconstruction of the Fatimid city. From Pauty, "Palais et maisons" (1923).

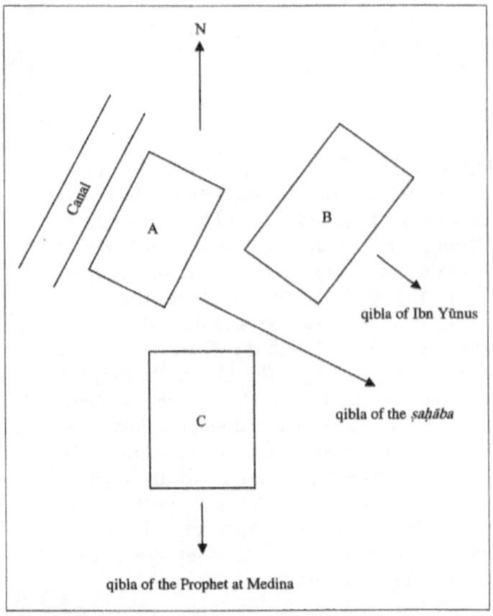

Fig. 11 The three main qiblas in medieval Cairo. The Fatimid city (A) is oriented in the qibla of the *ṣaḥāba*, that is, towards, winter sunrise. The City of the Dead (B) is in the qibla of Ibn Yūnus. The suburb of al-Qarāfa (C) faces due south, which was the qibla of the Shāfiʿīs, in imitation of the qibla of the Prophet when he was in Medina, a qibla used by certain interest groups from al-Andalus to Central Asia. After King, "Orientations in Cairo", p. 114.

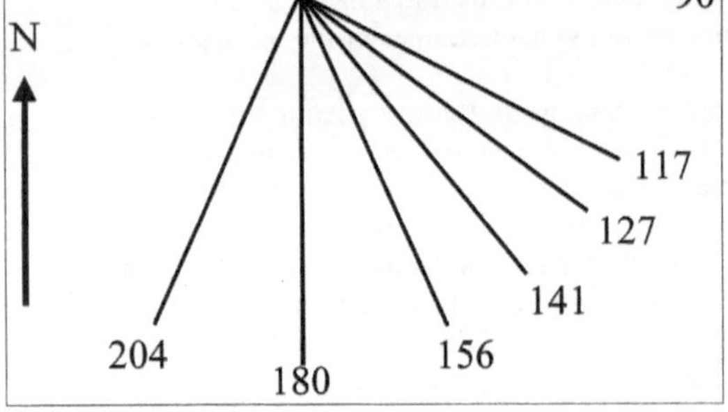

Fig. 12 The palette of qiblas accepted in medieval Cairo at one time or another by different interest groups. These are: (1) due east, (2) winter sunrise, the qibla of the *ṣaḥāba*, at 117°; (3) the qibla of Ibn Yūnus and most later astronomers, at 127°; (4) the qibla of the Mosque of Ibn Ṭūlūn, at 141°; (5) directions between the rising and setting of Canopus, from *ca.* 160° to *ca.* 200°; and (6) due south. This information was recorded by al-Dimyāṭī and later taken over by al-Maqrīzī. It is essentially confirmed by the orientation of surviving mosques, although these have yet to be measured properly. After King, "Orientations in Medieval Cairo", p. 114.

The overall situation in Cairo is considerably more complicated than I have indicated here.[27] However, none of the orientations of the mosques in Cairo have been measured properly yet, that is, according to the standards of the new discipline of archaeo-astronomy. The orientation of Islamic religious architecture, a substantial proportion of which is astronomically aligned,[28] constitutes the only chapter in that discipline in which there is a corresponding corpus of textual evidence.[29] The most important Egyptian work on the qibla, a lengthy treatise by the 12th-century legal scholar Zayn al-Dīn al-Dimyāṭī, which was the source of al-Maqrīzī's discussion of mosque orientations in Cairo (see Fig. 12), is still unpublished.[30]

The end of Ibn Yūnus

One day in the year 1009, when he was in good health, Ibn Yūnus predicted his own death in seven days. He attended to his own business, locked himself in his house, and washed the ink off his manuscripts. He then recited the *Qur'ān* until he died – on the day he had predicted. According to his biographer, Ibn Yūnus' son was so stupid that he sold his father's papers by the pound in the soap market.

The influence of Ibn Yūnus

The *Ḥākimī Zij* was influential in later Egypt, Syria, Iran and the Yemen (it was in the Yemen that some of the lost parts of the *Ḥākimī Zij* have been recovered). Numerous later *zīj*es compiled in Egypt, Persia, and Yemen are extant and contain material ultimately due to Ibn Yūnus. For example, the 13th-century Egyptian *Muṣṭalaḥ Zīj*, as well the *Īlkhānī Zīj* of Naṣīr al-Dīn al-Ṭūsī and the *Zīj* of Muḥyi 'l-Dīn al-Maghribī, both compiled at the observatory in Maragha in the 13th-century, relied on the *Ḥākimī Zīj*. Likewise, the *Mukhtār Zīj* by the 13th-century Yemeni astronomer Abu 'l-ᶜUqūl is based mainly on a *zīj* by Ibn Yūnus other than the *Ḥākimī*; and an anonymous 14th-century Yemeni *zīj* is adapted from the *Ḥākimī Zīj*. Ibn Yūnus' influence in astronomical time-keeping was more far-reaching. Indeed, his tables were ultimately responsible for the impressive developments which took place over the ensuing centuries in Cairo, Damascus, Taiz, Jerusalem, Tunis and Istanbul.[31]

The Leiden manuscript of the *Ḥākimī Zīj* bears a title-folio with biographical information in the hand of the leading astronomer in Cairo at the end of the 15th-century. Within a century this manuscript came into the possession of Taqiyyu 'l-Dīn, the director of the short-lived observatory in Istanbul – see already Fig. 2. Thence it was brought to Leiden by the Dutch orientalist Jacob Golius (1596-1667). By this time Ibn Yūnus had been virtually forgotten in Egypt.[32] His writings were replaced by a rich collection of Mamluk and Ottoman treatises, all competent but all lacking the brilliance of the master. The rare references to him in this literature show that the authors had no conception of his achievements.

---

27 Christel Kessler (see her early studies listed as "Mecca-Oriented Architecture", A-C) takes issue with some of my conclusions. Before more is said the orientations should be measured properly and the relevant texts should be published (see note 30).

28 But all of it is qibla-oriented, and different qiblas were used by different interest-groups. Only in the modern literature on Islamic architecture does one find misguided remarks to the effect that the qibla of this or that mosque is "wrong".

29 See King, "Orientations", for an overview. Rius, *Orientaciónes en al-Andalus y el Norte de África*, based mainly on medieval legal texts, takes the subject a major step forward for the Western region of the Islamic world, already well-served by the field-work reported in Bonine, "The Sacred Direction in Morocco".

30 The discovery of the Oxford manuscript of this treatise is announced in *idem*, "Orientations in Medieval Cairo", p. 115, n. 71. What is needed is a study such as that of Mònica Rius cited in the previous note.

31 See n. 9 above.

32 See, for example, the contents of the library of a 17th-century Cairo *muwaqqit* from the Mosque of Ibn Ṭūlūn, partially reconstructed in King, "Muezzins and Muwaqqits", p. 313-317.

The rediscovery of Ibn Yūnus

The importance of Ibn Yūnus was realized in the West when the Leiden manuscript was acquired. Already in 1666 a Latin translation by Golius of three eclipse records was published. In 1804, A.-P. Caussin de Perceval published the text of Ibn Yūnus' observational reports with a French translation. He also included the introduction to the *Zīj*, which contains the titles of the 81 chapters. J.-J. Sédillot's translation (now lost) of the Leiden and Paris manuscripts was summarized by J.-B. Delambre in 1819. The German scholar Carl Schoy published several articles containing translations and analyses of individual chapters of the *Zīj* relating to spherical astronomy and sundial theory. The present author in his youth surveyed the entire *zīj*. My doctoral thesis was restricted to the spherical astronomy therein, and various early writings of mine contain analyses of Ibn Yūnus' tables for time-keeping, his lunar equation tables, as well as his lunar crescent visibility theory, recovered in a Yemeni source. With this foundation it was possible to write the first overviews of astronomy in Mamluk Egypt and Rasulid Yemen.[33]

Concluding remarks

An overview of astronomy and astrology in Fatimid Egypt is, however, not possible at the present time. For one, the most important Fatimid astrological work is not known to have survived. This is the *Kitāb al-Qaḍāya 'l-ṣā'iba fī maᶜānī aḥkām al-nujūm*, a monumental work in 3000 folios, by the distinguished scholar al-Musabbiḥī (977-1029),[34] whom we have already mentioned. But we do have some control over Ibn Yūnus, as others have control over his late contemporary Ibn al-Haytham.[35] We have seen here how the leading astronomer in Egypt in Fatimid times was influential in various parts of the Islamic world for several centuries, even if he was not known in medieval Europe. His work added a distinctive Islamic flavour to Islamic astronomy, and it is he who laid the foundations for the activities of the *muwaqqit*s after the 13th century. In addition, his contributions to the qibla-problem influenced the orientation of mosques in Cairo from the Fatimid to the Ottoman period and are thus reflected in the city-plan of medieval Cairo.

---

33 King, "Mamluk Astronomy", and *idem*, *Yemeni Astronomy*.

34 Sezgin, *GAS*, VII, p. 185-186.

35 See the article by A. I. Sabra in *DSB*, and the contributions of R. Rashed and R. Morelon to this volume.

Bibliography and bibliographical abbreviations

Bonine, "The Sacred Direction and City Structure: A Preliminary Analysis of the Islamic Cities of Morocco", *Muqarnas: An Annual on Islamic Art and Architecture* 7 (1990), p. 50-72.

Caussin: A. P. Caussin de Perceval, "Le livre de la grande table Hakémite", in *Notices et extraits des manuscrits de la Bibliothèque Nationale*, 7 (An XII [= 1804]), p. 16-240, with separate pagination in the separatum.

Charette, *Orientalisme et histoire des sciences*: François Charette, *Orientalisme et histoire des sciences : l'historiographie européenne des sciences islamiques et hindoues, 1784-1900*, Master's thesis, Départment d'histoire, Université de Montréal, 1995.

Daffa & Stroyls, *Studies*: Ali A. Al-Daffa and John S. Stroyls, *Studies in the Exact Sciences in Medieval Islam*, New York: John Wiley & Sons, 1984. [To be used with caution.]

DSB: *Dictionary of Scientific Biography*, 16 vols., New York: Charles Scribner's Sons, 1970-80.

EHAS: Roshdi Rashed, with Régis Morelon, eds., *Encyclopedia of the History of Arabic Science*, 3 vols., London: Routledge, 1996; with French version *Histoire de la science arabe*, Paris: Seuil, 1997.

EI[1]: *The Encyclopaedia of Islam*, 1st edn., 4 vols., Leiden: E. J. Brill, 1913-38, reprinted in 9 vols., Leiden: E. J. Brill, 1987. [Also available in French and German.]

EI[2]: *The Encyclopaedia of Islam*, new edn., 9 vols. to date, Leiden: E. J. Brill, 1960 onwards. [Also available in French.]

Hill, transl., *al-Jazarī*: Donald Routledge Hill, *The Book of Knowledge of Ingenious Mechanical Devices by Ibn al-Razzāz al-Jazarī*, Dordrecht (NL), *etc.*: Reidel, 1974, reprinted Islamabad: Pakistan Hijra Council, 1989.

Islamic Geography: Fuat Sezgin, with Mazen Amawi, Carl Ehrig-Eggert and Eckhard Neubauer *et al.*, eds., *Islamic Geography*, Frankfurt am Main: Institut für Geschichte der Arabisch-Islamischen Wissenschaften, 276 vols. to date, 1992 to present (1998). [Facsimile reprints of early writings, mainly 19th and early-20th century.]

Kennedy, "[Islamic] Mathematical Geography": Edward Stewart Kennedy, "Mathematical Geography", in *EHAS*, I, p. 185-201 (I, p. 217-232 of the French version).

– , "*Zīj* Survey": *idem*, "A Survey of Islamic Astronomical Tables", *Transactions of the American Philosophical Society*, N. S., 42:2 (1956), p. 123-177, reprinted *ibid.*, n. d. [*ca.* 1990], with separate pagination.

Kennedy & Kennedy, *Islamic Geographical Coordinates*: *idem* and Mary Helen Kennedy, *Geographical Coordinates of Localities from Islamic Sources*, Frankfurt am Main: Institut für Geschichte der Arabisch-Islamischen Wissenschaften, 1987.

Kennedy & Ukashah, "IY's Candle-Clock": *idem* and Walid Ukashah, "The Chandelier Clock of Ibn Yūnis", *Isis* 60 (1969), p. 543-545, reprinted in *idem et al.*, *Studies*, p. 499-501.

Kennedy *et al.*, *Studies*: E. S. Kennedy, Colleagues and Former Students, *Studies in the Islamic Exact Sciences*, ed. by Mary Helen Kennedy and David A. King, Beirut: American University of Beirut, 1983.

Kessler, "Mecca-Oriented Architecture", A-C: Christel Kessler, "Mecca-oriented Architecture and Urban Growth of Cairo", *Atti del Terzo Congresso di Studi Arabi e Islamici* (Ravello, 1966), Naples: Istituto Universitario Orientale, 1967, p. 425 (A); *eadem*, "Funerary Architecture within the City", *Colloque International sur l'Histoire du Caire, 27 mars – 5 avril 1969*, Cairo: General Egyptian Book Organisation, 1969, p. 257-267 (B): and *eadem*, "Mecca-Oriented Urban Architecture in Mamluk Cairo: The Madrasa-Mausoleum of Sultan Sha^cbān II", in Arnold H. Green, ed., *In Quest of an Islamic Humanism: Arab and Islamic Studies in Memory of Mohamed Nowaihi*, Cairo: American University of Cairo Press, 1984, p. 97-108 (C).

King, *Ibn Yūnus*: David A. King, *The Astronomical Works of Ibn Yūnus*, Ph.D. dissertation, Yale University, 1972.

–, "IY on Lunar Crescent Visibility": *idem*, "Ibn Yūnus on Lunar Crescent Visibility", *Journal for the History of Astronomy* 19 (1988), p. 155-168, reprinted in *idem, Studies*, C-III.

–, "IY and the Pendulum": *idem*, "Ibn Yūnus and the Pendulum: A History of Errors", *Archives Internationales d'Histoire des Sciences* 29 (1979), p. 35-52.

–, "IY's Lunar Equation Table": *idem*, "A Double-Argument Table for the Lunar Equation Attributed to Ibn Yūnus", *Centaurus* 18 (1974), p. 129-146, reprinted in *idem, Studies*, A-V.

–, "IY's *Very Useful Tables*": *idem*, "Ibn Yūnus' *Very Useful Tables* for Timekeeping by the Sun", *Archive for History of Exact Science* 10 (1973), p. 342-394, reprinted in *idem, Studies*, A-IX.

–, "Mamluk Astronomy": *idem*, "The Astronomy of the Mamluks", *Isis* 74 (1983), p. 531-555, reprinted in *idem, Studies*, A-III.

–, "Muezzins and Muwaqqits": *idem*, "On the Role of the Muezzin and the Muwaqqit in Medieval Islamic Society", in F. Jamil Ragep & Sally P. Ragep, with Steven J. Livesey, eds., *Tradition, Transmission, Transformation: Proceedings of Two Conferences on Premodern Science Held at the University of Oklahoma*, Leiden, New York, N.Y., & Cologne: E. J. Brill, 1996, p. 285-346.

–, "Orientations": *idem*, "The Orientation of Medieval Islamic Religious Architecture and Cities", *Journal of the History of Astronomy* 26 (1995), p. 253-274.

–, " Orientations in Medieval Cairo": *idem*, "Architecture and Astronomy: the Ventilators of Medieval Cairo and their Secrets", *Journal of the American Oriental Society* 104 (1984), p. 97-133.

–, *Studies*, A-C: *idem, Islamic Mathematical Astronomy*, London: Variorum, 1986, 2nd rev. edn., Aldershot: Variorum, 1993 (A); *Islamic Astronomical Instruments*, London: Variorum, 1987, reprinted Aldershot: Variorum, 1995 (B); and *Astronomy in the Service of Islam*, Aldershot: Variorum, 1993 (C).

–, *World-Maps*: *idem, World-Maps for Finding the Direction and Distance to Mecca: Innovation and Tradition in Islamic Science*, Leiden: E. J. Brill, & London: Al-Furqan Islamic Heritage Foundation, 1999.

–, *Yemeni Astronomy*: *idem, Mathematical Astronomy in Medieval Yemen – A Bio-Bibliographical Survey*, (Publications of the American Research Center in Egypt), Malibu, Ca.: Undena, 1983.

King & Hawkins, "Orientation of the Kaᶜba": *idem* and Gerald S. Hawkins, "On the Orientation of the Kaᶜba", *Journal for the History of Astronomy* 13 (1982), p. 102-109, reprinted in *idem, Studies*, C-XII.

Lelewel, *Géographie du MA*: Joachim Lelewel, *La géographie du moyen âge*, with an *Atlas composé de cinquante planches*, 4 tomes (in 3 volumes) and an *Épilogue*, Brussels: Ve. et J. Pilliet, 1850 (atlas), 1852 (tomes I-IV) and 1857 (épilogue), reprinted in *Islamic Geography*, vols. 129-132 (the three volumes and the epilogue) and 133 (the atlas).

Nasr, *Islamic Science*: Seyyed Hossein Nasr, *Islamic Science – An Illustrated Study*, London: World of Islam Festival Publishing Co., Ltd., 1976. [To be used with caution.]

Paris ML 1998 Exhibition Catalogue: Sophie Makariou, ed., *L'apparence des cieux – Astronomie et astrologie en terre d'Islam*, Paris: Éditions de la Réunion des musées nationaux, 1998. [Catalogue of an exhibition held at the Musée du Louvre during 18.6.-21.9.1998.]

Pauty, "Palais et maisons": E. Pauty, "Les palais et maisons d'époque musulmane au Caire", *Mémoires publiés par les membres de l'Institut Français d'Archéologie Orientale du Caire* 62 (1932).

Rius, *Orientaciónes en al-Andalus y el Norte de África*: Mónica Rius, *Orientación de las mezquitas en al-Andalus y el Norte de África entre los siglos XI y XVIII*, thesis submitted to the Department of Semitic Philology, University of Barcelona, 1999.

Savage-Smith, *Islamicate Globes*: Emilie Savage-Smith, *Islamicate Celestial Globes – Their History, Construction, and Use*, Washington, D.C.: Smithsonian Institution Press, 1985.

Sayılı, *The Observatory in Islam*: Aydın Sayılı, *The Observatory in Islam*, (*Publications of the Turkish Historical Society*, Series VII, No. 38), Ankara: Türk Tarih Kurumu Basımevi, 1960, reprinted New York, N.Y.: Arno Press, 1981.

Sezgin, *GAS*: Fuat Sezgin, *Geschichte des arabischen Schrifttums*, 9 vols. to date, Leiden: E. J. Brill, 1967 onwards.

Shawkat, "Kharā'iṭ": Ibrahim Shawkat, "Kharā'iṭ jughrāfiyyi [?] al-ᶜArab al-awwal [?]", *The Professor = al-Ustādh* (College of Education, University of Baghdad) 10 (1962), p. 37-68.

Stephenson & Said, "Medieval Islamic Eclipse Observations", A-C: F. R. Stephenson and S. S. Said, "Precision of Islamic Eclipse Measurements", *Journal for the History of Astronomy* 22 (1991), p. 195-207 (A); "Solar and Lunar Eclipse Measurements by Medieval Muslim Astronomers. i. Background, and ii. Observations", *ibid.* 27 (1996), p. 259-273 (B), and 28 (1997), p. 29-48 (C).

Wiedemann, "Pendulum", A-B: Eilhard Wiedemann, "Über die angebliche Verwendung des Pendels zur Zeitbestimmung bei den Arabern", *Verhandlungen der Deutschen Physikalischen Gesellschaft* 21 (1919), p. 663-664 (A); and "Über die angebliche Verwendung des Pendels bei den Arabern", *Zeitschrift für Physik* 10 (1922), p. 267-268 (B); reprinted in *Eilhard Wiedemann: Gesammelte Schriften zur arabisch-islamischen Wissenschaftsgeschichte*, 3 vols., Frankfurt am Main: Institut für Geschichte der Arabisch-Islamischen Wissenschaften, 1984, II, p. 912-913 and 1033-1034.

V

Mamluk astronomy and the institution of the *muwaqqit*

Dedicated to the memory of my friend Salah Abdel-Khalek

Introduction

The contents of this chapter are not in the main new to the literature. I have already written preliminary surveys of Mamluk astronomy[1] and of the institutions of the muezzin and of the *muwaqqit*, the latter being a professional astronomer associated with mosques and *madrasas*.[2] In addition I have shown how various Mamluk treatises on astronomy can be of interest to a far greater audience than historians of science: Mamluk treatises on the *qibla* and on the ventilators that graced both the religious and secular architecture of Mamluk Cairo provide the answer to the questions surrounding not only the different orientations attested in Cairene religious architecture but also the orientations of the various parts of the medieval city.[3] In this brief overview I shall try to suppress technical details and to concentrate on mainstream developments and on issues that might be of interest to colleagues outside the history of science.

In the 1970s a distinguished historian of Islamic art, whose name need not be mentioned here, wrote that the Mamluks made no contributions whatsoever to science. This would be the opinion of most Islamists (as well as of most Egyptians and Syrians).[4] That scholar was, however, unaware that in the

[1] D. A. King, 'The astronomy of the Mamluks', *Isis*, 74 (1983), 531–55 (hereafter 'Mamluk astronomy'; reprinted in D. A. King, *Islamic Mathematical Astronomy* (London, 1986; 2nd edn 1993) (hereafter *Studies*, A), III. See also D. A. King, *Islamic Astronomical Instruments* (London, 1987) (hereafter *Studies*, B) and D. A. King, *Astronomy in the Service of Islam* (Aldershot, 1993) (hereafter *Studies* C). These books contain several articles on aspects of Mamluk astronomy, including, in addition to A–III, the overview of Mamluk astronomy, A-IX–XII, descriptions of the corpora of tables used for time-keeping in Mamluk Cairo, Damascus and Jerusalem, B-VIII–X, XII, and XVI–XVII, descriptions of various Mamluk astrolabes, quadrants, sundials and compendia, and C-VI–VII, universal solutions to problems of spherical astronomy.

[2] D. A. King, 'On the role of the muezzin and the muwaqqit in medieval Islamic society', in: F.J. and S. Ragep (eds.), *Tradition, Transmission, Transformation* (Leiden, 1996), 285–346.

[3] D. A. King, 'Architecture and astronomy: the ventilators of medieval Cairo and their secrets', *Journal of the American Oriental Society*, 104 (1984), 97–133.

[4] The exception to the latter would be those who have seen the chapter 'L 'astronomie en Syrie à l'époque islamique', in *Paris IMA Catalogue,* 386–95.

1950s E. S. Kennedy had discovered that the solar, lunar and planetary models proposed by a scholar of fourteenth-century Damascus were mathematically identical to those of Copernicus 150 years later. He can perhaps be forgiven for not knowing that a young historian of Islamic science had just prepared the first description of some thirteenth- and fourteenth- century Egyptian and Syrian tables for time-keeping and had labelled one of these 'the most sophisticated trigonometric tables known to me from the entire Middle Ages'. And in 1979 when I first described a spectacular astrolabe from early fourteenth-century Aleppo I labelled it 'the most sophisticated astrolabe ever made'.

I was soon to reap the rewards of having catalogued the 2,500 scientific manuscripts in the Egyptian National Library[5] as well as having studied Mamluk manuscripts in libraries elsewhere (notably in the Chester Beatty Library), and having catalogued all available Mamluk astronomical instruments.[6] This was the activity that provided enough momentum to produce the first overview history of Mamluk astronomy and a preliminary study of the institution of the *muwaqqit*. We can now say with confidence that Mamluk contributions to astronomy were among the most remarkable of their time, and that they were certainly as impressive as, and even more colourful than, those of the better-known regional schools in Iran and Central Asia (contemporaneous activity in the Islamic West and in Islamic India were simply not on a comparable level).[7]

Who were the Mamluk astronomers? Were the astronomers of Mamluk Egypt and Syria themselves Mamluks or did they simply work for the Mamluks? Or were they regular Egyptians and Syrians who were active during the Mamluk period? My earlier survey, and perhaps also the present chapter, would in fact be better entitled 'Astronomy under the Mamluks'. Nevertheless, a small minority of the astronomers were indeed *mamlūks*, if not (as was the

[5] See D. A. King, *A Survey of Scientific Manuscripts in the Egyptian National Library* American Research Center in Egypt, catalogs, vol. V (Winona Lake, Ind., 1986) (hereafter Cairo Survey; (see especially section C: Egyptian and Syrian authors from c. 1100 to c. 1600). This is based on the 2–volume Arabic catalogue published by the General Egyptian Book Organization in 1980–86.

[6] See D. A. King, *A Catalogue of Medieval Astronomical Instruments* (forthcoming: sections will deal with Mamluk astrolabes, quadrants and sundials – see also D. A. King, 'Medieval astronomical instruments: a catalogue in preparation', *Bulletin of the Scientific Instrument Society*, 31 (Dec. 1991), 3–7; and D. A. King, '1992 – a good year for medieval astronomical instruments', *Bulletin of the Scientific Instrument Society*, 36 (March 1993), 17–18), and the Mamluk Syrian examples in S. Cluzan, E. Delpont and J. Mouliérac (eds.), *Syrie, mémoire et civilisation* (Paris, 1993), 434–41.

[7] 'Better-known' is used advisedly. The *Zījes* of al-Ṭūsī and Ulugh Beg are still available only in manuscript form and no editor (or researcher) is in sight; at least the influential *Tadhkira* of al-Ṭūsī is available in a masterly study by F. J. Ragep (1993). There is no survey of astronomy in Iran and Central Asia beyond E. S. Kennedy's chapters on the exact sciences under the Seljuqs and Mongols and then the Timurids in *The Cambridge History of Iran* (Cambridge, 1968 to present), 6 vols. to date (vol. V, 659–79, and vol. VI, 568–80), which are restricted to general trends. Some of the literature published in 1994 to celebrate the 600th anniversary of Ulugh Beg is indicative of the poor understanding of his achievements.

case in Rasulid Yemen) the sultans themselves.[8] On the other hand, the institution of the *muwaqqit* seems very much to have been inaugurated during the Mamluk period, encouraged by the same rulers who founded and endowed the splendid mosques and *madrasas* of Mamluk Cairo. It is not surprising that this institution appeared for the first time in Mamluk times, because much of astronomy under the Mamluks was 'astronomy in the service of Islam', dedicated to the three areas of concern relating to the very pillars of Islam: the lunar calendar, the times of prayer and the sacred direction.[9] This having been said, it should be added that a great deal of energy was also expended by astronomers in Mamluk times on planetary astronomy, theoretical and practical, the most obvious application being mathematical astrology. When the sultan al-Nāṣir Muḥammad ibn Qalāwūn became ill with diarrhoea, astrologers and geomancers were consulted as well as his doctors.[10] But this anecdote, recorded by Ibn Abī 'l-Faḍā'il, is the only one of its kind known to me. There is in fact remarkably little astrology in the known Mamluk scientific treatises known to me. Perhaps colleagues can help by drawing my attention to other references in historical works.

An overview of Mamluk astronomy

First a general overview:[11] Let us, for the sake of convenience, begin in mid-thirteenth-century Syria. There we find an intensive programme of astronomy being pursued by various scholars (mentioned by Ibn Abī Uṣaybiʿa and Ibn al-Qifṭī), but these appear to have been mainly involved in teaching and few of them compiled treatises. At least three astronomers, on the other hand, produced zījes, that is, astronomical handbooks with tables, the emphasis being very much on planetary astronomy. For example, Muḥyī 'l-dīn al-Maghribī, an immigrant, compiled a couple of zījes in Damascus before moving on to participate in the observation programme of Naṣīr al-dīn al-Ṭūsī in Marāgha.[12]

[8] See D. A. King, *Mathematical Astronomy in Medieval Yemen – A Biobibliographical Survey*, American Research Center in Egypt, Catalogs, vol. IV, (Malibu, Ca., 1983), especially at 9, and 23–6, 27–9, 33, 34, 37 and 38 on individual sultans.

[9] For a survey see the study reprinted in *Studies*, C-I.

[10] 'Mamluk astronomy', 535.

[11] For details the reader is referred to my articles cited in notes 1 and 2 above, as well as to E. S. Kennedy and I. Ghanem (eds.), *The Life and Work of Ibn al-Shāṭir, an Arab Astronomer of the Fourteenth Century* (Aleppo, 1976) and G. Saliba, *A History of Arabic Astronomy* [sic] – *Planetary Theories during the Golden Age of Islam* (New York, 1994). (This deals only with non-Ptolemaic planetary theory, including that of Ibn al-Shāṭir; hereafter Saliba, *Studies.*). The major figures mentioned here, such as Ibn Yūnus, al-Ṭūsī, Ibn al-Shāṭir, al-Khalīlī and Ulugh Beg, are treated in *The Dictionary of Scientific Biography* (New York, 1970–80), 16 vols. (See generally articles 'Ibn al-Shāṭir', 'al-Khalīlī and 'Ulugh Beg'.) See also the article on al-Marrākushī in *The Encyclopaedia of Islam* (Leiden, 1960 to present), 8 vols. to date. (See generally articles 'Mīḳāt' (astronomical time-keeping); 'Ru'yat al hilāl' (lunar crescent visibility); and (forthcoming) 'Taḳwīm' (ephemerides) and 'Zīdj' (astronomical tables).

[12] A. Sayılı, *The Observatory in Islam* (Ankara, 1960; repr. New York, 1981), 187–223.

We continue with late thirteenth-century Egypt, where the emphasis was very much more on spherical astronomy and astronomical time-keeping. The most popular *zīj* in use in Egypt at that time was the anonymous *Muṣṭalaḥ Zīj*, compiled from various sources, not least from the monumental *Ḥākimī Zīj* of the illustrious Fatimid astronomer Ibn Yūnus (*c.* 990).[13] The most important astronomers were Shihāb al-dīn al-Maqsī, Najm al-dīn al-Miṣrī and Abū ʿAlī al-Marrākushī. Al-Maqsī compiled a corpus of tables for time-keeping for the latitude of Cairo, with some 30,000 entries. For each degree of solar longitude, corresponding roughly to each day of the year, the main set of tables gives the time since sunrise, the time till midday and the azimuth (direction on the horizon) of the sun. Smaller tables enable the user to find the time of the *ʿaṣr* prayer and the time when the sun is in the direction of Mecca. Others display the duration of morning and evening twilight to facilitate the determination of the *fajr* and *ʿishā'* prayers.[14] This corpus, which inspired the later ones for Damascus and Jerusalem (see below), was used in Cairo in one form or another until the nineteenth century. Najm al-dīn compiled a set of tables for time-keeping by the sun and stars that would work for all latitudes, with a grand total of 440,000 entries. This monumental table exists in a single copy, doubtless in the hand of Najm al-dīn himself, for it was not calculated to attract many copyists. The table was in a tradition much favoured by Mamluk astronomers, namely the compilation of tables and the development of instruments for solving the problems of spherical astronomy and time-keeping for all latitudes. They achieved some rather spectacular solutions.[15]

In 1280 Abū ʿAlī al-Marrākushī, another immigrant, compiled in Cairo a monumental compendium of spherical astronomy and instrumentation in several hundred folios. This was to form the basis of all later developments in the subject in Mamluk and Ottoman times (it remained unknown in the Maghrib and in the Islamic East).[16] At the same time we find the first mention of a family of *muwaqqit*s at the Mosque of ʿAmr in Fusṭāṭ, who served there from the middle of the thirteenth century to the first decades of the fourteenth. Their contributions are mentioned in a treatise on spherical astronomy which appears to be written by a later member of the family, Naṣīr al-dīn ibn Simʿūn (d. 1337).[17] If the office of the *muwaqqit* was introduced first in Egypt it did not take long to spread to Palestine and Syria: a manuscript of a treatise by Naṣīr al-dīn was copied in 1306/7 by Ibrāhīm ibn Aḥmad, *muwaqqit* at the Sacred Mosque in Hebron.[18] We should mention that before this time, in both Egypt and Syria, various muezzins had produced sophisticated astronomical instruments. As we know, however, from such works as the treatise on law

13 On the *Muṣṭalaḥ Zīj* see *Studies*, A-III, 535–36. On Ibn Yūnus see the article in *DSB*.

14 See the analysis in *Studies*, A-IX. 15 Surveyed in *Studies*, C-VI–VII.

16 The first part of this work was translated by Sédillot-*père* in 1834–5, and the second part was summarized by Sédillot-*fils* in 1844. For references see the article on al-Marrākushī in *EI*².

17 King, 'Muwaqqit', section 6, quoting MS Leiden Or. 468.

18 *Cairo Survey*, Pl. CIIIb at 323, with caption at 215.

enforcement (*iḥtisāb*) by Ibn al-Ukhuwwa (Cairo, *c.* 1300), the astronomical knowledge expected of muezzins was restricted to the basics of folk astronomy (simple schemes involving shadows and the lunar mansions for time-keeping by day and night, respectively).[19]

Moving back to Syria we find in about 1325 a lone astronomer, Ibn al-Sarrāj, in Aleppo. Not only did he devise and make instruments, he also wrote treatises on their construction and use. Ibn al-Sarrāj developed a universal astrolabe that could be used for all latitudes, apparently independently of eleventh-century Andalusian instruments of the same kind. He then devised an astrolabe that was universal in five different ways, an example of which happily survives in the Benaki Museum in Athens.[20] In one of his treatises, yet to be studied properly, Ibn al-Sarrāj describes all of the different kinds of instruments known to him and all those invented by himself. The instruments and writings of Ibn al-Sarrāj represent the culmination of astronomical instrumentation in the Islamic world.

A slightly later contemporary was al-Mizzī of Damascus who, in addition to making instruments, compiled a set of tables for time-keeping for the latitude of his home city. Al-Mizzī had studied astronomy in Alexandria under Ibn al-Akfānī, celebrated author of the bibliographical work *Irshād al-qāṣid*. So had Ibn al-Shāṭir, who was to become the chief *muwaqqit* at the Umayyad Mosque in Damascus. He wrote two *zīj*es and a treatise on theoretical astronomy in which he refined the non-Ptolemaic planetary models developed by the Marāgha school and produced geocentric models mathematically identical to those of Copernicus. Ibn al-Shāṭir's writings represent the culmination of the Islamic tradition of planetary astronomy.[21] He also made instruments, portable ones like the astrolabe and the compendium known as the *ṣandūq al-yawāqīt*, and fixed ones like an astronomical clock and a sundial 2 × 1 metres, which is the most spectacular sundial made between late Antiquity and the late Renaissance.[22] A colleague of Ibn al-Shāṭir in Damascus was Shams al-dīn al-Khalīlī, *muwaqqit* first at the Yalbughā Mosque and then at the Umayyad mosque. He compiled a new set of tables for time-keeping for Damascus (based on Ibn al-Shāṭir's new parameters), as well as a universal table for finding the *qibla* for any locality, and a set of tables for solving any problem of spherical astronomy for any latitude[23]. His works represent the culmination of the Islamic science of astronomical time-keeping (*ʿilm al-mīqāt*).

In the fifteenth century the main activity is in Egypt again, with

[19] King, 'Muwaqqit', section 4. On folk astronomy see ibid., section 3 and also D. A. King, 'A survey of medieval Islamic shadow schemes for simple time-reckoning', *Oriens*, 32 (1990), 199–249, esp. 192–6. [20] *Studies*, B-IX, and *Paris IMA Catalogue*, 434–5.

[21] See the article in *DSB* and Saliba, *Studies*.

[22] On the compendium see *Studies*, B-XII (with L. Janin). On an astrolabe by him see *Paris IMA Catalogue*, 435. On his sundial see a study by L. Janin in *Centaurus* 16 (1972) reprinted in Kennedy and Ghanem (eds.), *Ibn al-Shāṭir*, 107–21, and *Paris IMA Catalogue*, 439.

[23] See the article in *DSB*, supp., and the various analyses in *Studies*, A-XI and XIII. A complete analysis of his tables has been prepared but is not yet published.

astronomers such as Ibn al-Majdī, not known to have been associated with any mosque, interested in planetary astronomy and time-keeping, and Sibṭ al-Māridīnī, *muwaqqit* at the Azhar Mosque, interested mainly in time-keeping and small instrumentation. Ibn Abī 'l-Fatḥ al-Ṣūfī, also not known to have been associated with any mosque, prepared a recension for Cairo of the *Zīj* of Ulugh Beg (Samarqand, *c.* 1430).

In addition there are a host of minor figures, all competent astronomers, who were active as *muwaqqit*s in various mosques in Syria and Egypt. Of these we may mention Aḥmad al-Kawm al-Rīshī, *muwaqqit* at the Mosque of al-Mu'ayyad in Cairo, who in about 1410 compiled a recension for Cairo of the *Zīj* of Ibn al-Shāṭir; ʿIzz al-dīn al-Wafāʾī, *muwaqqit* at the same mosque around 1450, who wrote on small instrumentation and made sundials and compendia; and ʿAbd al-Raḥmān al-Ṣāliḥī, *muwaqqit* at the Umayyad Mosque in Damascus, who in about 1500 prepared a recension of the *Zīj* of Ulugh Beg, modifying the planetary tables for the longitude of his own city.

Although astrological topics are not featured in most known Mamluk astronomical treatises we can be sure that there was an active interest in astrology in Mamluk society. A magnificent copy of a *malḥama* or handbook of prognostications for each month of the Syrian year written for the treasury of the Sultan Qāytbāy (1468–96) survives as a witness to this interest.[24] What one can expect if a solar or lunar eclipse, a shooting star, a halo around the sun or moon, or various meteorological phenomena, occur during a given month is clearly outlined.

It was Sibṭ al-Māridīnī who, unwittingly and in total innocence, contributed most to the decline of astronomy in Egypt, and by extension in Syria, after 1500. He wrote several short handbooks on the basics of time-keeping and the use of the instruments (mainly on the quadrant, which had come to replace the astrolabe as the most popular instrument). Likewise Ibn al-Hāʾim (Jerusalem, *c.* 1400) wrote several works on the algebra of inheritance.[25] Commentaries and supercommentaries on these works clutter the shelves of manuscript libraries to this day. About the year 1800, as we know from the titles mentioned by ʿAbd al-Raḥmān al-Jabartī in his *History of Egypt*, it was the works of Ibn al-Hāʾim and Sibṭ al-Māridīnī which predominated in Egyptian scientific circles,[26] and writings such as those of al-Marrākushī, let alone Ibn al-Sarrāj, Ibn al-Shāṭir and al-Khalīlī, had been virtually, if not

24 MS Dublin Chester Beatty 4041.

25 H. Suter, 'Die Mathematiker und Astronomen der Araber und ihre Werke', *Abhandlungen zur Geschichte der mathematischen Wissenschaften,* 10 (1900); and 'Nachträge und Berichtigungen', *Abhandlungen zur Geschichte der mathematischen Wissenschaften,* 14 (1902), 157–85, repr. Amsterdam, 1982, and again in H. Suter, *Beiträge zur Geschichte der Mathematik und Astronomie im Islam* (Frankfurt am Main, 1986), 2 vols., no. 423, and *Cairo Survey*, no. C58.

26 J. Heyworth-Dunne, *An Introduction to the History of Education in Modern Egypt* (London, 1940), 62–64, and King, 'Mamluk astronomy', 552. The only work on theoretical planetary astronomy mentioned by al-Jabartī was the very elementary treatise of al-Jaghmīnī (Central Asia, *c.* 1220: see Suter, *MAA,* no. 403, and *Cairo Survey,* no. G17). See also note 36 below.

completely, forgotten. At the same time the centre of astronomical activity had moved from Cairo and Damascus (and from Samarqand) to Istanbul.[27] It is not surprising that the director of the short-lived observatory in Istanbul around 1577 was a Syrian who had worked both in Cairo and Nablus.

On *muwaqqits* and their role in Mamluk society

The function of the *muwaqqits* in their mosques and *madrasas* was to provide reliable information not only about the astronomically determined times of prayer, but also about the regulation of the lunar calendar and the determination of the sacred direction towards Mecca. Some of these men made instruments, such as sundials or quadrants; many of them wrote treatises on the use of this instrument or that, or on time-keeping in general. In Cairo by the thirteenth century, as mentioned above, there was a corpus of tables available for time-keeping by the sun (if not by the stars). Similar corpora followed in the fourteenth century for Damascus and Jerusalem. Some *muwaqqits* compiled new tables of this function or that, or 'published' new arrangements of existing tables. The *muwaqqits* do not seem to have been involved in the compilation of annual ephemerides giving the positions of the sun, moon and stars for each day of the year. (No Mamluk examples of such ephemerides exist, but we do have fragments of various Fatimid examples, two complete Rasulid Yemeni examples[28] and numerous Ottoman examples.[29]) Likewise they do not seem to have been involved in astrology.

On the organization of the activities of the *muwaqqits* in the mosques we have no information beyond the lists of numbers of such individuals and their salaries as found in the *waqf* documents of various Cairene *madrasas*.[30] As noted above, the earliest references to *muwaqqits* are from the second half of the thirteenth century and it seems that we are dealing with a very small group of individuals at the Mosque of ʿAmr in Fustat. I know of no references to this occupation at that time or indeed from the entire fourteenth century from Cairo itself. Nevertheless in the mid-fourteenth century we are confronted with an apparently well-organized team of *muwaqqit*s at the Umayyad Mosque in Damascus, headed by Ibn al- Shāṭir. (The activity there lasted until the nineteenth century.) In fifteenth-century Egypt we have a series of *muwaqqits* employed at different mosques and *madrasas* in Cairo. (This activity likewise lasted through the Ottoman period.)

The *muwaqqits* of Mamluk Egypt and Syria were known already by name

[27] On Ottoman astronomy see *Studies*, A-XII (on astronomical time-keeping), and a forthcoming bio-bibliographical survey edited by E. İhsanoğlu (to be published by IRCICA, Istanbul).

[28] *Cairo Survey*, no. E11, and King, *Yemen*, 33 (no. 11) and 39 (no. 22). These ephemerides are currently being investigated by Michael Hofelich, Frankfurt.

[29] About 140 examples are listed in *Cairo Survey*, no. H78.

[30] King, 'Muwaqqit', section 7, based mainly on E. Strauss, 'Prix et salaires à l'epoque mamlouke: une étude sur l'état économique de l'Égypte et de la Syrie à la fin du moyen âge', *Revue des Études Islamiques*, 17 (1949), 49–94. Unfortunately no information on the salaries of *muwaqqits* is contained in E. Ashtor's survey of prices and salaries in the medieval Islamic world (1969).

and by the titles of their works to H. Suter in about 1900,[31] but the writings of Ibn al-Shāṭir were not investigated before the 1950s, and the writings of the others not before the 1970s and 1980s. Now that we have an overview of them and can appreciate their scientific merit it is time to investigate the backgrounds of these men and their role in society. This is the difficult part. We have as yet no information on the rise and development of the institution of the *muwaqqit* other than the scraps that can be gathered from the biographies of these men (no doctoral student ready to undertake such a task is available yet), from their libraries, as far as these can be reconstructed (only one has been investigated but with surprising results), and from the *waqfiyyas* of the buildings in which they were employed (information is available for Egypt, that is, Cairo, but not, as far as I am aware, for Damascus or elsewhere). Thus, for example, the *waqfiyya* of the Mosque of the amir Qānim (dated 1446) mentions 900 dirhams per month for the imam, 500 for the preacher, 200 for the muezzin, 200 for the *muwaqqit* and 300 for the servant (!). On the other hand, the *waqfiyya* of the Mosque of al-Mu'ayyad (dated 1420) mentions 650 for the Ḥanafī shaykh, 150 for the Shāfiʿī shaykh, . . . and 15 for the muezzin; no *muwaqqits* are mentioned, which is curious because two well-known astronomers (al-Kawm al-Rīshī, *c.* 1410 and al-Wafāʾī, *c.* 1450) were *muwaqqits* in that mosque.

It has been possible – using marks of ownership in various manuscripts – to reconstruct part of the private library of ʿAbd al-Raḥmān al-Ṭūlūnī, who was imam and *muwaqqit* at the Mosque of Ibn Ṭūlūn in about 1650.[32] Al-Ṭūlūnī compiled a few treatises himself, which are fairly unimpressive, but his library contained, in addition to many of the more recent sophisticated works on time-keeping and instrumentation, a set of highly impressive treatises by various major scientists from earlier centuries, including al-Farghānī, al-Khujandī, al-Bīrūnī and Sharaf al-dīn al-Ṭūsī. But if al-Ṭūlūnī had read any of these there is no evidence of that in his own writings.

The fate of the instruments of the *muwaqqits* is rather sad. A sundial constructed in 1296/7 for the Mosque of Ibn Ṭūlūn was broken into pieces and used to fill a column in the mosque. It was recovered by Marcel, one of the scholars of Napoleon, put back together and copied in ink. One of the workers noticed the French scholars' interest in the pieces and liberated them. Marcel's copy was published in the *Description de l'Egypte*.[33] Al-Ṭanṭāwī, the last *muwaqqit* of any consequence at the Umayyad Mosque in Damascus, who was active around 1875, tried to realign the magnificent sundial of Ibn al-Shāṭir, made in 1371, and broke it. He was able to make a new copy which is still *in situ* on the main minaret of the mosque, and the pieces of the original sundial were used to line various water-conduits under the main courtyard of the

[31] Suter, *MMA*, still the main source for late Islamic astronomy, but to be supplemented by *Cairo Survey* for the late ('post-Sezgin') period. See King, 'Mamluk astronomy', 553–5, for a list of Mamluk astronomers with bibliographical references.

[32] *Cairo Survey*, no. D16. On his library, see King, 'Muwaqqit', section 10.

[33] *Studies*, B-XVI.

mosque. During excavations in 1958 some of the pieces were recovered and they have been set up in the garden of the Archaeological Museum in Damascus.[34] Most of the sundials still found in mosques in Cairo, Damascus and Jerusalem are non-functional, having lost their gnomons. I have seen one in Damascus covered by a drain-pipe.

Tasks accomplished and tasks for the future

There are so few researchers in the field that it is the duty of institutions with the necessary endowments (there are only two at present: the Institut für Geschichte der Arabisch–Islamischen Wissenschaften in Frankfurt and the Furqan Foundation in London) to produce facsimile editions of the best available manuscripts of works of major importance.[35] The treatise of al-Marrākushī was partly translated and partly summarized by the Sédillots *père et fils* in the mid-nineteenth century. With the exception of C. Schoy, who investigated the sundial theory, and myself, no one has worked on it since. G. Saliba is preparing an edition and translation of, and a commentary on the treatise on planetary astronomy by Ibn al-Shāṭir. The one extant *zīj* of the latter should be published in facsimile; no one is working on it at present. I have published a description and analysis of the corpus of tables for timekeeping used in Mamluk Cairo. Together with E. S. Kennedy I have published an analysis of the unusual planetary tables of Ibn al-Majdī. I have also written a detailed analysis of al-Khalīlī's corpus of tables for Damascus. Numerous treatises on instruments are available and merit study; foremost amongst these is the treatise by Ibn al-Sarrāj. The few available astrological writings have yet to be studied. And finally, the biographical dictionaries and chronicles have yet to be exploited for the background information that they could provide relating to Mamluk astronomy.

Interest in astronomy continued apace during the Ottoman period, and materials, both scientific texts and biographical information, are not lacking to continue a survey of Egyptian astronomy down to, say, 1900.[36]

Conclusion

In brief, there is a lot still to be done. If I have been able to convince colleagues that there is such a subject as Mamluk astronomy, that it is perhaps the most colourful chapter in the history of Islamic astronomy, and that historians as

[34] Kennedy and Ghanem (eds.), *Ibn al-Shāṭir*, 69–71, and *Paris IMA Catalogue*, 439.

[35] D. A. King, 'Some remarks on Islamic scientific manuscripts and instruments and past, present and future research', in J. Cooper (ed.), *The Significance of Islamic Manuscripts* (London, 1992), 115–44, esp. 128–30 and 134.

[36] al-Jabartī 's monumental *History of Egypt*, now available in translation with commentary (3 vols., Stuttgart, 1994), provides a remarkable amount of information on several of the astronomers whose works are so well represented in the manuscripts of the Egyptian National Library. These materials merit a separate study in the light of our recently acquired knowledge of the works themselves (see note 5 above).

well as historians of science can contribute to this topic, my present goal will have been achieved. We now know what the *muwaqqit*s wrote, if not how precisely they used their tables and instruments. How they received their training, how they were regarded in the society, what their attitude was to astrology, we do not know. Only when more research has been done will it be possible to say more on the relations of the astronomers, both the *muwaqqit*s and the independent scholars, to the ruling class and on the interest of that class in astronomy and astrology.

On the History of Astronomy in the Medieval Maghrib

Table of contents

Important notice [see now Preface, p. ix, n. 8.]

This article contains the unaltered text of my paper "On the History of Astronomy in the Medieval Maghrib" which appeared in *Études Philosophiques et Sociologiques Dédiées à Jamal ed-Dine Alaoui*, Université Sidi Mohamed Ben Abdallah, Publications de la Faculté des Lettres et des Sciences Humaines Dhar El Mahraz - Fès, N° Spécial 14 (Département de Philosophie, Sociologie et Psychologie), Fez, 1998 [published 1999], pp. 27-61. That, in turn, was an expanded and updated version of my paper "An Overview of the Sources for the History of Astronomy in the Medieval Maghrib", *Actes du 2. Colloque Maghrébin de l'histoire des mathématiques arabes (Tunis, 1-3 Dec. 1988)*, Tunis: Institut supérieur de l'Éducation et de la Formation continue, n.d. [*ca.* 1990], pp. 125-157, which was printed with only half of the bibliography. I have not attempted to bring the text up-to-date, so the reader must consult various important publications by the Barcelona school that have appeared in recent years, namely:

Àngel Mestres, *Materials andalusins en el Zīj d'Ibn Isḥāq al-Tūnisī* (edited text and tables, with introduction and commentary in English), doctoral thesis, University of Barcelona, 2000.

Mónica Rius, *La Alquibla en al-Andalus y al-Magrib al-Aqṣā*, (Anuari de Filologia (Universitat de Barcelona) XXI (1998-99) B-3), Barcelona: Institut "Millás Vallicrosa" de Història de la Ciència Àrab, 2000.

Julio Samsó, "An Outline of the History of Maghribī *Zījes* from the End of the Thirteenth Century", *Journal for the History of Astronomy* 29 (1998), 93-102, and "Astronomical Observations in the Maghrib in the Fourteenth and Fifteenth Centuries", *Science in Context* 14 (2001), pp. 165-178.

Also, David A. King and Julio Samsó, with a contribution by Bernard R. Goldstein, "Astronomical Handbooks and Tables from the Islamic World (750-1900): An Interim Report", *Suhayl – Journal for the History of the Exact and Natural Sciences in Islamic Civilisation* (Barcelona) 2 (2001), pp. 9-105.

These studies take our knowledge of this subject a substantial step further.

1. Introduction

During the "classical" period of astronomy in the Islamic Near East (from *ca.* 750 to *ca.* 1050) the discipline acquired a distinct Islamic character. Thereafter, and until the introduction of Western ideas, schools of astronomy with their own traditions, authorities and specific interests, developed in the different regions of the Islamic world. While the schools of Egypt, Syria and the Yemen have been studied recently, that of the Maghrib has remained largely untouched. Several of these sources have been rediscovered only during the past few years, so the time is ripe for a preliminary overview, and this is the aim of the present paper. As we shall see, there are still many gaps in our knowledge of the subject, but even if this survey provokes only criticism my purpose will have been served in drawing attention to the subject. A bio-bibliographical survey of the sources for the history of astronomy (and mathematics) in the Maghrib would be a useful contribution to the field but remains a task for the future.

Before we proceed we should point out that in the Muslim world in medieval times *astronomy was conducted at two different levels.* The handy term 'exact sciences', referring to both mathematical astronomy and mathematics proper, fails us here, not least because some of the medieval astronomical material is far from exact and represents a popular folk tradition, virtually independent of the mathematical one. I shall therefore take pains to distinguish between the *scientific or mathematical tradition* and the *popular or folk tradition.*

The topics treated in the medieval Arabic scientific texts dealing with astronomy (see Section 2) include:

- chronology,
- sexagesimal arithmetic,
- trigonometry,
- mathematical geography,
- spherical astronomy (the mathematics of the apparent daily rotation of the celestial sphere) and time-keeping (see also Section 4),
- solar, lunar and planetary means motions and equations,
- planetary latitudes and stations,

- parallax and eclipse theory,
- visibility of the planets and of the moon,

and –last but not least–

- theoretical astrology and the mathematics involved in its applications (see also Section 3).

The astronomical handbooks with tables known as *zījes* cover all of these topics. We know of about two hundred such works compiled by Muslim astronomers between the eighth and the nineteenth centuries. For individual years the daily solar, lunar and planetary positions were computed from the mean motion and equation tables in *zījes* and tabulated in ephemerides (*taqāwīm)*. Other independent treatises deal with planetary theory (not commonly found in the Maghrib) and instrumentation (mainly astrolabes, quadrants and sundials).

A different corpus of texts (see Section 6) deals with folk astronomy, a discipline devoid of theories, calculations and tables, which found considerably more favour with the religious authorities (*fuqahāʾ*) than the scientific tradition. The topics dealt with in these texts (classified as *kutub al-anwāʾ*, *kutub al-mawaqīt* and *kutub dalāʾil al-qibla*) include :

- shadow-lengths and time-keeping by day,
- the lunar mansions and time-keeping by night,
- the winds and their astronomically-defined limits,
- astronomical risings and settings, and
- information on agriculture arranged in the form of an almanach for the solar calendar.

Both the scientific and the folk scientific texts deal with the aspects of astronomy involved in religious life:

- the regulation of the lunar calendar,
- the organization of the astronomically-defined prayer-times, and
- the determination of the sacred direction towards Mecca (*qibla*).

Inevitably these problems were treated in two different ways, in the former texts mathematical methods are advocated, and in the latter the simpler procedures of folk astronomy.[1] The scholars of the religious law (*fuqahā'*) advocated only the procedures of folk astronomy in their writings on these subjects.

Since the first version of this paper was prepared for a conference on medieval Islamic *mathematics* (Tunis, 1988), I could not resist pointing out that much of our source material is based on mathematical procedures and is often highly technical; it merits our consideration as applied mathematics. The history of Islamic mathematics has suffered severely in the past at the hands of those who considered the subject restricted to texts on arithmetic, algebra and geometry.[2] For it is in *astronomical* sources that we find :

- discussions of plane and spherical trigonometry,
- applications of mathematical techniques such as stereographic and orthogonal projection and the analemma,
- tabulation of double – and triple-argument functions,
- graphical representations of double-argument functions,
- second-order interpolation schemes, and
- computational devices.

Our principal sources are manuscripts and instruments. A total of several hundred Maghribi astronomical manuscripts survive. The main libraries in which these are preserved are, of course, those in the Maghrib itself –Tunis, Algiers, Fez and Rabat, as well as other smaller collections, some private– but also the Deutsche Staatsbibliothek in Berlin, the Biblioteca de El Escorial near Madrid, the Egyptian National Library in Cairo, and the Süleymaniye and Topkapi Libraries in Istanbul. Also about ten dozen Maghribi astronomical instruments are found in various

[1]. My overview of these topics published as "Science in the Service of Religion: The Case of Islam", *Impact of Science on Society* (UNESCO), no. 159, vol. 40 (1990): no. 3, pp. 245-262, reprinted in D. King, *ASI* (see the bibliography), is available in French as "La science au service de la religion : le cas de l'islam", *Impact science et société*, n° 159, vol. 40 (1990): n° 3, pp. 283-302.

[2]. A broader approach, in which astronomical sources are used, is in Berggren 1986.

museums around the world. Some of the sources discussed below, however, are now preserved in places as far apart as Hyderabad and Chicago. Newly-identified sources offer far more scope to the writing of the history of astronomy in the Maghrib and promise to yield a far more accurate historical picture than has been possible hitherto. Nevertheless credit must be given at the outset to the labours of two Islamicists and historians of science who have done more than any other scholars in this century to document our subject. A glance at the bibliography following this paper will indicate the extent of the contributions of Henri P.-J. Renaud and Juan Vernet.

2. Mathematical astronomy

How little we know about the earliest Islamic astronomy in the Maghrib is shown by a treatise on the armillary sphere by Dūnāš ibn Tamīm, which survives in a unique manuscript in Istanbul (and which has been studied by S. M. Stern). The author was a well-known Jewish scholar of the middle of the tenth century and he was active in Kairouan. It is clear from this treatise that Dūnaš was *au fait* with the latest writings on astronomical instruments from Baghdad, and also that he was not working in a vacuum. One of his teachers was Abū ᶜAlī Aḥmad ibn ᶜUthman al-Asṭurlābī, which means that astrolabes were being made in Kairaouan even before his time. His other writings on astronomy, now lost, included materials on theoretical astronomy, mathematical astronomy and astrology; these are mentioned in a commentary on the *Book of Creation*, extant only in a very corrupt Hebrew version (studied by G. Vajda). In his treatise on the armillary sphere Dūnaš mentions a conversation on astronomy and astronomical instruments that he had with Abū 'l-Ḥasan Muḥammad ibn al-Ḥusayn, Fatimid Governor of the city of Mahdiyya, and he also states that he made an armillary sphere himself with the help of craftsmen organized by the Governor.

The earliest Maghribi *zīj* on which we have any information is the *Ḥall al-ᶜuqad wa-bayān al-raṣad* of the early eleventh century Kairouan astrologer Ibn Abī 'l-Rijāl (see Section 3), a work known to us only by title. This was no doubt an eclectic work and would be of great interest, but alas it appears to have been lost. (Compare the similar situation regarding the lost *Zīj* of Hamdānī in the history of Yemeni astronomy).

Considerable obscurity surrounds the activities in the thirteenth and early fourteenth century of scholars such as Ibn Hilāl of Ceuta, possibly the author of a commentary on Ptolemy's *Almagest*, and al-Mirrīkh in Marrakesh. Our knowledge of them we owe entirely to remarks recorded in passing in various later works (see below).

The fact that there was an active tradition of serious mathematical astronomy in the medieval Maghrib is recorded by the celebrated historian Ibn Khaldūn (*fl. ca.* 1400):

> "Contemporary Maghribi scholars are using as their reference work the astronomical handbook with tables (*zīj*) ascribed to Ibn Isḥāq. It is thought that Ibn Isḥāq based his work on astronomical observations. A Jew in Sicily who was skilled in astronomy and the mathematical sciences, and who occupied himself with astronomical observations, sent (Ibn Isḥāq) information on the conditions and motions of the planets which he had ascertained. Thus the Maghribis have been using Ibn Isḥāq's work because they assume that it is based upon reliable information. Ibn al-Bannā wrote an abridgment (of Ibn Isḥāq's *Zīj*) which he entitled *al-Minhāj*. People have been very eager to use the *Minhāj* because the operations described in it are easy."

The *Zīj* of Ibn al-Bannā of Marrakesh (1256-1321), until recently the most significant Maghribi scientific work known to us, contains about 15 folios of text and some 55 tables. It is extant in several manuscripts, attesting to its widespread use in the Maghrib over the centuries. J. Vernet published the text of the introduction and a Spanish translation in 1952 but the tables constituting the most important part of the work have never been properly studied (though see below).

In 1978, whilst working in a manuscript library in Hyderabad, the present author came by chance upon a copy of the *Zīj* of Ibn Isḥāq. It is a magnificent work, comprising some 400 pages, with some 360 astronomical tables relating to a wide variety of topics. It is far more substantial than the *Zīj* of Ibn al-Bannā (and indeed than most known *zījes* from other regions) and represents the missing link in the Maghribi astronomy. All future studies of the history of astronomy in the Islamic West will have to take it into consideration.

The *Zīj* of Ibn Isḥāq is important not least because it is a highly eclectic work and the author carefully identifies his sources, most of which are no longer extant in their original form. Its organization is unimpressive; for example, several topics are treated more than once because the author's sources had different procedures for handling them. This of course in no way detracts from the historical interest of the work. The *Zīj* was apparently completed in 1218 (the Hijra date is written in the highly unusual form w-y-d = 6(00) + 14 but there is a problem with the corresponding dates given in other calendars) in Tunis, and some of the tables for spherical astronomy are specifically computed for that city. However, at least one set of planetary tables is for the longitude of Toledo, and this was simply lifted from an Andalusian work. Ibn Isḥāq asserts that the structure of the tables –which is by no means standard– makes them universal so that the choice of a meridian is arbitrary; he therefore did not feel obliged to recompute them for Tunis. Our author adhered to the (false) notion of the trepidation of the equinoxes, and his work contains a great deal of material on this topic. He lists as his major inspiration the works of Zarqāllu (Azarquiel in the Latin West, *fl.* Toledo, *ca.* 1070), Ibn al-Kammād (*fl.* Seville, *ca.* 1150?), and Ibn al-Hāʾim (*fl.* Seville, *ca.* 1200). Only future research will establish the correspondence with the surviving works of these individuals.

Only parts of one of the *zījes* of Ibn al-Kammād survive in Latin, and of Ibn al-Hāʾim's *Zīj*, we have only a fragment of the introduction. In addition, some of the tables for lunar crescent visibility are said to be taken from the *Zīj* of Ibn Muʿādh (*fl.* Andalusia, late eleventh century), a work otherwise known only from a Latin translation of the introduction printed in Germany in the sixteenth century. One of Ibn Isḥāq's planetary latitude tables is said to be "according to the opinion of Ibn al-Bayṭār," perhaps to be identified with the celebrated contemporary Andalusian botanist and pharmacologist (who is not otherwise known to have concerned himself with astronomy and who must have compiled the table in his youth). The only observation mentioned by Ibn Isḥāq as having been his own is a determination of the obliquity of the ecliptic in Tunis in 588 Hijra [= 1192]. Likewise the only reference of Ibn Isḥāq's *Zīj* to the Jew in Sicily mentioned by Ibn Khaldūn is to one *Gh-'-l-m ben y-ḥ-'-r* (??) who made a similar determination in the year 584 Hijra [= 1188/89]. Thus Ibn Khaldūn seems to be quite misinformed (unless it was the Jewish colleague who communicated the Andalusian *zījes* to Ibn Isḥāq).

It should be already apparent that the *Zīj* of Ibn Isḥāq is a gold-mine for future investigations of the history of both Maghribi and Andalusian astronomy. Two further examples confirm this. Amidst the material on lunar crescent visibility is a table attributed to an individual named al-Qallās. This table is otherwise known to us only from a Latin translation of a recension by al-Majrīṭī (*fl.* Madrid, *ca.* 1000) of the *zīj* of al-Khawārizmī (*fl.* Baghdad, *ca.* 825). Now its authorship is established, but unfortunately we have no further information on al-Qallās. Secondly, there is an astrological table for predicting the length of an individual's life from the astronomical circumstances at his birth. This is attributed to the second century Hellenistic astrologer Vettius Valens (Wālīs) and came to Ibn Isḥāq by way of Ibn al-Kammād. It is also known to us from a Byzantine copy of the original Greek text, and there is textual evidence that it was available in its Arabic form in Baghdad in the ninth century. Furthermore several later (fifteenth century) Egyptian copies exist. But the table was apparently forgotten in the central lands of Islam despite its obvious utility for the gullible.

The Hyderabad manuscript was copied not in the Maghrib but in Egypt or Syria (the script is *naskhī*). A marginal note explains how to adapt some of the planetary tables from the longitude of Toledo to that of Damascus, which could be taken as indicating a Syrian provenance for the manuscript. The copy is undated but might well be from the thirteenth or fourteenth century. If it is indeed so early, then there is a possible explanation for its presence in Damascus: it is known that an astronomer of Maghribi origin worked there in the mid-thirteenth century, namely Muḥyī 'l-Dīn Yaḥyā ibn Muḥammad ibn Abī 'l-Shukr al-Maghribī. In the Biblioteca de El Escorial there is a unique copy in a Maghribi hand of a *zīj* for Damascus prepared by him in the year 1249. He cites no sources of any kind, does not mention trepidation, and appears to have made independent observations to determine the obliquity of the ecliptic and the latitude of Damascus. This manuscript needs to be investigated to see whether the parameters are related to any of those used by Ibn Isḥāq. Ibn Abī 'l-Shukr was taken as a prisoner by the Mongols from Damascus to Maragha in N. W. Iran, and there he compiled two more *zījes* and participated in the observations under Naṣīr al-Dīn al-Ṭūsī.

From a note on the title folio we learn that the Hyderabad manuscript of the *Zīj* of Ibn Isḥāq was in Istanbul about the year 1700; how it later came to Hyderabad I do not know. Miscellaneous tables attributed to Ibn Isḥāq have also turned up in later Egyptian and Yemeni *zījes*, attesting to the influence of his work outside the Maghrib. Research on Ibn Isḥāq's remarkable *Zīj* is currently being conducted by Professor Julio Samsó and his team in Barcelona.

Abū ᶜAlī al-Marrākushī was another Maghribi astronomer who admired the tradition of al-Zarqāllu and Ibn al-Kammād, although when mentioning them he made no reference to Ibn Isḥāq. He is better known for his work in spherical astronomy and instrumentation (see Section 4a).

One of the commentaries on the astronomical poem of al-Jādarī (see Section 6) records a measurement of the obliquity undertaken in Meknes by an unnamed observer in the year 1205/06. That was shortly before the observation of Ibn Isḥāq and about the same time as the activity in astrolabe construction in Marrakesh (see Section 5a). The commentary also mentions measurements by one Ibn Hilāl in Ceuta and by al-Mirrīkh in Marrakesh in 1304/05. The former is probably Abū ᶜAbdallāh Muḥammad ibn Hilāl, who was possibly the only Maghribi to write a commentary on Ptolemy's *Almagest.* Ibn Isḥāq and his followers such as Ibn al-Bannā and al-Marrākushī (see Section 4a) accepted the notion of the trepidation of the equinoxes, and there is substantial new material on this subject now available. The Maghribi interest in the changing obliquity can be seen as part of their preoccupation with this topic. A confused historical overview of Maghribi activity relating to trepidation is also preserved in the commentary on al-Jadarī's poem.

At first sight the *Zīj* of Ibn al-Bannā is not a much-simplified abridgment of the work of Ibn Isḥāq, although the earlier astronomer is cited in the introduction. The tables serve Marrakesh. Ibn Khaldūn was right in his assessment of its potential as a popular handbook, but Ibn al-Bannā was not completely convinced: he also prepared an almost trivial mini-*zīj* of a few pages entitled *al-Yasāra*, which he later expanded as *Tashīl al-ʿibāra fī takmīl mā naqaṣa min al-Yasāra*. Research on the tables in the larger *Zīj* is currently in progress in Barcelona.

Another small *zīj* for Tlemcen based on the *Yasāra* was compiled by Ibn Qunfudh in the fourteenth century and is available in several copies. The contemporary *zīj* for Fez by ʿAlī ibn Abī ʿAlī al-Qusanṭīnī also relies on Ibn al-Bannā, and has an introduction in verse. The planetary equation tables, taken from an unidentified work, are of particular interest in so far as they are based on a very simple Indian theory.

Two fairly substantial *zījes* by Muḥammad ibn al-Raqqām (*fl.* Tunis, fourteenth century) survive as manuscripts in Istanbul and Madrid/Rabat. Their relationship to each other and to other Maghribi works of the genre needs to be investigated. Ibn al-Raqqām is cited in a fourteenth century Hebrew *zīj* compiled by Isaac al-Ḥadib in Sicily.

A Maghribi recension (in Arabic) of the universal almanac of the Spanish Jew Abraham Zacuto (*fl.* Salamanca, late fifteenth century), together with several Arabic treatises on its use, has been located in Cairo and remains to be studied. A late recension for Algiers of the *Zīj* of Ibn al-Shāṭir (*fl.* Damascus, *ca.* 1350) also survives in Cairo, and two Tunisian recensions of the *Zīj* of Ulugh Beg (*fl.* Samarqand, *ca.* 1425) by ʿAbdallāh Ḥusayn Quṣʿa and Muḥammad al-Sharīf al-Tūnisī known as Sanjaqdār survive in several copies. It is clear that these more recent works were used in the Maghrib alongside the *Zīj* of Ibn al-Bannā and its offshoots. Their investigation is not as urgent as proper study of the original *zījes* of Ibn al-Shāṭir and Ulugh Beg, neither of which has yet found an editor.

At the time of writing I am unaware of the existence of any Maghribi ephemerides with solar, lunar and planetary positions for each day of the year common in other regions of the Islamic world.

Nevertheless such ephemerides were doubtless prepared annually in the Maghrib too.

3. Astrology

Most Islamic *zījes*, and particularly that of Ibn Isḥāq, contain material on astrology, and it was mainly such an association of mathematical astronomy with its notorious partner that made the entire scientific tradition suspect in the eyes of the religious authorities.

Ibn Abī 'l-Rijāl (see Section 2 above) was the tutor and astrologer of the Prince al-Muᶜizz ibn Bādīs at the Zirid court in Kairouan. His fame rests on an extensive treatise on astrology entitled *al-Bāriᶜ fī aḥkām al-nujūm* which was popular not only in the Islamic world but also in Europe, where it was translated into Castilian, Latin, Hebrew, Portuguese, French and English. It was not a work of great originality, being largely lifted from earlier Iraqi works, but then fate rather than originality always played a greater role in the transmission of Islamic scientific and pseudo-scientific knowledge to Europe. The treatise consists of eight books: I-III on interrogations (*masāʾil*), IV-VI on nativities (*mawālīd*), VII on catarchic astrology (*ikhtiyārāt*), and VIII on general (including political and historical) astrology. The author also compiled a poem on astrology which was very popular and was commented upon by Ibn Qunfudh (see Section 2) in 1372.

The historical sources record that in the year 1132, when the Almoravid Sultan ᶜAlī ibn Yūsuf ibn Tāshufīn wanted to build a wall around the city of Marrakesh, founded by his father some seventy years before, he consulted his astrologers. The venture was not entirely successful, in part at least as the result of the intervention of a crow. A later commentator exclaimed: "this is the reason why the walls of Marrakesh have never been solid and require constant maintenance".

The city of Bône was the birthplace of the prolific writer Abū l-ᶜAbbās Aḥmad ibn ᶜAlī al-Būnī (d. 1225) who wrote on magic and also dabbled in astrology. Other than two short treatises on astrology by Ibn al-Bannā and Ibn Hayḍūr al-Tādilī (*fl.* Fez, *ca.* 1400) which have never been studied, I know of no independent

Maghribi works on the subject or of any nativity books or astrological almanacs such as exist elsewhere in the Islamic world.

Ibn Khaldūn presented an eloquent refutation of astrology which he concluded as follows:

> "Thus the worthlessness of astrology from the point of view of the religious law, as well as the weakness of its achievements from the rational point of view, are evident. In addition, astrology does harm to human civilization. It hurts the faith of the common people when an astrological judgment occasionally happens to come true in some unexplainable and unverifiable manner. Ignorant people are taken in by that and suppose that all the other (astrological) judgments must be true, which is not the case. Thus, they are led to attribute things to some (being) other than their Creator.
>
> Further, astrology often produces the expectation that signs of crisis will appear in a dynasty. This encourages the enemies and rivals of the dynasty to attack (it) and revolt (against it). We have (personally) observed much of the sort. It is, therefore, necessary that astrology be forbidden to all civilized people, because it may cause harm to religion and dynasty. The fact that it exists as a natural part of human perceptions and knowledge does not speak against (the need to forbid it). Good and evil exist side by side in the world and cannot be removed. Responsibility comes in connection with the things that cause good and evil. It is (our) duty to try to acquire goodness with the help of the things that cause it, and to avoid the causes of evil and harm. That is what those who realize the corruption and harmfulness of this science must do..."

But there was no inquisition against the astrologers, and astrology's influence amongst the credulous continued.

Also on the subject of astrology, Renaud (1942) recorded the following story from contemporary documents. In 1715 the marabout of Tasaft was disturbed to learn that the army of the Sultan Mawlay Ismāʿīl was approaching his *zāwiya*, and so he consulted with one Muḥammad ibn ʿAlī al-Ṣanhājī, a *muwaqqit* or professional mosque astronomer who was currently visiting the *zāwiya*:

> "(Al-Ṣanhājī) had in his hand an astrolabe for measuring celestial altitudes and for determining the ascendant, on which are based the procedures for finding the horoscope. He asked for the following books to be brought to him: (1) *Minhāj aṭ-ṭālib fī taᶜdīl al-kawākib* (the *Zīj* of Ibn al-Bannā); (2) the poem in *rajaz* metre (on astrology by Ibn Abī 'l-Rijāl) and its commentary by Ibn Qunfudh; (3) the *al-Bāriᶜ fī aḥkām al-nujūm* by Ibn Abī 'l-Rijāl; and (4+5) the *Taḥṣīl al-maṭālib* and *al-Yasāra* (the small *zījes* of Ibn Qunfudh and Ibn al-Bannā). Then he said: 'The situation is serious and time is too short to ameliorate it. My mind is troubled by the celestial bodies which feature in the horoscope. Let us wait until tomorrow'."

Each of these sources is known to us.

4. Spherical astronomy and astronomical time-keeping

a) Treatises on spherical astronomy

We have already mentioned the celebrated astronomer of late thirteenth-century Cairo, Abū ᶜAlī al-Marrākushī; he was largely responsible for the development of astronomical time-keeping in the central lands of Islam. He compiled an extensive compendium on spherical astronomy and instrumentation –mainly astrolabes, quadrants and sundials– which during the ensuing centuries was highly influential in Egypt, Syria and Turkey, but not at all –as far as is indicated by the sources known to me– in the Maghrib (see Sédillot *père et fils*). Al-Marrākushī mentions that he personally conducted measurements of the latitudes and longitudes of several localities in the Maghrib before emigrating to Egypt. He was clearly familiar with the Andalusian-Maghribi traditions of astronomy since he quotes al-Zarqāllu and Ibn al-Kammād (if not Ibn Isḥāq); however, most of the other sources for his highly eclectic work are of Eastern Islamic provenance, and the majority of his tables are computed by himself for the latitude of Cairo.

The theoretical problems of spherical astronomy are treated in the various Maghribi *zījes* mentioned in Section 2 above. The only independent Maghribi work on the subject known to me is anonymous and was compiled in Sfax about the sixteenth century; it is now preserved in Cairo.

b) Tables for time-keeping

The extensive tables compiled by Muslim astronomers for time-keeping by day and night generally served specific latitudes. In Berlin there is a unique manuscript containing a set of tables for time-keeping computed in the fourteenth century for the latitude of Tunis. The times of midday, the *ẓuhr* and the *ʿaṣr* are displayed, as well as the time of an institution called the *taʾhīb* (see below and Section 5c), for each degree of solar longitude (corresponding roughly to each day of the solar year). The underlying latitude is 37°0', corresponding to Tunis, and in the same manuscript there is another contemporaneous set for 36°40', another medieval value for the latitude of Tunis.

A unique manuscript preserved in Cairo displays similar tables for the latitudes of Sijilmasa, Fez and Marrakesh. The time-keeping tables that were used in medieval Cairo and Damascus are extant in dozens of copies, but here it should be pointed out that most tables of this kind are miscatalogued (if they are catalogued at all), and it is quite probable that other copies of the Maghribi tables are simply waiting to be identified in libraries in the Maghrib.

Another Cairo manuscript contains a Tunisian recension of the minor universal auxiliary tables compiled by the Syrian astronomer al-Khalīlī (*fl.* Damascus, *ca.* 1360). The purpose of these tables was to facilitate the various operations of time-keeping by the sun for all terrestrial latitudes. In London there is a Maghribi copy of al-Khalīlī's major universal auxiliary tables – the culmination of Muslim achievement in this area – showing that these too were known in the Islamic West.

Yet another Cairo manuscript contains an Ottoman set of prayer-tables for Algiers, displaying for each day of the year the times of the *imsāk* in Ramadan, the *ʿaṣr* and an institution apparently called the *ṣanjāq*, probably related to the *ḍuḥā*. Although all these tables are of as much interest to the history of Islamic religious institutions as they are to the history of science, there are also other relevant sources (see Section 5c below).

c) Astrolabic clocks

Amongst the most remarkable historical objects in the Maghrib are the astronomical clocks in Fez (see Price, 1962). One of these, a water-clock operated by levers and strings and without any complicated gear mechanisms, was located in a room in the minaret of the Qarawiyyīn Mosque. It was made in 1286/87 by Ibn al-Ḥabbāk al-Tilimsānī, and when it was restored in 1346-48 by Abū ʿAbdallāh al-ʿArabī it was fitted with an astrolabic rete. Alas, the driving mechanism behind the clock is lost without trace, and it is not clear what changes have been made to the front of the clock. But fortunately the astrolabic part survives to this day. It is housed in a cabinet 2.4 metres high and 1.2 metres square; the rete is about 40 cm in diameter, and would have rotated once every 24 hours. It thus could imitate the apparent daily rotation of the heavens about the horizon of Fez, a kind of model of the universe in two dimensions. In addition, metal balls would fall through the doors above the clock every hour.

Another water-clock was constructed in the fourteenth century next to the Abū ʿInāniyya Madrasa in Fez. Its maker was Abū l-Ḥasan al-Tilimsānī, a *muwaqqit* at the Madrasa. The clock itself has now disappeared but the basic device for announcing the hours – operating in a housing about 12 metres long– is still *in situ* overlooking a street in the *sūq*. Each hour the clock would eject metal balls which would roll through the device and fall through the little door appropriate to the hour in question onto a series of gongs. Thus the clock would chime every hour, and in between the hours one could tell the time by checking which door was open.

Such clocks were known in Antiquity, the most spectacular being the one in the so-called "Tower of the Winds" in Athens, which dates from the first century before the Christian era. The Syrian engineer al-Jazarī described their construction in the early thirteenth century, but the Fez clocks are the sole surviving Islamic examples. Such Islamic astrolabic clocks were not common even in the Middle Ages: we know of other examples only from Cairo, Damascus and Toledo, and there is a description of one in the thirteenth-century Andalusian astronomical compendium entitled *Libros del saber de astronomia*, compiled by order of King Alfonso X el Sabio. We are thus singularly fortunate to have some remains of two clocks in Fez, as well as details about their construction recorded by a contemporary historian (al-Jaznāʾī in his *Zahrat al-ās*). The Fez clocks probably fell into disuse because they do not indicate the times of prayer.

5. Astronomical instruments

a) Astrolabes

The earliest known Maghribi astronomical instruments –apart from the various ones mentioned by Dūnaš ibn Tamīm (see Section 2), none of which survive– are two astrolabes made by Ibrāhīm ibn ʿAbd al-Karīm in Fez in the middle of the eleventh century. Earlier astrolabes are known from Andalusia, and these two pieces were clearly inspired by the tradition there.

Thereafter come ten astrolabes by a certain Abū Bakr ibn Yūsuf in Marrakesh at the beginning of the early thirteenth century. These incorporate a distinctive new design for the rete and are of exceptionally high quality (see, for example, Sarrus, 1853). One of the instruments made by Abū Bakr, which unfortunately does not survive, was a universal astrolabe of the kind invented by ʿAlī ibn Khalaf al-Shajjār in Toledo in the late eleventh century.

The next Maghribi astrolabe to survive is a unique example of a so-called "mixed" astrolabe made by ʿAlī ibn Ibrāhīm al-Jazzār, a muezzin in Tāza in 1327/28 and now preserved at Oxford. This is the only known example of a type of unusual astrolabe devised mainly in Iraq and Iran in the ninth and tenth centuries : its existence proves that the craftsmen in the Maghrib were well informed of the most sophisticated developments in the field.

A much later Maghribi astrolabe maker was the very prolific Muḥammad ibn Aḥmad al-Baṭṭūṭī. He worked in the early eighteenth century, and some twenty high-quality instruments made by him are known to us. His activities remind one of the astrolabe-makers of Lahore during the same period.

The late eighteenth-century "astrolabe" described by Delphin (1891) is nothing more than a sine quadrant (see below), supplemented with various data for time-keeping by the fixed stars and a table of ascensions for the latitude of Fez: most astronomers preferred to have information on paper rather than engraved on their instruments.

No Maghribi treatises of consequence on the construction or use of the astrolabe are known, although several poems on the operation of the instrument were written.

One of the most recent astrolabes to be made in the Maghrib (*ca.* 1980 ?) bears a dedication (in very dubious Arabic) to the famous traveler Ibn Baṭṭūṭa and the date 723 H [= 1323]. In Fez a few years ago it was offered for sale to a friend of mine for US$ 70,000. More recently it was purchased for a lower, but still substantial, price, and the unlucky purchaser has since tried to get rid of it through the leading auction houses in London, who are, however, now alerted to the problem (several dozen fake Indian and Persian instruments appear on the market each year). Until there is a catalogue of medieval instruments available, fakers will continue to be able to flourish and make money from unsuspecting tourists. Even this piece found "experts" in Jerusalem and Oxford who did not recognize it as a fake.

b) Quadrants

A quadrant made by Aḥmad ibn ʿAbd al-Raḥmān al-Dahmānī in 854 Hijra [= 1450/51) is preserved at Madrid (see Millás Vallicrosa, 1947). On one side there are the astrolabic markings (like one half of an astrolabe plate) for a specific latitude, namely 36° 40', which uniquely identifies the provenance of the instrument as Tunis. On the other side there is the grid of a sine quadrant for performing all kinds of trigonometric operations without calculation. The markings are carefully and elegantly engraved but they are standard, and the instrument is typical of the quadrants which were being produced in Syria and Egypt in large numbers in the fourteenth and fifteenth centuries. (This kind of instrument, with astrolabic markings on one side and a trigonometric grid on the other, was invented in Egypt in the eleventh or twelfth century, although it is often erroneously attributed in the modern literature to the Jewish astronomer Profeit Tibbón (Prophatius) who worked in Southern France at the end of the thirteenth century).

The only substantial Maghribi treatise on the use of the quadrant deals with the sine quadrant and was compiled by Abū Jaʿfar ʿUmar ibn ʿAbd al-Raḥmān al-Tūnisī al-Tūzarī about the year 1450. Again several poems are known to us, some of which were lithographed in Tunis about a century ago.

There are various Maghribi quadrants from the eighteenth and nineteenth centuries which survive, and perhaps the most interesting of these is one preserved in a museum in Chicago. It indicates the times of a prayer for the whole year and also the *qibla*, and its markings correspond to the latitude and longitude of Marrakesh.

c) Sundials

A horizontal sundial constructed in Tunis in the fourteenth century is now preserved in the Musée national in Carthage (see King, 1977). The sundial bears an inscription "constructed by Abū l-Qāsim ibn Ḥasan al-Shaddād in the year 746 Hijra [= 1345/46] in Tunis." Most Islamic sundials bear markings for the hours of daylight, but not this one: it bears only markings for the times of day having religious significance. The *ẓuhr* and the *ʿaṣr* markings are standard, but the sundial also bears a curve for the *ḍuḥā* at mid-morning and the *taʾhīb* one hour before midday.

It is well known that the institution of a prayer at mid-morning – that is, a sixth prayer – was favoured in some circles in early Islam, but it was known to have been practiced in Tunis in the fourteenth century: the sundial is proof that it was practiced there. The institution called the *taʾhīb*, in other sources also *taʾahhub*, relates to the communal prayers on Fridays. But not only does this sundial give us information on Tunisian religious practice, it also provides a key to understanding the distinctive definitions of the times of the daylight prayers in Islam (which, involving shadow *increases*, are *not* the same as those featured in the *ḥadīth*, which involve shadow *lengths*).

Two simple cylindrical dials with markings for astronomical midday, the *ẓuhr* and the *ʿaṣr*, have been noted in the literature (Marçais, 1903, and Bel, 1905). One, dated 1346, was found in Tlemcen, the other, doubtless earlier, in al-Manṣūra.

The sundial in the courtyard of the Mosque of Sīdī ʿUqba in Kairouan is of rather recent construction (see Janin, 1977). It consists of four sundials superposed one on top of the other. The four gnomons and the related markings serve to measure time from sunrise, time until sunset, and the time relative to each of the prayers. Technically it is not impressive; indeed, the curve for the *ʿaṣr* has been drawn twice, which hardly inspires confidence.

Various other sundials are still *in situ* in mosques around the Maghrib but they are nowhere catalogued and only those mentioned above have been studied.

A treatise on sundial theory –without any of the tables for sundial construction which are common in other Islamic works on the subject– was compiled by Ibn al-Raqqām of Tunis in the fourteenth century and survives in a unique manuscript in the Escorial. It has recently been published and analyzed for the first time (Carandell, 1988). A treatise by Ibn al-Raqqām's contemporary Ibn al-Najjār is known only by citation. Other less substantial Maghribi treatises on gnomonics survive, some with tables for the latitudes of Fez and Tunis.

6. Folk Astronomy

In the Islamic world in medieval times there were two virtually distinct traditions of astronomy: the mathematical tradition, represented in the Maghrib by the sources we have discussed in Sections 2-5 above, and a popular folk tradition devoid of any theory and innocent of any tables or instruments. The latter was particularly favoured by the scholars of religious law (*fuqahāʾ*) and was thus far more widely practiced than the former. In particular, folk astronomy was applied to the three aspects of daily life involving astronomy: the organization of the calendar through the actual sighting of the crescent (that is, without prediction by means of complicated mathematical procedures), the regulation of the prayer-times using simple shadow schemes by day and the lunar mansions by night (that is, without tables or instruments), and the determination of the *qibla* by astronomical risings and settings (rather than by using trigonometric procedures or tables).

In the Maghrib folk astronomy was represented mainly by the doctrines of Abū Miqraʿ and al-Jādarī. Abū Miqraʿ lived in Morocco in the early thirteenth century, and his opinions were formulated in verse by al-Marghīthī in the seventeenth century and subjected to numerous commentaries and super-commentaries thereafter (see Colin & Renaud, 1933). Al-Jādarī worked in Fez in the fourteenth century and compiled his own poem, which was also much discussed. Some of the commentaries on these verses contain interesting historical information: one provides us with a history of Maghribi involvement with the obliquity of the ecliptic and the theory of trepidation (see Section 2 above), as well as with parameters for morning and evening twilight. Others contain tables computed for the latitude of Tarudant and Tlemcen.

The early-thirteenth-century Maghribi savant Ibn Tūmart, author of the widely-used treatise *Kanz al-ʿulūm* on religion, law and the sciences, also dealt with astronomy at the popular level (see Vajda, 1957). He presented a shadow-scheme for time-keeping which is valid for the Yemen (then the centre of folk astronomy) rather than the Maghrib, as well as procedures for time-keeping by night using the lunar mansions.

An almanac of the same kind as the earlier and better-known *Calendar of Cordova* is attributed in several manuscripts to Ibn al-Bannā and was published by H. P.-J. Renaud (1948); more recently it has been studied by M. Forcada (1992). The simple tables in this work serve a latitude in Andalusia rather than Marrakesh, and the attribution should not be taken too seriously. A late Tunisian almanac has also been shown to contain Andalusian materials.

A fourteenth(?)-century writer Abū ʿAbdallāh al-Muʾaddib compiled a rather sophisticated treatise on folk astronomy which contains a few simple, but nevertheless historically interesting, tables. The treatise survives in a unique copy at Oxford. In the navigational atlas of the sixteenth-century scholar ʿAlī al-Sharafī of Sfax, extant in two fine copies in Oxford and Paris, there are likewise simple shadow-schemes for regulating the *ẓuhr* and *ʿaṣr* prayers, as well as a visually spectacular chart displaying the locations of various cities relative to the Kaʿba in Mecca – according to popular rather than precise geographical knowledge, and part of a tradition of Islamic sacred geography which started in the ninth century (see the article "Makka, iv: As centre of the world" in *EI*[2]).

H. P.-J. Renaud has surveyed various Maghribi treatises on folk astronomy for information on popular techniques of regulating the calendar and determining the *qibla*. A recent publication by the Maghribi scholar Muḥammad ibn ʿAbd al-Wahhāb on legal and astronomical aspects of lunar crescent visibility attests to continued local interest in this subject.

The first modern discussions of the often curious orientations of Maghribi mosques were by G. Sarton (1933) and H. Terrasse (1935), and the topic has been pursued more recently, and in greater depth, by M. Philibert (1972). However, the problems of mosque orientation (which are not restricted to the Maghrib) are mentioned already in various historical texts, and the answer to why there is such a problem is also found in textual sources. It was Renaud (1942a) who first considered these sources, albeit without relating the information they contain about finding the *qibla* by traditional procedures to the orientation of mosques. In this context the present writer has conducted an armchair survey of these orientations from the available literature on Maghribi religious architecture. In some cities of the Maghrib the orientations range from due east to due south, and to a certain extent these variations can now be explained.

More recently, M. Bonine (1990) has conducted the first field survey of the orientation of Maghribi cities and religious architecture relative to their topographical environment. His results can now be compared not with the modern *qibla* directions (which are irrelevant in any discussion of medieval orientations) but rather with the various *qibla* directions known from textual evidence to have been popular in medieval times amongst different interest groups.

Finally we should mention the lengthy poem entitled *al-Uqnūm* by ʿAbd al-Raḥmān al-Fāsī, which contains sections on various scientific subjects. This represents a particularly Maghribi trend to popularize astronomy – recall the poems of Ibn Abī 'l-Rijāl, al-Jādarī and al-Marghīthī, and also the versified *Zīj* of al-Qusanṭīnī. Unfortunately the trend also contributed to the stagnation of astronomy in the Maghrib. For by the nineteenth century it was mainly such works which were being memorized by those who were counted as astronomers.

7. Concluding Remarks

In 1932 H. P.-J. Renaud listed the various Maghribi astronomical and mathematical works which were published in lithograph in Morocco during the second half of the nineteenth century and the first few decades of the twentieth. When one knows what is in these works, the list is not impressive. Traditional astronomy had declined in the sense that the major Maghribi works had been forgotten and there was no-one left who could have understood them anyway. The situation was analogous in other regions of the Islamic world.

Now that we have better control over the sources in a way Delphin and Renaud could never have dreamed possible, what should we do with them? There are enough materials from the Maghrib to keep several researchers busy for many years. The *Zīj* of Ibn Isḥāq alone offers all manner of rewards. Only when it and the other sources I have mentioned have been investigated properly can we hope to gain a satisfactory overview of Maghribi activity in astronomy. In the mean-time it would be very convenient to have facsimile "editions" of the most important sources:

(1) the Hyderabad unicum of the *Zīj* of Ibn Isḥāq (publication would, I suspect, require international collaboration at a high level);

(2) the Escorial copy of the *zījes* of Ibn al-Bannā and al-Qusanṭīnī and Ibn Qunfudh's commentary on the astrological poem of Ibn Abī l-Rijāl;

(3) the Rabat manuscript of the *Qawīm Zīj* of Ibn al-Raqqām and the Istanbul fragments of his *Shāmil Zīj*;

(4) the Paris and Oxford manuscripts of al-Sharafī's navigational atlas; and

(5) the Oxford manuscript of Abū ʿAbdallāh al-Muʾaddib's treatise on folk astronomy.

Critical editions of various important texts are essential. Other worthwhile ventures could be detailed investigations of the surviving instruments of particular individuals such as Abū Bakr ibn Yūsuf and al-Baṭṭūṭī, or the preparation of a catalogue of Maghribi sundials. The analysis of the major Maghribi *zījes* has now begun, but it will be some time before we can assess them adequately. Until then it is hoped that by representing this preliminary survey some justice to a much-neglected scientific tradition has been done.

Primary sources – Manuscripts (selected)

Note: Not all available copies are listed; for more information the bio-bibliographical sources, such as Suter, Brockelmann, Renaud and King (*Cairo Survey*), should be consulted. The abbreviation MN refers to the Museo Naval in Madrid which houses an unnumbered manuscript, on the contents of which see Vernet, 1956 and 1980. The abbreviation ENL stands for the Egyptian National Library.

- Abū ᶜAbdallāh al-Muʾaddib, treatise on folk astronomy: Ms Oxford Pococke 249.
- Abū Miqraᶜ: see al-Marghīthī.
- Dūnaš ibn Tamīm, treatise on the armillary sphere: Ms Istanbul Aya Sofya 4861;
- Ḥusayn Quṣᶜa, recension of *Zīj* of Ulugh Beg: Ms Cairo ENL DM 814, *etc.*
- Ibn Abī 'l-Rijāl, *Kitāb al-Bāriᶜ*: Over 30 Mss are listed in Sezgin, VII, p. 187.
- Ibn Abī l-Shukr al-Maghribī, *Zīj* for Damascus: Ms Escorial ar. 932.
- Ibn al-Bannā, *Almanac*: published by Renaud, 1948.
- *Zīj*: Ms Escorial 9090, 1, Dublin Chester Beatty 4087, *etc.*
- *Yasāra* and Supplement: Ms London B. L. Add. 9599.
- Ibn Isḥāq, *Zīj*: Ms Hyderabad Andra Pradesh State Library 298.
- Ibn Qunfudh, *Zīj*: Ms Madrid MN.
- Commentary on Ibn Abī 'l-Rijāl's poem: Ms Escorial ar. 909, 4, *etc.*
- Ibn al-Raqqām, *al-Zīj al-Qawīm*: Ms Rabat Awqāf 260, 1, Madrid MN (fragments).
- *Al-Zīj al-Shāmil*: Ms Istanbul Kandilli 249.
- Treatise on sundial theory: Ms Escorial 918, 11 (published by Carandell, 1988).
- Ibn Tūmart, *Kanz al-ᶜulūm*: Ms Cairo ENL *falsafa* 411, *etc.*

- Al-Jādarī, poem on folk astronomy - commentaries: Mss London B. L. 411, 2, Cairo ENL K4311, *etc.*
- Al-Khalīlī, universal auxiliary tables (Maghribi copy): Ms London B. L. Add. 9599, 31.
- Al-Marrākushī, *Kitāb al-mabādi' wa-l-ghāyāt*: available in numerous copies (but none of Maghribi provenance ?)
- Al-Marghīthī, commentary on his poem on Abū Miqraᶜ: Ms Cairo ENL DM 415, *etc.*
- Al-Qusanṭīnī, *Zīj*: Ms Escorial 909, 2.
- Al-Ṣafāqusī = al-Sharafī, navigational atlas: Mss Paris B. N. ar. 2273, Oxford Bodleian Marsh 294.
- Anonymous (Algiers), prayers-tables: Ms Cairo ENL ṬFT 9, 1.
- Anonymous (Algiers), recension of the *Zīj* of Ibn al-Shāṭir: Ms Cairo ENL DM 533.
- Anonymous (Marrakesh ?), prayer-tables for the Maghrib: Ms Cairo ENL TR 338, 2.
- Anonymous (Sfax), treatise on spherical astronomy: Ms Cairo ENL K 7584, 1.
- Anonymous (Tunis), universal auxiliary tables for time-keeping by the sun: Ms Cairo ENL DM 689.
- Anonymous (Tunis), prayer-tables and tables for time-keeping by the sun and stars: Ms Berlin Ahlwardt 5724.

Primary sources – Instruments

Note: The following provisional list of Maghribi astronomical instruments is extracted from the table of contents of the catalogue of all medieval Islamic (and European) instruments currently in preparation in Frankfurt.[3] The order is roughly chronological, and some unsigned instruments may be Andalusian rather than

[3]. See King, "Medieval Astronomical Instruments: A Catalogue in Preparation", *Bulletin of the Scientific Instrument Society*, no. 31 (Dec., 1991), pp. 3-7 and "1992 – A Good Year for Medieval Astronomical Instruments", *ibid.*, no. 36 (March, 1993), pp. 17-18.

Maghribi. Information on later instruments is printed in a smaller font. The numbers preceded by the symbol # are those of the International Instrument Checklist. I have deliberately refrained from including the references to the catalogue sections where many of these instruments are already described because these are still not finalized. For abbreviations of museums the reader must consult the *Astrolabe Checklist* by Price *et al.*; PLU, however, stands for "present location unknown" and PC "private collection". The author will appreciate receiving from colleagues information on Maghribi instruments not recorded here.

Astrolabes

Two astrolabes by Ibrāhīm ibn ʿAbd al-Karīm of Fez

A mater dated 458 H (#3714 - Fez MA)

An undated astrolabe (#1079 - Palermo MN - stolen)

Seven astrolabes by Abū Bakr ibn Yūsuf of Marrakesh

A universal astrolabe in the tradition of ʿAlī ibn Khalaf dated 584 H mentioned in a medieval manuscript (#4039 - reference in Ms Cairo DM 291,1)

An astrolabe dated 605 H (#124 - Strasbourg OA)

An astrolabe dated 610 H (?) with later modifications dated 776 H (#2709 - Rabat KO)

An astrolabe dated 613 H (#1090 - Toulouse MPD)

An astrolabe dated 615 H (#125 - PLU, formerly Baron de Larrey)

An astrolabe reworked by a European with maker's name and date obliterated (#1069 - Madrid IVDJ)

A rete and set of plates (#1057 - London SM)

An astrolabe by al-Ḥusayn ibn ʿAlī, probably of Tunis, dated 709 H, with an Eastern Islamic replacement rete (#1204 - Cambridge WMHS)

An astrolabe by Muḥammad ibn Qāsim al-Qurṭubī of Fez dated 719 H (is this genuine?) (#4182 - PLU < Paris PC)

A mater and a universal myrtle astrolabe by ʿAlī ibn Ibrāhīm al-Jazzār, muezzin of Taza

A mater dated 724 H (#4041 - Geneva MHS)

A universal myrtle astrolabe dated 728 H (#3579 - Oxford MHS)

An astrolabe by Aḥmad ibn Muḥammad ibn Hārūn al-Hadamī of Marrakesh, dated 728 H, with a replacement rete (#1121 - Fez DB)

The astrolabic clock of Fez (#4042 - Fez QM)

An unsigned astrolabe in the tradition of Abū Bakr ibn Yūsuf dated 779 H (#1061 - Rome OA - stolen in 1984)

Unsigned, undated astrolabes in the tradition of al-Khamāʾirī of Seville (*ca.* 1215)

An astrolabe with a *ṣafīḥa shakkāziyya* on the back (#4044 - Rockford TM)

An astrolabe with a *ṣafīḥa shakkāziyya* on the back (#139 - Oxford MHS)

"Marcel's Egyptian astrolabe" (#1160 - PLU)

(#4045 - Bouillon MB)

An unsigned Andalusian/Maghribi astrolabe - "The Imola astrolabe" (#4184 - Bologna)

"Dr. Knuthsen's astrolabe" (#3640 - PLU)

With a replacement throne and Eastern Islamic rete (#3643 - Washington NMAH)

An unsigned Maghribi astrolabe (#4046 - Geneva MAH)

(#3601 or #3602 - Haifa NMM)

(#4047 - Point Lookout LC)

(#4048 - Halle DMG)

John Selden's Maghribi astrolabe (#149 - Oxford MHS)

(#1001 - Oxford MHS)

(#1067 - Madrid MA)

(#1066 - Madrid MN)

(#151 = #1080 - Genoa SLSP)

An astrolabe with eight plates serving latitudes from the Yemen to Provence (#4051 - Istanbul DM)

A Maghribi astrolabe for latitudes in the Yemen (#4170 - Cambridge, Mass. HU)

Three other undated, unsigned astrolabes not in the tradition of al-Khamāʾirī of Seville

With a *shakkāziyya* grid on the back (#1078 - Florence MSS)

An astrolabe with a "counter-ecliptic" frame on the rete (#1068 - Madrid IVDJ)

"Dorn's Aleppo astrolabe" (#126 - PLU)

An astrolabe by Aḥmad ibn ʿUmar al-Kabbī dated 933 H (#1188 - Cambridge WMHS)

An astrolabe by ʿAlī and Muḥammad al-Azdī dated 950 H ("the Valdagno astrolabe") (#148 - Rome OA - stolen)

An astrolabe by Yaʿqūb ibn Mūsā Tapiero of Fez supposedly dated 716 H but (more probably) 1016 H (#1122 - Morocco PC)

Miscellaneous bits and pieces

A single plate (#154 - Chicago AP)

Three astrolabes by al-Ḥasan ibn Aḥmad al-Baṭṭūṭī (precise location uncertain)

Dated 1097 H (#2707 - Point Lookout LC)

Dated 1103 H (#2568 - Washington NMAH)

Dated 1106 H (#3804 - Cairo MFI)

Some astrolabes by Muḥammad ibn Aḥmad al- Baṭṭūṭī (precise location uncertain) (more are known than are listed here)

Dated 1128 H (#3663 - Point Lookout LC)

Dated 1133 H (#2559 - Chicago AP)

Dated 1134 H (#1041 - Greenwich NMM)

Dated 1136 H (#1062 - PLU)

Dated 1138 H (#1143 - Fez MM)

Dated 1141 H (#1129 - Fez DM)

Dated 1146 H (#1023 - Oxford MHS)

Dated 1148 H (#1210 - Paris PC)

Dated 1151 H (#1027 - Greenwich NMM)

Undated (#1024 - Oxford MHS)

An unsigned Maghribi astrolabe with a transversal alidade (#4049 - London SM)

Two astrolabes by Sharaf Allāh of Essaouira

Undated and dedicated to the Sharīf Muḥammad ibn ʿAbdallāh (#3544 - Paris IMA)

Unsigned and dated 1179 H (#1070 - Madrid IGE)

Some late, signed and dated Maghribi astrolabes

An astrolabe by ʿAbdallāh ibn Sāsī dated 1099 H (#145 - Oxford MHS)

An astrolabe by Muḥammad ibn Aḥmad ibn Ibrāhīm of Fez, dated 1104 H (#3709 - PLU)

An astrolabe by Aḥmad ibn Muḥammad ibn Ibrāhīm of Fez dated 1116 H (#1162 - PLU)

An undated (?) astrolabe by Aḥmad ibn Ibrāhīm al-Jazzār (#3547 - Tunis MB)

An astrolabe by Qāsim ibn ʿAbd al-Salām al-Tamlī of Rabat in 1192 H (#1180 - Tetuan)

An astrolabe by Muḥammad ibn ʿAlī al-Ghazāwī dated 1314 H (#1181 - PLU)

Three astrolabes by Muḥammad ibn al-Mufaḍḍal ibn Aḥmad ibn Kīrān

Dated 1126 H (#3918 - Rockford TM)

Dated 1264 H (#4134 - Chicago AP)

Unsigned and undated (#155 - Oxford MHS)

Some late, unsigned Maghribi astrolabes

"Cowson" (#4185 - PLU)

(#4186 - Madrid MN)

(#4187 - Madrid MAN)

"Dorn II" (#3671 - St. Petersburg MVLM)

(#143 - Oxford MHS)

(#133 - London BM)

(#4188 - Rockford TM)

(#4135 - Paris ML)

(#4139 - PC)

An undated astrolabe with a European-type rete (#1034 - Greenwich NMM)

(#1048 - Greenwich NMM) (repeated ?)

A degenerate astrolabe bearing the date 1184 H (#1046 - Greenwich NMM)

(#4136 - Amsterdam SVM)

(#4189 - Cambridge WMHS)

A composite astrolabe (#1058 - London SM)

(#1083 - Paris MM)

Brussels ? (#4190 - PLU)

Two unsigned astrolabic plates with incomplete retes

(#4191 - London IO)

(#4137 - PC)

Miscellaneous bits and pieces

Two plates (#4192 - Bologna OA)

A single plate (#4193 - Paris IMA)

Two plates (#4194 - Haifa NMM)

A rete (#3539 = #3805 - Paris IMA)

An Ibn Bāṣo-type plate (#4195 - PLU)

A single plate (#3663 - Point Lookout LC)

A mater with a *ṣafīḥa shakkāziyya* on the back made to accommodate a rete and set of plates by [Abū Bakr ibn Yūsuf] (#1057 - London SM)

An astrolabe made in France in 1789 with Arabic inscriptions (#4197 - Fez MA)

A Maghribi astrolabe made by Oskar Heide (?) in Tunis [for Prof. Eilhard Wiedemann] (#4196 - Munich DM)

A fake astrolabe bearing an inscription mentioning Ibn Baṭṭūṭa and a date 732 H (#9021 - PC)

Quadrants

An astrolabic quadrant for [Tunis] by Aḥmad ibn ʿAbd al-Raḥmān al-Daḥmanī, dated 854 H (#5021 - Madrid MAN)

An unusual quadrant for [Marrakesh] (#5022 - Chicago AP)

An astrolabic / sine quadrant for Meknes by al-Baṭṭūṭī (#5023 - Tetuan)

Some typical late, attributed quadrants bearing astrolabic markings on one side and a sexagesimal sine grid on the other

For Meknes by Aḥmad ibn ʿAlī al-Andalusī, dated 1219 H (#5024 = #3701 - Oxford MHS)

For Oran (#5025 - PLU)

Some unsigned, undated quadrants

(#5026 - Chicago AP)

For Meknes (#5027 - PLU - London Christie's 1991)

For latitude 37°, Tunis (#5028 - Paris BN)

An altitude quadrant by ʿAbd al-Salām al-ʿAlamī of Fez, dated 1291 H (#5029 - Cairo ENL)

Two unsigned sine quadrants

Dated 1280 H (#5030 - Paris IMA)

Undated (#5031 - Paris IMA)

Sundials

A sundial for Tunis by al-Qāsim ibn Ḥasan ibn al-Shaddād dated 746 H (#7341 - Carthage MN)

A sundial from al-Mansura (#7345 - al-Mansura)

A column sundial by Aḥmad ibn Muḥammad al-Lamṭī dated 747 H (#7342 - Tlemcen SHM)

A sundial for Kairouan (#7343 - Kairouan)

A sundial from an oasis in Southern Tunisia (#7344)

Miscellaneous

Fragments of a spherical astrolabe for the latitude of [Tunis] (#8002 - Milan PC)

A late Maghribi nocturnal (#8091 - Oxford MHS)

Major Bibliographical Works on Islamic Astronomy

[Abbreviation: IGAIW= Institut für Geschichte der Arabischen-Islamischen Wissenschaften, Frankfurt am Main]

Astrolabe Checklist: S. Gibbs, J.A. Henderson and D. de Solla Price, *A Computerized Checklist of Astrolabes*, New Haven, Ct.: Yale University, Department of History of Science and Medicine, 1973.

A. El-Azzawi, *History of Astronomy in Iraq and its Relations with Islamic and Arab Countries in the Post-Abbasid Periods* (in Arabic), Baghdad: Iraq Academy Press, 1959. [See especially pp. 209-221 and 330-337 on Maghribi astronomers].

A. Brieux and F. Maddison, *Répertoire des facteurs d'astrolabes et de leurs œuvres*, Pt. I: *Islam*, as yet unpublished. [Supplements the work of Mayer with considerable new information.]

C. Brockelman, *Geschichte der arabischen Litteratur*, 2 vols., 2nd ed., Leiden: E. J. Brill, 1943-49, and Supplementbände, 3 vols., Leiden: E. J. Brill, 1937-42. [See especially II, pp. 330-332, 615-616, and SII, pp. 363-365, 707-709, on Maghribi astronomers and their works].

R.T. Gunther, *The Astrolabes of the World*, 2 vols., Oxford: Oxford University Press, 1932, reprinted in 1 vol., London: Holland

Press, 1976. [See especially vol. I, pp. 248-301 on Maghribi astrolabes].

E.S. Kennedy, "A Survey of Islamic Astronomical Tables", *Transactions of the American Philosophical Society*, N. S., 46:2 (1956), pp. 123-177, reprinted *ca.* 1990. [Contains information on about 120 *zījes*].

D.A. King, *A Survey of the Scientific Manuscripts in the Egyptian National Library*, (*Publications of the American Research Center in Egypt*), Winona Lake, Indiana: Eisenbraus, 1987. [See especially Section F on Maghribi and Andalusian astronomers].

M. Krause, "Stambuler Handschriften islamischer Mathematiker", *Quellen und Studien zur Geschichte der Mathematik Astronomie und Physik*, Abt. B: Studien, Band 3, Heft 4, (1936), pp. 437-532. [See especially n° 363 for manuscripts of al-Marrākushī's *magnum opus*].

L.A. Mayer, *Islamic Astrolabists and Their Works*, Geneva: Albert Kundig, 1956, with a supplement in R. Ettinghausen, ed., *Aus der Welt der islamichen Kunst*, Berlin: Gebr. Mann, 1959, pp. 293-296. [The fundamental work on the subject].

Price: See *Astrolabe Cheklist.*

Henri P.-J. Renaud, "Additions et corrections à Suter *Die Mathematiker und Astronomen der Araber*", *Isis*, 18 (1932), pp. 166-183. [Based mainly on manuscripts in El Escorial and Rabat].

H.P.-J. Renaud, *Les manuscrits arabes de l'Escorial*, tome II, fasc. 3: *Sciences exactes et sciences occultes*, Paris: Paul Geuthner, 1941.

H.P.-J. Renaud, "Quelques constructeurs d'astrolabes en Occident Musulman", *Isis* 34 (1942), pp. 20-23. [Corrections to Gunther and some new information].

A. Sayılı, *The Observatory in Islam*, (Publications of the Turkish Historical Society, Series VII, no. 38), Ankara, 1960, reprinted New York, N. Y.: Arno Press, 1981. [The fundamental work on the subject].

F. Sezgin, *Geschichte des arabischen Schriftums* (hereafter GAS), Band V: *Mathematik*, Band VI: *Astronomie*, Band VII:

Astrologie, Meterologie und Verwandtes, Leiden: E. J. Brill, 1974, 1978 and 1979. [The major bio-bibliographical reference work for authors before *ca.* 1050].

H. Suter, "Die Mathematiker und Astronomen der Araber und ihre Werke", *Abhandlungen zur Geschichte der mathematischen Wissenchaften*, 10 (1900), and "Nachträge une Berichtigungen", *ibid.*, 14(1902), pp. 157-185, reprinted Amsterdam: Oriental Press, 1982, and again in his *Beiträge* (see below). [The major reference work on Islamic astronomy up to the XVIth century, replaced by Sezgin, *GAS*, for the early period, supplemented by King, *Cairo Survey*, for the later period].

Major Books and Collections of Articles on Islamic Astronomy

J.L. Berggren, *Episodes in the Mathematics of Medieval Islam*, New York, *etc.*: Springer, 1986. [The first overview of the history of mathematics in the Islamic world to consider some astronomical material].

B.R. Goldstein, *Theory and Observation in Ancient and Medieval Astronomy*, London: Variorum Reprints, 1985. [Reprints of numerous articles].

W. Hartner, *Oriens-Occidens*, and *Oriens-occidens II*, Hildesheim: Georg Olms, 1968 and 1984. [Reprints of numerous articles, many dealing with Islamic astronomy.]

Hartner Festschrift: W. Saltzer and Y. Mayema, eds., *ΠΡΙΣΜΑΤΑ. Naturwissenschaftsgeschichtliche Studien - Festschrift für Willy Hartner*, Wiesbaden: Franz Steiner, 1977. [Contains several articles on Islamic astronomy].

Kennedy *et al.*: E.S. Kennedy, Colleagues and Former Students, *Studies in the Islamic Exact Sciences*, Beirut: American University of Beirut, 1983. [Reprints of numerous articles on all aspects of Islamic astronomy].

Kennedy Festschrift: D.A. King and G. Saliba, eds., *From Deferent to Equant: Studies in the History of History of Science in the Ancient and Medieval Near East in Honor of E. S. Kennedy*, Annals of the New York Academy of Science, 500 (1983). [Contains 35 contributions, several dealing with Islamic astronomy].

D.A. King, *Islamic Mathematical Astronomy*, London: Variorum Reprints, 1987. Abbreviated *IMA*. [Reprints of various articles].

D.A. King, *Islamic Astronomical Instruments*, London: Variorum Reprints, 1987. Abbreviated *IAI*. [Reprints of various articles].

D.A. King, *Astronomy in the Service of Islam*, Aldershot: Variorum Reprints, 1993. Abbreviated *ASI*. [Reprints of various articles].

D.A. King, Chapter "Islamic Astronomy", in C. Walker, ed., *Astronomy before the Telescope*, London: The British Museum, 1996, pp. 143-174.

J. Millás Vallicrosa, *Estudios sobre historia de la ciencia española*, and *Nuevos estudios sobre historia de la ciencia española*, Barcelona: Consejo Superior de Investigaciones Cientificas, 1949 and 1960 [Numerous articles].

C.A. Nallino, *Raccolta di scritti editi e inediti*, vol. V: Astrologia-Astronomia-Geografia, Rome: Istituto per l'Oriente, 1944. [Includes very valuable sections on folk astronomy].

D. Pingree, *Collected papers* to be published by Variorum Reprints, Aldershot.

J. Samsó, *Las ciencias de los antiguos en al-Andalus*, Madrid: Mapfre, 1992.

J. Samsó, *Islamic Astronomy and Medieval Spain*, Aldershot : Variorum Reprints, 1994.

G. Sarton, *An introduction to the History of Science*, 3 vols., Md.: Williams and Wilkins Co. for the Carnegie Institution, Washington, D.C., 1927. [Contains useful information, but should be used with some caution].

E. Savage-Smith, *Islamicate Celestial Globes: Their History, Construction, and Use*, Washington, D.C.: Smithsonian Institution Press, 1985.

C. Schoy, *Beiträge zur arabisch-islamichen Mathematik und Astronomie*, 2 vols., Frankfurt am Main: IGAIW, 1988. [Reprints of many monographs and articles].

H. Suter, *Beiträge zur Geschichte der Mathematik und Astronomie im Islam*, 2 vols., Frankfurt am Main: IGAIW, 1986. [Reprints of all monographs and articles].

J. Vernet, *Estudios sobre historia de la ciencia medieval*, Barcelona: Universidad de Barcelona & Bellaterra: Universidad Autónoma de Barcelona, 1979. [Reprints of numerous articles, several dealing with astronomy in Andalusia and the Maghrib].

Vernet Festschrift: Josep Casulleras and Julio Samsó, eds., *From Baghdad to Barcelona. Studies in the Islamic Exact Sciences in Honour of Prof. Juan Vernet*, 2 vols., Barcelona: Instituto "Millás Vallicrosa" de Historia de la Ciencia Arabe, 1996.

[J. Vernet ed.], *Historia de la ciencia arabe*, Madrid: Real Academia de Ciencias Exactas, Fisicas y Naturales, 1981. [Contains some important contributions].

[J. Vernet, J. Samsó, *et al.*], *El legado cientifico andalusi*, Madrid: Ministerio de Cultura, 1992. [Catalogue of an exhibition held at the Museo Arqueológico Nacional, Madrid, during April-June 1992].

E. Wiedemann, *Aufsätze zur arabischen Wissenschaftsgeschichte*, 2 vols., Hildesheim: G. Olms, 1970. [Numerous articles originally published in the *Sitzungsberichte der physikalisch-medizinischen Sozietät zu Erlangen*, 1902-1928.]

E. Wiedemann, *Gesammelte Schriften zur arabisch-islamischen Wissenschaftsgeschichte*, 3 vols., Frankfurt am Main: IGAIW, 1984. [Numerous articles].

Also articles in *EI*[1]: *Encyclopaedia of Islam*, 1st ed., 4 vols., Leiden: E. J. Brill, 1913-1934, *EI*[2]: *Encyclopaedia of Islam*, new ed., 9 vols. to date, Leiden: E. J. Brill, 1960 to present (especially articles 'Anwāʾ', 'Ibn al-Bannā', 'ʿIlm al-hayʾa', 'Ḳibla', 'al-Maghrib' (vii: Intellectual life; especially vol. V, p. 1209, where al-Marrākushī and Ibn al-Bannā are mentioned), 'Makka' (iv: as centre of the world), 'al-Marrākushī', 'Mīḳāt', 'Mizwala', 'Nudjūm', 'Rīḥ', 'Rubʿ', 'Ruʾyat al-hilāl' and 'Samt'. And *DSB*: *Dictionary of Scientific Biography*, New York: Charles Scribner's Sons, 16 vols., 1970-1980 (especially the article 'Ibn al-Bannā').

Published Material on the History of Astronomy in the Maghrib

M. Arribas Palau, "Un cuadrante árabe de Marruecos", *Al-Andalus*, 30 (1965), pp. 309-317.

A. Bel, "Un cadran solaire arabe", *Revue africaine*, 49 (1905), pp. 228-231. [Describes a cylindrical sundial discovered in al-Manṣūra].

M. E. Bonine, "The Sacred Direction and City Structure: A Preliminary Analysis of the Islamic Cities of Morocco", *Muqarnas: An Annual on Islamic Art and Architecture*, 7 (1990), pp. 50-72.

J. Carandell, "An Analemma for the Determination of the Azimuth of the *Qibla* in the *Risāla fī ʿilm al-ẓilāl* of Ibn al-Raqqām", *Zeitschrift für Geschichte der arabisch-islamischen Wissenschaften*, 1 (1984), pp. 61-72.

J. Carandell, *Risāla fī ʿilm al-ẓilāl de Muḥammad Ibn al-Raqqām al-Andalusī*, Barcelona: Universidad de Barcelona, Instituto "Millás Vallicrosa" de Historia de la Ciencia Árabe, 1988. [Text, translation and commentary on the most significant Maghribi treatise on sundials].

G.S. Colin, "Un juif marocain du XIVe siècle, constructeur d'astrolabe", *Hespéris*, 22 (1936), pp. 183-184. [Minimal details of an astrolabe made in Fez].

G.S. Colin & H.P.-J. Renaud, "Note sur le *muwaqqit* marocain Abū Muqriʿ – ou mieux Abū Miqraʿ – al-Baṭṭīwī (XIIIe s. J. C.)", *Hespéris*, 25 (1933), pp. 94-96.

M. Delphin, "L'astronomie au Maroc", *Journal Asiatique*, 8ème série, 17 (1891), pp. 177-201. [Deals only with a late XVIIIth-century "astrolabe" and the titles of some simple treatises].

M. Forcada Nogués, "Les sources andalouses du calendrier d'Ibn al-Bannā", *Actas del II Coloquio hispano-marroquí de ciencias históricas, "Historia, Ciencia y Sociedad", Granada, 6-10 noviembre, 1989*, Madrid: Instituto de Cooperación con el Mundo Árabe, 1992, pp. 183-196.

H. Goerseke, "Ein maghribinisches Astrolab aus dem Besitz der Bibliotek der Deutschen Morgenländischen Gesellschaft in Halle (Saale)", in *Studia Orientalia. In memoriam Caroli Brockelmann*, Halle, 1968, pp. 87-93. [A brief description of a fine anonymous (XVth-century ?) Maghribi astrolabe].

G. Hoest, *Nachrichten von Marokos und Fes im Lande selbst gesammelt, in den Jahren 1760 bis 1768*, Copenhagen: Christian Gottlob Prost, 1781. [Contains a popular calendar (pp. 250-258) but disappointingly little information on the activities of contemporary muezzins (p. 211)].

J.P. Hogendijk, "Three Islamic Lunar Crescent Visibility Tables", *Journal for the History of Astronomy*, 19 (1988), pp. 29-44. [Deals with the table of al-Qallās].

M. Hosotte-Reynaud, "Identification d'un des astrolabes de la Mosquée des Andalous", *Hespéris*, 44 (1957), and 3 plates. [Deals with an astrolabe with Arabic inscriptions made in France in 1789].

L. Janin, "Quelques aspects récents de la gnomonique tunisienne", *Revue de l'Occident musulman et de la Méditérranée*, 24 (1977), pp. 207-221.

Al-Jaznāʾī, *Zahrat al-ās*, ed. A. Bel, Algiers: Editions Jules Carbonel, 1923. [See especially pp. 38-39 (Arabic) and pp. 92-93 (trans.) on the Fez clocks].

B. Joly, "Un calendrier agricole marocain", *Archives Marocaines*, 3 (1905), pp. 301-319.

Justinard: see al-Zarhūnī below.

E.S. and M.H. Kennedy, *Geographical Coordinates of Localities from Islamic Sources*, Frankfurt am Main: IGAIW, 1987. [A computerized sorting of some 10,000 sets of coordinates according to locality, source, latitude and longitude; including those of al-Marrākushī and Ibn al-Bannā].

E.S. Kennedy and D.A. King, "Indian Astronomy in Fourteenth-Century Fez: The Versified *Zīj* of al-Qusanṭīnī", *Journal for the History of Arabic Science*, 6 (1982), pp. 3-45. [Reprinted in King, *IMA*, VIII].

D.A. King, "A Fourteenth-Century Tunisian Sundial for Regulating the Times of Muslim Prayer", *Hartner Festschrift* (1977), pp. 187-202. [Reprinted in *idem*, *IAI*, XVIII].

D.A. King, "On the Early History of the Universal Astrolabe in Islam and the Origin of the Term *Shakkāziya* in Medieval Scientific Arabic", *Journal for the History of Arabic Science, 3:2* (1979), pp. 244-257; reprinted in *idem*, *IAI*, VII. [Based partly on Andalusian materials].

D.A. King, "Some Early Islamic Tables for Determining Lunar Crescent Visibility", in *Kennedy Festschrift* (1983), pp. 185-225; reprinted in King, *ASI*, II. [Includes material from the *Zīj* of Ibn Isḥāq – now supplemented by Hogendijk (1988) – and the treatise on folk astronomy by Abū ʿAbdallāh al-Muʾaddib].

D.A. King, *Mathematical Astronomy in Medieval Yemen: A Bio-Bibliographical Survey*, Malibu, Ca.: Undena Publications, 1983. [Contains references to Yemeni fragments of the *Zīj* of Ibn Isḥāq].

D.A. King, "The Astronomy of the Mamluks", *Isis*, 74 (1984), pp. 531-555; reprinted in *idem*, *IMA*, III. [A survey of the astronomical tradition in Egypt and Syria, in which al-Marrākushī, Ibn al-Shāṭir and al-Khalīlī played major roles].

D.A. King, "A Survey of Medieval Islamic Shadow-Schemes for Simple Time-Reckoning", *Oriens*, 32 (1990), pp. 191-249. [Deals with numerous Maghribi schemes].

D.A. King, "The Orientation of Medieval Islamic Religious Architecture and Cities", *Journal for the History of Astronomy*, 26 (1995), pp. 253-274.

D.A. King, "Studies on Fake Astronomical Instruments. I: An Astrolabe for Ibn Baṭṭūṭa to take around the world", to appear in the *Bulletin of the Scientific Instrument Society*.

W. & G. Marçais, *Monuments arabes de Tlemcen*, Paris: Ancienne Librairie Thorin et fils, 1903. [Describes (pp. 291-293) a cylindrical sundial from Tlemcen].

J. Millás Vallicrosa, "Un ejemplar de azafea árabe de Azarquiel", *Al-Andalus*, 9 (1944), pp. 111-119.

J. Millás Vallicrosa, "Tres instrumentos astronómicos árabes de los museos de Tetuán y Madrid", *Al-Andalus*, 12 (1947), pp. 49-64.

F. Moreau & G. Ryckmans, "Un gnomon arabe du XIVème siècle", *Le Muséon*, 39 (1926), pp. 33-40. [Superseded by King, "Tunisian Sundial"].

A. de Motylinski, *Les mansions lunaires des Arabes*, Algiers, 1899. [Contains text and annotated translation of the poem on folk astronomy by Abū Miqraᶜ]. (Not consulted).

Muḥammad ibn ᶜAbd al Wahhāb ibn ᶜAbd al-Rāziq al-Marrākushī, *al-ᶜAdhb al-zulāl fi mabāḥith ruʾyat al-hilāl*, Qatar: Idārat al-shuʾūn al-dīniyya, 1397 Hijra = 1977. [A remarkable collection of medieval sources, which awaits proper analysis].

M. Philibert, *La qibla et le miḥrab*, Algiers (privately distributed), 1972.

D.J. de Solla Price, "Mechanical Water Clocks of the XIVth Century in Fez, Morocco", *Proceedings of the Tenth International Congress for the History of Science (Ithaca, N. Y., 1962)*, pp. 599-602.

H.P.-J. Renaud, "L'enseignement des sciences exactes et l'édition d'ouvrages scientifiques au Maroc avant l'occupation européenne", *Hespéris*, 14 (1932), pp. 78-89.

H.P.-J. Renaud, "Un prétendu catalogue de la Bibliothèque de la Grande Mosquée de Fès", *Hespéris*, 18 (1934), pp. 76-99. [Contains a handlist of the scientific manuscripts].

H.P.-J. Renaud, "Les Ibn Bāṣo", *Hespéris*, 24 (1937), pp. 1-12. [On the family of an Andalusian instrument designer also known in the Maghrib].

H.P.-J. Renaud, "Ibn al-Bannā de Marrakech: sufi et mathématicien", *Hespéris*, 25 (1938), pp. 13-42.

H.P.-J. Renaud, "Sur une tablette d'astrolabe appartenant à M. H. Terrasse", *Hespéris*, 26 (1939), pp. 157-169. [Deals with an instrument of Maghribi provenance].

H.P.-J. Renaud, "Histoire des sciences chez les Musulmans – Déterminations marocaines de l'obliquité de l'écliptique",

Bulletin de l'Enseignement Public (Morocco), no. 170, 28ème année, octobre-décembre 1941, pp. 321-335.

H.P.-J. Renaud, "Astronomie et astrologie marocaines", *Hespéris*, 29 (1942), pp. 41-63. [Deals mainly with relevant material from the *Riḥla* of the Marabout of Tasaft (*ca.* 1715) and from various popular treatises on the *qibla*].

H.P.-J. Renaud, "Sur les lunes du Ramadan", *Hespéris*, 32 (1945), pp. 51-68. [Based mainly on Maghribi sources].

H.P.-J. Renaud, "L'origine du mot *almanac*", *Isis*, 37 (1947), pp. 44-46. [Based on Maghribi sources].

H.P.-J. Renaud, *Le calendrier d'Ibn al-Bannā de Marrakech (1256-1321 J. C.)*, (Publications de l'Institut des Hautes-Etudes Marocaines, tome 34), Paris: Larose, 1948.

P. Ricard, "L'Horloge de la Médersa Bou-Anania de Fès", *Bulletin de la Société de Géographie d'Alger et de l'Afrique du Nord*, 25 (1924), pp. 248-254.

L. Richter-Bernburg, "Sāᶜid, the *Toledan Tables*, and Andalusi Science", in *Kennedy Festschrift*, pp. 373-401. [A very important study].

F. Rosenthal, *Ibn Khaldūn: The Muqaddimah –An Introduction to History*, 2nd ed., 3 vols., Princeton, N. J.: Princeton University Press, 1967. [See especially vol. III, pp. 133-137, on astronomy and pp. 385-392 for his refutation of astrology].

J. Samsó, "Nota acerca de cinco manuscritos sobre astrolabio", *Al-Andalus*, 31 (1966), pp. 385-392. [Based on manuscripts preserved in Tunis].

J. Samsó, "A propos de quelques manuscrits astronomiques des bibliothèques de Tunis: Contribution à une étude de l'astrolabe dans l'Espagne musulmane", *Il Coloquio Hispano-Tunecino*, n. d., pp. 171-190.

J. Samsó, "Un calendrier tunisien –d'origine andalouse ?– du XIXème siècle", *Cahiers de Tunisie*, 26 (1978), pp. 67-84.

J. Samsó, "La tradición clásica en los calendarios agricolas hispano-árabes y norteafricanos", *Segundo Congreso*

Internacional de Estudios sobre las Culturas del Mediterráneo Occidental, (Barcelona, 1978), pp. 177-186.

J. Samsó, "Azarquiel e Ibn al-Bannā", *Relaciones de la Peninsula Iberica con el Magreb (siglos XIII-XVI)*, Madrid, 1988, pp. 361-375.

J. Samsó and H. Mielgo, "Ibn Isḥāq al-Tūnisī and Ibn Muᶜādh al-Jayyānī on the *Qibla*", in Samsó, *Islamic Astronomy and Medieval Spain*, Aldershot: Variorum Reprints, 1994, VI.

J. Samsó and E. Millás, "Ibn al-Bannā, Ibn Isḥāq and Ibn al-Zarqālluh's Solar Theory", in Samsó, *Islamic Astronomy and Medieval Spain*, Aldershot : Variorum Reprints, 1994, X.

F. Sarrus, "Description d'un astrolabe construit à Maroc en l'an 1208", *Mémoires de la Société du Musée d'Histoire Naturelle de Strasbourg*, 1853, tome 4, pp. 1-32, pls.1-6. [Remarkable description of an astrolabe by Abū Bakr ibn Yūsuf].

G. Sarton, "Orientation of the *miḥrāb* in mosques", *Isis*, 20 (1933), pp. 262-264. [Deals with the problem in the Maghrib –see also Terrasse below].

J.-J. Sédillot, *Traité des instruments astronomiques des Arabes composé au treizième siècle par Aboul Hassan Ali de Maroc...*, 2 vols., Paris: Imprimerie Royale, 1834-1835; reprinted Frankfurt am Main: IGAIW, 1985. [A translation of the first half of al-Marrākushī's treatise].

L.A. Sédillot, "Mémoire sur les instruments astronomiques des Arabes", *Mémoires de l'Académie Royale des Inscriptions et Belles-Lettres de l'Institut de France*, 1 (1844), pp. 1-229; reprinted Frankfurt am Main: IGAIW, 1989. [Summarizes the second half of al-Marrākushī's treatise].

M. Souissi, "Présentation et analyse du traité *Somme des principes et des conclusions* par le savant marocain Al-Ḥassan al Marrākušī (était vivant en 1281)", *Cahiers de Tunisie*, 30 (1982), and an Arabic version in *Majallat Maᶜhad al-Makhṭūṭāt al-ᶜArabiyya*, [N. S.], 1:1 (1982), pp. 63-71. [Brief discussions, dealing mainly with al-Marrākushī's geographical table].

S. M. Stern, "A Treatise on the Armillary Sphere by Dūnaš ibn Tamīm", in *Homenaje a Millás Vallicrosa*, 2 vols., Barcelona: Consejo Superior de Investigaciones Cientificas, 1956, vol. 2, pp. 373-382; reprinted in *idem*, *Medieval Arabic and Hebrew Thought* (edited by F. W. Zimmerman), London: Variorum Reprints, 1983, XII.

H. Terrasse, Response to Sarton [see above], *Isis*, 24 (1935), pp. 109-110.

G. J. Toomer, "The Solar Theory of al-Zarqāl: A History of Errors", *Centaurus*, 14 (1969), pp. 306-336. [Demonstrates the importance of Maghribi sources for the history of Andalusian Astronomy].

G. Vajda, "Le commentaire kairouanais sur le *Livre de la Création*", *Revue des Etudes Juives*, 107 (1946-1947), pp. 99-156, and subsequent volumes. [Mentions the astronomical writings of Dūnaš ibn Tamīm].

G. Vajda, "Une synthèse peu connue de la révélation et de la philosophie: le *Kanz al-ᶜulūm* de Muḥammad b. ᶜAlī Ibn Tūmart al-Andalusī", in *Mélanges L. Massignon*, III, Damascus, 1957, pp. 359-374.

J. Vernet, *Contribución al estudio de la labor astrónomica de Ibn al-Bannā*, Tetuan: Editora Marroquí, 1952. [Contains text and translation to the introduction of his *zīj*].

J. Vernet, "Los manuscritos astrónomicos de Ibn al-Bannā", *Actes du VIIIème Congrès International d'Histoire des Sciences (Florence, 1956)*, pp. 297-298.

J. Vernet, "L'astronomie dans l'Islam occidental", *Archives Internationales d'Histoire des Sciences*, 16 (1963), pp. 225-240. [An overview].

J. Vernet, "La supervivencia de la astronomia de Ibn al-Bannā", *al-Qantara*, (Madrid), 1 (1980), pp. 447-451. [Description of an important Maghribi manuscript in the Museo Naval de Madrid].

[J. Vernet, J. Samsó, *et al.*], *Instrumentos astrónomicos en la España medieval*, Santa Cruz de la Palma, 1985. [A catalogue of an important exhibition; in Spanish and English].

Al-Zarhuni, *Riḥla*: Col. Justinard, transl., *La Riḥla du Marabout de Tasaft Sidi Mohammed ... ez Zerhouni*, Paris: P. Geuthner, 1940. [Discusses *inter alia* the problems of mosque orientation in the Maghrib].

Note: Other articles and books by J Millás, J. Samsó and J. Vernet dealing with astronomy in Muslim Spain are not listed here but should also be consulted: see the two new books by Samsó for references.

VII

A Hellenistic Astrological Table deemed Worthy of being Penned in Gold Ink: The Arabic Tradition of Vettius Valens' Auxiliary Function for Finding the Length of Life*

To the memory of David Pingree

1. Vettius Valens' table for determining the length of life

A basic concern of classical astrology was the prediction of the circumstances of an individual's life from the celestial configuration at the time of birth, and the prediction of his appearance and characteristics from the celestial configuration at the time of conception. The time of birth can be measured and the instantaneous celestial configurations relative to the local horizon determined by mathematical procedures; the time of conception is in general somewhat more tricky for an outsider to pinpoint, but rules were formulated to determine it from the time of birth and the period of gestation.

One aspect of the individual's circumstances which was of particular interest is the length of his life,[1] and already in classical Antiquity it was proposed that the maximum length of life in years for a specific location is the maximum rising time of a quadrant of the ecliptic over the local horizon, that is, one-half of the maximum length of daylight in equatorial degrees.[2] The late-2nd-century astrologer Vettius Valens, of Antioch and, later, Alexandria, contributed to this doctrine a scheme relating the length of an individual's life, subject to this condition for the maximum, to the longitude of the horoscopus or ascendant, that is, the point of the ecliptic instantaneously rising over the eastern horizon, at the time of birth.

The *Anthology* of Vettius Valens was very popular in late Antiquity, not least for the obscurity of its formulations.[3] The available Greek text in nine books is in bad shape; it was published by Wilhelm Kroll in 1908 on the basis of three Byzantine manuscripts (four others are derivative from these). Over 100 horoscopes in the *Anthology* were

* This study was first published in *Studies in the History of the Exact Sciences in Honour of David Pingree*, edited by Charles Burnett, Jan Hogendijk, Kim Plofker and Michio Yano, Leiden & Boston: Brill, 2004, pp. 666–714. It has here been reset, and the bibliography has been updated.

[1] A brief summary of this paper was published as 'Some Arabic Copies of Vettius Valens' Table for Calculating the Duration of Life', in *Symposium Graeco-Arabicum II*, Gerhard Endress, ed., Amsterdam: B.R. Grüner, 1989, pp. 25–8.

[2] Neugebauer, *HAMA*, II, p. 721.

[3] On Vettius Valens in Antiquity see the outdated article by Emilie Boer in Pauly-Wissowa, *Realencyclopädie*, 2–VIII,2, cols. 1871–73, and more especially the reference in the next note.

investigated by Otto Neugebauer and H.B. van Hoesen in 1959.[4] A more authoritative text was published by David Pingree in 1986 using a few more manuscripts than were available to Kroll.[5] Only one of these copies, used already by Kroll, contains the table whose Arabic descendant we shall be presenting in this paper, namely, MS Oxford Bodleian Seld. 22 & 20, originally a single codex of 286 folios. This is a late and corrupt Byzantine copy which was already in the possession of the French lawyer Christophe Longueil, who died in Padua in 1522, and then passed in 1556 to the library of John Dee and thence to John Selden (1585–1654).[6]

The fate of the *Anthology* in the East was first investigated by Carlo Alfonso Nallino in the 1890s and more recently has been pursued by Fuat Sezgin.[7] A Middle Persian commentary on the *Anthology* was prepared in the 6th century by the Sasanid minister Buzurjmihr. This commentary, now lost, was apparently arranged in ten books and was called *Vizīdhak* (Arabic, *Bizīdhaj*) which means simply 'Anthology'.[8] It was in this form, rather than the original Greek, that Vettius Valens' work became available to the astrologers of late-8th- and early-9th-century Baghdad. This commentary was translated into Arabic as *Kitāb al-Bizīdhaj*, but it is no longer extant in this form either. Vettius Valens was known to the Arabs as Wālīs, sometimes as *Wālinṭiyus*, and sometimes with the appellation al-Rūmī, 'the Byzantine', which would correspond to his accepted provenance Antioch, but also sometimes as al-Miṣrī, 'the Egyptian', or, more specifically, al-Iskandarānī, 'the Alexandrian' (see below). Two fragments directly attributed to Wālīs in the manuscript sources have been identified by Fuat Sezgin and remain to be properly studied.[9] There are numerous references to Wālīs and Buzurjmihr in several major Arabic astrological works, notably the *Kitāb al-Mughnī* of the 10th-century Christian Arab astrologer Ibn Hibintā[10] and the *Safīnat al-aḥkām*, an astrological compendium by the 13th-century polymath Naṣīr al-Dīn al-Ṭūsī.[11] But none of these, as far as I am aware, deals with the table which is the subject of this paper. As I shall show (see Section 3, *sub* MS A), there is some possibility that the table was known in medieval Europe, at least in Palermo in the 13th century.

The material of Vettius Valens on the length of life is found at the end of Book VIII of the *Anthology* in the form of two tables, the only ones in the entire work. As noted above, this material survives in only one late Byzantine manuscript. The first table (see Figs 1a–b), which has already been discussed by Otto Neugebauer,[12] bears the title *Kanónion*

4 See Neugebauer & van Hoesen, *Greek Horoscopes*, pp. 176–85.

5 Kroll, *Vettius Valens*, pp. 321–8; and Pingree, *Vettius Valens*, pp. 308–11.

6 *Ibid.*, p. x.

7 Nallino, '*Ilm al-falak*, pp. 192–6 (Arabic) and *idem*, *Scritti*, VI, esp. pp. 291–6; and Sezgin, *GAS*, VII, pp. 38–41 and 80. See also Ullmann, *NGWI*, pp. 281–2 and 297. (MS Istanbul Nurosmaniye 2920,3 of part of the *Kitāb al-Asrār* attributed to Wālīs does not contain material on our subject.) See also T. Fahd in the article '(Aḥkām al-)Nudjūm' [= astrology] in EI_2, esp. p. 106b.

8 Sezgin, *GAS*, VII, p. 80; and Ullmann, *NGWI*, p. 297.

9 Sezgin, *GAS*, VII, p. 41, and Ullmann, *NGWI*, pp. 281–2, esp. n. 4 on p. 282. See also Pingree, 'Astronomy and Astrology in Iran', pp. 241–2, on MS London BL Add. 23400 (n. 66).

10 See Sezgin, GAS, VII, pp. 162–4. The two manuscripts of the main fragments of this work are published in facsimile in Ibn Hibintā, *al-Mughnī*.

11 *Ibid.*, pp. 22–4.

12 Neugebauer & van Hoesen, *Greek Horoscopes*, pp. 174–5.

a(lpha) kaì plinthíon and variations thereon on subsequent pages of subtables, essentially meaning 'small table no. 1 also (known as) *plinthíon*'.[13] *Plinthíon* literally means 'small brick' but by extension 'a lattice of bricks' and we shall return to the use of this term when we consider the internal structure of the table. The latter displays values of an auxiliary function ζ, labelled simply *arithmos*, 'number', and the length of life, L, in years and months, with occasional entries of an additional 15 days. Both quantities are tabulated as a function of the longitude of the horoscopus λ_H at the time of birth, and the second being dependent on terrestrial latitude ϕ, in this case that of Alexandria, although this is not stated. Values are given for each degree of each sign beginning with Libra (see Figs. 1a–b). Each group of 5 entries is associated with the sun, moon or one of the five planets, but these were not used in known Greek horoscopes. Neugebauer has shown that the tabulated function is defined by:

$$L(\lambda_H) \;=\; \zeta(\lambda_H)/60 \times 2D(\lambda_H) \;..............\; (1)$$

where $2D$ is the length of daylight corresponding to a solar longitude equal to λ_H. The equinox was taken at Aries 8° as in Babylonian System B solar theory,[14] and the length of daylight in the surviving Greek table was computed using a linear zigzag function having the traditional extremal values 210° and 150° (ratio 7:5), a standard scheme for Alexandria.[15] The latitude of Alexandria corresponds to the first climate in the scheme of climates proposed elsewhere in the *Anthology* by Vettius Valens.[16] Only the table of years for Alexandria is mentioned in the published text, and Vettius Valens writes that this table is 'for the sake of example', and that intelligent people will be able to calculate such tables for each *klima*.[17] In only two of the many Greek horoscopes investigated by Neugebauer and van Hoesen is the table of life used. For two nativities in 419 A.D., both in the climate of Spain, the *arithmoí* are taken from Vettius Valens' first table and the lengths of daylight are computed because the climate is different from that of Alexandria.[18] Again the arithmetical schemes typical of the Babylonian tradition are used rather than values for the length of daylight computed trigonometrically, such as could easily have been derived from Ptolemy's tables of rising times for the climates in the *Almagest*.

---

13 *Plinthíon* is the diminutive of *plinthos* and means literally 'little brick'. By extension it was used to refer to rectangular boxes, checker-boards, squares of tartar, *etc.* (Liddell & Scott, *Lexicon*, p. 1421b) and as such corresponds to the Middle Persian *zik*, meaning thread or cord and by extension the warp of a fabric, which came to be used in Arabic as *zīj* for 'astronomical handbook with tables' (Kennedy, '*Zīj* Survey', pp. 123–4, and King & Samsó, 'Islamic Astronomical Tables', p. 12, n. 2).

14 Neugebauer, *HAMA*, I, pp. 369–72, and II, pp. 594–8. See also Cumont, *Astrology and Religion ...*, p. 36.

15 See, for example, Neugebauer & van Hoesen, *Greek Horoscopes*, p. 3.

16 Neugebauer, *HAMA*, II, p. 728, and also n. 10 on p. 718.

17 Pingree, ed., *Vettius Valens*, p. 283 (end of 8.1). I owe this reference to Alexander Jones.

18 Neugebauer & van Hoesen, *Greek Horoscopes*, pp. 136–8 (L 419) and 140 (L 431). Unfortunately (at least for me), most of the 65-odd horoscopes out of the 200-odd astronomical and astrological fragments from Roman Egypt (more precisely, from Oxyrhynicus, near Bahnasa some 150 km south of modern Cairo) so masterfully studied by Alexander Jones (*Astronomical Papyri from Oxyrhynicus*) predate the compilation of the *Anthology*, but even those from after *ca.* 300 (about 20 in number) show no traces of calculations of the length of life, despite the relative proximity of Alexandria.

VETTIVS VALENS

| ***Κανόνιον α′· πλινθίον***

	Ζυγοῦ					Σκορπίου					Τοξότου				
μοῖραι		ἀριθμοί	ἔτη	μῆνες	ἡμέραι		ἀριθμοί	ἔτη	μῆνες	ἡμέραι		ἀριθμοί	ἔτη	μῆνες	ἡμέραι
α	☾	β	ς	α	ιε	♂	ιδ	λθ	ο		♃	κς	ξζ	ς	
β		δ	ιβ	γ			ις	μδ	ε			κη	οβ	ς	
γ		ς	ιη	γ			ιη	μθ	α̲			λ	οζ	⟨ς⟩	
δ		η	κδ	δ			κ	νε	β			β	ε	β	
ε		ι	λ	γ	ιε		κβ	ξ	ς			δ	ι[ε]	γ	ιε
ς	♂	ιβ	λς	ζ̲		⟨♃⟩	κδ	ξε	α̲		♄	ς	ιε	ε	
ζ		κς	οη	α̲			η	κα	ι[α]			κ	να	β	
η		κη	πδ	ο			ι	κζ	β			κβ	νς	β	
θ		λ	πθ	ι̲α̲			ιβ	λβ	ζ			κδ	ξα	γ	ιε
ι		β	ς	ο			ιδ	λη	ο			κς	ξς	γ	
ια	♃	δ	ια	θ̲		⟨♄⟩	ις	μγ	γ		⟨☿⟩	κη	οα	δ	
ιβ		ς	ιζ	α̲			ιη	μη	ζ			λ	⟨ο⟩ς	ς	
ιγ		κ	νθ	[ι]α			β	ε	δ			ιδ	λε	ζ	ιε
ιδ		κβ	ξδ	ε̲	ιε		δ	ι	θ			ις	μ	ζ̲	ιε
ιε		κδ	ο	ς			ς	ις	α			ιη	με	η̲	
ις	♄	κς	ος	η̲		⟨☿⟩	η	κα	ε		♀	κ	ν	θ	
ιζ		κη	πα	γ			ι	κς	θ			κβ	νε	θ̲	
ιη		λ	πζ	δ̲			ιβ	λβ	ο			κδ	ξ	ι	
ιθ		ιδ	μ	ι̲α̲			κς	ξθ	β			η	κ	ι̲	
κ		ις	μς	θ̲			κη	οδ	δ			ι	κε	δ	
κα	☿	ιη	να	ε		⟨♀⟩	λ	οθ	ς		☾	ιβ	λ	ε	
κβ		κ	νζ	ι̲α̲			β	ε	γ			ιδ	λε	ε	
κγ		κβ	ξβ	ε̲	ιε		δ	ι	ς			ις	μ	ε	ι
κδ		κδ	ξη	η̲			ς	ιε	θ			ιη	με	ε	ιε
κε		η	κβ	θ			κ	νβ	ς			β	ε	γ	
κς	⟨♀⟩	ι	κη	ε	ι̲ε̲	☾	κβ	νζ	α̲		♂	δ	ι	α	
κζ		ιβ	λγ	ι⟨α⟩			κδ	ξβ	α̲			ς	ιε	α	
κη		ιδ	λθ	ε			κς	ξζ	ια			η	κ	γ	
κθ		ις	μδ	ι	ιε		κη	οβ	ια			ι	κε	β	
λ		ιη	ν	δ			λ	ο̲ζ̲	ζ̲			ιβ	λ	β	

Fig. 1a: An extract from the first table in the edition of the *Anthology* of Vettius Valens in the new edition by David Pingree. [From Pingree, *Vettius Valens*, p. 308.]

Sign Argument	VII	VIII	IX
1	2	14	26
2	4	16	28
3	6	18	30
4	8	20	2
5	10	22	4
6	12	24	6
7	26	8	20
8	28	10	22
9	30	12	24
10	2	14	26
11	4	16	28
12	6	18	30
13	20	2	14
14	22	4	16
15	24	6	18
16	26	8	20
17	28	10	22
18	30	12	24
19	14	26	8
20	16	28	10
21	18	30	12
22	20	2	14
23	22	4	16
24	24	6	18
25	8	20	2
26	10	22	4
27	12	24	6
28	14	26	8
29	16	28	10
30	18	30	12

Fig. 1b: The values of the auxiliary function given in Fig. 1a, which are sufficient to give an idea of the underlying structure of the entire table.

Neugebauer also showed how the function $\zeta(\lambda_H)$ could be generated from a scheme of the even numbers from 2 to 30, which, he wrote, 'has no astronomical significance whatever'. Indeed, he added that the theory well illustrates 'the combination of serious astronomical concepts with arbitrary manipulation of numbers in order to obtain new complex astrological rules'.[19] Two of the numerous Greek horoscopes investigated by Neugebauer and H.B. van Hoesen used Vettius Valens' first table and actually mentioned the associated length of life as displayed in the surviving table; however, since they were not computed for Alexandria they used different values for 2D to compute the proper life-span.[20]

The following description of the function ζ differs but slightly from that of Neugebauer but takes into consideration Vettius Valens' second table which Neugebauer did not discuss (Fig. 1c)[21] and perhaps casts just a little light on the inherent mystery. All values are *modulo* 30 with the convention that 30 ≡ 30 and not 0, and the value of ζ at Libra 0° (= Virgo 30°) is 30. The increase for each single degree is 2 (that is, 6 for each 3° or 12 for each 6°) with an additional 12 after each 6° and an extra 12 after each 30° (except at the end of Virgo). Vettius Valens' second table has different values for the first and second increases. This having been said, I confess that I was until recently at a loss to explain the motivation for these schemes. Certainly the tables are easier to use than the procedures for finding the life-span advocated by, say, Dorotheos of Sidon.[22]

A possible solution was suggested to me by José Chabás. His solution lies in the use of the term '*plinthíon*' meaning a 'lattice of bricks', or, more precisely, a 'square matrix of bricks'. Let *brick A* be the sequence of the first six even numbers:

$$A = 2, 4, 6, 8, 10, 12.$$

Now, let *brick* $B = A + 6$, *brick* $C = B + 6$, and so on. Thus,

$$B = 8, 10, 12, 14, 16, 18$$
$$C = 14, 16, 18, 20, 22, 24$$
$$D = 20, 22, 24, 26, 28, 30$$
$$E = 26, 28, 30, 2, 4, 6$$

---

19 *Ibid.*, pp. 174–6. See also Neugebauer & Saliba, 'Greek Numerology', on Hellenistic schemes for predicting the chances for the survival or death of a sick person.

20 See n. 18 above.

21 The second table is of the same kind and format (but without the planetary associations) and immediately follows the first table: see Kroll, *Vettius Valens*, pp. 325–8; and Pingree, *Vettius Valens*, pp. 312–15, and Neugebauer & van Hoesen, *Greek Horoscopes*, p. 174, n. 93. It is labelled simply *kanónion deuteron kai plinthíon* and variations thereon. Different values of the auxiliary function are given, now to two sexagesimal digits, although the second digit is always 0, 20 or 40. The underlying scheme is similar to that in the first table with the following changes: the value for Libra 0° = Virgo 30° is again 30 but the value for Libra 1° has been fiddled to be 0;20; the increase for each single degree of argument is now 2;20 (that is, 7 for each 3° or 14 for each 6°) with an additional increase of 10 after each 6° and an extra 12 after each 30° (except at the end of Virgo). By virtue of the choice of increments, the resulting values are of the same order as those in the first table. The length of life in years and months in this second table is based on these values of the auxiliary function and on the same scheme for the length of daylight as in the first table.

22 See Pingree, *Dorotheos*, pp. 237–41. See also *ibid.*, pp. 246, 264, on Vettius Valens in Dorotheos. On Dorotheos in the Arabic tradition see Sezgin, *GAS*, VII, pp. 32–8, esp. p. 37.

VETTIVS VALENS

| ***Κανόνιον δεύτερον καὶ πλινθίον***

	Ζυγοῦ				Σκορπίου				Τοξότου			
μοῖραι	ἀριθμοί	συγ	ἔτη	μῆνες	ἀριθμοί	συγ	ἔτη	μῆνες	ἀριθμοί	συγ	ἔτη	μῆνες
α	ο	κ	α	ο	ιβ	κ	λδ	δ	κδ	κ	ξγ	α
β	β	μ	η	β	ιδ	μ	μ	γ	κς	μ	ξθ	ο
γ	ε	ο	ιε	γ	ιζ	ο	μη	θ	κθ	ο	οδ	ια
δ	ζ	κ	κβ	δ	ιθ	κ	νγ	δ	α	κ	γ	α
ε	θ	μ	κθ	δ	κα	μ	νθ	δ	γ	μ	θ	α
ς	ιβ	ο	λς	γ	κδ	ο	ξε	ια	ς	ο	ιε	δ
ζ	κδ	κ	ογ	γ	ς	κ	ιζ	δ	ιη	κ	μς	ια
η	κς	μ	π	ο	η	μ	κγ	η	κ	μ	να	α
θ	κθ	ο	πζ	θ	ια	ο	κθ	⟨ια⟩	κγ	ο	νη	η
ι	α	κ	δ	ια	ιγ	κ	λς	α	κε	κ	ξδ	ζ
ια	γ	μ	ια	θ	ιε	μ	μβ	⟨ε⟩	κζ	μ	ο	ε
ιβ	ς	ο	ιη	ι[β]	ιη	ο	μη	ζ	λ	ο	ος	α
ιγ	ιη	κ	νε	ι	ο	κ	ο	ια	ιβ	κ	λα	α
ιδ	κ	μ	ξβ	ς	β	μ	ζ	β	ιδ	μ	λζ	β
ιε	κγ	ο	ξθ	θ	ε	ο	ιγ	γ	ιζ	ο	μγ	α
ις	κε	κ	ος	η	ζ	κ	ιθ	ζ	ιθ	κ	μθ	ζ
ιζ	κζ	μ	πγ	η	θ	μ	κε	ι	κα	μ	νε	ο
ιη	λ	ο	ς	θ	ιβ	ο	λβ	ο	κδ	ο	ξ	⟨ι⟩α
ιθ	ιβ	κ	λζ	δ	κδ	κ	ξδ	θ	ς	κ	ις	α
κ	ιδ	μ	μδ	ια	κς	μ	ο[β]	ι	η	μ	κα	ια
κα	ιζ	ο	να	ς	κθ	ο	ος	ι	ια	ο	κζ	ι
κβ	ιθ	κ	νη	α	α	κ	γ	ς	ιγ	κ	λγ	η
κγ	κα	μ	ξε	ς	γ	μ	θ	η	ιε	μ	λθ	γ
κδ	κδ	ο	οβ	γ	ς	ο	ιε	ι	ιη	ο	μς	δ
κε	ς	κ	ιθ	ια	ιη	κ	μη	α	ο	κ	ο	ι
κς	η	μ	κς	α	κ	μ	νδ	α	β	μ	ς	η
κζ	ια	ο	λγ	α	κγ	ο	ξ	α	ε	ο	ιβ	ζ
κη	ιγ	κ	μ	ς	κε	κ	ξε	ια	ζ	κ	ιη	⟨ς⟩
κθ	ιε	μ	μζ	ια	κζ	μ	οα	ια	θ	μ	κδ	δ
λ	ιη	ο	νδ	δ	λ	ο	οη	ο	ιβ	ο	λ[β]	β

Fig. 1c: An extract from the second table in the *Anthology*. [From Pingree, *Vettius Valens*, p. 312.]

Then the 'Table of Life', as some of the Arabic sources call it, reduces to the matrix:

$$\begin{array}{ccccc} C & E & B & D & A \\ B & D & A & C & E \\ A & C & E & B & D \\ E & B & D & A & C \\ D & A & C & E & B \end{array}$$

As José Chabás further pointed out to me, this matrix has different interesting symmetries, and changing the order of the columns does not modify the argument. Each brick appears only once in each row and each column. The sum of the entries in row 1 is 70; in row 2, 80; and in row 3, 90; and the basic pattern (70, 80, 90) repeats over and over for the following rows. Similarly, each column has exactly 2 'twos', 2 'fours', 2 'sixes', *etc.* Thus, the sum of all entries in any row is 2 x (2 + 4 + ... + 30) = 480. These and other features indicate that the matrix is 'well balanced' and 'fair' to all individuals!

During the course of my survey of the medieval scientific manuscripts in the Egyptian National Library in Cairo in the 1970s, I came across a number of Arabic copies of Vettius Valens' table of the auxiliary function $\zeta(\lambda_H)$ not previously known in the Islamic sources.[23] The copies were all of late medieval Egyptian provenance but for one copied from an Egyptian manuscript in Istanbul. Related research in other manuscript libraries in Europe, the Near East, and India, yielded only two other copies of this table, namely, one in another Egyptian manuscript in the Escorial and the second in a Syrian recension of a 13th-century Tunisian *zīj* now preserved in Hyderabad. I do not doubt that other copies will come to light now that attention has been drawn to the table.

2. The Arabic tradition

The table was clearly first copied in Arabic in the Abbasid period, more specifically in early-9th-century Baghdad. Our sources maintain that it was copied in gold ink, which must have been a rare occurrence, but not unique. Indeed, in this case, as in the case of the surviving fragments of the 'Blue *Qur'ān*' from 10th-century (Fatimid) Kairouan also written in gold ink, the inspiration may have come from Byzantine practice.[24] The legitimacy of writing or decorating the text of the *Qur'ān* in gold was discussed by some early Muslim authorities, most of whom did not approve of this.[25] However, in Fatimid Egypt there were dozens of *Qur'ān* copies in gold ink.[26]

In view of the purported utility of the table I find it surprising that it was not incorporated into the major Arabic astrological treatises and widely used all over the Muslim world thereafter. The fact that the underlying theory is absurd is of course no reason why the table should not have been popular.

The appointed term of a man's life or the date of his death (Arabic, *ajal*) was a topic discussed in the early Muslim literature on scholastic theology (*kalām*), inevitably without reference to astrological sources. The *Qur'ān* often emphasises the *ajal* as the irrevocable period of life assigned by God, and, according to Muslim tradition, *ajal* is determined for

---

23 The various Cairo manuscripts are listed in *Cairo ENL Survey*, no. A21.

24 See Bloom, 'The Blue Koran', especially p. 98b. On ink in Islamic society see the article 'Midād' in *EI*$_2$. I owe these references to the kindness of Dr. Jan-Just Witkam.

25 Jeffery, *Materials ...* , pp. 150–52, presents opinions recorded by Ibn Abī Dā'ūd (d. 928) in his *Kitāb al-Maṣāḥif.*

26 Bloom, *op. cit.* On the preparation of gold ink as described in the treatise by the Zīrid ruler of Ifrīqiya, al-Mu'izz ibn Bādīs (*ca.* 1025), see Levey, *Mediaeval Arabic Bookmaking*, pp. 9, 25, and 32–3.

a man whilst still in the womb.[27] It would be interesting to know how the 'Table of Life' was received when it was first made available in Arabic.[28]

As far as the later history of the table is concerned, the manuscripts I have located prove only that it was known in Tunis in the 13th century and then in Cairo from the 15th century onwards, as well as in Istanbul in the 19th century. I have a hunch that it was known in Seville or Cordova in the 12th century; if this is so, then it was probably also known in Palermo in the 13th century (see MS A in Section 3).

There appear to be at least three different Arabic traditions of the table. In the first, represented by the Escorial manuscript (MS B) and one Cairo copy made from it (MS C), the table itself is accompanied by a text – see Appendix B for a translation – in which the table is attributed to Wālīs, further identified as Wālinṭiyus (*W-'-l-x-ṭ-y-s*, where *x* represents an undotted carrier for a medial *nūn*, *etc.*), and is labelled *al-ha<u>sh</u>taq*, a word new to the modern literature. Now the Middle Persian for 'brick' is *i<u>sh</u>ti-*, and the New Persian, *<u>kh</u>i<u>sh</u>t*; a small brick in New Persian is *<u>kh</u>i<u>sh</u>tak*.[29] I suspect that *plinthion* was translated into Middle Persian as *i<u>sh</u>tak* and that this was rendered into Arabic as *ha<u>sh</u>taq*. The author of this text, which is clearly part of a larger work on astrology, had seen the table of the auxiliary function in its original form since he mentions that Vettius Valens had begun with entries for Libra. He had also seen the accompanying text in which Vettius Valens generously offered the table to his readers.[30] The table is also labelled *jadwal al-ḥayā wa-kamiyyat 'umr al-mawlūd*, 'table of life and the length of life of the newborn'. In the Arabic text there are two examples for the latitude of Baghdad, which, as we shall see, are based on parameters from the 9th century.

In the second tradition, represented only by the Hyderabad manuscript (MS A), the table is labelled *jadwal al-t-n-w-r,* which I read as *al-tanawwur*, 'table of illumination',[31] a curious development from 'little brick', and it is attributed to Wālīs. There is no accompanying text (at least not in MS A). The same title for the table and a corrupt version occur in two of the Egyptian manuscripts (MSS D and J).

---

[27] See the article 'A<u>dj</u>al' by Ignaz Goldziher and W. Montgomery Watt in *EI*$_2$. On some medieval Jewish notions on this subject see Weil, *Maimonides über die Lebensdauer*, and Kaufmann, 'R. Haja über die Lebensdauer'.

[28] It would not surprise me if a discussion was found in one or other of two works of al-Bīrūnī entitled *Kitāb al-Tanbīh 'alā ṣinā'at al-tamwīh*, 'An Exposé of the Art of Deception', in which he criticises astrologers who use the planets to determine the length of life, and *Kitāb al-Shumūs al-shāfiya li-'l-nufūs*, 'The Suns which Heal the Soul', in which he presents 'the best method'. Alas neither of these works is extant. See further al-Bīrūnī, *Chronology*, p. 79 (text) and p. 92 (trans.), and Sezgin, *GAS*, VII, p. 191, no. 9, and also n. 37 below.

[29] Buck, *Dictionary*, pp. 603–4; Steingass, *Dictionary*, p. 461b. I am grateful to Prof. Lutz Richter-Bernburg for his identification of the Persian connection.

[30] This appears to be a reference to a remarkable passage in the *Anthology* (8.5, ed. Pingree, p. 288). The following is an extract from a translation kindly provided by Alexander Jones:

> 'Hence for those who most desire to adhere to every methodical system, since each of those who have written treatises has, after working out the system in a distinctive and complicated manner, secretively and grudgingly omitted the solutions without presenting them, I who have sought (them) through much labour and much trial have set them out. … … if I very often remind you of my generosity and straightforwardness, what I say should be forgiven. … … .'

[31] The expression *al-jadwal al-mutanawwar* would be more acceptable. I have preferred the reading *al-tanawwur* to *al-tannūr*, 'baking oven', which makes little sense. Note that in MS J *al-tanawwur* becomes *al-tanwīr*. In my first study of these materials (see n. 1) I chose *al-tannūr*.

In the third tradition, represented by the remaining Cairo manuscripts, the table, now labelled *jadwal al-ḥayā*, 'table of life', is accompanied by a short text with a numerical example for an unspecified latitude (actually about 36°). There is no attribution to Wālīs. In one copy (MS G) there is an additional example for an unspecified latitude, this time for Cairo (see Appendix C).

In the Egyptian manuscript in the Escorial, as in six of the seven Cairo manuscripts, it is stated that the table 'was found in a work entitled *Kitāb al-Kāmil fi 'l-nujūm* (= *The Complete / Perfect Book on Astronomy / Astrology*), in the treasury of 'Abdallāh ibn Ṭāhir ibn al-Ḥusayn (sent to him) by (*'an*) al-Ma'mūn, (and) written in gold'. 'Abdallāh ibn Ṭāhir (798–844) was a well-known general, governor, and confidant of the Caliphs of Baghdad,[32] including al-Ma'mūn (reigned from 813 to 833), who was particulary interested in astronomy.[33] 'Abdallāh ibn Ṭāhir was governor of Khurasan, with its capital in Nishapur, and had a palace in Baghdad. It seems more likely that his treasury would have been in Baghdad. However, we should keep in mind that in the first and third traditions identified above we have worked examples for the use of the table respectively for the latitude of Baghdad and for an unspecified location with latitude 36°, which could serve Nishapur.[34] The *Kitāb al-Kāmil fi 'l-nujūm* remains unidentified; it is not one of the various titles attributed to Wālīs or his commentator by the 10th-century bibliographer Ibn al-Nadīm,[35] and it is not attested as a title of any known astrological work from the early Abbasid period.[36]

In the text accompanying the table in the Escorial manuscript it is stated that Wālīs wrote another tract of the variety known as *namūdār*: a procedure for finding the longitude of the horoscopus at birth when the time of birth is not known exactly.[37] I have come across two copies of this text and am confident that others exist in the vast sources available for the further study of Islamic astrology. This method is outlined in Section 4 and the text is translated in Appendix A.

---

[32] See the article "Abd Allāh b. Ṭāhir' by E. Marin in EI_2.

[33] See Sayılı, *Observatory in Islam*, pp. 50–87, on al-Ma'mūn's astronomical activity, and also pp. 8–49 on astrology in Islam. A new account of the geodetic activities sponsored by him is in King, 'Earliest Muslim Geodetic Measurements'.

[34] See Kennedy & Kennedy, *Islamic Geographical Coordinates*, p. 245, and also n. 78 below.

[35] Cited in Sezgin, *GAS*, VII, p. 41.

[36] A certain *al-Kitāb al-Kāmil* is attributed to Abū Ma'shar (787–886, active in Baghdad) by Ibn al-Nadīm: see Sezgin, *GAS*, VII, p. 151, and David Pingree in the *DSB* article 'Abū Ma'shar' (I, p. 38, nos. 22 and 24). Another work with the same title was penned by Ibn Nawbakht in 994 (Sezgin, *GAS*, VII, p. 172); it is an astrological history, published as Labarta, *Horóscopos históricos*. The other titles of the same kind associated with al-Farghānī (Sezgin, *GAS*, VI, pp. 150–51) and Abu 'l-Wafā' (*ibid.*, p. 223) are not astrological works.

[37] On the concept of the *namūdār* see al-Bīrūnī, *Astrology*, pp. 328–9; Sédillot-*fils*, *Ouloug Beg*, II, pp. 201–5 (Ulugh Beg on Ptolemy, Hermes and Zoroaster); North, *Horoscopes and History*, pp. 51–2 (Ptolemy, from *Tetrabiblos* III.2); and Elwell-Sutton, *Horoscope*, p. 83 (on a late Persian work). For a short discussion of other methods for finding the length of life see al-Bīrūnī, *Chronology*, pp. 78–80 (text) and pp. 90–92 (trans.). See also Tokan, 'al-Bīrūnī on the Length of Life' (abridged Arabic text with facsimile of the unique Konya manuscript), on Indian methods (Sezgin, GAS, VII, p. 92, and p. 190, no. 2), and n. 28 above.

3. The Arabic sources

Before investigating the table and the accompanying instructions I list the ten copies of the table (A–K), two copies of the *namūdār* (L–M), and various other relevant sources (N–Q):

A: MS Hyderabad Andra Pradesh State Central Library 298, *ca.* 200 folios, copied *ca.* 800 Hijra [≈ 1400 A.D.] in Homs. This is a unique copy of a recension of the *zīj* by the 13th-century Tunisian astronomer Ibn Isḥāq, which is perhaps the most important source for the history of post-classical Islamic astronomy to be discovered in the past few years.[38] Of the 360 tables in this work, no. 171, squeezed in between other astrological tables, is entitled *Jadwal al-tanawwur li-Wālīs li-ma'rifat 'umr al-mawlūd*, 'Table of illumination by Wālīs for finding the length of life of the newborn (*i.e.* the age to which a newborn will live)'. In the 188 pages of the introduction to the tables there is no explanatory text relating to this particular table.

There are, however, various tables for determining the period of gestation in the *Zīj* of Ibn Isḥāq,[39] and in the text the 12th-century astronomer Ibn al-Kammād, who hailed from Seville but was later active in Cordova,[40] is extensively cited. The writings of Ibn al-Kammād on this subject survive in an incomplete version of his *Kitāb Mafātīḥ al-asrār* – see MS Q – and have been studied in an article by Juan Vernet.[41] In this treatise Ibn al-Kammād cites Wālīs on the subject of gestation but does not mention the 'Table of Life'.[42] I conjecture that Ibn Isḥāq found the table in the missing part of the *Mafātīḥ al-asrār*.

As noted by Vernet, this treatise of Ibn al-Kammād was excerpted in Latin by Johannes de Dumpno in Palermo in 1260.[43] A copy survives in MS Madrid Biblioteca Nacional 10023, but this does not contain the 'Table of Life'. Ibn al-Kammād's tables for gestation, on the other hand, had a life of their own in the corpus of tables associated with Abraham Zacut.[44] As far as I am aware, and the same holds for various colleagues consulted, there is no mention of the table in other European sources.

---

38 On the author see Suter, *MAA*, no. 356; and, since the rediscovery of the Hyderabad manuscript, King, 'Astronomy in the Maghrib', pp. 181–4, Samsó, 'Maghribi *Zījes*', p. 93, and now Mestres, 'Hyderabad Manuscript', and *idem*, Zīj *of Ibn Isḥāq*, esp. p. 294, where the table is mentioned in passing.

39 Mestres, Zīj *of Ibn Isḥāq*, p. 292. See also n. 85.

40 Suter, *MAA*, no. 487, and 'Nachträge', no. 487; Brockelmann, *GAL*, S I, p. 864; Millás, *Estudios sobre Azarquiel*, pp. 345–7; Sarton, *IHS*, III:2, pp. 1514–15; and Vernet, 'Tractat', pp. 72–3.

41 *Ibid.* See also Chabás & Goldstein, '*Zīj* of Ibn al-Kammād', p. 37, and n. 44 below.

42 Vernet, 'Tractat', p. 86. Ibn al-Kammād states that Wālīs (al-Rūmī!) had compiled a work on the subject of gestation but that this had become corrupted over the centuries through continuous copying.

43 See also Carmody, *Bibliography*, p. 165. Johannes de Dumpno is apparently not mentioned in Sarton, *IHS*, and I have not encountered him elsewhere.

44 Chabás & Goldstein, *Zacut*, pp. 152–3.

B: MS Escorial ar. 961, copied in Cairo in 863 H [= 1459] by ʿAlī ibn Muḥammad al-Dalāmī, known as a copyist of several significant astronomical manuscripts,[45] is an important compendium of Abbasid and Mamluk treatises on astronomical instruments.[46] Here the table, with title *jadwal al-ḥayā wa-kamiyyat al-ʿumr li-ʾl-mawlūd*, occurs on fol. 44r, amidst an introduction on fols. 43r–44v (see Fig. 3a). The preceding and subsequent texts are quite unrelated to this section, which was inadvertently overlooked by H.-P.-J. Renaud when he catalogued the scientific manuscripts in the Escorial. (This explains how it remained unknown until the Cairo manuscript survey.)

C: MS Cairo Ṭalʿat *mīqāt* 102, 89 fols., copied *ca.* 1200 H [≈ 1800] apparently in Istanbul from the very manuscript now in the Escorial (B), contains (fols. 83v–85v) the same table and treatise with numerous copyist's errors. Here the treatise bears the spurious title *Risāla fī Ṭāliʿ al-mawlūd mina ʾl-asṭurlāb*, 'Treatise on (finding) the ascendant of the newborn with the astrolabe'.[47] See also MS K.

D: MS Cairo Muṣṭafā Fāḍil *mīqāt* 58, 15 fols., copied in Cairo *ca.* 850 H [≈ 1450], contains some anonymous tables for sundial construction (fols. 1v–2v) and timekeeping (fols. 2v–12v), amidst the latter of which (fol. 6r) occurs the table, here entitled *jadwal al-tanawwur alladhī yustakhraj minhu ʿumr al-mawlūd*.[48] It is stated that the table was found in the treasury of ʿAbdallāh ibn Ṭāhir: see Fig. 3b. There are some notes in a hand which is recognizable as that of the Cairo astronomer Ibn Abi ʾl-Fatḥ al-Ṣūfī (*ca.* 1500),[49] and it may be that MS G was copied from this.

E: MS Cairo Dār al-Kutub *mīqāt* 157, 7 fols., copied *ca.* 850 H [≈ 1450], contains a set of tables of the ascensions at the time of the *salām* (shortly before daybreak, when praises on the name of the Prophet are called out by the muezzin) by the 15th-century Cairo astronomer ʿIzz al-Dīn al-Wafāʾī.[50] On fol. 1r of this manuscript our table reappears copied upside down in a different, much later, hand, and with brief instructions on its use: see Fig. 3c.

F: MS Cairo Dār al-Kutub *mīqāt* 637, 48 fols., copied 868 H [= 1463/64], consists mainly of a copy of the *Zīj* called *al-Lumʿa* compiled by the 15th-century Cairo astronomer Aḥmad al-Kawm al-Rīshī.[51] Appended to this in different hands are various magical tables and qibla diagrams and our table (fols. 47v–48r) which now occurs together with another astrological table, illustrated in Fig. 3d.

45 *Cairo ENL Catalogue*, I, p. 704.

46 *Escorial Catalogue*, pp. 97–100.

47 *Cairo ENL Survey*, no. A21.

48 *Ibid.*, no. C141.

49 Suter, *MAA*, nos. 447 and 460 (confused); and *Cairo ENL Survey*, no. C98.

50 *Ibid.*, no. C61 (*ad* 3.1.24).

51 *Ibid.*, no. C41.

Fig. 3a-1: The text in MS B, f. 43r. [Courtesy of the Biblioteca de El Escorial.]

الرابع والعشرين كد وزادها هنا ١٤ وصار بازا الخامس
والعشرين ح وعاد الى الدرجتين وصار بازا السته والعشرين
يب وبازا السبعة والعشرين سه وبازا الثامنة والعشرين
ند وبازا التسعة والعشرين سو وبازا الثلاثين كح وجعل
الاسد والحوت مثله وبازا اول درجه من العقرب نه والحق
به الزيادة على هذا الترتيب وجعل الحمل والسنبله مثله
وبازا اول درجه من القوس كو والحق به الزيادة لذلك وجعل
الثور مثله وبازا اول درجه من الجدي كح والحق به الزيادة
على الرسم وجعل الجوزا مثله وبازا اول درجه من الدلو ك
والسرطان مثله وكان يرتب زيادته ان يزيد اسن اسن فاذا
مضت خمس درج زاد ١٤ ثم عاد فزاد اسن اسن ايضا لذلك
حتى يستوفي جميع البروج ودرج الزيادة في اوايل البروج سته
لانه ليس للدرجه الاولى اشتراك والباقيه مشتركه فهي خمس
درج ثم ما بين كل ١٤ مرتين وبين اختها وقال والبس ايضا
اذا زدت على ما بازا اول درجه من كل برج ١٢ خرج ما يجب
ان يكون بازا اول درجه من البرج الذي يليه على نسقه وامتحنا
ذلك فكان صحيحا ولم يذكر العلة في عمل سته من هذه

Fig. 3a-2: The text in MS B, f. 43v. [Courtesy of the Biblioteca de El Escorial.]

Fig. 3a-3: The text in MS B, f. 44r. [Courtesy of the Biblioteca de El Escorial.]

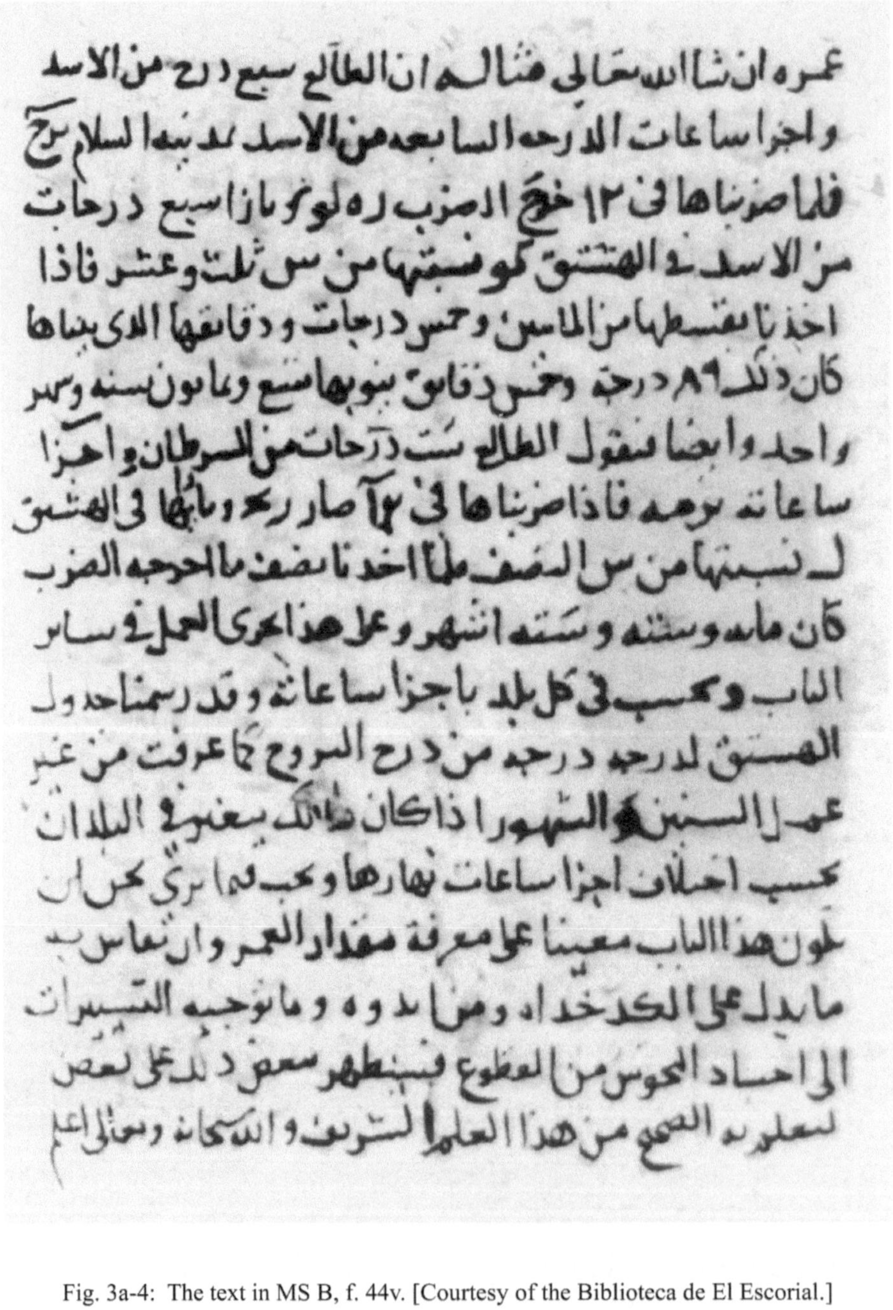

Fig. 3a-4: The text in MS B, f. 44v. [Courtesy of the Biblioteca de El Escorial.]

Fig. 3b: The table in MS D. [Courtesy of the Egyptian National Library.]

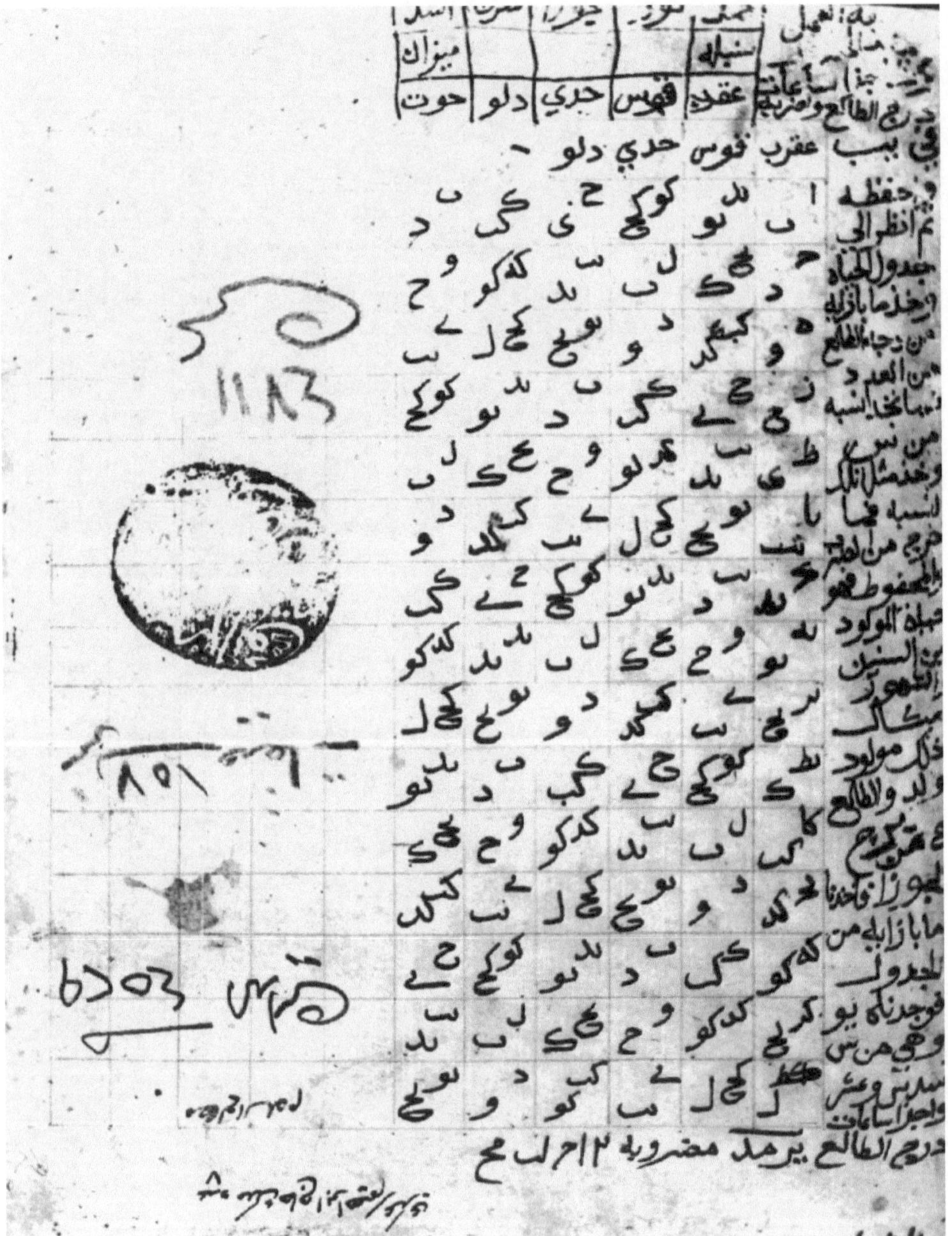

Fig. 3c: The table in MS E. [Courtesy of the Egyptian National Library.]

جدول طول البلاد والاقاليم بالدرجة الواقعة

جدول الحياه ومكث العمر للمولود والله اعلم

Fig. 3d: The table in MS F. [Courtesy of the Egyptian National Library.]

G: MS Cairo Dār al-Kutub *mīqāt* 1108, 59 fols., copied 1053 H [= 1643/44] by ʿAlī ibn Muḥammad, a copyist of two important astronomical manuscripts in the Egyptian National Library,[52] contains a collection of treatises and tables relating to timekeeping and lunar crescent visibility.[53] Amongst these (fol. 20v) is the *jadwal al-ḥayā* (see Fig. 3e) with some notes copied from the hand of Ibn Abi ʾl-Fatḥ al-Ṣūfī (see MS D).

H: MS Cairo Dār al-Kutub *mīqāt* 1205, 215 fols., copied *ca.* 1250 H [≈ 1850] in different hands, contains the *zīj* called *al-Zīj al-Mufīd* of the 17th-century Cairo astronomer Riḍwān Efendī.[54] At the end of this work there is appended (fol. 215v) the *jadwal al-ḥayā*.

J: MS Cairo Ṭalʿat *mīqāt* 241, 43 fols., copied *ca.* 1250 H [≈ 1850] in different hands, contains prayer-tables for Damietta by the 17th-century astronomer Quṭb al-Dīn al-Maḥallī and other anonymous tables for Cairo.[55] Amidst these (fol. 34r) is our table, here entitled *jadwal al-tanwīr*, 'Table of illumination', a nice corruption of *jadwal al-tanawwur*.

K: MS Cairo Ṭalʿat *mīqāt* 227, 163 fols., copied 1145 H [= 1732/33] in legible *naskhī* script (provenance uncertain), contains seven astrological treatises, including Abbasid, Mamluk, and Rasulid Yemeni works,[56] amidst which (fols. 87r–87v) there occurs the *jadwal al-ḥayā*, taken from a copy in the hand of ʿAlī ibn Ḥasan al-Baḥṭīṭī, a well-known copyist of astronomical works,[57] and a student of the Cairo astronomer Sibṭ al-Māridīnī *ca.* 900 H [≈ 1500],[58] this in turn based on a copy by al-Dalāmī, perhaps our source B!

L: MS Berlin Ahlwardt 5751 (= Mq. 101), 221 pp., copied *ca.* 700 H [≈ 1300],[59] is a fine copy of the astronomical handbook (*zīj*) of the early-11th-century Iranian astronomer Kūshyār ibn Labbān.[60] At the end of this copy occur various interesting very important spurious additions, including text and tables, amongst which[61] we find a *namūdār* attributed to Wālīs – see Fig. 3f and Section 4 – along with others attributed to Hermes, unidentified 'early scholars', and the 9th-century philosopher and astrologer al-Kindī on the technique associated with the Indians.[62] In addition there are some tables for finding the period of gestation.[63]

---

52 *Cairo ENL Catalogue*, I, p. 704.

53 *Ibid.*, I, pp. 188–9, and *Cairo ENL Survey*, no. D226.

54 *Ibid.*, no. D58.

55 *Ibid*, no. D37.

56 On the manuscript see *Cairo ENL Catalogue*, I, *sub* ṬM 227.

57 On al-Baḥṭīṭī see *ibid.*, I, p. 703.

58 On Sibṭ al-Māridīnī see Suter, *MAA*, no. 445, and *Cairo ENL Survey*, no. C97.

59 On the manuscript see *Berlin Catalogue*, pp. 203–6, esp. p. 206.

60 Kennedy, '*Zīj* Survey', nos. 7 and 9; Sezgin, *GAS*, VI, pp. 246–9; and King & Samsó, 'Islamic Astronomical Tables', p. 41.

61 One example is a set of tables for timekeeping by the stars, computed for the latitude of Qandahar *ca.* 1000: see King, *SATMI*, vol. 1, I–3.2.1.

62 No such treatise is listed in Sezgin, *GAS*, VII, pp. 259–61.

63 See n. 85 below.

Fig. 3e: The table in MS G. [Courtesy of the Egyptian National Library.]

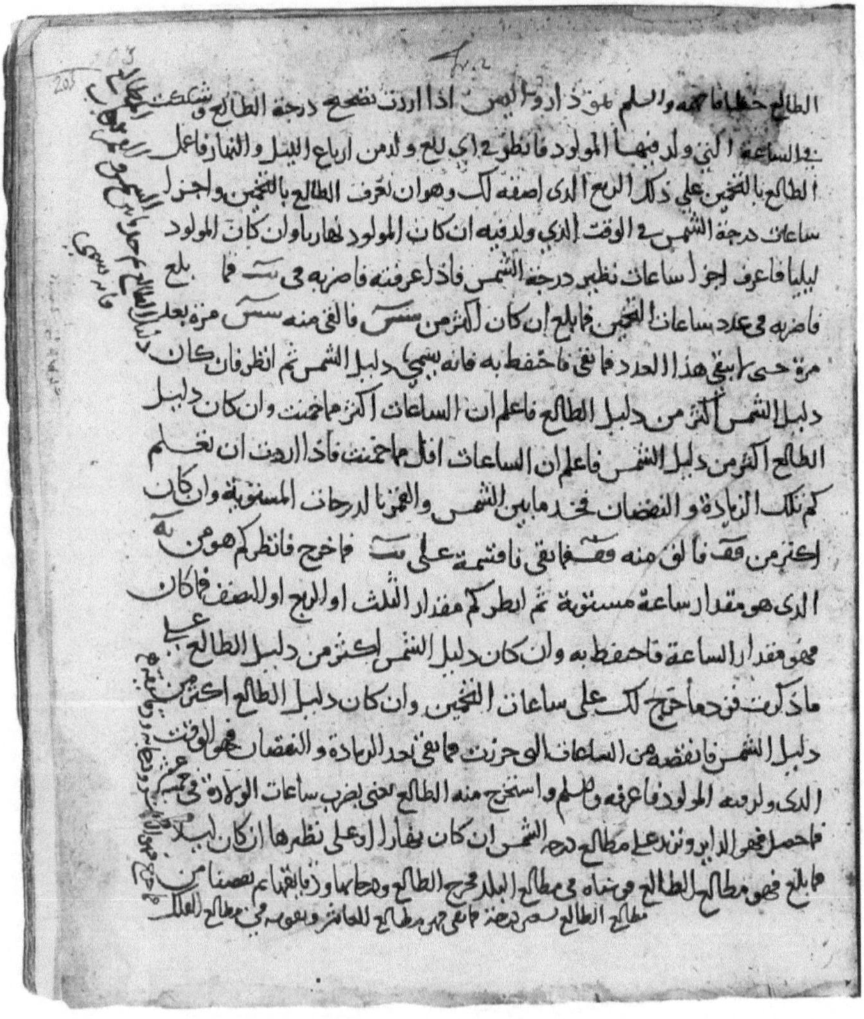

Fig. 3f: The text on the *namūdār* of Wālīs in MS L.
[Courtesy of the Deutsche Staatsbibliothek (Preussischer Kulturbesitz), Berlin.]

M: MS Damascus Ẓāhiriyya *'āmm* 9354, 181 fols., copied 641 H [= 1243/44], is a substantial fragment of the astrological compendium *Kitāb al-Mughnī* by Ibn Hibintā.[64] In this (fols. 22v–23r) is preserved the *namūdār* of Wālīs – see Section 4 – along with others. See also MS N below.

N: MS Munich ar. 852, 240 fols., copied *ca.* 650 H [≈ 1250], contains a second substantial fragment of the *Kitāb al-Mughnī* – see MS M above.

O: MS Dublin Chester Beatty 3640, 111 fols., copied *ca.* 670 H [≈ 1270] is a copy of the astrological compendium *Safīnat al-aḥkām* of al-Ṭūsī.[65]

P: MS London British Library Add. 23400, 103 fols., copied in the 10th/16th century, is another copy of the *Safīnat al-aḥkām.*[66]

Q: MS Escorial ar. 939,1, fols. 1r–9v, copied *ca.* 900 H (?) [≈ 1500], contains Chs. 10–15 of the astrological treatise *Mafātīḥ al-asrār* by Ibn al-Kammād – see MS A above.[67]

4. Finding the horoscopus at birth

The longitude (λ_H) of the horoscopus at birth can be found by measuring the time of the nativity and using tables of right and oblique ascensions $\alpha(\lambda)$ and $\alpha_\phi(\lambda)$ which were standard equipment of astronomers in late Antiquity and throughout the Middle Ages.[68] If the birth occurs at time T (in degrees) after sunrise on a day when the solar longitude is λ_S then:

$$\alpha_\phi(\lambda_H) = \alpha_\phi(\lambda_S) + T \text{ .... } (2)$$

If it occurs when a star with known mediation (co-culminating ecliptic longitude) μ is observed to be culminating, then:

$$\alpha_\phi(\lambda_H) = \alpha(\mu) + 90° = \alpha'(\mu) \text{ ..... } (3)$$

where $\alpha'(\lambda) = \alpha(\lambda) + 90°$ are the normed right ascensions, in use since Antiquity for determining λ_H from μ. (Various Islamic tables are known in which λ_H is tabulated for each degree of altitude for various stars, and for each degree of both altitude and longitude for the sun.[69] One table is known where λ_H is tabulated for various significant times of

---

64 See n. 10 above.

65 Sezgin, *GAS*, VII, p. 24.

66 *Ibid.* See also Pingree, 'Astronomy and Astrology in Iran', pp. 241–2.

67 *Escorial Catalogue*, pp. 54–5.

68 See my article '*Maṭāli'* [= ascensions]' in EI_2, and King, *SATMI*, vol. 1, pp. 37–8.

69 King, *SATMI*, vol. 1, I–3.2.1–2 and 3.1.1.

day – significant, that is, for religious and/or astronomical reasons – throughout the year, the values having been determined using an astrolabe.[70])

Since the horoscopus cannot be directly observed and its computation involves these subtleties, various solutions were developed in Antiquity which purported to enable the user in possession of only vague details of a nativity to determine the horoscopus accurately: these are known as *namūdārs*.[71] Frequently cited in the Islamic sources are the *namūdārs* of the Indians, Ptolemy, Hermes, Zoroaster, and Māshā'allāh, to which we may now add that of Vettius Valens – see Appendix A for a translation of the text in MSS L, p. 203 (I have not used MS M, fols. 22v–23r, or any other copies). The procedure outlined is by no means clear to me, and differs from what E.S. Kennedy has published recently from various other Arabic manuscripts.[72]

One is to guess the time of birth, say T' after sunrise, as best one can. Then using the lunar longitude λ_M, the length of daylight $2D(\lambda_S)$ corresponding to the solar longitude λ_S, and the oblique ascensions, one determines the 'indicators' (*dalīl*) of the horoscopus and of the sun, denoted by i_H and i_S, thus:

$$i_H \equiv [(2D(\lambda_S) \,/\, 12) \times 12 \times T'\,]\ [mod\ 360] \text{ ....(4)}$$

$$i_S = \alpha_\phi(\lambda_S) - \alpha_\phi(\lambda_M) \text{ .... (5)}$$

Then one is to form the quantity:

$$\Delta\lambda \equiv (\lambda_S - \lambda_M)\ [mod\ 180] \text{ .... (6a)}$$

with which the time correction is:

$$\Delta T = [\,\Delta\lambda \,/\, 12\,] \,/\, 15 \text{ .... (6b)}$$

and the true time of the nativity T is:

$$T = T' \pm \Delta T \quad (\text{as } i_S > \text{or} < i_H) \text{ .... (7)}$$

The text of the *namūdār* concludes with statements equivalent to (2) and (3). These procedures apparently defy rational explanation. There is perhaps a need for further investigation of all the available sources. This material is of interest not least because it represents an Arabic rendering of a very obscure Greek text[73]

---

[70] *Ibid.*, vol. 1, II–14.1, and IV–5.3.

[71] See n. 37 above.

[72] Kennedy, 'Kāshī on the Ascendent', pp. 143–4.

[73] Alexander Jones has kindly provided the following notes from the published Greek text. The 'indicators' of sun and the ascendant are what Vettius Valens calls 'gnomons'. There are several passages in the *Anthology* that discuss these. The one used in the Arabic version is apparently that found in 8.3 (Pingree, pp. 285–6):

'A different way. We found also in another way the magnitude corresponding to the sun's degree and multiplied it by 12, or at night the diametrically opposite (degree). Again we will multiply the same by the given hours of the giving birth; and after casting out 360s, regard the remainder as the horoscopic gnomon. Then,

5 The table and its format

The treatise accompanying the table in MS B and the derivative copy C (see Fig. 3a and also Appendix B for a full translation) begins with the words *tafsīr al-hashtaq li-Wālīs* ... , 'an explanation of (the table called) *hashtaq* due to Vettius Valens ... ', and continues with the remark that he 'made a *namūdār* for rectifying the longitude of the horoscopus and the hour of the nativity so that there should be no doubt about either' – see Section 4. The anonymous author then states that 'we have mentioned and explained this in the chapter on *namūdār*s in this book of ours (presumably the *Kitāb al-Kāmil fi 'l-nujūm* ?)' and that 'he (Vettius Valens) (made) a table which he called *al-hashtaq*'.

The author then describes in detail the structure of the table and mentions the significant entries therein, which correspond precisely to those in the various copies of the table:

> '(Vettius Valens) began in (the table) with Libra and put 2 for the first degree of it, 4 for the second, 6 for the third, 8 for the fourth, 10 for the fifth, 12 for the sixth, then he added to these twelve 14, and so there was 26 for the seventh, and then he again added 2 for the eighth making 28, and 30 for the ninth, and 2 for the tenth, *etc., etc.*'

The author explains that Vettius Valens made the entries for Leo and Pisces the same as those for Libra, and began the entries for Scorpio, as well as Aries and Virgo, with 14, and the entries for Sagittarius, and also Taurus, with 26, and the entries for Capricorn, as well as Gemini, with 8, and the entries for Aquarius, as well as Cancer, with 20. The rule is that one adds 2 for each degree and after each 5° adds 14, both operations being performed *modulo* 30, and that the 'degrees of increase at the beginnings of the signs are six (degrees), because the first degree has no contribution (*ishtirāk*) and the remaining

---

taking the interval from the sun to the moon in ascension (*pros anaphoran*), compare with the first, horoscopic, gnomon. For if the solar gnomon is greater, add to the hour, and if it is less, subtract. One should add or subtract as much as the difference of the solar magnitude indicates. For the complete (what?) before the comparison will be obtained from the addition and subtraction at the hour or also fraction of an hour.' [Last sentence unclear, obviously corrupt.]

Inserted in the Arabic version of this is a method found in the 5th-century '*additamenta*', ch. 1 (Pingree, p. 350):

'Another way. Reckoning the interval from conjunction to the moon or the (interval) from full-moon to the moon, and if it is within 180 degrees, multiplying by 12, see what fraction results of 15 days, and if it is found more than 180 degrees, subtracting the 180 degrees, compare the remainder, what fraction it is of the moon's (daily) course, and deduct this from the hourly magnitude.' (Apparently Vettius Valens uses 15 as a round figure for the moon's daily motion here.)

Another version of this last rule is in the *Anthology* (8.3, Pingree, p. 285), but without using 15 explicitly for the moon's daily course:

'Another way. Reckoning the interval from conjunction to the moon or the (interval) from full moon to the moon – and if it is found inside 180 degrees, use it in the demonstrated way, but if over 180 degrees, subtracting the 180 degrees, compare the remainder, what fraction it is of the course, and subtract this from the hourly magnitude.'

THE TABLE OF LIFE

	Sign	I	II	III	IV	V
		VI				VII
		VIII	IX	X	XI	XII
Degree						
1		14	26	8	20	2
2		16	28	10	22	4
3		18	30	12	24	6
4		20	2	14	26	8
5		22	4	16	28	10
6		24	6	18	30	12
7		8	20	2	14	26
8		10	22	4	16	28
9		12	24	6	18	30
10		14	26	8	20	2
11		16	28	10	22	4
12		18	30	12	24	6
13		2	14	26	8	20
14		4	16	28	10	22
15		6	18	30	12	24
16		8	20	2	14	26
17		10	22	4	16	28
18		12	24	6	18	30
19		26	8	20	2	14
20		28	10	22	4	16
21		30	12	24	6	18
22		2	14	26	8	20
23		4	16	28	10	22
24		6	18	30	12	24
25		20	2	14	26	8
26		22	4	16	28	10
27		24	6	18	30	12
28		26	8	20	2	14
29		28	10	22	4	16
30		30	12	24	6	18

Fig. 5a: The format of the 'Table of Life' in the Arabic sources, the order of the columns now reversed.

(degrees) are contributing (*mushtarika*).' Here we should surely read 26 rather than 6 for the amounts added (again *modulo* 30) at the first degree of each sign to the value at the end of the previous sign. Vettius Valens is quoted as saying that there is a difference of 12° between the values for the first degree of consecutive signs, a remark confirmed by inspection of the table. Furthermore, the treatise indicates that the entries in the table begin from the left hand side, as one would expect in a Greek astronomical table but not in an Arabic table, and also confirms the unusual distribution of the zodiacal signs at the head of the columns of the table (see Fig. 5a), which caused some confusion (see Fig. 3c) amongst copyists used to transcribing tables of spherical astronomical functions symmetrical about the solstices such as those in the main Cairo corpus of tables for timekeeping.[74]

The instructions on the use of the table indicate that one should first take the parts of the hours of the degree of the ecliptic which is instantaneously rising over the local horizon (*i.e.*, the horoscopus) at the time of birth, and multiply them by 12. The 'parts of the hours of the horoscopus' are then $2D(\lambda_H,\phi)/12$, so that multiplication by 12 gives us back the quantity $2D(\lambda_H,\phi)$. (This part of the instructions is absurd in that the first quantity $2D/12$ would have had to have been computed from $2D$ anyway.) Then one takes the number in the table corresponding to the horoscopus and divides it by 60, and takes this fraction of the result of the first multiplication. (For the latitude of Cairo, say, which is 30°, $2D$ varies between about 151° and 209° – since ζ varies between 2 and 30, for the latitude of Cairo the length of life varies between a minimum of about 5 years and a maximum of about 105 years.) Next one takes the quantity $\zeta(\lambda_H)$, and determines the length of life $L(\lambda_H,\phi)$ in years using formula (1) above. A marginal note in MS D in hand of the Cairo astronomer Ibn Abi 'l-Fatḥ al-Ṣūfī[75] states that to find the life expectancy one should simply multiply the entry in the table by the 'degrees of daylight (of the horoscopus)', taking the entry as sexagesimal minutes; in MS G, the copyist records a note by al-Ṣūfī in which the operation is described for the entries taken as units.

6. Some worked examples

An example is given in the treatise in MSS B and C, namely, for the case where the horoscopus is Leo 7° and the locality is Baghdad. The 'degrees of hours of the horoscopus' are stated to be 17;8° which when multiplied by 12 yields 205;36°. The appropriate entry in the *hashtaq* table is 26, the ratio of which to 60 is given as 1/3 + 1/10. The product of these two quantities is correctly calculated to be 89;5 (by truncation rather than rounding), that is, 89 years and 1 month. Another example is given in which the horoscopus is Cancer 6° and the degrees of its hours are 17;45°. The length of life is computed as 106 years and six months, a ripe old age and as old as one can get since the value of $\zeta(\lambda_H)$ in the table is 30. Values of $2D/12$ to two digits do not reflect small changes in latitude ϕ and obliquity of the ecliptic ε. However, the value 17;8° corresponds exactly to the pair: ϕ = 33° and ε = 23;51°, which are elsewhere associated with the early-9th-century astronomer

74 King, 'Cairo Corpus', pp. 351–3, and *idem*, *SATMI*, vol. 1, pp. 40–41, 43, and 45–7 for some examples,

75 See n. 49 above.

al-Khwārizmī.[76] For the parameters of the astronomers of al-Ma'mūn,[77] namely: ϕ = 33;27° and ε = 23;33°, I compute 17;9°. For the second example both sets of parameters yield 17;46° instead of 17;45° in the text. So it is not possible to determine the underlying parameters uniquely.

In MSS D and E, a different example is given. The horoscopus is now Gemini 5°, and the 'degrees of the hours' are 17;44°. Since the appropriate entry in the table is 16, the life expectancy is computed as 56 years and about 9 months. The latitude underlying the value of the 'degrees of the hours' is neither that of Baghdad nor that of Cairo, where both MSS B and D were copied, but rather about 36°, serving Raqqa, Aleppo or Rayy, or more significantly perhaps, the middle of the fourth climate.[78]

In MS G, the copyist has added another example for a baby named Muḥammad, again not his own addition (*hākadhā wajadtuhu*). The date of birth is given (according to the astronomical reckoning for the Hijra calendar) as Wednesday, Shawwāl 29, 1006 Hijra [= June 3, 1598] and the solar longitude is given as 13;13° which must be in Gemini so that λ_S = 73;13°. At midday in Cairo on that day the longitude of the sun was, by modern computation (to the nearest 0.1°) in fact about 72;3°, which says a lot for the solar tables in use at the time.[79] The time since sunrise T is given as 63;13°. This could have been computed directly from an observation of the solar altitude or found from tables of $T(h,\lambda_S)$ for each degree of both arguments which formed part of the corpus of tables for timekeeping used in Cairo throughout the medieval period.[80] Then the oblique ascensions of the horoscopus at sunrise (*i.e.*, of the solar longitude) are stated to be 57;50°. A table of oblique ascensions for Cairo with values to degrees and minutes for each degree of ecliptic longitude was standard equipment of the Cairo astronomers.[81] Accurately for ϕ = 30;0° and ε = 23;35°, the parameters of the Cairo corpus and originally due to Ibn Yūnus, the value should be 57;56°. (For the same latitude and the more recent value of the obliquity found by Ulugh Beg, namely ε = 23;30,17°, which was popular amongst the Cairo astronomers from the late 15th century onwards, I compute 58;0°, which shows yet greater divergence.) Then the oblique ascensions of the horoscopus at birth are simply 57;50° + 63;13° = 121;3°, so that the longitude of the horoscopus is Leo 9;20° = 129;20°. Using accurate tables and an oblique ascension of 121;3° I obtain 129;24°. The corresponding entry in the 'Table of Life' is 30, and the arc of daylight is given as 201;44°. For the first set of parameters I compute that it should be 201;40°. Then the length of life is correctly given as 1,40;52 (= 30/60 x 201;44) years, which corresponds to 100 years, 10 months and 12 days, although this equivalent is not given. But our copyist writes that the age of the boy when he died (*'umruhu*) was reported to be (*'alā mā dhukira*) 6 years,

---

76 King, 'al-Khwārizmī', p. 2.

77 These parameters were used in the table of oblique ascensions in MS Escorial ar. 927, fols. 49b–50a, of a recension of the *Mumtaḥan Zīj*. The pair ϕ = 33;21° and ε = 23;33° used in other spherical astronomical tables in this source is apparently due to the 10th-century astronomer Ibn al-A'lam. On early values for the latitude of Baghdad see further King, 'Earliest Muslim Geodetic Measurements', pp. 225–7.

78 See King, 'Geography of Astrolabes', pp. 6–9, also p. 10 above.

79 On the compilation of the solar ephemerides from which this value of the solar longitude was probably taken see Kennedy & King, 'Ibn al-Majdī's Auxiliary Tables'.

80 King, 'Cairo Corpus', pp. 359–62; and *idem*, *SATMI*, vol. 1, pp. 44–9 and 267–8.

81 King, 'Cairo Corpus', p. 359, and *idem*, *SATMI*, vol. 1, p. 264 and 283. See also n. 68 above.

5 months and 6 days. It is not completely clear that the boy died at this early age, not least because the usual formulae for the deceased are missing but also because the fractional parts of the computed age and actual age are suspiciously related in the ratio 2:1. It would be tempting to restore the '6' (*wāw* in *abjad* notation) years to '50' (*nūn*), but that would mean that the now mature Muḥammad was still alive when MS G was copied (1053 H), some 47 lunar years after he was born.

7. Finding the time of conception

Two notes (*fā'ida*) are found in MS B (and hence also MS C), the second of which is also found in MS D. In the first is stated the well-known rule of thumb that '(the longitude of) the moon at birth is the horoscopus at conception (*masqaṭ al-nuṭfa*), and (the longitude of) the moon at conception is the horoscopus at (the time of) birth'. This is found already in the *Centiloquium* attributed to Ptolemy (and actually due to the 10th-century Egyptian scholar Aḥmad ibn Yūsuf) as well as in numerous later Arabic sources.[82] It is invariably cited in the *namūdār* for finding the time of conception attributed to Hermes.[83]

In the second note, the number of days that the newborn will remain in its mother's womb is related to the astrological house in which the moon is to be found at the time of conception. If the moon is in the ascendant this time is 273 days (about 10 months). Otherwise the time of gestation is modified by the time taken by the moon to move from the position of the ascendant on the ecliptic to its instantaneous ecliptic position. The values given for the 12 houses starting with the ascendant are as follows:

273 275 278 280;30 283 285 288 260 263 265 268 270

The underlying scheme appears to be: take 273^d for the first house and add 2 $1/2^d$ for each house below the horizon and subtract 2 $1/2^d$ for each house above the horizon.[84] If we drop the resulting halves we obtain all of the values in the text except for the anomalous 280;30, which should perhaps be 280. As noted above, the value 273 is the number of days in ten months (actually $273^d\ 5^h$). One could argue that one should reckon with 30 / 13;11 = $2^d\ 6.6^h$ per house, this being the time the moon moves through 30° of the ecliptic from the horoscopus, corresponding roughly (if we ignore effects of latitude and obliquity!) to each house, but the text seems to use 12°/d for the lunar motion, corresponding to a month of 30^d and also roughly equal to the daily elongation of the moon from the sun. Several other Islamic tables for determining the length of gestation are available, but an investigation is beyond the scope of the present study.[85]

---

82 See Lemay, 'Centiloquium', especially pp. 97–8. For some later sources see, for example, al-Bīrūnī, *Astrology*, pp. 329–31, and Kennedy, 'Kāshī on the Ascendent', pp. 140–41.

83 Vernet, 'Tractat', p. 81.

84 *Ibid.*, pp. 91–4.

85 See n. 39 above. In MS L, pp. 199–200, there are tables giving the number of days the fetus remains in the mother's womb (*ayyām makth al-janīn fī baṭn ummihi*). On pp. 204 and 216 of the same manuscript there is a 'table of the *namūdār* of conception' displaying the 'correction of the moon' (*ta'dīl al-qamar*) for the two situations where the moon is above and below the horizon. Future researchers should also investigate the tables

The connection between the various notes and the 'Table of Life' is immediately apparent. Firstly, in the more likely situation, we consider the case of a newborn for whom the longitude of the horoscopus λ_H and the longitude of the moon λ_M have been determined at the instant of birth. The position of the moon relative to the horizon enables us to find the length of gestation and hence the date of conception. The longitude of the moon at birth is the longitude of the horoscopus on the day of conception, and this enables us to compute the precise time of conception. Secondly, we consider the situation of a couple desirous of producing a child destined to enjoy long life. The second part of the rule in the first note means that the length of life is entirely determined by the lunar longitude at conception, for this can replace the longitude of the horoscopus at birth as argument in the 'Table of Life'. One should certainly be careful about computing the lunar longitude at that time, and rounding procedures are also all-important, because if λ_H is such that the entry in the table for ζ is 30, the entry for the next degree of longitude will be just 2. In general, Muslims accept that *al-'ajala mina 'l-Shayṭān – al-ta'annī mina 'l-Raḥmān*, 'Haste is from the Devil, procrastination is from (God) the Merciful', in this particular case, *al-ta'annī* is not to be recommended. Or perhaps one would just need to get a better astrologer: a clever one could usually provide the kind of result his audience would want to hear.[86]

By considering the astrological house of the moon at the time of conception, the date of birth can be calculated using the information in the second note, and the precise time of birth is given by the first part of the rule in the first note: the lunar longitude at birth will be the same as the horoscopus at the time of conception. In either case, our material enables one to solve all of the astrological problems involved.

Appendix A: The *namūdār* of Vettius Valens

MS Berlin Ahlwardt 5751 (= L), p. 203 – see Fig. 3f

'The *namūdār* of Wālīs. If you want to rectify (the value of) the longitude of the horoscopus and you have doubts about the hour when the newborn was born, investigate in which quarter of the day and night he was born and determine [the time of day (!)] and (the approximate value of) the horoscopus by guessing (*al-takhmīn*) for that quadrant as I shall describe to you. You find the horoscopus and the degrees of the hours corresponding to the solar longitude for the time when he was born if the birth was during the day and corresponding to the opposite point of the ecliptic if the birth was during the night. When you have found it multiply by 12 and multiply the result by the number of hours you guessed. If the result is greater than 360 subtract 360s until the remainder is (less than 360): keep the result in mind; it is called the 'indicator of the horoscopus' (*dalīl al-ṭāli'*). Then take the difference between the sun and moon in degrees of the ascensions and this is

associated with al-Sijzī (*ca.* 980) in MS Cairo ṬFF 18,3 (see *Cairo Survey*, no. B56–2.1.5), as well as those in the *Zīj*es of al-Qāsim ibn Maḥfūẓ al-Baghdādī (1285), extant in MS Paris BNF ar. 2486, and Shams al-Dīn al-Wābkanawī al-Bukhārī (*ca.* 1320), extant in MS Istanbul Aya Sofia 2694.

[86] On the occasional necessity to fudge astronomical data in order to produce acceptable astrological predictions see North, *Horoscopes and History*, p. 52.

called the 'indicator of the sun' (*dalīl al-shams*). If the indicator of the sun is greater than the indicator of the horoscopus then you know that the hours are more than you guessed, and if the indicator of the horoscopus is greater than the indicator of the sun you know that they are less. If you want to know by how much they are greater or less then take the number of equal degrees between the sun and moon: if it is greater than 180 subtract 180 and divide the result by 12: investigate what fraction the quotient is of 15, which is the amount of one equal hour (in degrees), and investigate whether it is one-third, one-quarter, one-half (or whatever), and it will be the amount of the hour (*miqdār al-sā'a*); keep it in mind. If the indicator of the sun is greater than the indicator of the horoscopus, add your result to the hour you guessed, and if the indicator of the horoscopus is greater than the indicator of the sun, subtract it. The result after addition or subtraction will be the time when the newborn was born. Know this and peace (be upon you).

Find the horoscopus from (the time) [illegible word] by multiplying the hours at the birth by 15: the result will be the time since sunrise (in equatorial degrees). Add this to the (oblique) ascensions of the longitude of the sun by day or to the ascensions of the opposite point of the ecliptic by night. The result will be the (oblique) ascensions of the horoscopus: we take the inverse oblique ascensions and the result in degrees and minutes will be the horoscopus. If we subtract 90° from the oblique ascensions of the horoscopus the remainder will be the (right) ascensions of the tenth house (that is, upper midheaven): we take the inverse right ascensions and the result will be the longitude in degrees and minutes of the tenth house.'

Appendix B: Translation of the anonymous treatise on Vettius Valens' table

MSS Escorial ar. 961,5a (B) – see Fig. 3a – and the derivative Cairo Ṭal'at *mīqāt* 102,9 (C)

'An explanation of the *hashtaq* of Wālīs who compiled it without meanness and with no precociousness for finding the length of life. Wālīs, who is (also called) Wālinṭiyus (*W-'-l-x-ṭ-y-s*), devised a *namūdār* for correcting the longitude of the ascendant and the hour of birth so that no doubt remains (concerning their accuracy). We mentioned it and explained it in the chapter on *namūdār*s in this book of ours and we refrain from repeating this here.

(Wālīs also compiled) a table which he called the *hashtaq* starting with Libra, and he put 2 for the 1st degree of (this sign), 4 for the 2nd, 6 for the 3rd, 8 for the 4th, 10 for the 5th, and 12 for the 6th. Then he added 14 to these 12 and so the entry for the 7th degree was 26. Then he added 2 again for the 8th and so for it the entry was 28, and then 30 for the 9th, 2 for the 10th, 4 for the 11th, and 6 for the 12th. Here he added 14 and so for the 13th the entry was 20, and then he started adding 2 degrees again so that the entry for the 14th degree was 22, for the 15th 24, for the 16th 26, for the 17th 28, and for the 18th 30. Here he added 14 and so for the 19th degree the entry was 14, for the 20th 16, for the 21st 18, for the 22nd 20, for the 23rd 22, and for the 24th 24. Here he added 14 and for the 25th the entry was 8, and then he started adding 2 degrees again so that for the 26th the entry was 10, for the 27th 12, for the 28th 14, for the 29th 16, and for the 30th 18. He made Leo and Pisces the same. For the 1st degree of Scorpio (he put) 14 and added to this in the same fashion, and he made Aries and Virgo the same. For the 1st degree of Sagittarius (he put) 26 and added to this in the same fashion, and he made Taurus the same. For the 1st degree of Capricorn (he put) 8 and added to this in the same fashion, and he made Gemini the same. For the 1st degree of Aquarius and also Cancer he put 20.

The way he increased was to add 2s and if 5 degrees had passed to add 14 and then to add 2s again until he had completed all of the signs [read: sign!]. The number of degrees of increase at the beginning of each sign is 6 [read: 26!], because the 1st degree has no contribution (*ishtirāk*) and the remaining (degrees) are contributing (*mushtarika*), and they are 5 degrees, (the value in the table) increasing by 14 between (each set) and the next (? *ukhtuhā*, lit., its sister). Wālīs also said that if you add 12 to the entry for the 1st degree of each sign you obtain the entry which must be for the 1st degree of the next sign. We checked this and it was correct. He did not mention the reason (*'illa*) for his procedures for obtaining these entries or for the increases which he applied.

Then (Wālīs described the following procedure) for determining the length of life of a newborn for whom 'the stars' have indicated that he will be healthy during the years of his early upbringing and that he will have a life (beyond childhood), if God Almighty wills. You take the degrees of the hours corresponding to the longitude of his ascendant in the locality where he was born, by day or night depending on the time of his birth, multiply these by 12 and obtain the product. Then investigate the entry in the *hashtaq* corresponding to an argument equal to the longitude of the ascendant, divide this by 60 and take this fraction of the product you obtained by multiplying the degrees of the hours by 12. The result will be the years and the fractional parts months, and this will be the length of his life, if God Almighty wills.

Example. The ascendant is Leo 7° and the corresponding degrees of the hours in Baghdad are 17;8°. When we multiply this by 12 the product is 205;36. In the *hashtaq* the entry for Leo 7° is 26, the ratio of which to 60 is as one-third plus one-tenth (to unity). If we take the share (of this fraction) of the 205 degrees and (the extra) minutes whose derivation we explained, the result is 89;5, which represents 89 years and one month.

Again suppose the ascendant is Cancer 6°. The degrees of its hours are 17;45 and if we multiply this by 12 the result is 213. (The entry) in the *hashtaq* is 30, the ratio of which to 60 is one-half (to unity). If we take one-half of the product we obtain 106 (years) and 6 months.

The procedure is the same as this in all other localities [reading *al-bilād* for *al-bāb*]: one (must) calculate with the (appropriate) degrees of the hours in each locality. We have compiled the *hashtaq* table for each degree of the signs (by the procedure) you know, without working out the years and months because these change for (different) localities according to the variation in the degrees of the hours of daylight in each. In our opinion this chapter (or this method) should be helpful for finding the length of life. It should be compared with the indications of the *katkhudāh* and its supporters and the interruptions to life (*quṭū'*) necessitated by the prorogations (*tasyīrāt*) towards the bodies of the planets. (The truth ??) will become apparent by (comparing) one of these (methods) with the others, so that we can know through it what is correct of this noble science. God – may He be praised and exalted – knows best.'

Title of table: 'Table of life and the span of life for a newborn.'

Note to the left side of the table: '(The author) said that this table was found in the *Kitāb al-Kāmil fī 'l-nujūm*, 'The Perfect Book on Astronomy/Astrology', in the treasury of 'Abdallāh ibn Ṭāhir ibn al-Ḥusayn, (sent to him) by al-Ma'mūn. (It was) written in gold.'

[On the other marginalia see Section 7.]

Appendix C: Shorter text to the 'Table of Life'

MS Cairo Dār al-Kutub *mīqāt* 1108,4b, fol. 20v (G) – see Fig. 3e

'Praise be to God. This is the Table of life and the lifespan of a newborn as I found it in the *Kitāb al-Kāmil fī 'l-nujūm* in the treasury of 'Abdallāh ibn Ṭāhir ibn al-Ḥusayn (sent to him) by al-Ma'mūn and written in gold. The way to use this table is (as follows). Take the degrees of the hours corresponding to the longitude of the ascendant, multiply them by 12, and keep (the product) in mind. Then look at the table of life and take the entry opposite the argument equal to the longitude of the ascendant. Make what you find a ratio to 60 and take that ratio of the product you kept in mind. It will be the (length of) life of the newborn in years and months.

Example. A baby is born and has ascendant Gemini 5°. We take the corresponding entry in the table, which we find to be 16, which is one-sixth plus one-tenth of 60. The degrees of the hours corresponding to the longitude of the ascendant are 17;44. (When we) multiply (this) by 12 the result is 212;48°. One-sixth plus one-tenth of this is 56;45, which is fifty six years and nine months approximately. God knows best the Unseen.

(The following note) was found in the handwriting of Ibn Abi 'l-Fatḥ al-Ṣūfī. Take the diurnal arc corresponding to the ascendant and keep it in mind. Enter with the longitude of the ascendant in the table, multiply what you find by the quantity you kept in mind and divide the product by 60. The quotient will be years (of the individual's life). God knows best. This is as I found it, and as I found it I copied it.

Another example. The blessed child Muḥammad was born on Wednesday, 29 Shawwāl, 1006 (Hijra). The position (of the sun) was (Gemini) 13;13°, the time since sunrise 63;13°, the (oblique) ascensions (of the sun) 57;50°, the (oblique) ascensions of the ascendant 121;3°, and the ascendant Leo 9;20°. The (corresponding) value in the table is 30. The arc of daylight corresponding to the ascendant is 201;44°, and the product is 1,47;52, the first (digit) being 60s (*awwaluhu marfū' marra*) on the basis that the entries in the table are minutes. His lifespan according to what was related was six years, five months, and six days. God knows best. This is as I found it.'

Bibliography and bibliographical abbreviations

Berlin Catalogue: Wilhelm Ahlwardt, *Die Handschriften-Verzeichnisse der Königlichen Bibliothek zu Berlin*, vol. 17: *Verzeichniss der arabischen Handschriften*, vol. 5, Berlin: A. Asher & Co., 1893.

al-Bīrūnī, *Astrology*: R. Ramsay Wright, trans., *The Book of Instruction in the Elements of the Art of Astrology by ... al-Bīrūnī*, London: Luzac & Co., 1934, repr. Baghdad: Maktabat al-Muthannā, n.d.

al-Bīrūnī, *Chronology*: C. Edward Sachau, ed., *Chronologie orientalischer Völker von Alberuni*, Leipzig: F.A. Brockhaus & Otto Harrassowitz, 1923, repr. Baghdad: al-Muthannā, n.d., transl. as C.E. Sachau, *The Chronology of Ancient Nations – An English Version of the Arabic Text of the Athár-ul-bákiya of Albírúní, or 'Vestiges of the Past,' collected and reduced to writing by the author in A.H. 390-1, A.D. 1000*, London, 1879, repr. Frankfurt am Main: Minerva, 1984.

al-Bīrūnī: see also Togan.

Bloom, 'The Blue Koran': Jonathan M. Bloom, 'The Blue Koran: An Early Fatimid Kufic Manuscript from the Maghrib', in *Les manuscripts du Moyen-Orient – Essais de codicologie et de paléographie. Actes du Colloque d'Istanbul (Istanbul, 26–29 mai 1986)*, François Déroche, ed., Istanbul: Institut Français d'Études Anatoliennes d'Istanbul, & Paris: Bibliothèque Nationale, pp. 95–9 and pl. XIIIA.

A. Bouché-Leclercq, *L'astrologie grecque*, Paris: Ernest Leroux, 1899.

Brockelmann, *GAL*: C. Brockelmann, *Geschichte der arabischen Litteratur*, 2nd edn., 2 vols., Leiden: E.J. Brill, 1943–49, and Supplementbände, 3 vols., Leiden: E.J. Brill, 1937–42.

Buck, *Dictionary*: C.D. Buck, *A Dictionary of Selected Synonyms in the Principal Indo-European Languages*, Chicago, Ill.: The University of Chicago Press, 1949.

Cairo ENL Catalogue and *Survey*: David A. King, *A Catalogue of the Scientific Manuscripts in the Egyptian National Library* (in Arabic), 2 vols., Cairo: General Egyptian Book Organization, 1981 and 1986; and *idem*, *A Survey of the Scientific Manuscripts in the Egyptian National Library* (in English), Winona Lake, Ind.: Eisenbrauns, 1986.

Carmody, *Bibliography*: Francis Carmody, *Arabic Astronomical and Astrological Sciences in Latin Translation: A Critical Bibliography*, Berkeley & Los Angeles: University of California Press, 1956.

Centiloquium: see Lemay.

Chabás & Goldstein, '*Zīj* of Ibn al-Kammād': J. Chabás and B.R. Goldstein, 'Andalusian Astronomy: *al-Zīj al-Muqtabis* of Ibn al-Kammād', *Archive for History of Exact Science* 48 (1994), pp. 1–41.

—— , *Zacut*: *eidem*, *Astronomy in the Iberian Peninsula: Abraham Zacut and the Transition from Manuscript to Print*, *Transactions of the American Philosophical Society*, vol. 90, pt. 2, Philadelphia: The American Philosophical Society, 2000.

F.H. Cramer, *Astrology in Roman Law and Politics*, Memoirs of the American Philosophical Society 37 (1954).

Franz Cumont, *Astrology and Religion among the Greeks and Romans*, first publ. 1912, repr. New York: Dover Reprints, 1960.

Dorotheos: see Pingree, *Dorotheus*.

EI_2: *The Encyclopedia of Islam*, new edn, 12 vols. to date, Leiden: E.J. Brill, 1960 onwards.

Elwell-Sutton, *Horoscope*: L.P. Elwell-Sutton, *The Horoscope of Asadullāh Mirzā – A Specimen of Nineteenth-Century Astrology*, Leiden: E.J. Brill, 1977.

Escorial Catalogue: H.-P.-J. Renaud, *Les manuscrits arabes de l'Escurial ...* , tome II, fasc. 3: *Sciences exactes et sciences occultes*, Paris: Paul Geuthner, 1941.

Goldstein: see Chabás & Goldstein.

Ibn Hibintā, *al-Mughnī*: Ibn Hibintā, *al-Mughnī fī aḥkām al-nujūm*, 2 vols., Fuat Sezgin *et al.*, eds., Frankfurt: Institut für Geschichte der Arabisch-Islamischen Wissenschaften, 1987. [A facsimile of the Damascus and Munich manuscripts (MSS M and N)].

Ibn Nawbakht: see Labarta .

Jeffery, *Materials ...* : Arthur Jeffery, *Materials for the History of the Text of the Qur'ān – The old codices – The Kitāb al-Maṣāḥif of Ibn Abī Dāwūd together with a collection of variant readings ...* , Leiden: E.J. Brill, 1937.

Jones, *Astronomical Papyri from Oxyrhynicus*: Alexander Jones, *Astronomical Papyri from Oxyrhynicus*, 2 vols. in 1, Philadelphia, Pa.: American Philosophical Society, 1999.

Kaufmann, 'R. Haja über die Lebensdauer': David Kaufmann, 'Ein Responsum des Go'ans R. Haja über Gottes Vorherwissen und über die Dauer des menschlichen Lebens', *Zeitschrift der Deutschen Morgenländischen Gesellschaft* 49 (1895), pp. 73–84.

Kennedy, 'Kāshī on the Ascendent': Edward Stewart Kennedy, 'Treatise V of Kāshī's *Zīj*: Determination of the Ascendent', *ZGAIW* 10 (1995/96), pp. 123–45.

——, 'Kāshī's Spherical Astronomy': *idem*, 'Spherical Astronomy in Kāshī's *Khāqānī Zīj*', *ZGAIW* 2 (1990), pp. 1–46, repr. in *idem*, *Studies*, VII.

——, *Studies*: *idem*, *Astronomy and Astrology in the Medieval Islamic World* (Variorum Collected Studies Series: CS600), Aldershot: Ashgate-Variorum, 1998.

——, '*Zīj* Survey': *idem*, 'A Survey of Islamic Astronomical Tables', *Transactions of the American Philosophical Society*, N.S., 46:2 (1956), repr. n.d. [*ca.* 1990].

Kennedy & Kennedy, *Islamic Geographical Coordinates*: *idem* and Mary Helen Kennedy, *Geographical Coordinates of Localities from Islamic Sources*, Frankfurt am Main: Institut für Geschichte der Arabisch-Islamischen Wissenschaften, 1987.

Kennedy & King, 'Ibn al-Majdī's Auxiliary Tables': *idem* and David A. King, 'Ibn al-Majdī's Tables for Calculating Ephemerides', in *Journal for the History of Arabic Science* (Aleppo) 4 (1980), pp. 48–68, repr. in King, *Studies*, A–VI.

King, 'Astronomy in the Maghrib': David A. King, 'On the history of astronomy in the medieval Maghrib', in Mohammed Abattouy, ed., *Études d'histoire des sciences arabes*, Casablanca: Fondation du Roi Abdul Aziz Al Sa'ud pour les Études Islamiques et les Sciences humaines, 2007, pp. 175–218, repr, in *idem*, *Studies*, E–VI.

——, 'Cairo Corpus': *idem*, 'Ibn Yūnus' *Very Useful Tables* for Reckoning Time by the Sun', *Archive for History of Exact Science* 10 (1973), pp. 342–94, repr. in *idem*, *Studies*, A–IX. See now *idem*, *SATMI*, vol. 2, pp. 43–9 and 247–98.

——, 'Earliest Muslim Geodetic Measurements': *idem*, 'Too Many Cooks ... – A Newly-Rediscovered Account of the First Islamic Geodetic Measurements', *Suhayl* 1 (2000), pp. 71–99, repr. in King, *Studies*, E–X.

——, 'Geography of Astrolabes': *idem*, 'Bringing Astronomical Instruments Back to Earth: The Geographical Data on Medieval Astrolabes (to *ca.* 1100)', in *North Festschrift*, pp. 3–53, repr. in *idem*, *SATMI*, vol. 1, XVI.

——, 'al-Khwārizmī': *idem*, 'al-Khwārizmī and New Developments in Mathematical Astronomy in the Ninth Century', *Occasional Papers on the Near East* (New York University, Hagop Kevorkian Center for Near Eastern Studies) 2 (1983), 47 pp.

——, *SATMI*: *idem*, *In Synchrony with the Heavens – Studies in Astronomical Timekeeping and Instrumentation in Medieval Islamic Civilization*, 2 vols., Leiden & Boston: Brill, 2004–05, especially vol. 1, part I: *A Survey of Tables for Regulating Time by the Sun and Stars*, and part II: *A Survey of Tables for Regulating the Times of Prayer*.

——, *Studies*, A: *idem*, *Islamic Mathematical Astronomy*, London: Variorum, 1986.

——, *Studies*, E: *idem*, *Islamic Astronomy and Geography*, Farnham: Ashgate-Variorum, 2012.

King & Samsó, 'Astronomical Handbooks and Tables': *idem* and Julio Samsó, with a contribution by Bernard R. Goldstein, 'Astronomical Handbooks and Tables from the Islamic World (750–1900): An Interim Report', *Suhayl* 2 (2001), pp. 9–105. [A summary is in the article 'Zīdj' in EI_2.]

Kroll, *Vettius Valens*: Wilhelm Kroll, *Vetii Valentis anthologiarum libri*, Berlin: Weidmann, 1908.

Labarta, *Horóscopos históricos*: Ana Labarta, *Mūsā ibn Nawbajt: al-Kitāb al-Kāmil – Horóscopos históricos*, Madrid & Bellaterra: Universidad Autónoma de Barcelona, Instituto Hispano-Árabe de Cultura, 1982.

Lane, *Lexicon*: Edward W. Lane, *An Arabic-English Lexicon*, 8 pts., London: Williams and Norgate, 1863, repr. Beirut: Librairie du Liban, 1968.

Lemay, '*Centiloquium*': Richard Lemay, 'Origin and Success of the *Kitāb* [*al-*]*Thamara* of Abū Ja'far Aḥmad ibn Yūsuf ibn Ibrāhīm from the Tenth to the Seventeenth Century in the World of Islam and the Latin West', *Proceedings of the First International Symposium for the History of Arabic Science (Aleppo, 1976)*, 2 vols., Aleppo: Institute for the History of Arabic Science, 1978, II, pp. 91–107.

Levey, *Medieval Arabic Bookmaking*: Martin Levey, 'Mediaeval Arabic Bookmaking and its Relation to Early Chemistry and Pharmacology', *Transactions of the American Philosophical Society* (Philadelphia), N.S., 52:4 (1962).

Liddell & Scott, *Lexicon*: H.G. Liddell and R. Scott, *A Greek-English Lexicon*, 9th edn., Oxford: Clarendon Press, 1953.

Mestres, 'Hyderabad Manuscript': Àngel Mestres, 'Maghribī Astronomy in the 13th Century: a Description of Manuscript Hyderabad Andra Pradesh State Library 298', in *Vernet Festschrift*, I, pp. 383–443.

——, Zīj *of Ibn Isḥāq*: *idem*, *Materials andalusins en el* Zīj *d'Ibn Isḥāq al-Tūnisī* (edited text and tables, with introduction and commentary in English), doctoral thesis, University of Barcelona, 2000.

Millás, *Estudios sobre Azarquiel*: José María Millás Vallicrosa, *Estudios sobre Azarquiel*, Madrid & Granada: Consejo Superior de Investigaciones Científicas and Instituto Miguel Asín, 1943–50.

Nallino, *'Ilm al-falak*: Carlo Alfonso Nallino, *'Ilm al-falak: ta'rīkhuhu 'ind 'l-'Arab fi 'l-qurūn al-wusṭā*, Rome, 1911, repr. Baghdad: al-Muthannā, ca. 1960.

——, *Scritti*, V–VI: *idem*, *Raccolta di scritti editi e inediti* (ed. Maria Nallino), vol. V: *Astrologia–Astronomia–Geografia*, and vol. VI: *Letteratura–Linguistica–Filosofia–Varia*, Rome: Istituto per l'Oriente, 1944 and 1948.

——, 'Tracce': *idem*, 'Tracce di opere greche giunte agli Arabi per trafila pehlevica', in *A Volume of Studies Presented to Professor E.G. Browne*, Cambridge, 1922, pp. 345–63, repr. in *idem*, *Scritti*, VI, pp. 285–303.

Neugebauer, *HAMA*: Otto Neugebauer, *A History of Ancient Mathematical Astronomy*, 3 vols., New York, *etc.*: Springer, 1975.

Neugebauer and H.B. van Hoesen, *Greek Horoscopes*, *Memoirs of the American Philosophical Society* 48 (1959).

Neugebauer & Saliba, 'Greek Numerology': *idem* and George Saliba, 'On Greek Numerology', *Centaurus* 31 (1989), pp. 189–206.

North, *Horoscopes and History*: John D. North, *Horoscopes and History* (Warburg Institute Surveys and Texts XIII), London: University of London, Warburg Institute, 1986.

North Festschrift: *Between Demonstration and Imagination: Essays in the History of Science and Philosophy Presented to John D. North*, Arjo Vanderjagt and Lodi Nauta, eds., Leiden: E.J. Brill, 1999.

Pauly-Wissowa, *Realencyclopädie*, 2–VIII,2: G. Wissowa *et al.*, *Paulys Realencyclopädie der classischen Altertumswissenschaften*, 2. Reihe, Band VIII,2, Stuttgart: Alfred Druckenmüller, 1958.

Pingree, 'Astronomy and Astrology in Iran': David Pingree, 'Astronomy and Astrology in Iran', *Isis* 54 (1963), pp. 229–46.

——, *Dorotheus*: *idem*, ed., *Dorotheus Sidonius carmen astrologicum*, Leipzig: B.G. Teubner, 1976.

——, *Vettius Valens*: *idem*, ed., *Vettii Valentis Antiocheni anthologiarum libri novem*, Leipzig: B.G. Teubner, 1986.

Renaud: see *Escorial Catalogue*.

Saliba: see Neugebauer & Saliba.

Samsó, *Astronomy and Astrology in al-Andalus and the Maghrib*: Julio Samsó, *Astronomy and Astrology in al-Andalus and the Maghrib*, Aldershot: Ashgate-Variorum, 2007.

——, 'Maghribi *Zījes*': *idem* 'An Outline of the History of Maghribī *Zīj*es from the End of the Thirteenth Century', *Journal for the History of Astronomy* 29 (1998), pp. 93–102, repr. in *idem*, *Astronomy and Astrology in al-Andalus and the Maghrib*, XI.

Samsó: see also King & Samsó.

Sarton, *IHS*: George Sarton, *Introduction to the History of Science*, 5 vols. in 3, Baltimore: Williams and Wilkins & Co. for the Carnegie Institution, Washington, D.C., 1927–48, repr. 1962.

Sayılı, *The Observatory in Islam*: Aydin Sayılı, *The Observatory in Islam* ... (Publications of the Turkish Historical Society, Series VII, No. 38), Ankara: Türk Tarih Kurumu Basımevi, 1960, repr. New York: Arno, 1981.

Sédillot-*fils*, *Oloug-Beg*: L.P.E.A. Sédillot, *Prolégomènes des tables astronomiques d'Oloug-Beg*, 2 vols., Paris: Firmin Didot Frères, 1847–53.

Sezgin, *GAS*: Fuat Sezgin, *Geschichte des arabischen Schrifttums*, 13 vols. to date, especially vol. 6: *Astronomie*, and vol. 7: *Astrologie, ...* , Leiden: E.J. Brill, 1978 and 1979.

Steingass, *Dictionary*: F. Steingass, *A Comprehensive Persian-English Dictionary*, London, 1892, repr. Beirut: Librairie du Liban, 1975.

Suhayl = *Suhayl – Journal for the History of the Exact and Natural Sciences in Islamic Civilisation* (Barcelona). [Available on the Internet.]

Suter, *MAA*: Heinrich Suter, 'Die Mathematiker und Astronomen der Araber und ihre Werke', *Abhandlungen zur Geschichte der mathematischen Wissenschaften* 10 (1900), and 'Nachträge und Berichtigungen ...', *ibid.* 14 (1902), pp. 157–85, repr. in *idem*, *Beiträge zur Geschichte der Mathematik und Astronomie im Islam*, 2 vols., Frankfurt: Institut für Geschichte der Arabisch-Islamischen Wissenschaften, 1986, I, pp. 1–314.

Togan, 'al-Bīrūnī on the Length of Life': Zeki V. Togan, 'Muḥammad ibn Aḥmad al-Bīrūnī: *Hikayat ṭariq ahl al-Hind fi istixrac al-'umr*', *Islām Tetkikleri Enstitüsü Dergisi* (Istanbul), 1 (1954), pp. 1–25 (appended with new pagination after p. 184 of the volume).

Ullmann, *NGWI*: Manfred Ullmann, *Die Natur- und Geheimwissenschaften im Islam*, Leiden: E.J. Brill, 1972.

Vernet, *Estudios*: Juan Vernet, *Estudios sobre historia de la ciencia medieval*, Barcelona: Universidad de Barcelona (Facultad de Filología) & Bellaterra: Universidad Autónoma de Barcelona (Facultad de Filosofía y Letras), 1979.

——, 'Tractat': *idem*, 'Un tractat d'obstetricia astrologica', *Boletín de la Real Academia de Buenas Letras de Barcelona* 22 (1949), pp. 69–96, repr. in *idem*, *Estudios*, pp. 273–300.

Vernet Festschrift: *From Baghdad to Barcelona–Studies in the Islamic Exact Sciences in Honour of Prof. Juan Vernet*, Josep Casulleras and Julio Samsó, eds. (*Anuari de Filologia* (Universitat de Barcelona) XX (1996) B–2), 2 vols., Barcelona: Instituto 'Millás Vallicrosa' de Historia de la Ciencia Árabe, 1996.

Vettius Valens: see Kroll and Pingree.

ZGAIW: *Zeitschrift für Geschichte der arabisch-islamischen Wissenschaften* (Frankfurt am Main).

Weil, *Maimonides über die Lebensdauer*: Gotthold Weil, *Maimonides: Über die Lebensdauer – Ein unediertes Responsum*, Basle: Karger, 1953.

VIII

The Sacred Geography of Islam

Contents

1. Introduction

In Islam, as in no other religion in human history, the performance of the most important ritual acts—prayer at specific times and in a specific direction, and fasting during a particular month—has been assisted by procedures involving astronomy and mathematics. This activity resulted from a distinctive conception of time and space in Islam. The organisation of the lunar calendar, the regulation of the astronomically-defined times of prayer, and the determination of the sacred direction towards the sacred Kaaba in Mecca—these are three topics of traditional Islamic science still of concern to Muslims today, and each has a history going back close to 1400 years. They form a small but highly significant part of a scientific tradition which knew no rival for several centuries. This paper deals with the third of these topics, the sacred direction in Islam and the corresponding conceptions of a sacred geography.

The Kaaba is a shrine of uncertain historical origin that served as a sanctuary and centre of pilgrimage for centuries before the advent of Islam in the early 7th century. In Muslim tradition the Kaaba was built by Abraham and Ishmael; otherwise practically nothing is known of its early history beyond what can be intimated from the nature and layout of the edifice. In any case, the *Qurân* advocates prayer towards the Kaaba, and the edifice was adopted as the focal point of the new religion, a physical pointer to the presence of God. Thus, since the early 7th century, Muslims have faced the Kaaba during their prayers. The sacred direction is called *qibla* in all languages of the Islamic commonwealth. Mosques are built with the prayer-wall facing the Kaaba, the direction being indicated by a *mihrâb* or prayer-niche. In addition, according to Islamic law, certain ritual acts—such as reciting the *Qurân*, announcing the call to prayer, and slaughtering animals for food—are to be performed facing the qibla. Also Muslim graves and tombs were laid out so that the body would lie on its side and face the qibla. Thus the qibla is of prime importance in numerous aspects of the life of every Muslim.

2. The dichotomy of science in Islamic civilisation

Islamic science functioned on two different levels throughout the period 750–1900. First, there was the folk scientific tradition, which had the backing of the religious authorities. Second, there was the mathematical tradition, which was cultivated by a small subset of society but whose prescriptions were generally too complicated to be put to into practice. Nowhere is this dichotomy more clear than in the situation regarding the obligation of Muslims to pray and to perform various other ritual acts facing the qibla.

The Muslims inherited the folk astronomical and folk cosmological notions of pre-Islamic Arabia, and when they expanded from the Arabian Peninsula they encountered the folk astronomy of the Hellenistic world. Arabic texts on folk astronomy, of which the most developed stem from the Yemen, deal with the changing seasons, with the stars and the lunar mansions, with the phases of the moon, with time-reckoning by shadow-lengths, as well as with the winds and the effects of the seasons on agriculture. In particular, directions were defined by astronomical horizon phenomena: the risings and settings either of the sun at the equinoxes and solstices or of the brightest fixed stars. As we shall see, the techniques of folk astronomy were employed by the legal scholars to determine the qibla, and a sacred geography defining the positions of various parts of the world in relation to the central shrine, the Kaaba, was developed (Section 3).

The Muslims were also heirs to the Hellenistic, Iranian and Indian traditions of mathematical astronomy and to the Hellenistic tradition of mathematical geography and cartography. They assimilated these with remarkable facility, and it was not long before they were in a position to criticise and improve upon them, and to select what they found most acceptable. Out of this amalgam of notions there developed a distinctly Islamic astronomy, which was without rival from the 9th to the 15th century. (The label "medieval" used for this period and the scientific tradition which it produced has no negative connotation since the tradition was unrivalled at the time; the term is here intended simply to distinguish between the scientific traditions of Antiquity and the modern age.) Muslim scientists

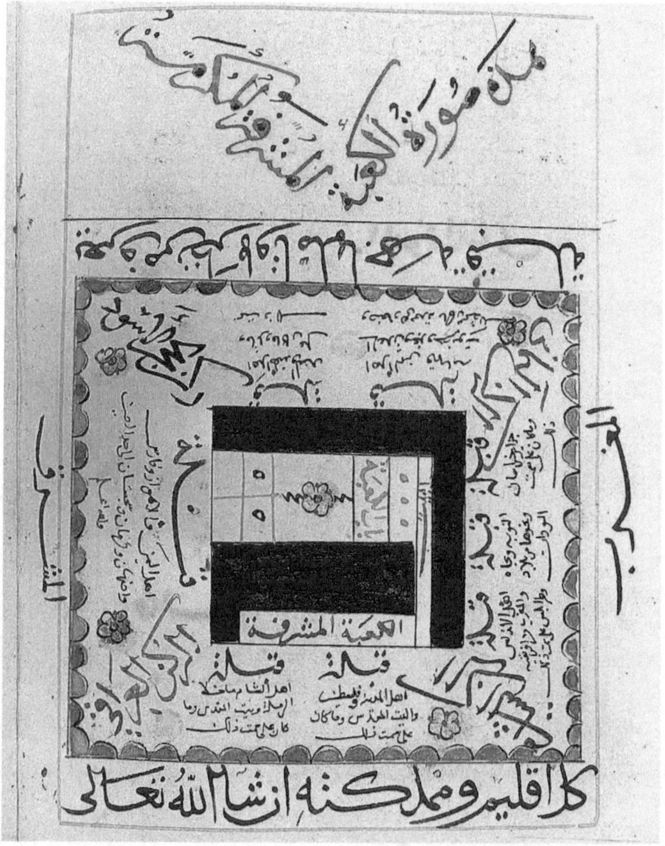

Fig. 1. A diagram of sacred geography with the qiblas of eight different regions of the Islamic commonwealth displayed around the four walls of the Kaaba. This is found in one of the copies of the cosmography of the early-15th-century scholar Pseudo-Ibn al-Wardî. [Taken from MS Istanbul Topkap AIII 3025, fol. 30v, courtesy of the Topkap Library.]

developed new trigonometric techniques and manifold tables for solving various astronomical problems. As we shall see (Section 4), they applied mathematical techniques to the available geographical data in order to calculate the qibla in various places. The scientists developed a sacred geography of their own, defining the directions of various cities to the centre, now Mecca, the city, rather than the Kaaba, the edifice. This distinction is significant and characterises the two different approaches.

3. The sacred geography of the legal scholars

The scholars of the sacred law of Islam created a sacred geography based on the notion that in order to face the Kaaba in any region of the world one should face the same direction in which one would be standing if one were directly in front of the appropriate segment

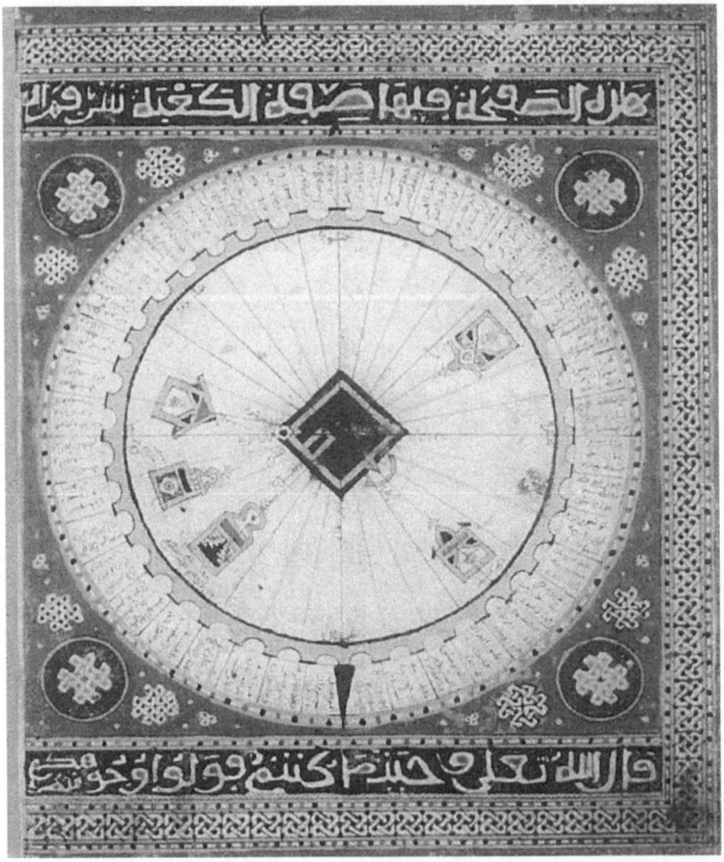

Fig. 2. A diagram of the world divided into sectors about the Kaaba, found in a copy of the navigational atlas of the 16th-century Tunisian scholar Ahmad al-Sharafî al-Safâqusî. [From MS Paris BNF ar. 2278, fol. 2v, courtesy of the Bibliothèque nationale de France.]

of the perimeter of the Kaaba. Since that sacred edifice is itself aligned in astronomically-significant directions, the directions adopted by the legal scholars for the qibla were towards the risings and settings of the sun at the equinoxes or the solstices or of various significant qibla-stars.

From the 9th to the 16th century Muslim scholars working exclusively in the folk astronomical tradition developed a tradition of sacred geography in which the qiblas of different regions of the world around the Kaaba were associated with particular astronomical horizon phenomena. Some 20 different schemes of this kind of sacred geography—sometimes illustrated in manuscripts, sometimes described in word—are known from some 30 different medieval sources, and there are surely more yet to be rediscovered. These directions adopted for the qibla obviously corresponded only roughly to the qiblas that were calculated by the Muslim astronomers. This tradition was particularly popular in the medieval Yemen, not least because it was a legal scholar of Yemeni origin named Ibn Surâqa who ca. 1000 first proposed various serious schemes with 8, 11 and 12 sectors around the Kaaba. In various later works, such as the geographical writings of Yâqût (Syria ca. 1200) and al-Qazwînî (Iraq ca. 1250), the information on the qibla is suppressed, and in yet later works such as the nautical atlas of Ahmad al-Sharafî al-Safâqusî (Tunis ca. 1575) and various other Ottoman compilations, the localities are uniformly distributed on a ring around the Kaaba with no specific qibla-values. Figures 1 and 2 show two examples of these schemes of sacred folk geography, involving no calculations whatsoever.

4. The sacred geography of the scientists

4.1. *Mathematical solutions to the qibla problem*

Muslim astronomers from the 8th century onwards concerned themselves with the determination of the qibla as a problem of mathematical geography. This activity required knowledge of geographical coordinates and involved the computation of the direction of one locality from another by procedures of geometry or trigonometry, such as the analemma (a sophisticated device for reducing problems on a sphere from three dimensions to two) or spherical trigonometry.

The qibla at any locality was defined as the direction to Mecca along the great-circle on the terrestrial sphere. The basic problem, illustrated in Figure 3, is to determine the direction to Mecca M from any locality X, given the latitudes of both localities, measured by MB ($= \phi_{\mathrm{M}}$) and XA ($= \phi$), and the longitude difference AB ($= \Delta L = L - L_{\mathrm{M}}$). The qibla is measured by the angle AXM ($= q$). We shall also have occasion to use the latitude difference XD ($= \Delta\phi = \phi - \phi_{\mathrm{M}}$) and the distance to Mecca XM ($= d$).

Once the geographical data are available, a mathematical procedure is necessary to determine the qibla. The first Muslim astronomers who considered this problem developed a series of approximate solutions, all adequate for most practical purposes. The approximate solutions were all inspired by cartographic considerations. However, also in the early 9th century, accurate solutions by solid trigonometry and projection methods were formulated, and in the 10th century solutions by spherical trigonometry. Over the centuries one

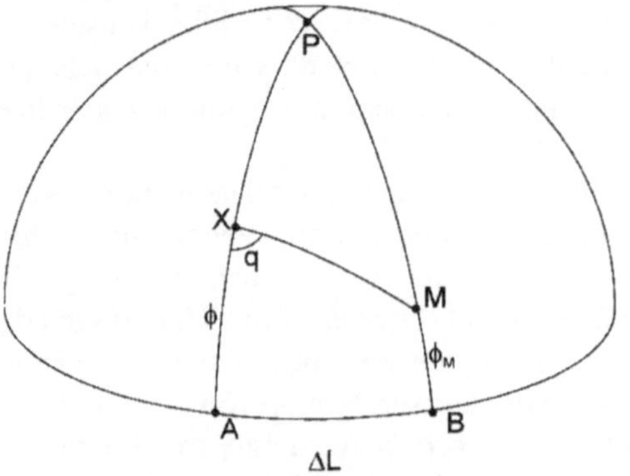

Fig. 3. The mathematical problem of determining the qibla.

"standard approximate method" was generally advocated in popular treatises on astronomy, and the most widely-known exact procedure amongst the experts was the "method of the *zîjes*". (A *zîj* is an astronomical handbook with tables, and some 225 works of this kind were compiled between the 8th and the 19th centuries.)

There is evidence that various scholars over the centuries favoured simple cartographic solutions to the qibla problem: in an anonymous treatise from 9th-century Baghdad three essentially cartographic methods are presented, together with an accurate solution by solid trigonometry and even a table displaying the qibla for a whole range of latitudes and longitudes. The author, as yet unidentified, was clearly capable of considering the qibla-problem in both cartographic and mathematical terms. The approximate procedures he presents are equivalent to:

$$q = \tan^{-1}\left\{\frac{\Delta L}{\Delta\phi}\right\}, \quad \tan^{-1}\left\{\frac{\Delta L \cos\phi_{\mathrm{M}}}{\Delta\phi}\right\}, \quad \tan^{-1}\left\{\frac{\sin\Delta L \cos\phi_{\mathrm{M}}}{\sin\Delta\phi}\right\}.$$

The corresponding basic triangles are shown in Figure 4(a)–(c):

Only the first of these had any following in later centuries, namely, on maps with square grids on which a given locality was simply joined to Mecca by a straight line.

Another approximate method first attested in the early 9th century but possibly earlier lends itself to the simple geometrical construction shown in Figure 4(d), and is equivalent to the formula:

$$q = \tan^{-1}\left\{\frac{\sin\Delta L}{\sin\Delta\phi}\right\}.$$

This, the "standard approximate method", was widely used until the 19th century but was generally not approved by the most serious astronomers.

The modern formula for the qibla is:

$$q = \cot^{-1}\left\{\frac{\cos\Delta L \sin\phi - \cos\phi \tan\phi_{\mathrm{M}}}{\sin\Delta L}\right\},$$

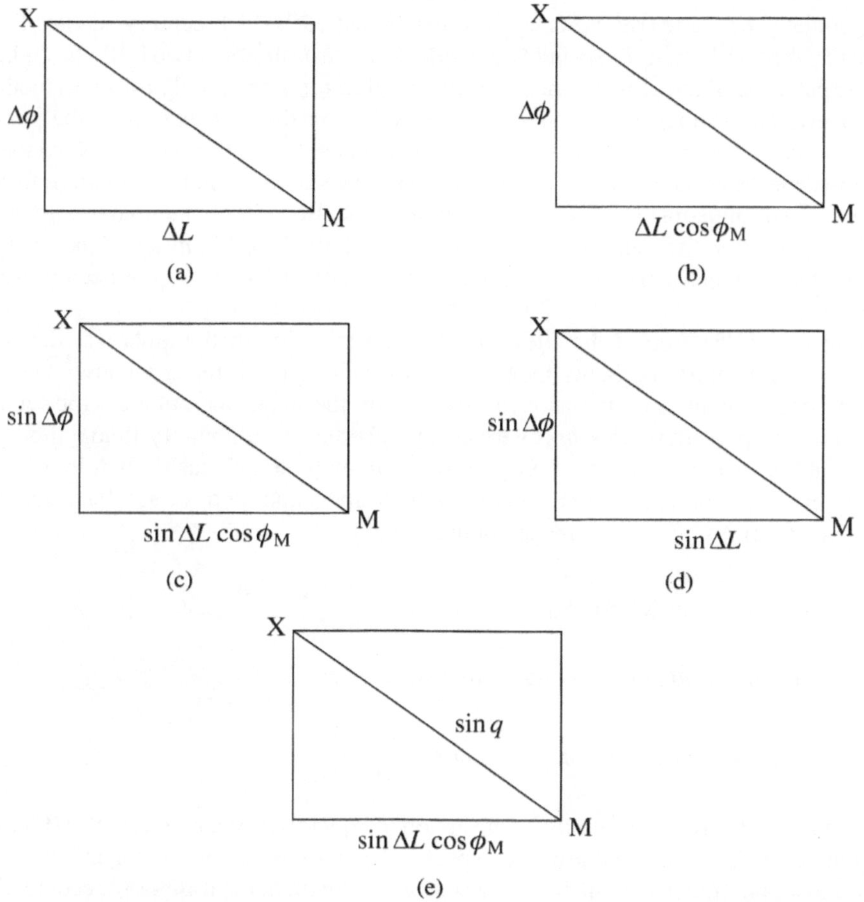

Fig. 4. (a)–(c) The underlying diagrams from which the 9th-century approximations can be derived. (d) The standard approximate construction for the qibla proposed by al-Battânî ca. 910. (e) The triangle from which the qibla can be derived from the distance to Mecca, according to the last procedure in the "method of the *zîjes*", first derived in the 9th century. This underlies the Mecca-centred maps discussed in 4(e).

and various accurate solutions were devised by Muslim astronomers from the 9th century onwards which were mathematically equivalent to this although they usually involved various steps. The distance to Mecca was also of interest. It may be derived using:

$$d = \sin^{-1}\left\{\frac{\sin \Delta L \cos \phi_M}{\sin q}\right\}.$$

As noted above, an anonymous astronomer in 9th-century Baghdad derived an accurate formula for the qibla by solid trigonometry. From that century onwards, numerous Muslim scientists discussed the qibla problem, presenting other accurate solutions by spherical trigonometry, or reducing the three-dimensional problem to two dimensions and solving it by geometry or plane trigonometry. The celebrated early-11th-century scholar Ibn al-Haytham, who was born in Basra but later worked in Cairo, authored two brilliant studies of the determination of the qibla by means of spherical trigonometry and projection methods, respectively. The former, written first, is remarkable for the fact that all of the possible cases ($0 < \Delta L < 360°$ and $-90° < \phi < +90°$) are considered. The second, which uses an analemma construction, is spectacular for its conciseness, and from it the modern formula can be derived immediately. Also Ibn al-Haytham mentions in his autobiography that he wrote a treatise "on the determination of the azimuth of the qibla in all of the inhabited world by tables which I compiled..."; unfortunately, this treatise has not been preserved for us.

The so-called "method of the *zîjes*" for the determination of the qibla was developed in the mid-9th century, probably by Habash al-Hâsib, who is the first known author to have proposed it explicitly. It was also favoured by the major astronomers for centuries thereafter. The procedure was first derived by spherical trigonometry (anonymous 9th-century author, maybe Habash), then by projection methods (Habash), then by spherical trigonometry (various 10th-century scholars). It involves finding three auxiliary quantities "modified" from ΔL, ϕ_{M}, and $\Delta\phi$ and defined using:

$$\sin \Delta L' = \sin \Delta L \cos \phi_{\mathrm{M}}; \qquad \sin \phi'_{\mathrm{M}} = \frac{\sin \phi_{\mathrm{M}}}{\cos \Delta L'}; \qquad \Delta\phi' = \phi - \phi'_{\mathrm{M}}.$$

With these auxiliary quantities d and q are determined by:

$$\cos d = \cos \Delta L' \cos \Delta\phi', \qquad \sin q = \frac{\sin \Delta L'}{\sin d}.$$

What is involved here is essentially a conversion of equatorial/polar coordinates (longitude and latitude) to coordinates centred on a specific point on the sphere (qibla and distance).

The most comprehensive medieval discussion of mathematical qibla-procedures is by the celebrated al-Bîrûnî, who worked in Central Asia ca. 1025. In his treatise *Tahdîd nihâyât al-amâkin*, "The Determination of the Locations of Cities", the leading scientist of medieval Islam set out to establish the geographical coordinates of Ghazna (now in Afghanistan) in order to compute the qibla there: the result was an involved discussion of determinations of terrestrial latitude, the length of a degree along the terrestrial meridian, the differences in terrestrial longitudes by eclipse observations, the distances between cities by measuring caravan routes, and the derivations of the qibla for a given locality by mathematics, *inter alia*. Having first determined the longitude difference between Mecca and Ghazna with remarkable accuracy, al-Bîrûnî then applied complicated techniques of spherical trigonometry and projection methods to derive the qibla. His treatise is the most valuable work on mathematical geography from the medieval period.

4.2. *Tables displaying the qibla*

Muslim astronomers compiled a series of tables displaying the qibla q for each degree (within a certain range) of longitude different from Mecca ΔL and latitude difference $\Delta\phi$

Fig. 5. An extract from the qibla-table of al-Khalîlî, showing qibla-values for the latitude range 28°–33°. The entries are written in standard Arabic alphanumerical notation. [From MS Paris BNF ar. 2558, fols. 55v–56r, courtesy of the Bibliothèque nationale de France.]

or latitude ϕ based on both approximate and exact formulae. The first of these was prepared in Baghdad, probably in the 9th century, and displays values of the qibla in degrees and minutes were given ostensibly for each 1° of ΔL and $\Delta\phi$ from 1° to 20°. Several other qibla-tables are known but the most impressive from the point of view of its scope and its accuracy is that of the 14th-century Damascus astronomer Shams al-Dîn al-Khalîlî, which displays $q(L, \phi)$ to degrees and minutes for each 1° of ΔL from 1° to 60° and each 1° of ϕ from 10° to 50°, with most of the 2 880 entries accurately computed: see Figure 5.

The extensive geographical tables that were a standard feature of Islamic *zîj*es rarely included qibla-values. The earliest table of this kind was by Abd al-Rahmân al-Khâzinî, who worked in Merw in Central Asia ca. 1125. It gives the qibla-values of about 250 localities along with their longitudes and latitudes. These basic coordinates appear to have been read from a map, for they are mainly those of his predecessor by one century, al-Bîrûnî, as found in his astronomical handbook, but generally rounded to the nearest 0; 10°, and occasionally there is a jump of 1° or 2° in one or other of the coordinates. al-Khâzinî's qibla-values are likewise given to the nearest 0; 10° but invariably they do not correspond to the longitudes and latitudes as well as one might expect; indeed, the errors can be as much as several degrees in places. Another table of this kind is found in a work by the Cairene astronomer Najm al-Dîn al-Misrî ca. 1325. This gives the longitudes, latitudes and qiblas (measured from the east or west) of some 150 localities, and whilst entries for both coordinates are given to the nearest degree, the qibla-values are computed to the nearest minute but do not correspond happily to the coordinates. Several other Egyptian tables of this kind present similar problems. A more successful table was that associated with al-Khalîlî (see above), serving 48 localities mostly in Syria and Palestine: here the qibla-values are remarkably accurate.

An enormous table very different from any of the others in its scope was compiled in 15th-century Timurid Central Asia. The author has not been identified, but he seems to have had a connection with the astronomical activity in Samarqand ca. 1425. In the available version of this table, values of the qibla and the distance to Mecca are given to the nearest minute, virtually without error, for 274 localities. But the original table had values of q and d to seconds, and these were accurate to within a few seconds. The information in this remarkable table, in its simplified form, was used in the gazetteers which were common on Iranian astrolabes and compasses from the 17th century onwards (see Section 4.3).

4.3. *Instruments for finding the qibla*

Several types of instruments were devised over the centuries for finding the qibla. One variety serves to actually determine the qibla of any locality: isolated examples of spherical devices for finding the qibla graphically are attested from the 12th century to the 14th. Another variety has markings based on computed qibla-values that provide a convenient practical means of displaying the qibla. From the 12th century onwards we find qibla directions for a specific locality marked on the back of an astrolabe or on a horizontal sundial. Iranian instrument-makers from the 16th century onwards were fond of representing graphically on the backs of astrolabes, for various localities, the solar altitude in the direction of the local qibla for all solar longitudes (that is, throughout the year).

Qibla-indicators in the form of magnetic compasses with information on the qiblas of various localities engraved around their rims are described in texts from the Yemen and from Egypt ca. 1300. A glazed ceramic bowl from Damascus ca. 1520 with qibla-values painted on it seems to attest to an earlier Iranian tradition. Brass qibla-indicators were particularly popular in Iran from the 17th century onwards, now using the qibla-values from the Timurid geographical table (see Section 4.2) in various degrees of corruption.

4.4. *Maps for finding the qibla*

Muslim scholars also prepared maps specifically for finding the qibla. One from early-13th-century Egypt is naive: on a standard coordinate grid fitted within a circular frame one simply joins one's locality to Mecca and measures the inclination to one's local meridian—see Figure 6. In spite of its simplicity, this approximate procedure works reasonably well

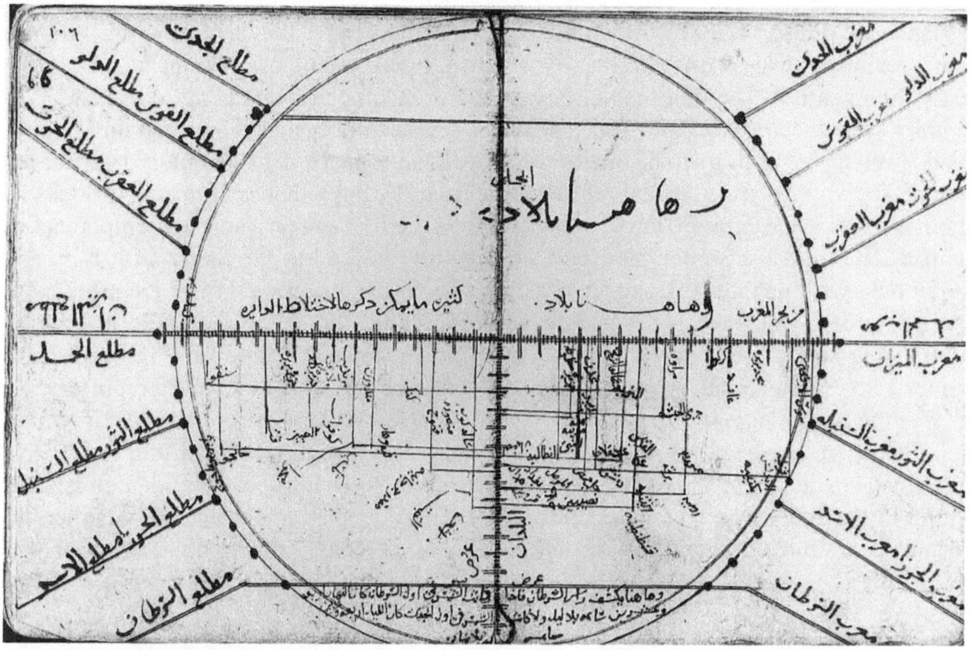

Fig. 6. A map of the world combining mathematical geography (localities plotted on a grid according to their coordinates) with folk geography (directions defined in terms of the rising and setting of the sun). The map occurs in a copy of an Egyptian treatise on folk astronomy compiled ca. 1210. The coordinates of the localities on the map are in the tradition of al-Bîrûnî (ca. 1025) and al-Khâzinî (ca. 1125), hence the map is of considerable historical interest. [From MS Princeton Yahuda 4657, fols. 65v–66r, courtesy of Princeton University Library.]

for localities in such regions as Egypt and Iran. Yet the map is not Mecca-centred and directions around the circumference are crudely associated with solar risings and settings. The map thus constitutes a unique example of the combination of mathematical cartography and folk astronomy. Another curious qibla-map with a grid based on an orthogonal projection is found on an astrolabe from Lahore dated 1666. A 'qibla-map' 'invented' by an Armenian for the Ottoman Grand Vizier in the late 1730s, is simply a European map of the landmass north of the equator between West Africa and Japan fitted with a magnetic compass and an additional pointer at Mecca, which, unhappily, is not at the centre. A later Ottoman world-map actually centred on Mecca is also known but has not been studied.

4.5. *Three Safavid world-maps centred on Mecca*

In 1989, 1995 and 2001, respectively, three remarkable world-maps from Safavid Iran came to light. Each is engraved on a circular brass plate some 22.5 cm in diameter and each is fitted with a highly sophisticated longitude and latitude grid centred on Mecca: see Figure 7. Some 150 localities are carefully marked on each map, the coordinates having being taken from the Timurid table mentioned above (Section 4.2). A non-uniform linear scale attached at the centre and a scale around the circumference enables the user to read off at a glance the direction of Mecca and the distance to Mecca. A grid achieving this purpose was previously unknown to the history of cartography, Islamic or European, before the early 20th century. The first map is unsigned and incomplete; its dating to the late 17th century seems secure on art-historical grounds. The second is signed by a certain Muhammad Husayn who appears to be one of the instrument-makers of 17th-century Isfahan, but who is known only from textual references and not by any other instruments: it looks as though it was made ca. 1800 but it is as genuine as the first and probably contemporaneous with it. The third is also complete and as elegantly finished as the first, and it is signed by an otherwise-unknown Hasan Husayn, who is, perhaps, however, to be identified with the maker of various unsigned sundials and qibla-indicators bearing a Shiite inscription mentioning both the imams Hasan and Husayn, which seem to date from the early 18th century. So even with the instruments themselves there are some unresolved problems.

Was the world-map grid designed by a 17th-century Safavid astronomer? I cannot accept that it was, since all of the contemporaneous writings from Safavid Iran that I have examined are devoid of scientific initiative. Now there were Europeans resident in Isfahan in the 17th century who were in contact with the Iranian instrument-makers. Also the instruments are fitted with a European-type magnetic compass (although the compass was

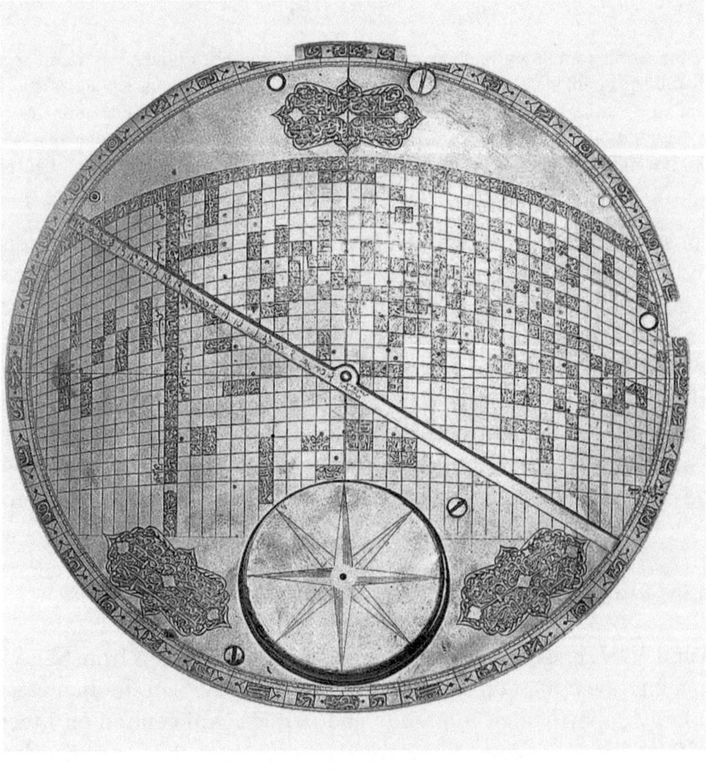

Fig. 7. The Safavid world-map rediscovered in 1989. [Dar al-Athar al-Islamiyyah, Kuwait, photo courtesy of Christie's of London.]

known in Islam at least since 1300) and a European-type inclining sundial (although such sundials were known in 14th-century Syria and 15th-century Egypt). If the grid was proposed to the Safavid instrument-makers by a European then that person would have had to have been exceptionally gifted. I could find no trace of European input on the grids, in spite of having considered what is known about the Europeans who were in Isfahan in the 17th century.

The problem with the maps is simply that they *look* very Islamic both from a geographical point of view and from a mathematical point of view: they feature essentially the world of medieval Islamic cartography and use somewhat outdated geographical data for regions of the world other than Greater Iran. Also, the underlying mathematics is in fact very simple when considered in terms of the well-known Islamic "method of the *zîj*es" for finding the qibla: see Figure 4(e). Furthermore, the latitude curves are approximated with arcs of circles (the centre of the arc with the smallest radius is still visible, the other centres would be off the map), and there is a well-attested tradition in Islamic instrument-making of using arcs of circles to approximate graphic representations of more complex curves (in this case, ellipses). Nevertheless, the idea of using such a method in this way to produce such a map-grid smacks of genius; also, even given the basic formula underlying the grid, the way to construct the grid is not self-evident. I have argued that the maps, all of which are without any doubt copies of an original, might have been inspired by a prototype developed alongside the monumental Timurid geographical table mentioned above. Concrete evidence of this is lacking. But I suspect that the original inspiration for such a grid might have come much earlier, and I have hypothesised that it was perhaps mentioned in some treatise now lost by al-Bîrûnî or his predecessor Habash. al-Bîrûnî wrote on mappings which preserve direction and distance to the centre, and Habash, some of whose works were available to al-Bîrûnî, devised an astrolabe based on such a projection (which is not the same as the one underlying the Safavid world-maps). I could show only that both of these early scholars and the later Timurid calculator had the necessary interest and ability to have devised a grid such as features on the Safavid world-maps, but not that they ever did.

The Safavid world-maps thus remain something of an enigma. Indeed, they have already become the object of some controversy, with some colleagues insisting, for example, that the possibility of European influence behind the grids should not be ruled out, or that one should not underestimate the Safavid astronomers. One colleague has proposed that the grids were added after the localities, which would be a first in the history of cartography, but he overlooked the fact that the deviations in the positions of the localities on the maps with respect to the longitudes and latitudes in the Timurid table clearly indicate that the grids were constructed first. Hence the inspiration for the Mecca-centred world-maps and their early development of remain clouded in obscurity.

Added in proof

In 2001, Professor Jan Hogendijk of Utrecht identified two Arabic treatises, one from 10th-century Baghdad and the other from 11th-century Isfahan, both dealing with the determination of the qibla by means of ellipse segments. The latitude curves on the maps are arcs of circles approximating segments of ellipses. We have thus come considerably closer to the origin of the Safavid world-maps, and the publication of Hogendijk's discoveries is awaited with anticipation.

5. The orientation of mosques and Islamic cities

The alignment of medieval mosques reflects the fact that the astronomers were not always consulted on their orientation. But since we now know from textual sources which directions were used as a qibla in each major locality we can better understand not only the mosque orientations but also can recognise numerous cities in the Islamic world that can be said to be qibla-oriented. In some, such as Taza in Morocco and Khiva in Central Asia, the orientation of the main mosque dominates the orientation of the entire city.

In the case of Cairo, various parts of the city and its suburbs are oriented in three different qibla directions: see Figure 8. The Fatimid city, founded in the 10th century, faces winter sunrise, which was the qibla of the Companions of the Prophet who erected the first mosque in nearby Fustat some three centuries previously. The Mamluk "City of the Dead" faces the qibla of the astronomers. The suburb of al-Qarâfa faces south, another popular qibla direction. The outer south-east walls of the splendid Mamluk mosques and madrasas on

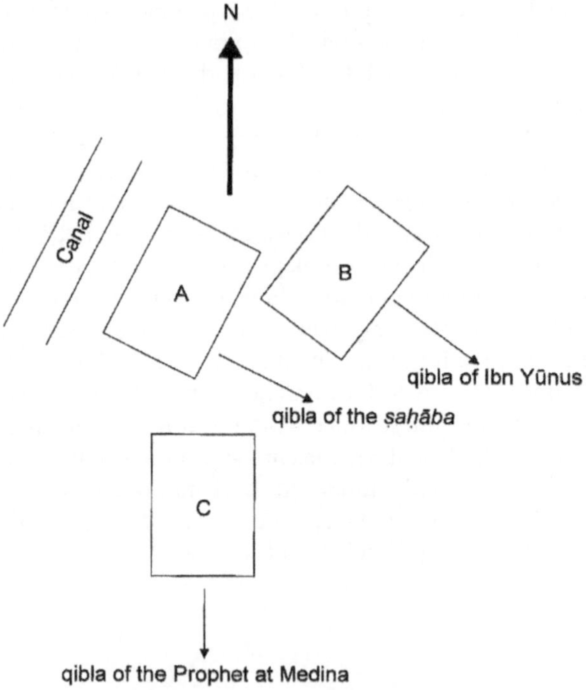

Fig. 8. The main orientations of three parts of medieval Cairo. Each part is qibla-oriented, the old city (A) in the qibla of the Companions of the Prophet (winter sunrise at ca. 27° S of E), the 'City of the Dead' (B) in the qibla of the astronomers (at 37° S of E), and the area around the tomb of al-Shâfi'î (C) in the qibla of the Shâfiite school, that is, the qibla of the Prophet when he was in Medina (due south).

the main thoroughfare of the city are aligned with the street plan, and the insides of the walls are aligned with the qibla of the astronomers; one can observe the varying thickness of the walls, which are at a 10° angle to each other, when standing in front of the windows inside the mosque overlooking the street outside. Figure 9 shows the qiblas used in Cairo and also Cordova and Samarqand.

Of all the regions of the Muslim world it is al-Andalus which has been best served by modern scholarship on orientations. A recent study by Mònica Rius on orientations in al-

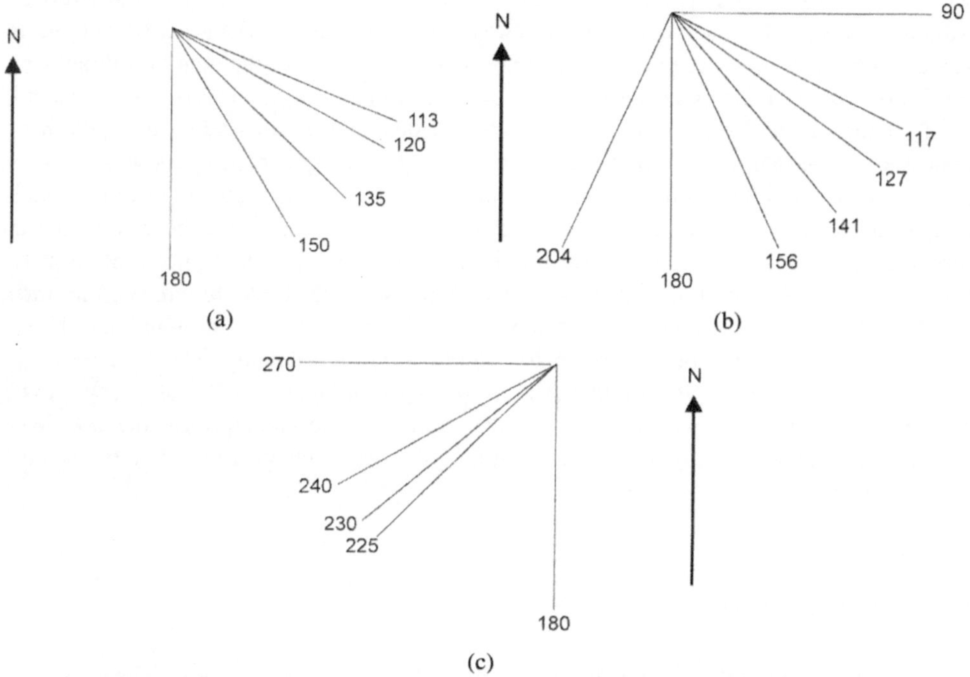

Fig. 9. Various qiblas accepted in Cordoba, Cairo-Fustat and Samarqand by different interest groups. (a) According to a 12th(?)-century Andalusi treatise on the astrolabe, mosques in Cordova were oriented in these different directions: 23° S of E, the qibla computed by the standard approximate formula (which works well for localities in the central regions of the Islamic world, but not for al-Andalus, where the error is more than 10°); 30° S of E, winter sunrise; 45° S of E, a compromise between due east and due south; 30° E of S, the direction of the Great Mosque (which is 'parallel' to the major axis of the Kaaba); and due south (not specifically mentioned in this text). (b) The Egyptian historian al-Maqrîzî (d. 1442) mentioned these qiblas as being used for mosque orientation in Cairo: due east (not specifically mentioned in this text); 27° S of E, winter sunrise, the qibla of the Companions of the Prophet; 37° S of E, the qibla of the astronomers, computed according to an exact procedure and first attested in the writings of the 10th-century Fatimid astronomer Ibn Yunus; 39° E of S, the qibla of the Mosque of Ibn Tûlûn, variously explained; and any direction between ca. 24° E of S and ca. 24° W of S, between the rising and setting of Canopus. (c) The legal scholar al-Bazdawî (d. 1089) reported these qiblas as being used for mosque orientation in Samarqand: due west, used by the Hanafite school of law and corresponding to the direction in which the road to Mecca left the city; 30° S of W, winter sunset, as used for the Great Mosque; 40° S of W, a value underlying a table for the altitude of the sun in the azimuth of the qibla, presented by al-Bazdawî but lifted from some earlier source; and due south, used by the Shâfiite legal school, corresponding to the qibla of the Prophet in Medina.

Andalus and the Maghrib compares the information in medieval folk astronomical texts with actual orientations. This is an area of the history of urban development in the Islamic world which has only recently been understood for the first time, not least because, prior to the discovery of the textual evidence, it was by no means clear which directions were used as qiblas, and even if a qibla direction at variance from the "true" qibla was clearly popular, it was not known why.

Only with the systematic scientific cartographic surveys of the 18th and 19th centuries did the first accurate measurements of geographical longitudes of localities in the Islamic world become available. Thus most of the accurately-computed qiblas of the Muslim astronomers were off by a few degrees anyway. Nowadays urban Muslims are content to use the qibla directions found by calculation from modern geographical coordinates. In rural areas where there is no mosque nearby, astronomical horizon phenomena are still used for the sacred direction. In recent years, various devices for finding the qibla have appeared on the market. The first were in the form of a magnetic compass with a list of directions for the qiblas in the major cities of the world. In some later electronic models one would feed in one's latitude and longitude, and the device would beep until it was turned to the appropriate direction for the qibla, upon which it would purr continuously; such devices were sometimes fitted to an electronic clock which would indicate the times of Muslim prayer. Most recently such devices are linked to a satellite which can locate the user and so does not need to be informed of his/her coordinates. If one were to take such a device into a mosque built in medieval times just before the call to prayer was to be announced, it is most unlikely that the device would show the same qibla as the *mihrâb* of the mosque, and also highly improbable that the thing would beep at the same time as the call to prayer.

6. Concluding remarks

The activities of both the legal scholars and the scientists are remarkable for the enthusiasm and ingenuity with which they were conducted, the sophistication and complexity they achieved, and not least for the occasional universality they attained, in both cases providing solutions to the qibla-problem for all localities. The legal scholars developed a sacred geography centred on the Kaaba, and the scientists a sacred geography centred on Mecca. In the long term it was irrelevant whether a particular qibla was "correct"; at the time it was important only that those who adopted it thought it was correct. Numerous historians of Islamic architecture have been misled into pronouncing the orientation of this or that mosque "incorrect"; it is they who stand to benefit most from our new insights about medieval notions of the qibla. Likewise, it is easy for us moderns with a mathematical bent to pour scorn on the legal scholars for not listening to the scientists and on some of the latter for not being able to compute the qibla properly. But one should not lose sight of the fact that the ultimate purpose of the adoption of a sacred direction in Islam was in obediance to a divine injunction that the devotion of the faithful should be focussed on a sacred shrine in the most sacred city at the heart of the Islamic world.

References

[1] J.L. Berggren, *Episodes in the Mathematics of Medieval Islam*, Springer, New York, 1986. [A useful introduction to the mathematical methods used by medieval Muslim scientists.]

[2] al-Bîrûnî, *Tahdîd*: J. Ali, *The Determination of the Coordinates of Cities … by al-Bîrûnî*, American University of Beirut Press, Beirut, 1967; and E.S. Kennedy, *A Commentary upon Bîrûnî's Kitâb Tahdîd [Nihâyât] al-Amâkin, An 11th Century Treatise on Mathematical Geography*, American University of Beirut, Beirut, 1973. [Translation of and commentary on the most significant surviving medieval treatise on mathematical geography.]

[3] M. Chebel, *Symboles de l'Islam* and *Symbols of Islam*, with photographs by L. Hamani, both Éditions Assouline, Paris, 1997. [Beautifully illustrated.]

[4] *The Encyclopaedia of Islam*, new edition, 10 vols. to date, E.J. Brill, Leiden, 1960 to present. [Contains various survey articles including "Anwâ" [aspects of folk astronomy]; "Ilm al-hay'a" [astronomy]; "Kibla" [sacred direction]; "Layl and nahâr" [aspects of folk astronomy]; "Makka: As Centre of the World" [sacred geography]; "Matâli" [ascensions]; "Matla" [horizon directions]; "Mîkât" [astronomical time-keeping]; "Samt" [direction]; "Tâsa" [magnetic compass]; and "Zîdj" [astronomical handbooks with tables].]

[5] D.A. King, *Astronomy in the Service of Islam*, Variorum, Aldershot, 1993. [Reprints of various articles, including an analysis of the earliest mathematical methods for finding the qibla and a study of the orientation of the Kaaba and its influence on mosque orientations.]

[6] D.A. King, *Islamic astronomy*, Astronomy Before the Telescope, Christopher B.F. Walker, ed., British Museum Press, London, 1996, repr. 1999, pp. 143–174. [An overview.]

[7] D.A. King, *The Sacred Geography of Islam*, to appear. [An analysis of the numerous medieval schemes of folk geography in which the world was divided into sectors around the Kaaba; a summary is in the article "Makka, iv" in the *Encyclopaedia of Islam*.]

[8] D.A. King, *In Synchrony with the Heavens. Studies in Astronomical Timekeeping and Instrumentation in Medieval Islamic Civilization*, 2 vols., 1: *The Call of the Muezzin (Studies I–IX)*, and 2: *Instruments of Mass Calculation (Studies X–XVIII)*, E.J. Brill, Leiden, 2004. [See especially VIIa: "The Orientation of Medieval Islamic Religious Architecture and Cities", first published in *Journal for the History of Astronomy* 26 (1995), 253–274, and VIIb: "Architecture and Astronomy: The Ventilators of Medieval Cairo and Their Secrets", first published in *Journal of the American Oriental Society* 104 (1984), 97–133. See also [9].]

[9] D.A. King, *World-Maps for Finding the Direction and Distance of Mecca: Innovation and Tradition in Islamic Science*, E.J. Brill, Leiden, and Al-Furqan Islamic Heritage Foundation, London, 1999. [An overview of the treatment of the qibla-problem by Muslim scholars and a detailed study of the two 17th-century Mecca-centred world-maps with speculations on the source of their inspiration. The third world-map and the newly-discovered 10th- and 11th-century treatises are introduced in King, *In Synchrony with the Heavens*, VIIc.]

[10] D.A. King, J. Samsó, *Astronomical handbooks and tables from the Islamic world (750–1900): An interim report*, Suhayl—Journal for the History of the Exact and Natural Sciences in Islamic Civilisation (Barcelona) **2** (2001), 9–105. [A supplement to E.S. Kennedy's 1956 survey of Islamic *zîjes*, with some remarks on qibla-tables and geographical tables displaying the qibla for numerous localities.]

[11] M. Rius, *La Alquibla en al-Andalus y al-Magrib al-Aqsà*, Anuari de Filologia (Universitat de Barcelona) **XXI** (1998–1999) B-3, Institut "Millás Vallicrosa" de Història de la Ciència Àrab, Barcelona, 2000. [The first investigation of mosque orientations in the light of medieval folk astronomical and legal texts on the qibla.]

IX

Al-Bazdawī on the Qibla in Early Islamic Transoxania

1 – *Introduction*

In the seventh century, within decades of the death of the Prophet Muḥammad, the Muslims conquered an area stretching from Andalusia to India. Wherever they settled they built mosques oriented in the qibla, that is, so that the prayer-niche or *miḥrāb* would be facing Mecca, in accordance with a Quranic injunction that Muslims should face the sacred compound in Mecca during prayer.[1] The orientations of the earliest mosques built by the *Ṣaḥāba*, the contemporaries of the Prophet, and the *Tābiᶜūn*, the next generation of Muslims, were established by non-mathematical procedures. The qibla was defined in terms of the cardinal directions, or by the rising and setting of the sun or stars, or by the wind directions.[2]

The determination of the qibla by mathematical means is a complicated problem of mathematical geography, which was pursued with enthusiasm by Muslim astronomers from the ninth century onwards. To solve the qibla problem, one requires a knowledge of terrestrial coordinates and a trigonometric formula giving the direction from one locality to another on a terrestrial sphere Lists of such coordinates and statements about the appropriate formulae are attested in numerous medieval sources.[3] Inevitably controversies arose in different localities about which qibla directions were legally acceptable, and the records of these discussions constitute documents of considerable interest to both the history of Islamic science and the history of Islamic architecture.[4]

Acknowledgements:

The research on medieval Islamic science conducted at the American Research Center in Egypt during the period 1972-1979 was sponsored mainly by the Smithsonian Institution (1971-79) and the National Science Foundation, Washington, D. C. (1972-80), the Ford Foundation (1976-79), and the American Research Center in Egypt (1979). This support is gratefully acknowledged.

It is a pleasure to thank the Egyptian National Library for providing a microfilm of the copy of al-Bazdawī's treatise on which this study is based. I owe many of the finer points of the text edition and translation of al-Bazdawī's treatise to discussions with my friend Prof. Richard Frank of the Catholic University of America, *khayr al-nudamā' wa-'andamuhum*, begun on board MTS Argonaut on the Red Sea and concluded in Cairo.

1. See the article "*Kibla* (ritual and legal aspects)" in EI_2 by A. J. Wensinck.
2. See *King* 5 and also *King* 7 (forthcoming).
3. See my article "*Kibla* (astronomical aspects)" in EI_2, and also *King* 7, Section 2.
4. See already *Renaud* 2 on the qibla in the Maghrib, *King* 2 on the qibla in Andalusia, and *King* 1, pp. 368, and 4 on the qibla in Egypt. More details are given in *King* 7, Section 4.

The treatise presented in this study deals with the determination of the qibla in early Islamic Transoxania.[5] It affords new light on qibla determinations in early Islamic practice and contains information which will be useful to historians of Islamic architecture when the religious architecture in Transoxania is properly surveyed for the first time. The author of the treatise, Abu'l-Yusr al-Bazdawī, was a *qāḍī* in Samarqand in the late eleventh century[6] and was a Ḥanafī scholar of some standing. He was the author of several works on law, including a commentary on the major work of Abu Ḥanīfa, after whom the Ḥanafī school is named, and a commentary on a work of Abu Ḥanīfa's student al-Shaybānī, who was one of the founders of the Ḥanafī school. The *nisba* al-Bazdawī indicates that our author or his family originated from Bazda or Bazdawa, a small town with a castle on the road between Nasaf and Bukhara.[7]

al-Bazdawī's treatise is extant in a manuscript preserved until recently in Sohag in the Nile Valley, but the manuscript is now no longer in the Library of Sohag. The authorities there say that the manuscript has been taken to Cairo, but I have been unable to ascertain its fate more precisely.[8] However, a hand copy of the Sohag manuscript, prepared in 1936, is preserved in the Egyptian National Library, numbered B 19385 and containing twelve pages of text.[9] This study is based entirely on this late copy:[10] I present a translation of al-Bazdawī's treatise (Section 2) and a commentary thereon (Section 3), as well as an assessment of al-Bazdawī's understanding of the qibla problem and of his suggestions for its solution (Section 4). The Arabic text edited from the Cairo copy is also presented (Appendix A).

5. A good introduction to the area is *Le Strange*.

6. On al-Bazdawī, see *Brockelmann*, I, p. 460, and SI, pp. 637-638, and *Sezgin*, I, pp. 412 and 428.

7. *Le Strange*, p. 471.

8. It is a pleasure to thank my friend Mr. Peter Mackenzie-Smith, formerly of the British Council in Cairo, and his friends amongst the British VSO teachers in Sohag, for making enquiries about the manuscript on my behalf.

9. This manuscript was first catalogued in *Sayyid*, I, p. 397. The letter B indicates that it belongs to the manuscripts dealing with religion that were acquired by the Egyptian National Library between 1936 and 1955.

10. The importance of the treatise was first recognized during my recent survey of the scientific manuscripts in the Egyptian National Library. See further *Cairo Cat.*, vol. I, *sub* B 19385; vol. II, Section 3.3.2; and *Survey*, no. B88.

2. *Translation of al-Bazdawī's Treatise*

There follows a free translation of al-Bazdawī's treatise. Words in parentheses have no counterpart in the original and have been added to clarify the meaning. In general the meaning of the text is clear. Page references in the margin relate to the pagination of the Cairo manuscript. The paragraph numbers accord with those used by me in the edition of the text presented in the Appendix.

p. 1 *Treatise on the Azimuth of the Qibla*[1]

p. 2 (1) In the Name of God, the Merciful and Compassionate. Praise be to God- the Most High, the Mighty, the Mild, the Benificent, the Wise, the Knowing, the Ruler, the Truth, the Revealer, the Possessor of strength and power, the Firm[2] – for the (different) kinds of virtues, high moral standards, and good qualities with which he has blessed us. Blessings upon His chosen, faithful, elected, and distinguished Prophet, and on all of his pure family, his companions, and his wives.

(2) The shaykh, imam, the gallant master, the leader of Islam, Abu'l-Yusr al-Bazdawī,[3] may God have mercy on him, said:

The greatest of the religious observances after faith in God – may He be exalted – is prayer,[4] for the Prophet – may (God) bless him and grant him salvation – said: "Prayer is the pillar of religion; whoever neglects it has destroyed (his) faith."[5]

He considered faith without prayer to be like a ruined house, and a ruined house is a house but is no (longer) of any use. This Prophetic statement shows us that prayer is one of the greatest of the religious observances, and that anyone who neglects it indeed invalidates his faith.[6]

(3) Now anyone who needs to perform the prayer several times each day cannot perform it except after he knows its basic elements and conditions. One of the conditions required in every prayer is that one should face the Ka^{c}ba,

1. The title *Risāla fi samt al-qibla* is probably spurious. al-Bazdawī nowhere mentions the word *samt*, "azimuth," in his treatise, although he does use the related word *musāmit*, "in the same direction as . . . ".

2. These are some of the ninety-nine names of God, on which see the article "*al-Asmā' al-ḥusnā*" in *EI*$_2$ by L. Gardet.

3. al-Bazdawī is nowhere else named in the Cairo manuscript. Later in the text he mentions that he was a judge in Samarqand (paragraph 9), and that he arrived in Samarqand in 473 Hijra (paragraph 9), and he also names his grand-father (paragraph 11).

4. The word *al-ṣalāt* is here omitted from the Arabic text ! The prayer intended is the ritual liturgical prayers, on which see the article "*Ṣalāt*" in *EI*$_1$ by A. J. Wensinck.

5. On the statements attributed to the Prophet Muḥammad see the article "*Ḥadīth*" in *EI*$_2$ by J. Robson. A concordance of the canonical *ḥadīth* literature is *Wensinck*, but this particular *ḥadīth* does not occur in the canonical collections.

6. Note that the text has *lā yubaṭṭilu/yubṭilu īmānahu*, "does not invalidate his faith," and I have emended this to *la-yubaṭṭilu/yubṭilu īmānahu*.

wherefore one has to know in which direction is the Ka^{c}ba. Imams in the first generations – may God have mercy on them – formulated many (legal) problems relating to each of the conditions (of prayer) and they expounded solutions demonstrating their correctness; they were not satisfied to follow the authority of others in these (matters). Likewise they formulated many solu-
p. 3 tions to problems relating to (obligatory) alms-giving // , even though most people do not need to know the solutions to such problems. They (also) formulated solutions to many problems relating to fasting, even though there was not much need to know the problems and solutions relating to it. They (further) formulated the solutions to many problems relating to marriage and divorce, sales and crimes, and to other problems as well, relating to every (conceivable) subject. Even though ordinary people have no need of solutions to such problems most of the time, they *(viz.* the imams*)* pursued the evidence that leads to the correct solution of these problems, thought about them, and were not content to follow the authority of others, speculating on behalf of the people as a whole, so that when an individual happened to need a particular response he should find it or something similar to it and not be perplexed about it. Abū Ḥanīfa – may God have mercy upon him – was our leader in formulating the solutions to problems, reasoning on the basis of their evidence, and his followers after him thought about those solutions, as did the other legal scholars of the community – may God have mercy upon them. Solutions for which they found provative evidence to indicate their validity they adhered to and asserted as valid, but where no provative evidence of the correctness (of a solution to a given problem) became apparent to them, they rejected it, and were not content to follow the authority of others.

(4) Now most of the imams of the first generations avoided thinking about the matter of the qibla and were content to follow the authority of others, even though the laying out of the qibla is not a matter in which one is obligated to follow the authority of another. They did this only because they did not have the means (*āla*) to know the qibla, since the qibla can be known only by the science of arithmetic, and they had no insight into calculation. Accordingly they followed the authority of others because of their inability to find it by using (the traditional method of invoking) legal evidence. Most of the imams of the first generations had not made any efforts (*ijtahadū*) in arithmetic, and
p. 4 some of the followers of our school had // insight into it and so sought solutions to the problems relating to the qibla and they got them right. But the (imams) did not compile any books on (the subject) because of the abstruse nature of arithmetic, and because most people, especially the legal scholars, avoided arithmetic. Some of them did compile books but they compiled them in such a way that they were so difficult to comprehend that only those thoroughly familiar with arithmetic could understand what was in them. So these books fell into disuse and everybody ended up relying on established authority.

(5) Some of the more recent followers of al-Shāfiʿī in Transoxania and Khurasan who had never smelled the fragrance of arithmetic found fault with the righteous men of the first generations and they made the qibla in the direction between the rising point (of the sun) at (mid-) winter and its setting point. They relied on two Prophetic statements, one of which is their relating from the Prophet – may (God) bless him and grant him salvation – that he said: "The qibla is between the east and west,"[7] and the second is their likewise relating from him – may (God) bless him and grant him salvation – that he said: "Do not face towards or away from the qibla when you are relieving yourself; rather face east or west."[8] The authenticity of these two Prophetic sayings is not recognized because the reliable authorities did not relate them in their books.[9] Furthermore even if the reports are true, it is plain for any intelligent person that the argument based on these reports is not valid, for every intelligent person knows by immediate intuition that the qibla of all localities is not between the east and the west. The qibla of some localities only is (in this direction). There is no (mention) in the statement of the Prophet – may (God) bless him and grant him salvation – that this qibla is the qibla of any (particular) place, and therefore one cannot rely on these two
p. 5 Prophetic statements. If they say // the Prophetic statement: "the qibla of the people of Iraq is between the east and the west" is related, then we reply that this addition (*viz.* "of the people of Iraq") is not authentic because Iraq had not been conquered in those days. Rather it is more probable that what is intended by these two Prophetic statements is the qibla of the people of Medina, for it is (indeed) between the east and the west.

(6) When the righteous first generations conquered the(se) lands, they made the qibla of Transoxania and Khurasan at the setting point of (the sun in) the autumn, which is the setting point when day and night are equal. When the sun has entered the sign of Libra and twenty days have passed [of the month of Aylūl (= September)!] so that (the sun) has moved about twenty degrees (during that month), then whoever faces the sun at sunset (is facing) the qibla laid out by the men of the first generation. (The season) is the time when the peasants of Bukhara finish sowing wheat and barley. The situation is the same when the sun has entered the sign of Aries.

(7) The distorted qibla which is accepted by the Shāfiʿite school is altogether erroneous. That it is erroneous is obvious to anyone of normal adult intelligence who has given a modicum of thought (to the matter), let alone having smelled the fragrance of arithmetic. It is correct that the qibla of the cities of

7. See *Wensinck*, V, p. 259b for references to this *ḥadīth*.
8. This *ḥadīth* is not in the canonical collections.
9. This is true only of the second *ḥadīth* quoted: see notes 7 and 8 above.

Transoxania and Khurasan should be between the rising and setting point of (the sun at mid-) winter only if these cities have the same longitude as Mecca, or if the longitude of these cities is close to the longitude of Mecca, like the longitude of Mecca (as compared) with the longitude of Medina. By "longitude" is meant the distance of the city from the sea in the west which encircles the earth (*i.e.* the Atlantic). The distance of Mecca from the sea in the west is sixty-
p. 6 seven degrees, and the distance of Bukhara // from the sea in the west is eighty-seven degrees, and the distance of Samarqand from it is eighty-nine degrees, and the distance of Nasaf from it is eighty-eight degrees. Thus the difference between the distance of Mecca and the distance of Bukhara is twenty degrees, and each degree is approximately twenty-five *farsakhs*, so that the longitudinal difference between Mecca and Bukhara is five hundred *farsakhs* or thereabouts. The differences in longitude between Samarqand and Mecca, and between Mecca and Nasaf, are greater than this. So whoever (in these parts) puts his face in the direction between the rising point (of the sun) in winter and its setting point will certainly not be facing Mecca. There is about five hundred (text has: fifty) *farsakhs*' distance in longitude between (Mecca and these parts), so whoever puts his face in the direction between the setting point (of the sun) in winter and its rising point knows with certainty that he is not facing Mecca because the distance in longitude between Mecca and Bukhara is five hundred *farsakhs*. Thus the error in what recent scholars of the Shāfiᶜī school introduced in the qibla is established with certainty.

(8) The qibla laid out by those who conquered this district, which is (the qibla) used for prayer by the followers of Abū Ḥanīfa – may God have mercy upon him, has a deviation to the right of the person praying (in the true qibla) and he is (facing) to the left of that qibla (*i. e.* due west). Thus the face (?) of the qibla . . . [LACUNA?] ... is to the east (!!), for we have stated above that this qibla is towards the setting point of (the sun at) the equinox, *viz.*, is at the middle of the setting points of the sun. The setting point of (the sun at) the equinox is not equal . . . [LACUNA?] . . . for these cities in latitude (?). This
p. 7 would be correct only if the latitude of these // cities and the latitude of Mecca were the same (or) their latitudes were close, so that the (person) facing the setting point of (the sun at) the equinox would be facing Mecca. But there is a large difference in latitude between Mecca and these cities, for the latitude of Mecca is twenty-one degrees, the latitude of Bukhara is thirty-eight degrees, the latitude of Samarqand is forty degrees, and the latitude of Nasaf is approximately thirty-six degrees. By "latitude" is meant the distance of the locality from the equator, which is mid-heaven [!]. Now all the inhabited part (of the earth) is in one of the two halves of the earth, namely, the half which is to the left of the qibla (??). The latitudinal difference between Mecca and Bukhara is sixteen degrees, each degree being close to twenty-five *farsakhs*, so that there are four hundred *farsakhs* between them, and between Mecca

and Samarqand the difference is greater. Thus it is not correct at all that the person facing the setting point of (the sun at) the equinox is facing Mecca, because of the large difference in latitude between the two, close to five hundred (text has: fifty) *farsakhs*. Because the sun sets at the equinox to the left of Mecca and not in a direct line with Mecca, the person facing west on that day will not be facing Mecca. This error occured on their part simply because they laid (the qibla) down in approximation, without having any knowledge of calculation. One who uses an approximation is acting without adequate evidence; he is in fact functioning by the dictates of the heart. The one who judges by approximation often errs, but despite this he may (validly) pray if he has no other evidence. The prayer of those who have no adequate
p. 8 evidence is permissable and likewise the prayer of those who follow them, but //
when someone has revealed their error with the correct evidence, they may no longer pray towards the qibla (that has been shown to be incorrect).

(9) During the time of my stay in Samarqand, when I was a judge at the *ḥaḍra*, I heard some people in whom I have confidence say that the religious scholars took it upon themselves to make a thorough investigation of the qibla laid out by those who had conquered the region of Transoxania, and that they agreed that it was inclined away from the Ka^{c}ba towards the left (?). So they had recourse to those who are experts in calculation and who have insight in this matter, and they laid out for them a qibla to the left of the person praying and to the right of the Ka^{c}ba (?), *i.e.* it is between the qibla of the followers of Abū Ḥanīfa – may God have mercy on him – and that introduced by the (more ?) recent followers of al-Shāfi^{c}ī – may God have mercy upon him. They laid out the qibla of the Jāmi^{c} Mosque in Samarqand in the same direction, and the followers of Abū Ḥanīfa and al-Shāfi^{c}ī at that time agreed on this, *i.e.* a qibla which has not inclination in any direction (from the true direction of Mecca). I checked it when I came to Samarqand in the year four hundred and seventy-three and the sun was in the sign of Gemini: I found it correct and proper.

(10) Afterwards when I realized that this vain and absurd disagreement concerning the qibla between the followers of Abū Ḥanīfa and the recent followers of al-Shāfi^{c}ī is a disagreement of ignorant men who do not really know – since anyone who really knows the qibla would say that both parties are wrong, while anyone who has no true knowledge but does have piety and normal adult intelligence will not enter into an argument over the qibla but will acknowledge that he does not know, so that the dispute will continue only between unlearned men who have neither piety nor normal adult intelligence together with precise reflection – I felt a desire to compose a short work on the question of the qibla in which I would show what is right and demonstrate it with brilliant reasoning from the evidence and also show in it the

method of knowing (the qibla) by the easiest means. I pray for God's ultimate
p. 9 reward and His guidance // in completing it, relying upon Him for protection
from slipping and falling into error, and hoping for His good pleasure.

(11) So I say: We explained at the beginning of the book that which proved the error of what was introduced by the more recent followers of al-Shāfi^{c}ī concerning the qibla, and we mentioned that the qibla laid out by those who conquered the(se) regions was erroneous. I heard from one of the upright legal scholars in whom I have confidence that anyone who leaves Mecca and gives careful consideration to the matter of the qibla and then goes back to Transoxania will recognize that the qibla (there) is inclined to the left of the (true) qibla. My grandfather, the shaykh, the ascetic imam, Abū Muḥammad ^{c}Abd al-Karīm ibn Mūsā, travelled to Mecca as a pilgrim and whilst there thought about the matter of the qibla. When he returned to his own town he changed the qibla of his mosque to the left of the person praying, *i.e.* to the right of the qibla (?), because he had become certain of the error of the (first) qibla. But then he annulled (the new qibla) and put (the qibla) back to what it had been previously, on account of the people's gossiping so much, because ignorance together with fanaticism prevailed amongst the unlearned of his town. I heard this (story) from honest and reliable people. Likewise, the experts in arithmetic were of one mind and changed the qibla of the Jāmi^{c} Mosque in Samarqand from what it had been to the left of the person praying, and the religious leaders of both schools agreed upon it. Their agreement indicated the error of both qiblas.

(12) The position of the followers of our school (*i.e.* the Ḥanafīs) that the
Pole Star should be aligned with a person's two earlobes when he wishes to
pray is incorrect, since this would be correct if the latitude of Mecca and the
latitude of that region were equal, which is not the case. Indeed, there is a
large difference in latitude between them, as we have shown. The statement
of some of the followers of al-Shāfi^{c}ī that the Pole Star should be at the neck
p. 10 of a person when he wants to pray is also completely erroneous. // It would be
correct only if the longitude of Mecca and the longitude of these cities were
equal, and we have shown that this is not the case. The opinion of the followers of al-Shāfi^{c}ī is more erroneous than the opinion of the followers of Abū Ḥanīfa and those who conquered this region, because the roads in this region (leading) to Mecca were laid out by the ancients towards the setting point of (the sun at) the equinox, and they did not lay them out towards [what is between] the setting point of (the sun at mid–) winter and its rising point.

(13) I shall now demonstrate a method for finding the qibla in the cities of Transoxania – Bukhara, Samarqand, and Nasaf, and the towns and villages which belong to them – in such a way that people can understand. The calcu-

lators and those who practice disciplines other than arithmetic agree that when the sun enters Gemini [and has moved through ten degrees of it], it becomes in the same direction as Mecca when it is midday (at Mecca), (the sun being then) like a shield over a man's head, so that there remains no shadow anywhere and so they have the expression "there is no shadow left in the wells." Similarly when the sun enters the sign of Cancer and has moved through twenty degrees of it, the sun is again in the same direction as Mecca when it is (in the zenith of Mecca) at midday, and in every place whose longitude is the same as the longitude of Mecca, so that if a person faces the sun at midday (in Mecca) when the sun has entered the sign of Gemini [and moved through ten degrees of it] or when the sun has moved through twenty degrees of the sign of Cancer, then that (direction) will be the qibla of that place. Midday at Mecca will be after midday in every place which differs in longitude from Mecca and whose longitude is greater than that of Mecca, such as Bukhara and Samarqand. We have already noted the difference between the longitude of these cities and that of Mecca; thus the declining of the sun from the meridian at Mecca takes place only after the midday at the cities of Bukhara, Samarqand, and Nasaf, by two and two-thirds hours. So when two and two-thirds hours have passed after midday in the cities of Bukhara, Samarqand,
p. 11 and Nasaf, // whoever faces the disc of the sun at that moment when the sun is at the beginning of the sign of Gemini or the end of the sign of Cancer in one of these cities, this will be the qibla of these cities. So if the person is one of those who understands the hours he will after finding the hours proceed according to what we have explained. If he does not know the hours precisely, then when the shadow of every object after midday becomes half (the length of the object), which is the time when the followers of Abū Ḥanīfa – may God have mercy upon them – perform the *ẓuhr* prayer at that time (of the year), *viz.* the summer, whoever faces the disc of the sun at that time in any of these cities, this (direction) will be the qibla of that city. If the person is one of those who knows the entry of the sun in the zodiacal signs, this is good, otherwise he should have recourse to someone who does know this, and inquire, and lay out the qibla at that time according to what we have stated. Likewise, when the sun has entered the sign of Capricorn, which is at the beginning of winter when the day-light is at its shortest, then when the sun has almost set, whoever faces the sun at that time in any of these cities of Samarqand, Bukhara, or Nasaf, this (direction) will be the qibla of those cities.

(14) However, the way to make the qibla in Bukhara and Nasaf the same as the qibla of the Jāmi^{c} Mosque in Samarqand is to look at the disc of the sun after midday in the city of Samarqand in the Jāmi^{c} Mosque until it is in the same direction as the qibla (of the mosque), and to see how long the shadow of any object is, and to measure it, and then to return to Bukhara or Nasaf or Karminīya or Kish with haste, and there face the sun after midday when the

shadow of any object is the same as it was in Samarqand: this (direction) will be the qibla of that city, for in a period of one or two, or up to ten, days, no more than a small difference (in shadow length) will occur, and such a difference need not be taken into consideration.

(15) The calculators have devised another method for finding the qibla and laying it out which can be used on any day (of the year). The sun is in the same direction as the Ka^{c}ba at some particular moment on every day (of the year), so whoever knows that moment and faces the sun at that moment in any of these cities, then that (direction) will be the qibla of those cities. But whoever wants to lay out the qibla on any day must learn how the solar altitude is found. When the sun is in the first degree of the sign of Aries and the altitude of the sun after midday is forty degrees, then anyone who faces the sun at that moment in one of these cities will be facing the qibla of that city. When the sun is in the second degree of the sign of Aries and the altitude of the sun after midday is forty-one degrees, then whoever faces the sun in one of these cities at that moment will be facing the qibla of that city. In this way (the required altitude of the sun) increases degree by degree every two days or more, until the sun enters the sign of Cancer and has moved through sixteen degrees of it [*sic*] and when it enters the sixteenth degree and the altitude of the sun after midday is sixty-six degrees, whoever faces the sun at that time in one of these cities will be facing the qibla of that city. Then (the required altitude of the sun) decreases in this way until the sun enters the sign of Capricorn and moves through sixteen degrees of it [*sic*]. When (the sun) enters the nineteenth [!] degree and you measure the altitude after midday in one of these cities and the altitude is eleven degrees, then whoever faces the sun in any one of these cities at that moment will be facing the qibla of that city. Then the degree(s of the altitude of the sun) start to increase every two days or more until (the sun) reaches Aries. I have written the increase and decrease of the degree (of altitude of the sun) throughout the year in a table which I prepared, and I have written this in alphanumerical notation which is *abcd* . . . (for 1234 . . .).[10]

[THERE IS NO TABLE IN THE CAIRO MANUSCRIPT !]

10. On the *abjad* notation for numbers in Islamic astronomical tables, see *Irani*.

3. *Technical Commentary on al-Bazdawī's Treatise*

The commentary relates to the numbered paragraphs in the translation in Section 2.

In the commentary I use the following notation freely:

h_q solar altitude in the azimuth of the qibla

L terrestrial longitude

L_M longitude of Mecca

n length of gnomon

q qibla (measured from the meridian)

s length of gnomon shadow

t_q hour angle when the sun is in the azimuth of the qibla

δ solar declination

ΔL difference in longitude from Mecca

Δs increase of gnomon shadow over midday minimum

$\Delta\varphi$ difference in latitude from Mecca

ϵ obliquity of the ecliptic

φ solar longitude

φ terrestrial latitude

φ_M latitude of Mecca

The mathematical problem of determining the qibla q for any locality in terms of the geographical coordinates of the locality (L,φ) and of Mecca $L_M, \varphi_M)$ has the solution:

$$q = \text{arc cot} \left\{ \frac{\sin \varphi \cos \Delta L - \cos \varphi \tan \varphi_M}{\sin \Delta L} \right\},$$

where $\Delta L = | L - L_M |$. Equivalent exact trigonometric or geometric solutions were known to Muslim astronomers from the ninth century onwards. Also, however, various approximate methods were devised in the ninth century, of which the most popular method, which has been used for over a millenium, was to use the construction:

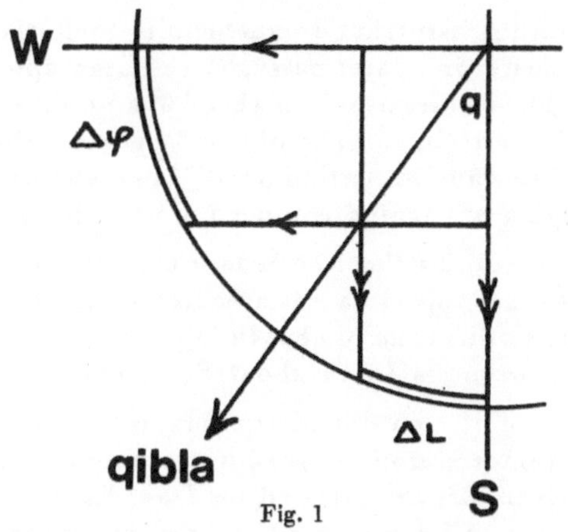

Fig. 1

which is equivalent to the approximate formula

$$q = \text{arc tan} \left\{ \frac{\sin \Delta L}{\sin \Delta\varphi} \right\}$$

For a survey of these methods, see my article *Kibla* in *EI*$_2$, and also *King* 5, Section 2.

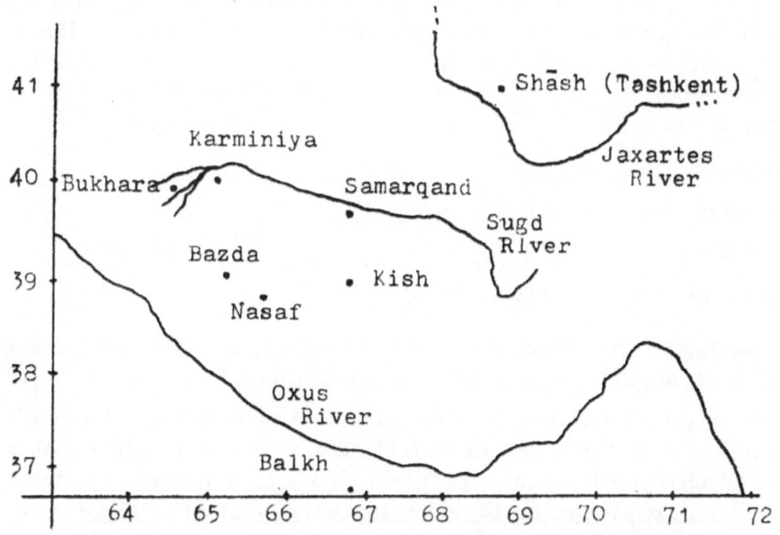

Fig. 2 : Map of Transoxania (from *Le Strange*, map IX facing p. 433

§ 5: al-Bazdawī is stating here that some Shāfiʿīs took the qibla to be due south: this is made clearer in a later passage (see Paragraph 7) when he states that their qibla would be correct only if there was no difference in longitude between Transoxania and Mecca. The phrase "between the rising point (of the sun) at (mid–) winter and its setting point" is an awkward way of refering to due south. The qibla at Medina is indeed almost due south.

§ 6: al-Bazdawī is informing us that the *Ṣaḥāba* took the qibla as due west. In Paragraph 8 he states that this qibla was accepted by the Ḥanafī school. When he implies that at the equinox the sun has been in the sign of Libra for 20 days, he means that the sun enters Libra about September 20.

The identification of the autumnal equinox with agricultural activity is typical of the kind of information recorded in the almanacs that were common in medieval times. Examples are attested for Iraq, Egypt, Andalusia and the Yemen, but not, as far as I know, for the Eastern provinces of the Muslim world. On the almanacs of Ibn Māsawayh (Baghdad, early 9th century) and ʿArīb b. Saʿd (Cordova, 10th century), see *Troupeau* and *Pellat*. On a nineteenth-century Tunisian almanac, based on a much earlier Andalusian tradition, see *Samsó* 4. On an anonymous Egyptian almanac of uncertain date (but certainly earlier than the 18th century), see *Munoz*. On a late Yemeni almanac, see *Serjeant*. On the origin of certain of these almanacs, see *Samsó* 1, 2, and 3.

According to *Taqizadeh*, p. 40, the Avestan time of harvest in Iran began on September 14th. The following dates for the sun's entry into Libra are given in *Troupeau*, p. 135; *Pellat*, pp. 140-142; and *Serjeant*, p. 455:

Ibn Māsawayh	September 22
al-Bīrūnī	16/17
al-Qazwīnī	18
al-Marzūqī	24
Ibn ʿArīb	18/23*
Muḥammad Ḥaydara (20th century)	23

(* according to the *Mumtaḥan Zīj* (*Kennedy* 2, no. 51) and *Sindhind Zīj* (*Kennedy* 2, no. 28), respectively; on this, see now *Samsó* 3, pp. 178-179.)

One might be forgiven for supposing that al-Bazdawī is suggesting that the equinox is at Libra 20°. Certain Muslim astronomers adhered to a (false) late–Hellenistic notion known as trepidation, which attributes to the equinox an oscillatory motion about the sidereally-fixed point Aries 0°. Other examples of specific values given to the current distance between the equinox and Aries/Libra 0° occur in later Maghribi sources, such as some prayer–tables for Morocco by Muḥamamd b. Muḥammad al-Jannād (MS Cairo Taymūr *riyāḍa* 238, where the value is 18°) and a later eighteenth–century Moroccan star catalogue (*cf. Delphin*, p. 181, where the value is 16°). There is a considerable amount of unstudied material relating to trepidation available in the Islamic sources, and more basic research is necessary in this field. See already *Goldstein* and the literature there cited.

§ 7: On the longitude values given by al-Bazdawī, see the commentary to Paragraph 8 below.

On the *farsakh*, see the article in EI_2 by W. Hinz. The value 1° = 25 *farsakhs* is attributed to Hermes (*cf.* the article *Hirmis* in EI_2 by M. Plessner) by al-Bīrūnī, quoting the eight-century astronomer al-Fazārī (*cf. Ali*, p. 177 and *Kennedy* 1, p. 132). Other values cited by al-Bīrūnī include three values around 18 or 19 *farsakhs* attributed to "the Indians," and to the early ninth–century astronomers al-Farghānī and Ḥabash. Yet other values are attested in the Islamic sources: for example, the seventeenth-century scholar Ḥājjī ᶜAbd al-ᶜAlī Tabrīzī of Hyderabad used 1° = 20 to 22 *farsangs* (*cf. Habib*, p. 25 of the author's typescript).

§ 8: The remark that the qibla of the Ḥanafis "has a devation to the right of the person praying (in the true qibla) and he is (facing) to the left of the qibla" is an obscure way of saying that the qibla of the Ḥanafīs, which was due west, was too far to the right. al-Bazdawī goes on to say that the qibla could only be due west if the latitudes of the cities in Transoxania were the same as the latitude of Mecca (this is in fact erroneous).

al-Bazdawī gives the following geographical coordinates in Paras. 7 and 8:

Locality	Latitude	Longitude
Mecca	21°	67°
Bukhara	38	87
Samarqand	40	89
Nasaf	36±	88

al-Bazdawī's coordinates for Samarqand are attested elsewhere only in the geographical work of the late thirteenth–/early fourteenth–century Syrian ruler and scholar Abu'l-Fidā', quoting the unidentified source *Kitāb al-Aṭwāl*, but this source has (39;20°, 87,50°) for Bukhara, (39;0°, 88;40°) for Nasaf, and (21;40°, 67;13°) for Mecca. The latitude 38° for Bukhara, which is in error by almost two degrees, is not attested in any known medieval source. Likewise, no known medieval sources list a latitude for Nasaf less than 39°. al-Bazdawī's values for Mecca are common to several early Islamic sources, including the geographical work of the ninth-century Baghdad scholar al-Khwārizmī. However, al-Khwārizmī has (37;30°, 89;30°) for Samarqand and (37;50°, 87;20°) for Bukhara. For this information I have relied on the computer printouts of medieval geographical coordinates described in *Kennedy & Haddad* which have been brought up–to–date by Dr. Kennedy in Cairo during 1978-79.

For al-Bazdawī's coordinates I compute that the *qiblas* (measured from the south) for Bukhara, Samarqand, and Nasaf are:

51;6° 51;8 56;23

according to the exact formula, and

49;28° 49;0 54;10

according to the standard approximate formula. (The modern coordinates are:

Mecca	21;27°	39;49°
Bukhara	39;48	64;25
Samarqand	39;40	66;58

so that the true *qiblas* for Bukhara and Samarqand are respectively 56;4° and 59;48° measured from the south, but these are, of course, irrelevant to any discussion of medieval qiblas or orientations.)

§ 9: The expressions "inclined away from the Ka^cba towards the left" referring to the qibla in the direction of due west and "to the left of a person praying (in the qibla of due west) and to the right of the Kaᶜba" referring to the new qibla (in the direction of winter sunset?) are extremely awkward. They *seem* to relate to the direction of the local qibla as seen from the Kaᶜba.

We know from modern excavations (according to a sketch attached to a letter from Prof. N. B. Nemtseva to Prof. L. Golombek of the University of Toronto and The Royal Ontario Museum dated June 5, 1977) that the qibla of the Jāmiᶜ Mosque at Samarqand is about 35° S. of W. This corresponds roughly to the azimuth of the setting sun at midwinter for latitude $\varphi = 40°$ and obliquity $\epsilon = 23\frac{1}{2}°$, which is about $31\frac{1}{2}°$ S. of W., but I have no information on the local horizon at the site to confirm whether this is indeed a solstitial alignment. In Paragraph 11 al-Bazdawī seems to imply that the qibla of the Jāmiᶜ Mosque was originally due west and was changed to the left, probably towards winter sunset. *Allāhu aᶜlam.* Finally, it is probably quite fortuitous that the alignment of the mosque corresponds fairly closely to the qibla that can be computed from al-Bazdawī's geographical data, namely, about 40° S of W.

It would be interesting to know how al-Bazdawī checked this qibla when he came to Samarqand in the year 473 Hijra. The sun was conveniently in Gemini (see below), and he probably checked that the sun was in the direction of the qibla of the mosque at a certain time after midday (see below).

§ 11: At first it seems strange that al-Bazdawī thinks one can gain more insight into the matter of the qibla by actually going to Mecca and then going back to one's own country. However, each side and corner of the Kaᶜba in Mecca was associated with different geographical regions (see *King* 6 and *King* 7, Section 3.3). Furthermore the Ka^{c}ba is itself astronomically aligned (see *Hawkins & King* and *King* 7, Section 3.5). Thus someone standing in front of the eastern wall of the Ka^{c}ba is facing winter sunset. al-Bazdawī and others, assumed that, since the qibla of Transoxania was towards the north-eastern wall of the Ka^{c}ba, it was appropriate to use winter sunset as the qibla in Transoxania – see Fig. 3.

One can only feel sympathy for al-Bazdawī's grandfather when he was forced to annul the new qibla (towards winter sunset?) that he found by actually going to Mecca, and reinstate the erroneous qibla (toward due west?) that was generally accepted.

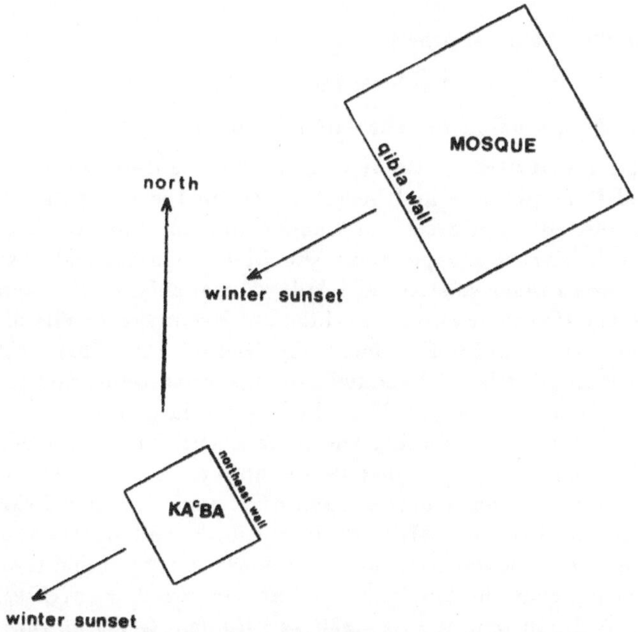

Fig. 3: al-Bazdawī's conception of the qibla in Transoxania. The qibla there is towards the northeastern wall of the Ka'ba. When standing in front of this wall one is facing winter sunset. Therefore the qibla in Transoxania is towards winter sunset.

§ 12: The mode of formulating these two statements that the qibla is due west and due south, respectively, is typical of the kind of information contained in the early *kutub dalā'il al-qibla*.

al-Bazdawī favors the western qibla of the Ḥanafīs against the southern qibla of the Shāfi'īs, and quite reasonably invokes the example of the roads from Transoxania to Mecca. The road from Samarqand to Bukhara is roughly due west. The road from Bukhara to Marw and Nisapur on the way to Mecca is in a direction of roughly 45° S. of W. Between Nisapur and Rayy (Tehran) the road is again roughly due west. (See *Le Strange*, Map 1 facing p. 1.) Note that the fact that the road from Samarqand to Bukhara is more or less due west did not lead al-Bazdawī to question why their latitudes should differ by 2°.

§ 13: al-Bazdawī is trying to say that the sun is in the direction of the qibla when it is in the zenith of Mecca, but his explanation is confused by his ignorance of spherical astronomy. The sun is in the zenith at Mecca at midday on two days of the year, namely, when its longitude λ is such that its meridian altitude

$$90^\circ - \varphi_M + \delta(\lambda)$$

is equal to 90°, that is, when

$$\delta(\lambda) = \varphi_M = 21^\circ,$$

using al-Bazdawī's value for the latitude of Mecca.

The solar longitudes corresponding to the situation $\delta(\lambda) = \varphi_M$ are described by al-Bazdawī here and below as "when the sun enters Gemini" and "when the sun enters the sign of Cancer and has moved through twenty degrees of it". Since these positions should be symmetrical with respect to Cancer 0°, the summer solstice, al-Bazdawī probably means Gemini 10° and Cancer 20°, the two solar positions which are 20° on either side of the solstice; however, he never specifically mentions Gemini 10°. These solar positions $\lambda = 70^\circ/110^\circ$ implied by al-Bazdawī are approximations, and the corresponding solar declination is about 22°, which is too large for φ_M.

The same method for finding the qibla occurs in other medieval Islamic sources, of which I shall cite just two examples. In the popular *Mulakhkhaṣ fi'l-hay'a* of Maḥmūd ibn ʿUmar al-Jaghmīnī, compiled in Khwārizm (modern Khiva in the U.S.S.R.) in 618H(=1221), the author advocates first the approximate geometrical construction outlined above and then mentions this second method, stating that the longitudes of the sun are $\lambda = 67;21^\circ/112;39^\circ$ since he uses $\varphi_M = 21;40^\circ$ and $\epsilon = 23;35^\circ$ (see *Rudloff & Hochheim*, p. 272, for a translation of this passage). In a Yemeni manuscript of this work copied *ca.* 1500, namely MS Cairo Dār al-Kutub *hay'a* 69, the two solar positions are copied incorrectly (fol. 18v) as Gemini 7;8° and Cancer 28°, *i.e.* $\lambda = 67;8^\circ/118^\circ$, which are not even symmetrical with respect to the solstice. Again, in an Egyptian treatise on the use of the sine quadrant in 24 chapters extant in MS Istanbul Topkapi A 3509,7 (copied 676H), the anonymous author solves $\delta(\lambda) = 21^\circ$ with $\epsilon = 24^\circ$ and obtains $\lambda = 64^\circ/116^\circ$ (fols. 318v–319r). In fact these values of λ correspond to $\epsilon = 23;35^\circ$, and the solution of $\delta(\lambda) = 21;30^\circ$ (another popular medieval value for the latitude of Mecca) with $\epsilon = 24^\circ$ is $\lambda = 64^\circ/116^\circ$! With $\delta(\lambda) = 21;30^\circ$ and $\epsilon = 23;35^\circ$ one obtains $\lambda = 66^\circ/114^\circ$.

The time-difference of $2\frac{2}{3}$ hours between Mecca and Transoxania given by al-Bazdawī is grossly inaccurate. In fact, it is double the correct amount for a longitude difference $\Delta L \approx 20^\circ$. The time difference should be determined by reckoning 1 hour for each 15 degrees of ΔL, since 24 hours corresponds to the 360° of the apparent daily rotation of the heavens.

It is unlikely that anyone other than an astronomer would "know the truth about the hours," as far as the determination of equinoctial hours from the instantaneous solar altitude is concerned. Simple arithmetical rules for regulating the seasonal hours (which, as twelfth divisions of the length of daylight, depend on terrestrial latitude and vary throughout the year) by means of shadows are attested in the non-technical literature of the Muslims. (A survey of these shadow-schemes is contained in my forth coming *Studies in Astronomical Timekeeping in Medieval Islam.*)

From al-Bazdawī's remark about the shadow at the *ẓuhr* prayer we may conclude that the Ḥanafīs in Transoxania performed the *ẓuhr* some time after midday, rather than as soon as the sun had declined from the meridian, which is the standard definition. When the horizontal shadow s of a vertical object of length n is $\frac{1}{2}n$ the solar altitude is 60°, and at midsummer in Transoxania this is about one hour after midday. Since the midday shadow would be about $\frac{1}{4}n$, it may be that the beginning of the *ẓuhr* was defined by $\Delta s = \frac{1}{4}n$, where Δs is the increase in the shadow beyond its midday minimum. This is the definition for the beginning of the *ẓuhr* usually associated with Andalusian practice (see, for example, *King* 3, pp. 191 and 193-194). Notice that the time defined by $s = \frac{1}{2}n$ is much earlier than the 2 2/3 hours after midday which al-Bazdawī has just prescribed.

Finally, al-Bazdawī remarks that the setting sun at midwinter defines the qibla in Transoxania. This direction, about $31\frac{1}{2}^\circ$ S. of W. for latitude 50°, is at variance with the direction he advocates in Paragraph 15 by about 15°. However, it may be that here we have the reasons behind the orientation of the Jāmi^c Mosque in Samarqand. The direction of the *rising* sun at midwinter was used as the qibla in early Islamic Egypt and also in Andalusia (cf. *King* 1, p. 372; *King* 2, p. 371; and also Appendix B of this paper).

§ 15: Given the local latitude φ, the azimuth of the qibla q, and the solar longitude λ or declination $\delta(\lambda)$, it is possible to determine the solar altitude h_q and the hour-angle t_q when the sun is the direction of Mecca. Thus, for example, the tenth-century Cairo astronomer Ibn Yūnus compiled a table of h_q for Cairo as a function of solar longitude, giving values in degrees and minutes for each degree of solar longitude, which corresponds roughly to each day of the year. The corpus of tables which was used in medieval Cairo contains tables of both h_q and t_q as well as of the corresponding functions in the direction perpendicular to the qibla (see further *King* 1, p. 368). Various later tables of these functions for different latitudes are attested in the Islamic sources, but none is known to have been compiled for the region of Transoxania other than the table of h_q which al-Bazdawī is about to present. (All such tables are discussed in my forthcoming *Studies in Astronomical Timekeeping in Medieval Islam.*)

Unfortunately the table is missing from the Cairo manuscript and I wonder whether it was also missing from the Sohag manuscript. However, al-Bazdawī gives sufficient information on his table to enable us to investigate its accuracy. He gives the following values of h_q for various solar positions:

Solar position	h_q
Aries 1°	40°
Aries 2°	41°
Cancer 16°	66° (maximum)
Capricorn 16°/19° ??	11° (minimum)

Firstly, it is not clear whether the table was arranged according to solar longitude or days of the Syrian calendar. al-Bazdawī confuses these systems and when he implies that the tabulated function reaches its maximum at Cancer 16° he is misinterpreting the fact that the sun enters Cancer on June 16th. Likewise when he implies that the function reaches its maximum at Capricorn 16° and, in the next sentence, 19°, he is misinterpreting the fact that the sun enters Capricorn on December 16th/19th. Since he earlier implies (Paragraph 4) that the autumnal equinox is on September 20th, the change from a figure 20 to 16 or 16/19 may indicate that he did not compute the table himself. Indeed, I doubt if he had the vaguest notion how the entries in such a table would be computed. The function $h_q(\lambda)$ for $\varphi = 40°$ and for $q = 45°$ does indeed assume a value of about 40° at $\lambda = 0°$ (and even 41° when rounded at $\lambda = 1°$!), a maximum of about 69° at the summer solstice $\lambda = 90°$, and a minimum of about 12° at the winter solstice $\lambda = 27°$. Selected recomputed values for $\varphi = 40°$, $\epsilon = 23;35°$, and $q = 44°$, 45°, and 46° are shown in Table 1.

From an inspection of these recomputed values, I think that it is safe to assume that whoever computed the table from which the four values mentioned in the text were taken, was using some approximate means of determining the solar altitude in the direction of south-west. Approximate values for the equinoxes and solstices could be found very easily (and more accurate values by taking additional care) by using an analogue computer device such as an astrolabe or by geometric construction (involving a technique known as the analemma).

The next question for us to consider is how the person who computed the table arrived at a value $q = 45°$ for the qibla at Samarqand. There are at least three possible answers to this. Perhaps the unknown compiler of the table considered that for Samarqand and Mecca, ΔL ($= 22°$ according to

Table 1
Sample Recomputed Values of $h_q(\lambda)$ for
$\varphi = 40°$, $\epsilon = 23;35°$, and $q = 44°, 45°$, and $46°$

λ \ q	44°	45°	46°
0°	40;36°	40; 7	39;37
1	41; 5	40;36	40; 6
2	41;33	41; 4	40;35
•			
•			
•			
90	68;48	68;33	68;16
•			
•			
•			
270	12;24	11;42	10;58

al-Bazdawī) was close enough to $\Delta\varphi$ ($= 19°$ according to al-Bazdawī) that one could take $\Delta L = \Delta\varphi$ and hence derive $q = 45°$ by the standard approximate method. Alternatively, he may have decided that $q = 45°$ was a happy compromise between the qibla of the Shāfiᶜīs (due south) and the qibla of the Ḥanafīs (due west). Similar situations occur in medieval Andalusian and Maghribi sources, where the qibla may be due south, due east, or conveniently south-east (see *King* 2, pp. 370-387, and 3, pp. 190-191). Another possibility is that the qibla was taken as south-west in order to "face" the north-east wall of the Kaᶜba (see further *King* 7, Section 3 and the commentary to Paragraph 11 above).

4. *Concluding remarks*

al-Bazdawī informs us that the *Ṣaḥāba* and later Ḥanafīs took the qibla in Transoxania as due west, and that the Shāfiᶜīs took it as due south. He rightly criticizes both traditions, the latter more than the former, since the road to Mecca from Samarqand goes due west rather than due south. He himself prefers, without presenting any valid reasons, to accept the orientation of the Jāmiᶜ Mosque in Samarqand.

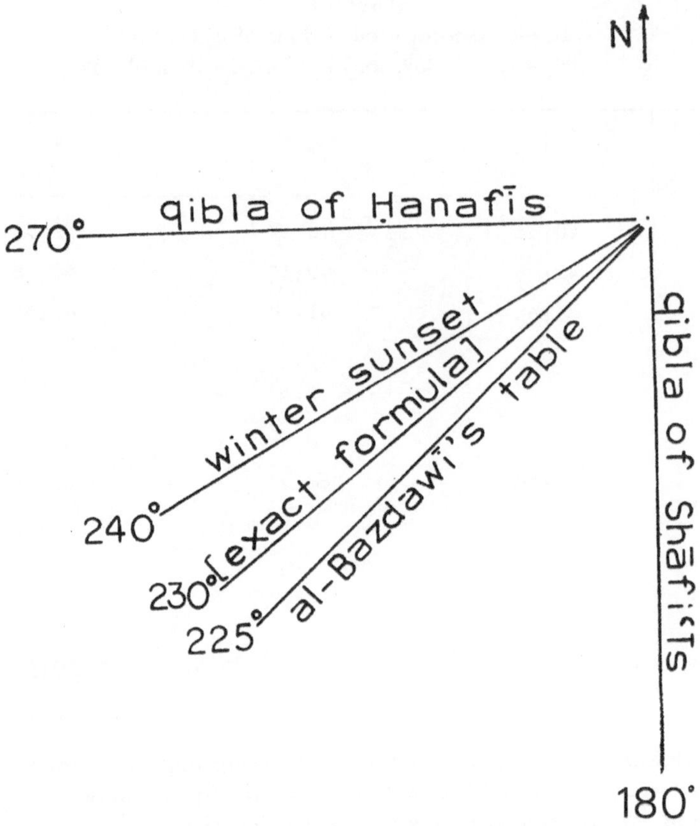

Fig. 4. The different qiblas for Samarqand mentioned by al-Bazdawī

al-Bazdawī's knowledge of astronomy was such that he confused months and zodiacal signs. He presented various geographical coordinates in his treatise, but was apparently incapable of using them to compute the qibla, even by the simple approximate method which was common knowledge amongst contemporary astronomers. The table that al-Bazdawī included in his treatise would enable one to lay out the qibla at Samarqand at 45° S. of W. It seems probable that whoever computed the table considered this a compromise between the Shāfiʿī qibla (due south) and the Ḥanafī qibla (due west). This is about 10° S. of the direction al-Bazdawī was aiming at, namely, the orientation of the Jāmiʿ Mosque of Samarqand, which is at about 35° S. of W. The reason for this orientation may well be that the qibla was taken as the direction of the setting sun at midwinter, which is at about 31½° S. of W. (The proximity of this direction to the modern qibla for Samarqand, which is about 30° S. of W., is fortuitous.) A less likely reason may be that the qibla was computed from available geographical data: al-Bazdawī's data can be used to derive a qibla of about 40° S. of W.

It is instructive to compare al-Bazdawī's treatise with a treatise by his earlier contemporary from further south, the celebrated scholar Abu'l-Rayḥān al Bīrūnī. In his work entitled *Taḥdīd nihāyāt al-amākin*, "The Determination of the Coordinates of Localities," al-Bīrūnī's ultimate purpose was to establish the geographical coordinates of Ghazna and hence determine the qibla there. As well as achieving these ends admirably, he provided us with a richly-documented essay on mathematical geography and spherical astronomy.[1] al-Bīrūnī would have been dismayed by the weakness of al-Bazdawī's reasoning and shallowness of his scientific knowledge. In the *Taḥdīd* he wrote as follows:[2]

"... Let us point out the great need for ascertaining the direction of the qibla in order to perform the prayer which is the pillar of Islam... It is known that this direction varies with the place at which the direction of the Kaᶜba is to be determined. This is witnessed in the Sacred Mosque itself, and should be more evident when considered from other places. If the distance from the Kaᶜba is small, its direction may be determined by a diligent seeker, but when the distance is great, only the astronomers can determine that direction. Every challenge calls for the right men........ Some scholars have been discussing completely irrelevant phenomena, like the directions from which the winds blow, and the risings of the lunar mansions. Even the professional astronomers find the qibla problem difficult to solve, so you can imagine how difficult it is for the non-astronomer."

al-Bazdawī was not the only medieval writer to discuss the qibla in Samarqand. Bābur, in *ca.* 1500, noted that the orientations of the qiblas in the Masjid al-Muqatta' and the madrasa of Ulugh Beg were different, and attributed the orientation of the latter to the astronomers.[3] No doubt other such discussions are to be found in the literature.

The state of documentation of orientations of Islamic architecture in

1. This treatise is published by Bulgakov in *al-Bīrūnī*, translated in *Ali*, and commented upon in *Kennedy* 1.

2. *cf. Ali*, pp. 11-12.

3. *Babur*, 1, p. 80, cited in *Sayili*, p. 24.

See also the valuable study *Nemtseva*. With regard to the qibla corresponding to Ulugh Beg's geographical coordinates for Samarqand and Mecca, see already E. S. Kennedy's comments in *Nemtseva*, p. 52. These coordinates are:

	φ	L
Samarqand	39;37°	99;16°
Mecca	21;40	77; 0

and the accurately computed qibla is 53;8,30° W. of S. According to the standard Islamic approximate formula the value would be 50;53°.

Transoxania is rather poor, and most modern plans purporting to give orientations are not to be trusted.[4] Also, aerial and satellite photographs of the architectural sites in this particular area are hardly likely to fall into the hands of scholars in the near future. It is to be hoped, therefore, that future researchers will make careful measurements of orientations of individual buildings and, when interpreting these, will take into consideration the kind of information presented by al-Bazdawī.

Added in proof:

(1) In September, 1983, I had the privilege of researching in the Library of the Oriental Institute in Tashkent. There I came across in manuscript no. 177 a work by the late–tenth–/early–eleventh–century legal scholar cum mathematician ᶜAbd al-Qāhir al-Baghdādī (*Brockelmann*, I, p. 482 and SI, pp. 666–667, and *Sezgin*, V, pp. 357) on the divergences of opinion on the qibla in early Islamic Iran. This treatise was not previously known to exist, and a study is in progress.

(2) In January, 1984, I was able to measure the orientations of various religious edifices in Samarqand. The magnetic declination there is currently ca. 4° E, and my readings with a pocket compass have been adjusted by ca. 5°. I anticipate that the values given below are correct to within ca. ±5°.

(a) The site identified as a mosque in Afrasiab faces roughly 245° (= 25° S. of W.), that is, towards winter sunset (ca. 250°). On the Jāmiᶜ Mosque in Samarqand (which faced roughly the same direction), see the commentary to Paragraph 9 above.

(b) The basic orientation of the Shāh-i-Zinde complex (most surviving monuments date from the fourteenth and fifteenth centuries) is north-south, with most mihrābs rom facing south, according with the Shāfiᶜī practice.

(c) The basic orientation of the Bībī Khānum Mosque (ca. 1400) is in the cardinal directions with the mihrab facing due west, thus corresponding to the Ḥanafī practice.

(d) The tombs of Timur and Ulugh Beg in the Gur-Emīr Mausoleum (fifteenth century) face roughly 255° (= 15° S. of W.). I cannot explain this orientation.

(e) The basic orientation of the Registan complex (built in the 15th–17th centuries) is ca. 200° (= 20° W. of S.). Again, this orientation corresponds to none of those mentioned by al-Bazdawī.

(f) The Mosque of Khōja Zulmurād in the quarter known as Chahār-rakh (built in the 19th (?) century and still in use) faces ca. 265° (= 5° S. of W.), probably intended as an alignment towards due west in accordance with Ḥanafī practice.

4. See *King* 7, Section 4.9, for a survey, based mainly on *Cohn-Wiener*, *Pugachenkova Pugachenkova & Rempel* 1 and 2, *Zasypkin*, and *Gafurov & Litvinskii* 1 and 2.

These results are interesting, but it is clear that a proper survey of the orientations of all of the principal medieval religious monuments in Central Asia would reveal much more.

APPENDIX A

The Arabic Text of al-Bazdawī's Treatise

See pp. 38-31 of this issue.

[Page 30 of this article was blank and has been omitted for the purposes of this volume.]

APPENDIX B

An Addition to *King* 2

H.P.J. Renaud, on p. 58 of his article on astronomy and astrology in medieval Morocco (see *Renaud* 2 in the bibliography), quotes the ninth-century Andalusian historian and jurist Ibn Ḥabīb (on whom see the article in *EI*$_2$ by A. Huici Miranda) as saying that the qibla at Cordova is the rising point of the star α Scorpio "because it rises at the corner of the Black Stone." This azimuth was about 30° south of east for the latitude of Cordova at Ibn Ḥabīb's time, and corresponds precisely to the azimuth of the rising sun at the winter solstice. See now the commentary on Paragraph 13 of al-Bazdawī's treatise above.

Bibliographical Abbreviations

Ali	J. Ali, *The Determination of the Coordinates of Cities: al-Bīrūnī's Taḥdīd al-amākin*, Beirut: American University of Beirut Publications, 1966.
Bābur	*Bâburnâma*, E. J. W. Gibb Memorial Series, vol. 1, 1905, trans. by L. King, 2 vols., London, 1921.
al-Bīrūnī	al-Bīrūnī, *Kitāb Taḥdīd nihāyāt al-amākin*, ed. P. Bulgakov, *Majallat Maᶜhad al-Makhṭūṭāt al-ᶜArabīya*, vol. 8 (1962).
Brockelmann	C. Brockelmann, *Geschichte der arabischen Litteratur*, 2 vols. (2nd. ed.), Leiden: E. J. Brill, 1943-49, and *Supplementbände*, 3 vols., Leiden: E. J. Brill, 1937-42.
Cairo Cat. and *Survey*	D. A. King, *A Catalogue of the Scientific Manuscripts in the Egyptian National Library* (in Arabic), 2 vols., Cairo: General Egyptian Book Organization in collaboration with the Smithsonian Institution and the American Research Center in Egypt, 1981 and 1985 (?), and *A Survey of the Scientific Manuscripts in the Egyptian National Library* (in English), Publications of the American Research Center in Egypt, Winona Lake, Ind.: Eisenbrauns 1985.
Cohn-Wiener	E. Cohn-Wiener, *Turan: Islamische Baukunst in Mittelasien*, Berlin: Ernest Wasmuth Verlag, 1930.
EI_1	*Encyclopaedia of Islam*, 1st. ed., 4 vols., Leiden: E. J. Brill, 1913-34.
EI_2	*Encyclopaedia of Islam*, 2nd. ed., 4 vols. to date, Leiden: E. J. Brill, 1960 onwards.
Gafurov & Litvinskii 1	B. G. Gafurov & B. A. Litvinskii, *Istoriya i Kultura Narodov Srednei Azii*, Moscow: Nauka, 1976.
2	————————, *Srednyaya Aziya v Drevnosti i Srednevekov'e*, Moscow: Nauka, 1977.
Goldstein	B. R. Goldstein, "On the Theory of Trepidation . . . ," *Centaurus*, 10 (1964), pp. 232-247.
Habib	I. Habib, "Cartography in Mughal India," in *Medieval India: A Miscellany*, IV (Aligarh Muslim University), Bombay: Asia Publishing House, 1977, pp. 122-134. (I have used Prof. Habib's typescript.)
Irani	R. A. K. Irani, "Arabic Numeral Forms," *Centaurus*, 4 (1955), pp. 1-12.
Kennedy 1	E. S. Kennedy, *A Commentary upon al-Bīrūnī's Kitāb Taḥdīd al-Amākin*, Beirut: American University of Beirut Press, 1973.
2	————————, "A Survey of Islamic Astronomical Tables," *Transactions of the American Philosophical Society*, N. S., 46:2 (1956), pp. 123-177.
Kennedy & Haddad	F. I. Haddad and E. S. Kennedy, "Geographical Tables ofMedieval Islam," *al-Abḥāth*, 24 (1971), pp. 87-102.
King 1	D. A. King, "Ibn Yūnus' *Very Useful Tables* for Reckoning Time by the Sun," *Archive for History of Exact Sciences*, 10 (1973), pp. 342-394.
2	————————, "Some Medieval Values for the Qibla at Cordova," in "Three Sundials from Islamic Andalusia," *Journal for the History of Arabic Science*, 2 (1978), pp. 358-392.

3 ————————, "A Fourteenth- Century Tunisian Sundial for Regulating the Times of Muslim Prayer," in W. G. Saltzer and Y. Maeyama, eds., *Prismata: Festschrift für Willy Hartner*, Wiesbaden: Franz Steiner Verlag, 1977, pp. 187-202.

4 ———————, "Architecture and Astronomy: The Ventilators of Medieval Cairo and Their Secrets," *Journal of the American Oriental Society*, 104:1 (1984), pp. 97-133.

5 ————————, "Astronomical Alignmnents in Medieval Islamic Religious Architecture," *Annals of the New York Academy of Sciences*, 385 (1982), pp. 303-312.

6 ————————, *The World About the Kaba: A Study of the Sacred Direction in Medieval Islam* (in preparation), to be published by Islamic Art Publications S. p. A. (Summaries are to appear in *Interdisciplinary Science Reviews*, 10 (1985), and *Proceedings of the Second International Quran Conference, (New Delhi*, 1982*)*.

Le Strange G. Le Strange, *The Lands of the Eastern Caliphate*, London: Frank Cass & Co. Ltd., 1966 (reprint of 1905 edition).

Muñoz R. Muñoz, "Un Calendario Egipcio del Siglo XVIII," *Awraq*, 1 (1978), pp. 67-81 (to be continued).

Nemtseva N. B. Nemtseva, "The Origins and Architectural Development of the Shāh-i Zinde," translated from the Russian by J.M. Rogers and A. Yasin, *Iran*, 15 (1977), pp. 51-73.

Pellat Ch. Pellat, *Le Calendrier de Cordoue*, Leiden: E. J. Brill, 1961.

Pugachenkova G. A. Pugachenkova, *Puti Razvitiya Arkhitektur'i Youjnogo Turkmenistana . . .*, Moscow: Izdatel'stvo Akademii Nauk S.S.S.R., 1958.

Pugachenkova & Rempel 1 G. A. Pugachenkova and L. I. Rempel, *V'idaioushiesya Pamyatniki Arkhitektur'i Uzbekistana*, Tashkent: Gosudarstvennoe Izdatel'stvo Khudozhestvennoi Literatury Uz. S. S. R., 1958.

2 ——————————————, *Istoriya Iskusstv Uzbekistana*, Moscow: Iskusstvo, 1965.

Renaud 1 H. P. J. Renaud, *Le Calendrier d'Ibn al-Bannā' de Marrakech* (1256-1321 *J. C.*), Publications de l'Institut des Hautes-Etudes Marocaines, tôme XXXIV, Paris: Larose Editeurs, 1948.

2 ———————, "Astronomie et Astrologie Marocaines," *Hespéris*, 29 (1942), pp. 41-63.

Rudloff & Hochheim G. Rudloff and A. Hochheim, "Die Astronomie des Maḥmūd ibn Muḥammad ibn 'Omar al-Ğagmīnī," *Zeitschrift der Deutschen Morgenländischen Gesellschaft*, 47 (1893), pp. 213-275.

Samsó 1 J. Samsó Moya, "Las "Pháseis" de Ptolomeo y el "Kitāb al-Anwā'" de Sinān b. Ṯābit," *al-Andalus*, 41 (1976), pp. 15-48.

2 ———————, "De nuevo sobre la traducción árabe de las "Pháseis" de Ptolomeo y la influencia clásica en los "Kutub al-Anwā'"," *al-Andalus*, 41 (1976) pp. 471-479.

3 —————————, "La tradicion clâsica en los calendarios agricolas hispanoarabes y norteafricanos," *Second International Congress of Studies on Cultures of the Western Mediterranean*, (Barcelona, 1975), pp. 177-186.

Sayili A. Sayili, *The Observatory in Islam*, Ankara: Turkish Historical Society (Series VII, No. 38), 1960.

Sayyid F. Sayyid, *Fihrist al-makhṭūṭāt*, 3 vols., Cairo: Dār al-Kutub, 1961-1963.

Sezgin F. Sezgin, *Geschichte des arabischen Schrifttums*, 8 vols. to date, Leiden: E.J. Brill, 1967 to present.

Taqizadeh S. H. Taqizadeh, *Old Iranian Calendars*, London: Royal Asiatic Society, 1938.

Troupeau G. Troupeau, "Le Livre des Temps de Jean ibn Māsawayh," *Arabica*, 15 (1968), pp. 113-142.

Wensinck A. J. Wensinck *et al.*, *Concordance et Indices de la Tradition Musulmane*, 7 vols., Leiden: E. J. Brill, 1936-1969.

Zasypkin B. N. Zasypkin, *Arkhitektura Srednei Azii*, Moscow: Izdatel'stvo Akademii Arkhitektury S.S.S.R., 1948.

في كل يوم ينبغي أن يتعلم ما يعرف به ارتفاع الشمس فإذا كانت الشمس في برج الحمل في أول درجة منه وكان ارتفاع الشمس بعد الزوال أربعين درجة فمن استقبل قرص الشمس في ذلك الوقت في بلدة من هذه البلاد فذلك قبلة تلك البلدة وإذا كانت الشمس في برج الحمل في الدرجة الثانية منه وكان ارتفاع الشمس بعد الزوال إحدى[1] وأربعين درجة فمن فمن استقبل الشمس في بلدة من هذه البلاد في ذلك الوقت فذلك قبلة تلك البلدة وهكذا يزداد في كل يومين وأكثر من ذلك درجة درجة حتى تنزل الشمس برج السرطان وتقطع منه ست عشرة درجة فإذا نزلت[2] في الدرجة السادسة عشر وكان ارتفاع الشمس بعد الزوال ست وستين درجة فمن استقبل الشمس في ذلك الوقت في بلدة من هذه البلاد فذلك قبلة تلك البلدة ثم ينقص في الدرج هكذا حتى تنزل الشمس برج الجدي وتقطع منه ست عشرة درجة فإذا نزلت[3] إلى الدرجة التاسعة[4] عشر منه وأخذت الارتفاع بعد الزوال في بلدة من هذه البلاد وكان الارتفاع إحدى[5] عشر درجة فمن استقبل الشمس في بلدة من هذه البلاد في ذلك الوقت فذلك قبلة تلك البلدة ثم تأخذ[6] الدرجة في الزيادة كل يومين وأكثر حتى تصير إلى الحمل وقد كتبت زيادة الدرجة ونقصانها في السنة كلها في جدول وضعته[7] وكتبت ذلك بحروف الجمل وهي حروف[8] ابجد .*

١ – آحاد ٢ – نزل (انظر ٣ – ٢) ٣ – نزل (انظر ٢)
٤ – التاسع ٥ – احد ٦ – يأخذ ٧ – وضعتها
٨ – حرف

* ملاحظة للناسخ : بعون الله تم نسخ هذه الرسالة في يوم الخميس ٢٤ صفر سنة ١٣٥٥ الموافق ١٥ مايو سنة ١٩٣٦ على نفقة دار الكتب المصرية نقلا عن النسخة الخطية المستحضرة من مجلس محلي سوهاج تحت نمرة ٢١ أصول وكتبها راجي عفو المتين محمود عبد اللطيف فخر الدين النساخ بدار الكتب المصرية العامرة .

طول مكة مثل بخارى[1] وسمرقند يكون زوال مكة بعد زوال تلك البلدة وقد ذكرنا التفاوت بين طول هذه البلاد وطول[2] مكة فإنما تزول الشمس بمكة عن كبد السماء بعد زوال بلدة بخارى وبلدة سمرقند وبلدة نسف بساعتين وثلثي[3] ساعة فإذا مضت ساعتان وثلثا ساعة بعد الزوال ببلدة بخارى وسمرقند
ونسف // فمن استقبل قرص الشمس في ذلك الوقت والشمس في أول برج ص ١١
الجوزاء أو في آخر برج السرطان في بلدة من هذه البلاد فذلك قبلة تلك البلدة فإن كان الإنسان ممن يقف على الساعات يعمل[4] بعد معرفة الساعات على ما بينا وإذا كان لايعرف حقيقة الساعات فإذا صار ظل كل شيء بعد الزوال مثل نصفه وهو وقت يؤدي صلاة الظهر أصحاب أبي حنيفة رحمهم الله في ذلك الزمان فإنه زمان الصيف فمن استقبل قرص الشمس في ذلك الوقت في بلدة من هذه البلاد فهو قبلة تلك البلدة فإن كان الإنسان ممن يعرف دخول الشمس في البروج وإلا يرجع إلى من يعرف ذلك فيسأل فيسوي القبلة حينئذ على ما ذكرنا وكذلك إذا نزلت الشمس برج الجدي وهو أول الشتاء حين يكون النهار في غاية القصر فإذا كادت الشمس تغرب فمن استقبل الشمس حينئذ في بلدة من هذه البلاد بلدة سمرقند وبخارى ونسف فذلك قبلة تلك البلدة .

(١٤) ولكن تسوية القبلة بقبلة مسجد الجامع بسمرقند في بلدة بخارى ونسف هو[5] أن ينظر إلى قرص الشمس بعد الزوال في بلدة سمرقند وفي مسجد الجامع حتى يصير محاذياً لقبلته وينظر كم ظل كل شيء فيأخذه ثم يعود إلى بخارى أو نسف أو كرمنية أو كش بالعجلة ويستقبل الشمس بها بعد الزوال إذا صار ظل كل شيء مثل ماصار بسمرقند فذلك قبلة تلك البلدة فإن في اليوم واليومين إلى عشرة لايقع إلا تفاوت يسير لايعتبر مثل ذلك التفاوت .

(١٥) وقد وضع الحسّاب[6] طريقاً آخر لمعرفة القبلة وتسويتها يمكن //
العمل به في كل يوم فإن الشمس تصير مسامتة[7] للكعبة في كل يوم في[8] ص ١٢
وقت من الأوقات فمن عرف ذلك الوقت واستقبل الشمس في بلدة من هذه البلاد في ذلك الوقت فذلك قبلة تلك البلدة ولكن من أراد تسوية القبلة

١ – بخارا ٢ – مكرر ٣ – وثلث ٤ – بمعنى : فليعمل
٥ – وهو ٦ – الخشاب ٧ – مسامتا ٨ – من

خطأ تلك القبلة إلا أنه نقضها وردها إلى ما كانت لكثرة قيل[١] الناس فإن الجهل مع الحمية غالبان في عامة بلده[٢] وقد سمعت هذا من قوم عدول ثقات وكذلك الحُسّاب اتفقوا وحولوا قبلة المسجد الجامع بسمرقند عما كان إلى يسار المصلي واتفق عليه الأئمة من الفريقين فدل اجتماعهم على خطأ كلتي[٣] القبلتين .

(١٢) وما قاله أصحابنا إن الجدي ينبغي أن يكون على شحمتي أذن الإنسان إذا أراد أن يصلي ليس بصحيح فإنه يستقيم هذا إذا كان عرض مكة وعرض هذه البلاد سواء وليس كذلك بل بينهما في العرض تفاوت عظيم كما بينا وما قاله بعض أصحاب الشافعي إنه[٤] ينبغي أن يكون الجدي على قفا الإنسان إذا أراد أن يصلي خطأ محض // أيضاً فإنه إنما يستقيم هذا إن لو[٥] كان ص ١٠ طول مكة وطول هذه البلاد سواء وقد بينا أنه ليس كذلك وما ذهب إليه أصحاب الشافعي أشد خطأ مما ذهب إليه أصحاب أبي حنيفة والذين فتحوا البلاد لأن طرق هذه البلاد إلى مكة وضعها المتقدمون إلى مغرب الاستواء وما وضعوها إلى مغرب الشتاء ومشرقه .

(١٣) فأبين الآن طريق معرفة القبلة في بلاد ما وراء النهر بخارى[٦] وسمرقند ونسف وما يتبعها من البلاد والقرى بقدر مايقع في أفهام الناس أجمع الحساب وأصحاب التجارب سوى الحساب على[٦] أن الشمس إذا نزلت برج الجوازاء فإن الشمس تصير مسامتة[٧] لمكة وقت الزوال كالترس على رأس الإنسان حتى لايبقى في موضع من المواضع ظل حتى قالوا لايبقى في الآبار ظل وكذلك إذا نزلت الشمس برج السرطان وقطعت[٨] منه عشرين درجة تصير الشمس أيضاً مسامتة[٩] لمكة وقت الزوال ولكل بلدة يكون طولها مثل طول مكة فإذا قابل الشمس إنسان وقت الزوال حين نزلت الشمس برج الجوزاء أو حين قطعت[١٠] الشمس عشرين درجة من برج السرطان يكون ذلك قبلة تلك البلدة وكل بلدة يخالف طولها[١١] طول مكة وكان طول تلك البلدة أكثر من

١ – بمعنى : قيل وقال ٢ – برده ٣ – كلف ٤ – ناقص
٥ – بخارا ٦ – ناقص ٧ – مسامتا ٨ – وقطع
٩ – مسامتا ١٠ – قطع ١١ – ناقص

خطأهم بالدليل لاتجوز[1] الصلاة إلى تلك القبلة بعد ذلك .

(٩) وقد سمعت أناساً أثق بهم أيام مقامي بسمرقند حين كنت قاضياً بالحضرة[2] يقولون إن[2] العلماء تكلفوا في القبلة التي وضعها من فتح بلاد ما وراء النهر واتفقوا أنها منحرفة عن الكعبة يساراً فرجعوا إلى أهل الحساب الذين لهم بصيرة في هذا الباب فوضعوا لهم قبلة إلى يسار المصلى ويمين الكعبة وهو ما بين قبلة أصحاب أبي حنيفة رحمه الله وما أحدثه المتأخرون من أصحاب الشافعي رحمه الله وقبلة مسجد الجامع بسمرقند وضعوها على ذلك وعليه اتفق أصحاب أبي حنيفة والشافعي يومئذ وهو قبلة لا انحراف فيها إلى جهة وقد امتحنتها حين دخلت سمرقند سنة ثلاث وسبعين وأربعمائة وكانت الشمس في برج الجوزاء فوجدتها مستقيمة قويمة .

(١٠) ثم إني لما رأيت هذا الاختلاف الفاسد الباطل في أمر القبلة بين أصحاب أبي حنيفة وأصحاب الشافعي المتأخرين منهم فإنه اختلاف الجهال فإن العالم بالقبلة يخطئهم جميعاً والذي لا علم له بالقبلة وله تقوى وعقل كامل لا ينازع في القبلة ويقر بالجهل فإنما تبقى المنازعة بين عوام لا تقوى لهم ولا كمال عقل مع حدة[3] تأمل أردت أن أصنف[4] كتاباً قصيراً في أمر القبلة وأبين فيه ماهو الصواب وأدل عليه بدلائل نيرة وأبين فيه طريق معرفته بأسهل الوجوه طالباً لثواب الله والتوفيق // لإتمامه معتصماً به من الزلل والخطأ راجياً مرضاته . ص ٩

(١١) فأقول قد بينا في أول الكتاب ما تبين به خطأ ما أحدثه المتأخرون من أصحاب الشافعي في أمر القبلة وذكرنا ما وضع الذين فتحوا البلاد من القبلة أنه ليس بمستقيم وقد سمعت من أثق به من الفقهاء الصلحاء أن من خرج من مكة وتأمل في أمر القبلة ثم رجع إلى بلاد ما وراء النهر عرف أن القبلة منحرفة[5] إلى يسار القبلة وكان جدنا الشيخ الإمام الزاهد أبو محمد عبد الكريم بن موسى خرج إلى مكة حاجاً وتأمل ثم في أمر القبلة فلما رجع إلى بلده حول قبلة مسجده إلى يسار المصلى وهو يمين القبلة لما أنه استيقن

١ – يجوز ٢ – إن : أن ٣ – حد ٤ – أضيف
٥ – منحرف

أنه لا يكون متوجها إلى مكة لأن بعد[1] ما بين مكة وبخارى في الطول خمسمائة فرسخ فإذا قد[2] وقع اليقين على خطأ ما أحدثه المتأخرون من أصحاب الشافعي في[3] القبلة .

(٨) وأما القبلة التي وضعها الذين فتحوا البلاد وهي التي يصلي عليها أصحاب أبي[4] حنيفة رحمه الله ففيه انحراف إلى يمين المصلى وهو يسار القبلة فإن وجه القبلة إلى المشرق فإنا قد ذكرنا أن هذه القبلة إلى مغرب الاستواء وهو وسط المغارب ومغرب الاستواء ليس بمساو لهذه البلاد في العرض وإنما يستقيم هذا
إن لو كان عرض هذه // البلاد وعرض مكة سواء وبينهما مقاربة في العرض ص ٧
حتى يكون المحاذي لمغرب الاستواء محاذياً لمكة وبين مكة وهذه البلاد تفاوت عظيم في العرض فإن عرض مكة إحدى وعشرون درجة وعرض بخارى ثمان[5] وثلاثون درجة وعرض سمرقند أربعون درجة وعرض نسف قريب من ست[6] وثلاثين درجة ويعني[7] بالعرض بعد البلدة عن خط الاستواء وهو وسط السماء[8] فإن العمران[8] كلها في أحد النصفين من الأرض وهو النصف الذي في يسار القبلة وكان بين مكة وبخارى من التفاوت في العرض[9] ست عشر[9] درجة كل درجة قريبة[10] من خمسة وعشرين فرسخا فيكون بينهما من الفراسخ أربعمائة فرسخ وبين مكة وسمرقند التفاوت أكبر[11] فلا يستقيم البتة أن يكون المحاذي لمغرب الاستواء متوجها إلى مكة لما بينهما من التفاوت الكبير[12] في العرض قريب من[13] خمسمائة فرسخ[13] ولأن الشمس تغرب عند الاستواء على يسار مكة لاموازياً لمكة فلا يكون المتوجه إلى المغرب يومئذ متوجهاً إلى مكة وإنما وقع[14] لهم هذا الغلط لأنهم وضعوها[15] بالتحري من غير أن تكون[16] لهم معرفة بالحساب والمتحري عامل بلا دليل بل هو عامل بتحكيم القلب وكثيراً ما يخطئ المتحري ولكن مع هذا تجوز[17] صلاة المتحري إذا لم يكن معه دليل آخر[18] ومن صلى ولا دليل
معه[18] فتكون صلاته جائزة وصلاة من تقلدهم كذلك ولكن // إذا بين إنسان ص ٨

١ — البعد ٢ — ناقص ٣ — من ٤ — أبو
٥ — ثمانية ٦ — ستة ٧ — وتعين ٨ — فإن العمران : قال العمراني
٩ — ست عشر : ستة عشر ١٠ — قريب ١١ — اكثر ١٢ — الكثير
١٣ — خمسين فرسخاً : خمسائة فرسخ ١٤ — رفع ١٥ — وصفوا
١٦ — يكون ١٧ — يجوز ١٨ — ومن صلى ولا معه دليل : ولا معه دليل

هذه القبلة قبلة أي ناحية فلا يصح التعلق بهذين الحديثين فإن قالوا // قد روى قبلة أهل العراق ما بين المشرق والمغرب فنقول[1] هـذه الزيادة ليست بصحيحة فإن العراق لم تكن فتحت يومئذ فالأولى[2] أن المراد من هذين الحديثين قبلة أهل المدينة فإنها بين المشرق والمغرب . ص ٥

(٦) وإن السلف الصالح وضعوا قبلة ما وراء النهر وخراسان حين فتحوا البلاد إلى مغرب الخريف وهو المغرب عند استواء النهار والليل فالشمس إذا نزلت في برج الميزان ومضى عشرون يوماً حتى قطعت قريباً مـن عشرين درجة فمن واجه الشمس عند الغروب فتلك القبلة التي وضعها أولئك السلف وهو وقت فراغ عامة بخارى مـن زراعة الحنطة والشعير وكذلك إذا نزلت الشمس في برج الحمل .

(٧) وما ذهب أصحاب الشافعي منهم إليه من تحريف-القبلة خطأ محض لا يخفى خطأه على من له عقل كامل وتأمل قليل تأمل فضلاً من أن يكون شم رائحة من الحساب فإنه إنما يستقيم أن تكون قبلة بلاد ما وراء النهر وخراسان ما[3] بين مشرق الشتاء ومغربه إذا كانت هـذه البلاد مساوية لمكة في الطول أو طول هذه البلاد قريباً مـن طول مكة كطول مكة مع طول المدينة يعني بالطول بعد البلدة مـن بحر المغرب المحيط بالأرض وبعد مكة من بحر المغرب سبع[4] وستون درجة وبعـد بخارى[5] // من بحـر المغرب سـبع[6] وثمانون درجة وبعد سمرقند منه تسع[7] وثمانون درجـة وبعد نسف منه ثمان[8] وثمانون درجة فيكون التفاوت بين بعد مكة وبعد بخارى عشرين درجة وكل درجة قريبة[9] من خمسة وعشرين فرسخا فيكون التفاوت بين مكة وبخارى في الطول خمسمائة فرسخ أو قريباً منه وبين سمرقند وبين مكة التفاوت أكثر من هذا في الطول وكذلك بين مكة ونسف فمن جعل وجهه إلى ما[10] بين مشرق الشتاء ومغربه لا يكون متوجهاً إلى مكة بيقين وبينهما من البعد في الطول قريب من[11] خمسمائة فرسخ[11] فمن جعل وجهه إلى ما بين مغرب الشتاء ومشرقه يعلم يقينا ص ٦

١ – نقول ٢ – على ٣ – با ٤ – سبعة ٥ – وبعد بخارى (مكرر) ٦ – سبعة ٧ – تسعة
٨ - ثمانية ٩ – قريب ١٠ – ناقص ١١ – خمسين فرسخا : خمسمائة فرسخ (انظر ص ٤ رقم ١٣)

فن واتبعوا دلائلها وتأملوا فيها وما اكتفوا فيها بالتقليد وإن كانت عامة الناس لايحتاجون إلى تلك المسائل في عامة الأزمان نظراً للناس حتى إذا وقع لإنسان حاجة يجدها أو يجد مثلها ولا يتحير فيها وكان أبو حنيفة رحمه الله رئيسنا في وضع المسائل بدلائل وأصحابه[1] بعده تأملوا في تلك المسائل وكذلك فقهاء الأمة من غيرهم رحمهم الله فما وجدوا له دليلاً يدل على صحته اعتقدوه وقالوا به وما لم يوضح لهم دليل صحته طرحوه ولم يكتفوا بالتقليد[2] .

(٤) ثم إن السلف من الأئمة أكثرهم أعرضوا عن التأمل في أمر القبلة واكتفوا بالتقليد وإن كان وضع القبلة ليس بواجب عليه[3] التقليد وإنما فعلوا ذلك لأنه لم يكن لهم آلة معرفة القبلة فإن القبلة لا تعرف إلا بعلم الحساب وما كان لهم بصر بالحساب فقلدوا غيرهم لعجزهم عن معرفتها بالدلائل فإن عامة السلف من الأئمة ما كانوا اجتهدوا في علم الحساب وقد كان لبعض أصحابنا // بصر في علم الحساب فطلبوا دلائل القبلة ووقفوا عليها ولكن ص ٤
لم يصنفوا فيه كتاباً لغموض علم الحساب ولإعراض أكثر الناس عن علم الحساب خصوصاً الفقهاء منهم وبعضهم صنفوا كتباً ولكن صنفوها غامضة لا يقف على ما فيها إلا المتبحرون في[4] علم الحساب فتعطلت تلك الكتب وبقي التقليد في الناس أجمع .

(٥) ثم إن بعض المتأخرين من أصحاب الشافعي ممن لم يشم رائحة الحساب بما وراء النهر وخراسان خطأوا السلف الصالح وحرفوا القبلة ما بين مشرق الشتاء ومغربه واعتمدوا على حديثين أحدهما ما[5] رووا عن النبي عليه الصلاة والسلام أنه قال القبلة ما بين المشرق والمغرب والثاني ما رووا عنه عليه الصلاة والسلام أيضاً أنه قال لا تستقبلوا القبلة عند الخلاء ولا تستدبروها ولكن شرقوا أو غربوا وهذان الحديثان لا تعرف صحتهما لأن ما رواهما[6] الثقات في كتبهم ثم وإن ثبت فلا يخفى على عاقل أن الاحتجاج به غير صحيح فإنه يعرف كل عاقل ببديهة عقله أن قبلة البلاد كلها لا تكون بين المشرق والمغرب وإنما تكون قبلة بعض البلاد وليس في حديث النبي عليه الصلاة والسلام أن

١ – با ٢ – واضح ٣ – ناقص ٤ – من ٥ – ناقص
٦ – روى

APPENDIX A
The Arabic text of al-Bazdawī's treatise edited from MS Cairo Dār al-Kutub B 19385

رســـالة في سمـــت القبلة ص ١

دافيد كينج

(١) // بسم الله الرحمن الرحيم الحمد لله العلي العظيم الحليم الكريم الحكيم العليم الملك الحق المبين ذي العزة والقوة المتين على ماأنعم علينا مـــن أنواع الفضائل ومكارم الأخلاق والشمائل والصلاة على رسوله المصطفى الأمين المجتبى المكين وعلى آله وأصحابه وأزواجه الطاهرين أجمعين . ص ٢

(٢) قال الشيخ الإمام والقرم الهمام صـــدر الإسلام أبو اليسر البزدوي[١] رحمه الله أما بعد فإن أعظم العبادات بعد الإيمـــان بالله تعالى الصلاة[٢] فإن النبي عليه الصلاة والسلام قال الصلاة عماد الدين فمن تركها فقد هدم الإيمان وجعل[٣] الإيمان بلا صلاة كالبيت المنهدم والدار المنهدمة[٤] دار إلا أنه لا يمكن الانتفاع بها فدلنا هـــذا الحديث أن الصلاة من أعظم العبادات وأن تاركها ليبطل[٥] إيمانـــه .

(٣) ثم كل مـــن هو يحتاج إلى أداء الصلاة في كل يوم مراراً كثيرة لا يقدر على أدائها إلا بعد معرفة أركانها وشروطها ومن شروطها التي تحتاج إليه في كل صلاة التوجه إلى الكعبة فلا بد من معرفة الكعبة انها في أي جهة فإن السلف من الأئمة رحمهم الله وضعوا مسائل كثيرة في كل شرط من شروطها وأظهروا دلائل صحتها وما اكتفوا فيها بالتقليد وكذلك وضعوا مسائل كثيرة في الزكـــاة // وإن كان أكـــثر الناس لايحتاج إلى معرفـــة مسائل الزكاة ص ٣
ووضعوا مسائل كثيرة في الصوم وإن كانت الحاجة تقـــل إلى معرفة مسائله[٦] ووضعوا مسائل النكاح والطلاق والبيوع والجنايات، ومسائل أخرى في[٧] كل

ملاحظة : كل ماورد في الحواشي هو وارد في الاصل ، إلا ماذكر خلافه .

١ – اليزدوى ٢ – ناقص ٣ – جعل ٤ – المنهدم
٥ – لايبطل ٦ – مسائلها ٧ – اخر

Too Many Cooks ...
A New Account of the Earliest Muslim Geodetic Measurements

Contents: Introduction. 1: Ibn Yūnus' account. 2: Al-Bīrūnī's account. 3: The account of Ḥabash. 4: The account of Yaḥyā ibn Aktham. 5: Concluding remarks. Appendix: Arabic texts. Bibliography and bibliographical abbreviations.

Introduction

The measurement of the length of one degree on the meridian by astronomers commissioned by the Abbasid Caliph al-Ma'mūn in Baghdad *ca.* 830 is in one sense well known, and in another, clouded in obscurity. Different versions of the observations are recorded by Ibn Yūnus (*fl.* Cairo *ca.* 990) and al-Bīrūnī (*fl.* Central Asia *ca.* 1025), as well as by various later writers. Particularly those versions in which two groups of astronomers laden with instruments are reported heading off in opposite directions along a meridian in the middle of the desert simply boggle the mind. There is a substantial secondary literature on these reports, including important contributions by C. A. Nallino, S. H. Barani, A. Sayılı, E. S.

Kennedy, R. P. Mercier, and F. J. Ragep.[1] In this paper these well-known accounts will be discussed, but only briefly, for my main purpose is to present a new contemporaneous account of the measurements.

The earliest Muslim latitude and longitude measurements in Mecca and Baghdad, also commissioned by al-Ma'mūn, are less well known, and they too are clouded in obscurity. Until recently they were known only from passing remarks by al-Bīrūnī,[2] and it was Aydın Sayılı who first drew attention to that scholar's brief mention of the simultaneous lunar eclipse observations in the two cities.[3] It was clear that the main purpose of the measurements was to derive the *qibla* at Baghdad, but details were lacking.[4]

In 1985 Dr. Y. Tzvi Langermann of the Hebrew University in Jerusalem published a newly-discovered treatise on both sets of observations by the contemporary astronomer Ḥabash,[5] and I have come across a different

[1] See Nallino, *Scritti*, V, pp. 408-457; Barani, "Muslim Geodesy"; Sayılı, *The Observatory in Islam*, pp. 85-86; Kennedy, *al-Bīrūnī's Taḥdīd*, pp. 131-136; and the recent critical survey of the evidence in Mercier, "Muslim Geodesy", pp. 178-181. On Ragep's astounding findings see n. 67 below. See also the forthcoming volumes of Sezgin, *GAS*, relating to geography. The various places mentioned in this article are shown on a map in Mercier, "Muslim Geodesy", p. 180.

[2] On al-Bīrūnī see the article by E. S. Kennedy in *DSB*. His treatise on mathematical geography has been published by P. Bulgakov, translated by J. Ali, and analyzed by E. S. Kennedy - see the bibliography for details.

[3] Sayılı, *The Observatory in Islam*, p. 85.

[4] For a general survey of the *qibla* problem see King, "Sacred Direction", and the articles "Ḳibla. ii. Astronomical aspects" and "Makka. iv. As centre of the world" in *EI²*, reprinted in King, *Studies*, C-IX and C-X, as well as King & Lorch, "Qibla Charts". See also n. 60 below on non-mathematical methods that were widely used in medieval Islam.

[5] On Ḥabash see the articles by W. Hartner in *EI²* and S. Tekeli in *DSB*, and the recent study of his *zīj* in Debarnot, "Ḥabash's Zīj (Istanbul MS)". (The later recension extant in a Berlin manuscript is currently being investigated by Dr. Benno van Dalen (Utrecht and Frankfurt).) On the title of his treatise, see Sezgin, *GAS*, V, p. 276, no. 5, where no manuscripts are listed: the treatise is now published in Langermann, "Ḥabash on Distances".

account by Yaḥyā ibn Aktham,[6] the judge appointed by al-Ma'mūn to oversee the observations. In this paper I shall compare both of these reports in the light of the later accounts of Ibn Yūnus and al-Bīrūnī. Mercier's recent overview of the evidence available to him, which included the treatise of Ḥabash, draws attention to the obvious inconsistencies in the sources. From the outset it must be borne in mind that this newly-discovered report of Yaḥyā ibn Aktham raises yet more questions and casts but little light on the confusion.

In the sequel I use the following notation freely:

- q *qibla* (local direction of Mecca, measured from the local meridian)
- L terrestrial longitude
- ΔL longitude difference from Mecca
- ΔL_B longitude difference between Baghdad and Mecca
- $\Delta\phi$ latitude difference from Mecca
- $\Delta\phi_B$ latitude difference between Baghdad and Mecca
- ϕ latitude
- ϕ_B latitude of Baghdad
- ϕ_M latitude of Mecca.

In fact, q is a trigonometric function of ΔL, ϕ and ϕ_M, but most of the earliest mathematical procedures, including the one used by al-Ma'mūn's astronomers, were approximate, and one popular method used for several centuries involved only ΔL and $\Delta\phi$.[7]

In the translations the distinction between numbers written in words and those expressed in Arabic numerals has been preserved. Numbers expressed in alphanumerical (*abjad*) notation are underlined. Insertions in the text to smooth the flow of the translation, as well as occasional citations of the Arabic original, are in parentheses; restorations are in square

6 On Ibn Aktham see, for example, the biographical notice (devoid of any reference to the observations commissioned by al-Ma'mūn) in al-Khaṭīb al-Baghdādī, *Ta'rīkh Baghdād*, XIV, pp. 191-204, and also Sayılı, *The Observatory in Islam*, pp. 56 and 69. On his son Muḥammad, who authored a treatise on number theory, see Sezgin, *GAS*, V, pp. 273-274.

7 See King, "Earliest Qibla Methods and Tables", especially pp. 103-107.

brackets; and my comments in curly brackets.

1. Ibn Yūnus' account

In Chapter 2 of his monumental work, a *zīj* or astronomical handbook with tables,[8] dedicated to the Fatimid Caliph al-Ḥākim and called the *Ḥākimī Zīj*, Ibn Yūnus[9] refers in general terms to the measurement of longitude differences by means of lunar eclipse observations,[10] and then briefly refers to the geodetic measurements:[11]

> Discussion of the (distances in) cubits between places: Sanad ibn ᶜAlī mentioned in a statement of his which I found that al-Ma'mūn ordered Khālid ibn ᶜAbd al-Malik al-Marwarrūdhī and him to measure the amount of one degree of a great circle on the surface of the earth. He said: "We both set off together for this (purpose). (al-Ma'mūn) ordered ᶜAlī ibn ᶜĪsā al-Asṭurlābī and ᶜAlī ibn al-Buḥturī to do the same and they went off in a different direction." Sanad ibn ᶜAlī said: "Khālid ibn ᶜAbd al-Malik and I travelled to (a place) between *W'mxh*[12] and Tadmur (= Palmyra) and there we measured the amount of a degree of a great circle on the surface of the earth. It was fifty-seven miles (*mīl*). ᶜAlī ibn ᶜĪsā and ᶜAlī ibn al-Buḥturī also made measurements and the two of them

[8] A *zīj* is a set of astronomical tables with introductory and explanatory materials. See Kennedy, "*Zīj* Survey", for an introduction, based on about 125 such works, many of which contain several hundred pages of tables and text. The number of known *zīj*es is now closer to 200.

[9] On Ibn Yūnus see my article in *DSB*, and also King, *Studies*, A-IX, pp. 343-345.

[10] See Schoy, "Muslim Geography", pp. 265-267, for a translation of the relevant passage from MS Leiden Or. 143, p. 80.

[11] The text, found in MS Leiden Or. 143, pp. 81-82, was first translated in Caussin, "Ibn Yūnus", pp. 78 and 80, n. 1. The text is published in Barani, "Muslim Geodesy", pp. 8-11, and a new translation is in Mercier, "Muslim Geodesy", p. 179.

[12] The letter *x* here denotes an unpointed carrier in Arabic, which could be read as a *bā'*, *yā'*, *nūn*, etc. See further the commentary to this passage.

found the same as this. The two reports from the two directions (telling of) the two measurements with the same result arrived at the same time."

Aḥmad ibn ᶜAbdallāh, known as Ḥabash, stated in his book in which he mentioned the observations (of the compilers) of the *Mumtaḥan* (*Zīj*) in Damascus, that al-Ma'mūn ordered that one degree of a great circle on the surface of the earth should be measured. He said: "They set off for this (purpose) in the desert of Sinjār, (travelling) until two measurements of the meridian altitude on the same day differed by one degree, and then they measured the distance between the two places, and it was 56 and one-quarter miles, where each mile is four thousand cubits, these being the "black cubits" that were adopted by al-Ma'mūn."

Ibn Yūnus continues with a description of the way in which the distance along the meridian can be measured. One takes two ropes, each 50 cubits long. Lay out the first one in the meridian. Then lay out the second one along the first one, starting at the mid-point of the first. And so on ... (for 56-odd miles!).[13] He concludes with a discussion of measuring rods, describing instruments of square cross-section and 24 times as long as they are wide, adding that these would cost 1000 dirhams if made of silver, or 1000 dinars in gold. Since, according to a 14th-century Egyptian source, Ibn Yūnus was paid 100 dinars a day for his services to the Fatimid Caliph al-Ḥākim,[14] we may speculate that he himself might have had such a rod in gold, if not several!

Commentary

The *Mumtaḥan Zīj* was the major production of the astronomers commissioned by al-Ma'mūn, who were under the direction of Yaḥyā ibn

[13] See Mercier, "Muslim Geodesy", p. 181, for a translation of this passage.

[14] MS Leiden Or. 468, fol. 84v. On the importance of this manuscript see King, *Studies*, B-VII, pp. 250-252. It is now being studied by Margarita Castells of Barcelona.

Abī Manṣūr.[15] This work is available only in a later recension that has still not been studied in detail. Groups of astronomers in Baghdad and Damascus were involved in observations for this work, but it is by no means clear what the input of the two groups was. Ibn Yūnus mentions that Ḥabash was writing about the Damascus observations, but the surviving manuscript of Ḥabash's treatise mentions Damascus only once, when talking about a large equatorial ring made by Khālid al-Marwarrūdhī, "who was in Damascus".[16]

Caussin de Perceval, who first translated the text of this passage, thought that the mysterious place-name *W'mxh* (MS Leiden Or. 143, p. 81, l. 19) might be the ancient Apamea,[17] but the name is nowhere else attested in the early Islamic sources. The name reappears as *W'myh* in the parallel passage in MS London B.L. Or. 3624, fol. 42v, l. 14, of the *Mukhtār Zīj* of Abū 'l-ᶜUqūl (*fl.* Taiz *ca.* 1300),[18] a work relying heavily on Ibn Yūnus. The best-known city with the name Apamea is near Homs and cannot be considered. There is, however, another Apamea due north of Palmyra, near Zeugma.[19] Others have suggested that the word is simply a corruption of (al-)Raqqa,[20] which has the advantage of lying roughly due north of Palmyra and is connected to it by a Roman road, the Via Diocletiana (which runs north-east leaving Palmyra in order to avoid the

[15] On this work see Vernet, "Tabulae Probatae", and Kennedy, "*Zīj* Survey", pp. 132 (no. 51) and 144-147. On its context see Sayılı, *The Observatory in Islam*, pp. 50-87.

[16] Langermann, "Ḥabash on Distances", p. 125.

[17] Caussin, "Ibn Yūnus", p. 80, note a.

[18] On this author and his work see King, *Astronomy in the Yemen*, pp. 30-32, and Varisco, *Yemeni Almanac*, pp. 10-11.

[19] Mercier, "Muslim Geodesy", p. 179, n. 27.

[20] See, for example, Barani, "Muslim Geodesy", p. 28, and Kennedy, *al-Bīrūnī's Taḥdīd*, p. 131. In fact Raqqa is not due north of Palmyra (it lies roughly NNE), but the two cities have the same longitude in various early Islamic geographical tables - see Mercier, "Muslim Geodesy", *loc. cit.*

mountains and then due north to meet the Euphrates just west of Raqqa).[21] But Ibn Yūnus, writing in Cairo, was well aware of the existence of distant Raqqa, not least because his predecessor al-Battānī had worked there.[22] Mercier has suggested a possible misreading of θεμα, the name of a locality north of Palmyra and just across the Euphrates, as οεμα, which could conceivably be rendered *Wāmiya* in Arabic.[23] The problem of this mysterious name is not yet solved.

Two different results, 57 and 56 1/4 miles, are reported for the two different expeditions. Al-Farghānī, a late contemporary of al-Ma'mūn's astronomers, mentioned 56 2/3,[24] which Nallino and Barani have seen as a mean value between these two results that subsequently passed as the "real finding" (Barani) of al-Ma'mūn's scientists.[25] But as we shall see, the more original reports mention a figure of 56 miles. Here the reader should be reminded that these findings were not exactly "new", even in the Islamic context: already al-Fazārī in the 8th century had recorded the circumference of the earth according to "the Indians", namely, 6,600 *farsakhs*, where a *farsakh* is 16,000 cubits. Thus, as al-Bīrūnī points out elsewhere in the *Taḥdīd*, the length of 1° is 18 1/3 *farsakhs*, that is, 55 miles, since one *farsakh* equals three miles.[26] What is new is the organized scientific measurements, but only in so far as we are prepared to believe the reports, for certainly some imaginary elements have been inserted. Also the value 56 1/4 was not new: it can be derived from the 75 miles attributed in numerous sources to al-Khwārizmī, another of al-

[21] *Ibid.*

[22] On al-Battānī see the article by W. Hartner in *DSB*.

[23] Mercier, *op. cit.*; Sédillot proposed Wāsiṭ - see Nallino, *Scritti*, V, p. 429 - but Arabic does not allow of such spelling-mistakes.

[24] Mercier, "Muslim Geodesy", p. 178.

[25] Nallino, *Scritti*, V, pp. 431-432; and Barani, "Muslim Geodesy", p. 31.

[26] See *al-Bīrūnī's Taḥdīd*, text, pp. 211-212, translation, p. 177, commentary, p. 132, as well as Pingree, "al-Fazārī", p. 114.

Ma'mūn's astronomers, by converting from Roman miles to Arab miles.[27]
On the black cubit see the commentary to the text of Ḥabash.

2. Al-Bīrūnī's account

We continue with the references of al-Bīrūnī to the observations, quoting his monumental work on mathematical geography, the *Taḥdīd nihāyāt al-amākin*.[28] The text is brief and no details are given:

> ... The *qibla* of Islam is the Sacred Mosque ... . The latitude of Mecca is known but there are various opinions about the number of minutes in the twenty-second degree, because (?) the astronomers (*al-ḥussāb*) take it to be twenty-one degrees {with no minutes}. It is related that Manṣūr ibn Ṭalḥa al-Ṭāhirī concerned himself with finding its correct magnitude: he determined that this was in excess of (twenty-one degrees) by two-thirds of a degree, which is in agreement with what Ḥabash reported about the latitude measurement commissioned by al-Ma'mūn (*raṣad al-Ma'mūn iyyāhu*). (But) others have claimed that the excess (over twenty-one degrees) is (only) one-third of a degree. The longitude (of Mecca) is also known, because all the reports agree that Manṣūr ibn Ṭalḥa found its longitude to be sixty-seven degrees, which agrees with what Ḥabash al-Ḥāsib stated in his *Kitāb al-Ab*ᶜ*ād wa-'l-ajrām*. (He said) that al-Ma'mūn had organized someone to observe lunar eclipses (at Mecca), and that (that person) {singular} had found a difference of three degrees between its meridian and that of Baghdad. So, if the longitude of Baghdad is seventy degrees, that of Mecca would be sixty-seven degrees. ...

27 Mercier, "Muslim Geodesy", *loc. cit.*

28 Bulgakov, ed., *al-Bīrūnī's Taḥdīd*, pp. 209-210. Ali's translation (p. 175) is unreliable here. See also Kennedy's commentary (p. 130).

Later in the same treatise al-Bīrūnī again cites the same work of Ḥabash:[29]

> The observations (commissioned) by al-Ma'mūn (took place) only when he (had) read in the books of the Greeks that the equivalent (*ḥiṣṣa*) of one degree was five hundred *stades*, which is the unit of theirs that they used for measuring distances, and when he did not find amongst the translators sufficient knowledge about its length compared with other known (units). At that time - according to what Ḥabash related on the authority of Khālid al-Marwarrūdhī and a group of scholars of (instrument) construction and expert constructors from amongst the carpenters and brass-workers - (al-Ma'mūn) ordered the construction of the instruments and the selection of a place for this survey. There was chosen a location in the desert of Sinjār in the area (*min ḥudūd*) of Mosul, nineteen *farsakhs* from the city (*min qaṣabatihi*) and forty-three *farsakhs* from Samarra, where they were satisfied that (the ground was) level. They transported the instruments there and they selected a place where they observed the solar altitude at midday. Then two groups set forth (in two different directions). Khālid and a group of surveyors and instrument-makers (*al-massāḥ wa-'l-ṣunnā*ᶜ) headed in the direction of the northern (celestial) pole, and ᶜAlī ibn ᶜĪsā al-Asṭurlābī and Aḥmad ibn al-Buḥturī the surveyor with a(nother) group towards the south (celestial) pole. Each of the two groups observed the altitude of the sun at midday until they found that it had changed by one degree, apart from the change that resulted in the solar declination (*siwa 'l-taghayyur al-ḥādith mina 'l-mayl*). They measured the track on their way out (*fī dhihābihim*) and set up markers (*sihām*, lit., arrows) as they went, and as they returned they investigated (*i*ᶜ*tabarū*) the distance (*al-misāḥa*) for the second time. The two groups met again at the place from which they had set out, and they found that one degree of the terrestrial meridian is equivalent to fifty-six miles. (Ḥabash) claimed that he had heard Khālid dictating that number to the *Qāḍī* Yaḥyā ibn Aktham, so (Ḥabash himself) heard it from the horse's mouth (*fa-'ltaqaṭahu minhu samā*ᶜ*an*).

29 *al-Bīrūnī's Taḥdīd*: text, pp. 213-214; translation, pp. 178-179; and commentary, p. 133. See also Barani, "Muslim Geodesy", pp. 11-14.

> Abū Ḥāmid al-Ṣaghānī, who heard it from Thābit ibn Qurra, told me the same thing.
> (On the other hand) it is related of al-Farghānī (that he reported) two-thirds of a mile in addition to the (56) miles mentioned above. Similarly I found all of the records confirming these (additional) two-thirds, and I may not attribute that to their having dropped out of the manuscript of the *Kitāb al-Abᶜād wa-'l-ajrām* because Ḥabash derived from that (value) the circumference and diameter of the earth and all of the (planetary) distances. When I investigated (these) I found that they result from fifty-six miles only {that is, without the two-thirds}. It is preferable to imagine that (these different results for the length of one degree) derive from two accounts by two teams. (Certainly) this is a subject surrounded with confusion, which should inspire renewed investigations and observations. Who is prepared to help me with this (project)? ... ...

Elsewhere al-Bīrūnī deals in general terms with the problem of determining longitude differences from simultaneous observations of a lunar eclipse,[30] but does not mention the Abbasid observations again.

Commentary

Manṣūr ibn Ṭalḥa was a philosopher, mathematician, astronomer and music theorist, who died *ca.* 855.[31] Khālid al-Marwarrūdhī was one of the leading astronomers of al-Ma'mūn: he apparently authored a *zīj* that has been lost without a trace.[32]

Of particular interest in this account are the coordinates of the locations in the plain of Sinjār where the two groups of observers set forth in two different directions. This information does not stem from the treatise of

[30] *al-Bīrūnī's Taḥdīd*: see text, pp. 166-169, translation, pp. 129-132, and commentary, pp. 100-101, and my review of the last in *Centaurus* 19 (1974), pp. 320-323, especially p. 322. See also Mercier, "Muslim Geodesy", p. 176.

[31] Sezgin, *GAS*, V, p. 245, and VI, p. 145.

[32] *Ibid.*, V, p. 244, and VI, p. 139.

Ḥabash as it has come down to us - see the next section. What is disappointing is that there is no record of two different results - such as one would expect from any scientific mission - which could then have been compared and used to produce a "final" result. Rather, in this case, the "final" result is presented as a *fait accompli*.

Having briefly discussed these two later accounts, both of which have been known to previous researchers, we are now in a position to investigate the more original accounts.

3. The account of Ḥabash

The first two parts of the treatise of Ḥabash, preserved in a precious 17th(?)-century astronomical compendium given to Rabbi Josef Kafaḥ by a Muslim Yemeni colleague before he left Sanaa for Jerusalem,[33] provide considerably more information - see Arabic text no. 1. The text translates:[34]

> In the Name of God the Merciful and Compassionate, from whom we seek help. *The Book of Sizes and Distances* by Ḥabash ibn ᶜAbdallāh the astronomer (*al-ḥāsib*).
> He said that the Commander of the Faithful, al-Ma'mūn, wanted to know the size of the earth, so he made some investigations about this and found that Ptolemy had stated in one of his books that the circumference of the earth was so many thousand *stades*. Thereupon he asked the interpreters about the meaning of *stades* and they gave different interpretations. [He said of their interpretations:] "They do not dispense with (?) what we wanted to know." {The text is corrupt here.}

[33] This is yet additional proof of the importance of Yemeni manuscripts for the history of Islamic astronomy. See King, *Astronomy in the Yemen*, for a first look at over 100 Yemeni astronomical manuscripts, now supplemented by Langermann, "Bibliographical Notes", based on Jewish sources.

[34] Langermann, "Ḥabash on Distances", pp. 115-116 and 122-123. Some problems in the published text have been fixed, and the translation is my own. I am grateful to Dr. Langermann for photocopies of the relevant sections of the original manuscript.

(Al-Ma'mūn) therefore sent out Khālid ibn ᶜAbd al-Malik al-Marwarrūdhī, ᶜAlī ibn ᶜĪsā al-Asṭurlābī and Aḥmad ibn al-Buḥturī the surveyor together with a group of surveyors and craftsmen including carpenters and brass-workers, to make correctly the instruments which they would need, transporting all of them to a place which they selected in the desert of Sinjār.
Next Khālid and one group headed towards the North Pole of the Banāt Naᶜsh (= Ursa Minor), and ᶜAlī and Aḥmad and a(nother) group headed towards the South (celestial) Pole. They {*i.e.*, the former group ?} continued until they finally found that the (solar) meridian altitude had decreased and changed from what they had found in the place where they had separated, by the amount of one degree, after they had subtracted from that (the change in) solar declination during the time it took (for them) to cover that distance. (As they went) they set up markers (*sihām*), and then they retraced their steps using the markers and checked the distance a second time. They found the amount of one degree of the earth's surface to be fifty-six miles, where one mile is four thousand black cubits, this being the cubit established by al-Ma'mūn for measuring clothes, surveying buildings and reckoning distances between stations on the pilgrimage road (*qismat al-manāzil*).
I heard this (information) which I have mentioned in my book {*sc.* in this book of mine?} from Khālid ibn ᶜAbd al-Malik al-Marwarrūdhī as he was conveying it to the *Qāḍī* Yaḥyā ibn Aktham. Yaḥyā [had been] ordered [by al-Ma'mūn] {the text appears to be corrupt here} to write down for him {sc. for al-Ma'mūn} all that Khālid told him, so he wrote (it) for him. I have written what I heard from Khālid himself.
The Commander of the Faithful al-Ma'mūn - may God be pleased with him - (also) wanted to measure the azimuth of the *qibla*. So he sent out (someone) at the time of a lunar eclipse to measure the longitude between Mecca and Baghdad. (That person) found that the meridian of Mecca was west of that of Baghdad by approximately three degrees. If we modify one degree of the celestial equator {the expression *al-falak al-mustaqīm*, meaning *sphaera recta* or right sphere, is usually used for

astronomical ascensions[35] - compare the terminology of Yaḥyā ibn Aktham} we find that the amount on the celestial day-circle corresponding to the parallel of Mecca (*ḥiṣṣatuhā fī* [*madār*] *al-falak al-makhṭūṭ ᶜalā ᶜarḍ Makka*) is approximately fifty-six minutes. So if we multiply the distance between the meridians of Mecca and Baghdad, (namely,) three degrees, by 56 minutes where 60 parts are one degree (*juz'*), we find that the corrected longitudinal distance between them on that parallel (*fī 'l-falak al-makhṭūṭ ᶜalā madār ᶜarḍ Makka*) is approximately two degrees and forty-eight minutes. (That same observer ?) also found the latitude of Mecca to be approximately twenty-one degrees and forty-two minutes and that of Baghdad to be approximately thirty-three degrees and twenty-seven minutes. (If) we subtract the latitude of Mecca from that of Baghdad the difference between the two is eleven degrees and forty-five minutes. If we multiply the difference between the two longitudes, namely <u>2</u> degrees and <u>48</u> minutes, by itself, the result is 7 degrees, 50 minutes and 24 seconds. Half of this is 3 degrees, 55 minutes and 12 seconds, or 235 minutes and 12 seconds. This we divide by 11 degrees and 45 minutes, which is the difference between the two latitudes, [the quotient is 0;20(,1,1). We add this to the 11;45 and obtain] for the distance between Baghdad and Mecca 12 degrees and 5 minutes. This we multiply by 56 miles and obtain approximately six hundred and seventy-six and two-thirds miles as the distance between Baghdad and Mecca as the crow flies (*ᶜala 'l-sahm al-mustawī*). When Khālid ibn ᶜAbd al-Malik al-Marwarrūdhī submitted (this) value of one degree of the earth('s circumference) to al-Ma'mūn, (the Caliph) wanted to check it and so he [sent out] {reading *wajjaha* for *wajada*} someone {singular! - compare the account of Yaḥyā ibn Aktham} to measure the road between Baghdad and Mecca by the shortest route and he {singular!} found it to be seven hundred and twelve miles. So we (?) found that the difference between the distance as the crow flies (*al-sahm alladhī yajrī fī 'l-hawā*) and the measure-ment of the road was thirty-five miles and one-third of a mile. Al-Ma'mūn said that this was not to be regarded as excessive because there must be (in addition to) the flat

[35] See the article "Maṭāliᶜ" in *EI²*.

parts, inclines up or down on (the road) amounting to this (difference).

Commentary

In his *Planetary Hypotheses*, Ptolemy stated that the circumference of the earth was 180,000 *stades* (*astādhiyya*).[36]

The determination of terrestrial latitude is best achieved by measuring the altitude of the Pole Star or by measuring the meridian altitudes of the sun. Our text asserts that the second method was used. As John Britton has shown, there are problems inherent in the use of the meridian quadrant which affect the accuracy of the altitude measurements.[37] Furthermore, given the problems associated with obtaining a reasonable result from such a crude procedure, it is hardly necessary to make a correction for the change in solar declination during the time the team moved the fifty-odd miles. The report attributed to Yaḥyā ibn Aktham - see below - states that the observations were made at the summer solstice - when the daily change in declination is a minimum - but adds that the declination at that time is zero, which is clearly absurd. Does the text mean that one can find the meridian by standing perpendicular to the direction of the setting sun at the equinoxes?

The information on the so-called "black" cubit is interesting: W. Hinz has stated in one place that this unit was standardized by al-Ma'mūn and elsewhere that it was not.[38] Now we know from a contemporary source that it was.

It is the account of the simultaneous lunar eclipse observations and latitude measurements in Baghdad and Mecca which adds to the information recorded by al-Bīrūnī. The simultaneous observation of a lunar eclipse in two localities is a very sensitive kind of observation even with

[36] See Goldstein, "Ptolemy's *Planetary Hypotheses*", p. 7.

[37] See Britton, "Ptolemy on the Obliquity", for details.

[38] See Hinz, *Maße und Gewichte*, pp. 60-61, and the same author's article "<u>Dh</u>irāᶜ" in *EI*², as well as other sources cited in Barani, "Muslim Geodesy", pp. 44-45. See also Mercier, "Muslim Geodesy", p. 177, n. 13, pointing to errors made by Hinz.

sophisticated instruments, and it is particularly difficult to obtain good results when the longitude difference is so small. Problems include the determination of the exact moment of a particular eclipse phase and the determination of the time at that moment. An error of 1° in the measurement of the altitude of a star at one end could of itself lead to an error of the same order in the time estimate, and hence in the longitude difference. The details of the necessary observations are recorded by Ibn Yūnus and al-Bīrūnī.[39] Ḥabash unfortunately gives no information of consequence on the simultaneous eclipse observations; he simply reports that the astronomers measured the distance between the meridian of the two localities as 3°, and that they then "adjusted" this to 2;48°. Indeed, the length of the arc corresponding to a longitude difference of ΔL on a parallel of latitude through Mecca is $\Delta L \cos \phi_M$ and $\mathrm{Cos}\ \phi_M / R = 0;56$ ($R = 60$), so that the 3° reduces to 2;48°.[40]

The latitudes found for Mecca and Baghdad, *viz.* 21;42° and 33;27°, are new to the modern literature, but again no information is provided on the way they were derived. (The accurate values are 21;26° and 33;21°.) Al-Bīrūnī records that Manṣūr ibn Ṭalḥa found $\phi_M = 21;40°$, and this is clearly rounded from 21;42°. The latter value is not attested to in any known source, but the approximation 21;40° was widely used over the centuries. The value 33;27° for ϕ_B is not attested elsewhere; the approximation 33;25° was, however, widely used. (The former value is attested as a latitude for Damascus,[41] but it is not known who first derived it.) See below on other values used for ϕ_M and ϕ_B.

Notice that the latitude difference (33;27° - 21;42°) is correctly derived as 11;45°. In the report attributed to Yaḥyā ibn Aktham, however, 12;15°

[39] Ibn Yūnus' account is translated in Schoy, "Muslim Geography", pp. 265-267. On al-Bīrūnī's discussion see n. 30 above.

[40] See King, "Earliest Qibla Methods and Tables", pp. 90-91.

[41] See King, *Studies*, A-X, p. 78, on the tables for time-keeping for Damascus prepared in the first half of the 14th century by al-Mizzī. It is also used on an astrolabic quadrant for Damascus dated 735 H [= 1334/35] (International Instrument Checklist No. 5008, now preserved in the British Museum, London, inventory no. 62 12-27 1), on which see Mayer, *Islamic Astrolabists*, p. 44, *sub* ʿAlī ibn al-Shihāb.

is used - see below. (The difference between the modern values is 11;55°.) To find the hypotenuse of a right-angled triangle on the sphere (where the longitude difference is measured on a small circle), Ḥabash not unreasonably applies Pythagoras' Theorem for a plane triangle. With Langermann, I have restored a line in the translation where Ḥabash uses the standard approximate rule:

$$\sqrt{(a^2 + b^2)} \approx a + b^2 / 2a \qquad (b << a^2)$$

for taking the square root, a rule known in ancient Babylonia, Egypt, India and China.[42] Thus he uses:

$$\sqrt{[(11;45)^2 + (2;48)^2]} \approx 11;45 + (2;48)^2 / (2 \times 11;45)$$

Ḥabash's results for both the calculated distance Baghdad - Mecca, namely, 12;5° = 676 2/3 miles, and for the measured distance, 712 miles, are different from those reported by Yaḥyā ibn Aktham - see below. (Note that the figure of 712 miles is confirmed by al-Bīrūnī.[43]) See below on the fact that Ḥabash presents no further information on the *qibla* at Baghdad.

4. The account of Yaḥyā ibn Aktham

I now present some new material, taken from MS Princeton Yahuda 4657 (fols. 71v-72r), a copy datable to about 1500 of an early-13th-century Egyptian treatise on folk astronomy by an unidentified author named Sirāj al-Dunyā wa-'l-Dīn.[44] This account of the observations, recorded by the *Qāḍī* Yaḥyā ibn Aktham, is often at variance with that recorded by Ḥabash

[42] van der Waerden, *Ancient Geometry and Algebra*, pp. 46-47,168 and 216. For similar approximations in Islamic sources see Youschkevitch, *Mathématiques arabes*, p. 77, corresponding to Juschkewitsch, *Mathematik im MA*, p. 194; and Berggren, *Episodes*, pp. 48-53.

[43] *al-Bīrūnī's Taḥdīd*: text, p. 234, and translation, p. 208.

[44] On this manuscript see *Princeton Catalogue*, no. 4983, and also *Cairo ENL Survey*, no. C4.

discussed above. A translation follows - see Arabic text no. 2:

> Now that I have finished my statement on the most widely used calendars, I should like to mention some aspects of geometry (*al-handasa*) relating to the size (*dawr*) of the earth.
> Chapter. The motive which inspired al-Ma'mūn to (promote) observations, taken from the *Qāḍī* Yaḥyā ibn Aktham. (He reported): "The Commander of the Faithful al-Ma'mūn became very excited about knowing the size of the earth. So he asked me and I told him that the astronomers familiar with geometry (*al-muhandisūn min aṣḥāb al-najāma*) had knowledge of these matters. So he summoned Khālid ibn ᶜAbd al-Malik al-Marwarrūdhī, Yaḥyā ibn Abī Manṣūr, ᶜAlī ibn ᶜĪsā and Aḥmad ibn al-Buḥturī, and they chose a group of brass-workers and carpenters (to assist them in making instruments). Al-Ma'mūn questioned them on the way to proceed and they answered him in unison that it was easy. (They would do) this by facing (*annahum tawajjahū*) ...
> ... {there appears to be a lacuna in the text} for a while (*waqtan*) when the sun was entering the sign of Cancer (*nuzūl al-shams ra's al-saraṭān*) {*i.e.*, at the time of the summer solstice}, for at this time the sun has no declination {!! - read: no change in declination}.
> They headed for the desert between Tadmur (= Palmyra) and Raqqa and observed the celestial pole (!) until they had accurately determined (*ḥaqqaqū*) its altitude. Next they moved (*tawajjahū*) in a straight line from north to south parallel to the meridian until the altitude of the pole had changed by one degree and then they measured the distance between the two localities (where the observations had been made). They found that one degree on the surface of the earth was (equal to) fifty-six miles, where one mile is four thousand "black" cubits (*dhirāᶜ bi-'l-sawād*), (one of which) is six *qaḍbas*, one of which is four *aṣbaᶜ*s, one of which is the width of six barley corns laid side by side in a row. Then they multiplied <u>56</u> by <u>360</u> which is the total circumference of the meridian (*dawr al-samā'*): the result was twenty thousand one hundred and sixty miles, which is the measure of the circumference of the earth (*dawr al-samā'*). (This) they divided by 3 and one-seventh, and the diameter of the earth came to 6,414 miles and 3 fifths {actually, 6/11} of a mile.
> Then al-Ma'mūn told them that he would like to know the azimuth of the

qibla and the distance between Baghdad and Mecca, and they replied that (they could do) this at the time of a lunar eclipse. So they {he?} sent out (*anfadhū*) a group of observers to Mecca to determine the precise time of the beginning and end of the lunar eclipse. Those who had stayed in Baghdad did the same. They {*sc.* the two groups} found that the meridian of Mecca was west of that of Baghdad by three degrees. They corrected (*ᶜaddalū*) these degrees by the right ascensions {!! *bi-maṭāliᶜ al-falak al-mustaqīm* - compare the terminology of Ḥabash} and they found them to be (equal to) two degrees and 48 minutes, which is the "corrected" longitude between Mecca and Baghdad.
They (also) found the altitude of the (celestial) pole at Mecca to be 21 degrees and at Baghdad to be 33 degrees {note: no minutes are given for either value}, so that the latitude of Baghdad is 12 degrees and 15 [minutes] north of that of Mecca. When they multiplied the longitude difference by itself (the result) was 7 degrees (*daraja*) and 50 minutes and two-fifths of a minute {*i.e.*, 7;50,24}, and when they multiplied the latitude difference by itself (the result) was 150 units (*juz'*) and [three minutes] {text: one-third of a unit} and fifty [seconds] {text: units} {*i.e.*, 150;3,50 - note that the square of 12;15 is 150;3,45}. When they added the two quantities and took the square root of the sum the result was 12 degrees and [30 minutes] {text: 20} to a close approximation. And when they multiplied that by 56 miles the result was 700 and 3 miles and one-half and one-quarter of a mile. {See the commentary on this calculation.}
Al-Ma'mūn wanted to check the calculations made by Yaḥyā {*sc.* Yaḥyā ibn Abī Manṣūr} and his colleagues, so he sent out someone to measure the road to Mecca, and they {plural! - compare the account of Ḥabash} found that the (number of) miles between Baghdad and Mecca by the shortest and straightest road was 700 and 10 miles, which was about 6 miles more than the results of the calculation. al-Ma'mūn pronounced that the calculation was more accurate and that the difference was due to the depression (*istifāl*) of the *wādī*s and the elevation (*irtifāᶜ*) of (hills on) the surface of the earth. When he had become convinced of this, he became excited to know the maximum distance of the moon from the surface of the earth, so he asked them about this and they answered ....
.... ."

Commentary

There are clearly some corruptions of the text, which, as it stands, does not reflect well on the *Qāḍī*'s competence to participate in the observational program. For example, it is not absolutely clear from the text that he knew the difference between an equinox and a solstice. If the sun was indeed at the summer solstice, then the daily change in its declination would be negligible even in more sophisticated observations. This con tradicts the statement of Ḥabash that the astronomers made a correction for the change in solar declination inbetween the two decisive solar altitude measurements. The next points to note are first that this text states that the observations were made by heading for the desert between Palmyra and Raqqa, and second that there is no mention whatsoever of Sanad ibn ᶜAlī.

This text mentions that the altitude of the celestial pole in Mecca was found to be 21° and in Baghdad, 33°. Yet the latitude difference stated is 12;15°, and the ensuing distance calculations are based on that value (not 11;45° as reported by Ḥabash). Since we have no information on the way in which the values were obtained, there is little more to say on them, although it may be useful to remark on other early values of these two parameters.

Excursus: Early values of the latitudes of Mecca and Baghdad

In addition to the values 21;0° and 21;40° for ϕ_M that are mentioned in our sources, we may cite various others used in the 9th and 10th centuries. Ptolemy's value of 22° for ϕ_M (for "Makoraba")[45] was used by the 10th-century astronomer Abū 'l-Wafā' (who used 33;25° for ϕ_B and 3° for ΔL_B).[46] The identity of the astronomers who derived the value 21;30° for ϕ_M within a few years of the observations commissioned by al-Ma'mūn is unknown. Likewise, we do not know who determined ϕ_M as 21;20° (the

[45] Kennedy & Kennedy, *Islamic Geographical Coordinates*, p. 225.

[46] Kennedy, "Abū'l-Wafā' on the Distance Baghdad-Mecca", p. 200.

accurate value is 21;26°).[47] This was used in one of the first tables displaying the *qibla* as a function of terrestrial coordinates, which was prepared in Baghdad in the 9th or perhaps the 10th century.[48]

We have already encountered the values 33° and 33;27° for ϕ_B in our sources. In various early Abbasid works, some to be associated with al-Khwārizmī, a contemporary of these observations, the value 33° is also used for ϕ_B (the accurate value is 33;20°).[49] Yet in al-Khwārizmī's and also al-Battānī's geographical tables the value given is 33;9°.[50] This distinctive value reappears on one of the plates of of an astrolabe made in 11th-century al-Andalus.[51] We need not assume that such a value was actually measured: it may have been calculated form the length of longest daylight.[52] The more accurate value 33;25° was used, for example, by al-Nayrīzī in Baghdad *ca.* 900,[53] and is also found in the geographical tables of Suhrāb, *ca.* 930.[54] The value value 33;21° appears to have been derived by Ibn al-Aᶜlam, a 10th-century Baghdad astronomer of some

[47] King, "Earliest Qibla Methods and Tables", pp. 126-127 and 129.

[48] *Ibid.*, pp. 118-126.

[49] King, "al-Khwārizmī", p. 2.

[50] Kennedy & Kennedy, *Islamic Geographical Coordinates*, p. 55.

[51] Namely, the astrolabe made by Ibrāhīm ibn al-Sahlī dated 478 H [= 1085/86] (International Instrument Checklist No. 121, preserved in the Staatliche Kunstsammlungen in Kassel, inventory no. A38) - see Mayer, *Islamic Astrolabists*, p. 52. The latitude 33;9° is also stated to serve Damascus.

[52] The length of longest daylight chosen could have been 14;15 hours, the obliquity of the ecliptic that of Ptolemy, 23;51,20°. (Muslim astrolabists used this latter parameter for centuries, unswayed by the more up-to-date values found by the Abbasid astronomers and their successors.)

[53] Schoy, "al-Nayrīzī über die Qibla", p. 57.

[54] Kennedy & Kennedy, *Islamic Geographical Coordinates*, p. 55.

renown.[55] It is used in several tables in the anonymous recension of the *Mumtaḥan Zīj*, the only available version of the major production of the astronomers of al-Ma'mūn.[56]

Returning to the text, it is not without interest to follow the calculations recorded by Yaḥyā of the distance Baghdad - Mecca. These do not correspond to those reported by Ḥabash, but it seems as though the latter were made by Ḥabash himself. The squares of the sides of the (plane) right-angled triangle of which this is the hypotenuse are given as 7;50,24 and 150;3,[45] (text has 50″). Their sum is 157;54,9. Using the standard approximation, the square root of this is:

$$\sqrt{(144 + 13;54,9)} \approx 12 + 13;54,9 / (2 \times 12)$$
$$\approx 12 + 0;34,45 \approx 12;34 \text{ (by truncation)} \approx 12;30 \text{ (text has 12;20)}.$$

Now we multiply 12;34 by 56: the product is 11,43;44, which is indeed about 703 + 1/2 + 1/4, as stated in the text. Note, however, that the results of al-Ma'mūn's final survey of the road between Mecca and Baghdad as reported here are also at variance with what was recorded by Ḥabash.

The reader will observe that no value is presented for the *qibla* at Baghdad. Clearly, though, the method used was equivalent to:

$$\sin q = \Delta L \cos \phi_M / \sqrt{[(\Delta\phi)^2 + (\Delta L \cos \phi_M)^2]} .$$

[55] On Ibn al-Aʿlam see Sezgin, *GAS*, V, p. 309, and VI, pp. 215-216, and on his lost astronomical tables see Kennedy, "Ibn al-Aʿlam", Mercier, "Ibn al-Aʿlam" and Tihon, "Alim".

His derivation of ϕ_B is described in a note on the flyleaf of an early copy of the *Zīj-i Īlkhānī* which passed through Christie's of London in 1986. The text translates: "Note (*fā'ida*): al-Sharīf {the standard epithet of Ibn al-Aʿlam} found that the highest (solar) altitude was 80;12° and the lowest 33;6°. From this it was established that the obliquity of the ecliptic was 23;33°, the latitude of Baghdad 33;21° and the altitude of (the sun at) the equinox(es) 56;39°. Copied from the handwriting of the author."

[56] See MS Escorial ar. 927, fols. 64r, 64v and 93v (pp. 125, 126 and 184 of the "facsimile" published in Frankfurt in 1986, which does not accurately reproduce the original manuscript).

This corresponds to one described in an early Abbasid text.[57] Its justification is simple: distances on the sphere are considered in a plane. And, in particular, distances on the parallel of latitude of Mecca are modified by a factor of cos ϕ_M = Cos ϕ_M / R (where R = 60).

For Ḥabash's values 2;48° and 12;5° for the numerator and the denominator, I compute 13;24° for *q*. For Ibn Aktham's values 2;45° and 12;34°, I compute 13;38°. (The value 13;13° for Baghdad is specifically mentioned in the instructions to an Abbasid *qibla* table, but the table is based on ϕ_M = 21;20°.[58])

The geographical parameters derived from these observations (ϕ_M = 21;40° and ΔL = 3°), but now with the better value ϕ_B = 33;25°, were used by al-Nayrīzī (*fl.* Baghdad *ca.* 900) in a numerical example appended to his treatise on the determination of the *qibla* by spherical trigonometry.[59]

5. Concluding remarks

It should be borne in mind that before these observations were conducted to establish the mathematically-computed *qibla*, the Muslims in the province of al-ᶜIrāq had been using astronomical alignments or cardinal directions for the *qibla*. A popular *qibla* direction for al-ᶜIrāq was winter sunset, chosen in order to be "facing" the north-east wall of the Kaᶜba.[60]

[57] King, "Earliest Qibla Methods and Tables", p. 102. See also Ahmedov *et al.*, "Istanbul Manuscripts", on the source.

[58] King, "Earliest Qibla Methods and Tables", pp. 127-129.

[59] Schoy, "al-Nayrīzī über die Qibla", p. 57.

[60] On the use of astronomical alignments for the *qibla* see King, *Studies*, C-X-XIII (where C-X is a reprint of the article "Makka. iv. As centre of the world" in *EI*²), and, more recently, *idem*, "Orientations".

It is not only for the history of Islamic architecture that the *qibla* is important. In an Arabic account of the supernova of 1006 A.D. the direction of the supernova is defined in terms of the "*qibla* of al-ᶜIrāq" (see Goldstein, "Supernova of 1006", p. 107). But which *qibla* is being referred to here? Winter sunset, a popular traditional direction, or the *qibla* determined by al-Ma'mūn's astronomers? Or yet another direction?

Others preferred due south, following the practice of the Prophet in Medina. In the year 847 the mosque in the new Abbasid city of Samarra was apparently (that is, according to Herzfeld and Sarre) laid out at about 15° W of S,[61] perhaps as a result of the observations described above. Surely other religious sites could be identified in Baghdad with an orientation of about 13° W of S. But very little reliable information on the orientations of Islamic religious architecture in Iraq is currently available, and little more should be said until they are measured properly.

As Mercier has already pointed out, all of these reports about the expeditions must be regarded with a certain amount of scepticism. He has expressed doubt that the Palmyra expedition ever took place, and his arguments are, given the evidence, sound. Without wishing to do injustice to Sanad ibn ᶜAlī, Mercier suspects that we are dealing with a legendary report based on a possible earlier (that is, pre-Islamic) expedition. But the main cause of scepticism remains the fact that the results are too good to be true, if indeed the procedures involving measuring solar meridian altitudes were used. Not least is the problem presented by the size of the solar disc. Again, an error in one minute in the solar altitude would yield an error of one mile in the distance, and measuring the solar altitude to the nearest minute is something that one might have been able to do with a good instrument in Baghdad, but anticipating that one could achieve the same accuracy in the wilderness during an expedition of several days requires considerable imagination. Small wonder, then, that not long thereafter someone thought of simply measuring the distance between Baghdad and Samarra, and the difference in solar meridian altitudes in the two places. Unfortunately, however, Samarra is by no means due north of Baghdad. The report of this minor expedition, cited by al-Bīrūnī from Abū 'l-Faḍl al-Harawī, confirmed the result 56 2/3, and was rightly regarded by al-Bīrūnī as rather fishy.[62]

Some later Muslim scholars ignored the results of al-Ma'mūn's

[61] Herzfeld & Sarre, *Archäologische Reise*, III, pl. XIX. (Note that in pl. XX no orientation is given.).

[62] *al-Bīrūnī's Taḥdīd*: text, pp. 212-213; translation, pp. 177-178; and commentary, pp. 132-133.

expeditions altogether. Thus, for example, the celebrated early-14th-century geographer-prince Abū 'l-Fidā' preferred Ptolemy's value of 66 2/3 miles per degree.[63] A report such as the one by Yaḥyā ibn Aktham would certainly have encouraged serious people to favour such a reliable authority from Antiquity (though see below).

Our problem is that we do not have access to the original reports by the astronomers who participated in the expedition(s). Were there really two different expeditions, one in the desert north of Palmyra that started from Damascus and another in the desert of Sinjār that started from Baghdad? Even the latter, which is reported to have taken the astronomers 200 km into the desert, must have been a major undertaking. And it would have been utter folly to schlepp instruments and personnel 500 km across inhospitable terrain from Baghdad to the Syrian desert. We are still, however, confronted with the report by Sanad ibn ᶜAlī, which is early enough that one cannot explain it away.

Other reports of different astronomical observations (mainly of conjunctions and eclipses, equinoxes and solstices) conducted by the Abbasid astronomers in the 9th century, such as we find quoted in the writings of Ibn Yūnus and al-Bīrūnī, are impressive by any standards and are not marred by the inconsistencies found in these various reports of the geodetic measurements. What has happened here confirms the adage: "Too many cooks spoil the broth". Ibn Yūnus made a similar criticism of al-Ma'mūn's team-work, stating that one good man could achieve more on his own.[64] He had, of course, himself in mind. Alas he did not tap his patron al-Ḥākim for a geodetic expedition up the Nile. Al-Bīrūnī, who was more convinced of the need for new geodetic measurements, did conduct the necessary research programme and was able to derive a reasonable value without involving two teams of astronomers - and one superfluous *qāḍī* completely innocent of what they were doing - wandering about in the desert.[65]

[63] Mercier, "Muslim Geodesy", p. 176, n. 12, and pp. 177-178.

[64] Caussin, "Ibn Yūnus", pp. 34-40.

[65] See the summary in Mercier, "Muslim Geodesy", pp. 182-184.

One may wonder why in later mathematical geography Muslim astronomers preferred to use the less accurate Ptolemaic value of 66 2/3 miles for one degree. For example, it is Ptolemy's value that is used on 17th-century Persian maps of the world centred on Mecca, from which one can simply read off the *qibla* and the distance to Mecca for any locality in the Islamic commonwealth.[66] The explanation is probably that it was thought that the parameter 66 2/3 actually was the result of al-Ma'mūn's astronomers. Even the great 13th-century polymath Naṣīr al-Dīn al-Ṭūsī, in his influential *Tadhkira fī ᶜilm al-hay'a*, attributes this value to them.[67] Thus the efforts and results of al-Ma'mūn's astronomers were not only not properly recorded at the time, these results were also not properly transmitted for posterity.

Acknowledgements: This study could not have been written without the cooperation of Princeton University Library and the generosity of Dr. Y. Tzvi Langermann (Jerusalem). I am much indebted to Kurt Maier (Frankfurt) and Dr. Muhsin Zakery (Halle) for their constructive criticism. Any remaining confusion is the responsibility of myself, and, in no small measure, the cooks and the *qāḍī*.

[66] On these remarkable maps (one discovered in 1989 and another in 1995) see Lorch & King, "Qibla Charts", pp. 198-200 and 202; my article "Samt" [= direction] in *EI²*, with corrections in the article "Ṭāsa" [= magnetic compass]; and King, *World-Maps*. A detailed study is in preparation.

[67] See Ragep, *al-Ṭūsī's* Tadhkira, I, pp. 310-314, and II, pp. 501-510. Ragep has shown that the value is recorded already in a report by Muḥammad ibn Mūsā ibn Shākir (d. 873).

النصوص العربية

١: مخطوطة الريان كفاح في القدس (ص. ١-٣ من رسالة حبش)

بسم الله الرحمن الرحيم وبه نستعين كتاب الأجرام والأبعاد لحبش بن عبد الله الحاسب قال إن أمير المؤمنين المأمون أحب أن يعرف مقدار الأرض فبحث عن ذلك فوجد بطليموس (!) يذكر في بعض كتبه أن دور الأرض كذا وكذا ألف اسطادنيوس[1] فسأل المفسرين عن عن تفسير الاسطادنيوس[2] فاختلفوا في تفسيره هذا ...[3] لا يغني[4] عما نريد فوجه خالد بن عبد الملك المروروذي[5] وعلي بن عيسى الأسطرلابي وأحمد بن البحتري الذارع مع جماعة من الذراع والصناع من النجارين والصفارين لتصحيح ما يحتاجون إليه من الآلات وحملهم بأجمعهم إلى موضع اختاروه من برية سنجار[6] ثم توجه خالد وطائفة معه قطب بنات نعش الشمالي وتوجه علي وأحمد وطائفة معهم إلى جهة القطب الجنوبي فمضوا حتى انتهوا إلى أن وجدوا غاية ارتفاع نصف النهار قد زال وتغير عن ارتفاع نصف النهار الذي وجدوه في الموضع الذي افترقوا منه مقدار درجة واحدة بعد أن نقص من ذلك مقدار ميل الشمس في مدة مسافة الطريق في ذهابهم ونصبوا السهام ثم رجعوا أيضا على تلك السهام وامتحنوا الذرع ثانية

[1] بغير نقط في الأصل.

[2] بغير نقط في الأصل.

[3] إنه يجوز أن تنقص كلمة هنا.

[4] في الأصل: يني.

[5] في الأصل: المروزودي.

[6] في الأصل: سيحان.

فوجدوا مقدار درجة واحدة من وجه الأرض ستة وخمسين ميلا والميل أربعة آلاف ذراع بالأسود وهو الذراع الذي وضعه المأمون لذرع الثياب ومسح البناء وقسمة المنازل وسمعت هذا الذي ذكرته في كتابي من خالد بن عبد الملك المرورروذي[7] يحدث به القاضي يحيى بن أكثم[8] فأمر يحيى أن يكتب له جميع ما ذكره خالد فكتب له وكتبته أنا من خالد سماعا وإن أمير المؤمنين المأمون رضي الله عنه أحب أن يمسح سمت القبلة فوجه في وقت كسوف قمري ليمتحن الطول بين مكة وبغداد فوجد نصف نهار مكة غربيا عن نصف نهار بغداد بثلث درج بالتقريب فإذا[9] عدلنا درجة واحدة من الفلك المستقيم نجد حصتها في مدار[10] الفلك المخطوط على عرض مكة ست وخمسين دقيقة بالتقريب فإذا ضربنا البعد بين نصف نهار مكة ونصف نهار بغداد ثلث درج في ٥٦ دقيقة وجعلنا كل س جزءا نجد البعد المعدل في الطول بين نصف نهار مكة ونصف نهار بغداد في الفلك المخطوط على مدار عرض مكة درجتين وثماني وأربعين دقيقة بالتقريب ووجد أيضا عرض مكة أحد وعشرين درجة واثنان وأربعون دقيقة بالتقريب وعرض بغداد ثلث وثلثون درجة وسبع وعشرين دقيقة بالتقريب فنقصنا عرض مكة من عرض بغداد فبقي الفضل بين العرضين أحد عشر درجة وخمسة وأربعين دقيقة فإذا ضربنا الفضل بين الطولين وهو ب درج مح دقائق في نفسه يجتمع ٧ درج و ٥٠ دقيقة ٢٤ ثانية ونصفه ٣ درج و ٥٥ دقيقة و ١٢ ثانية يكون ذلك ٢٣٥ دقيقة و ١٢ ثانية نقسمه على ١١ درجة و٤٥ دقيقة وهو الفضل بين العرضين فبلغ مسافة الطول بين بغداد ومكة ١٢ درجة و ٥ دقائق ضربناه في ٥٦

7 في الأصل: المروزودي.

8 في الأصل: اكتم.

9 في الأصل: فاد.

10 في الأصل: مقدار.

ميلا فبلغ المسافة بين بغداد ومكة على السهم المستوي ستمائة وستة وسبعين ميلا وثلثي ميل بالتقريب وقد كان المأمون عندما رفع إليه خالد بن عبد الملك مقدار درجة واحدة من الأرض أحب أن يمتحن ذلك فوجد من ذرع الطريق بين بغداد ومكة على أقرب الطرق فوجده سبع مائة واثني عشر ميلا فوجدنا الخلاف بين السهم الذي يجري في الهوى وبين ذرع الطريق خمسة وثلثين ميلا وثلث ميل فقال المأمون هذا ما لا يستكثر لأن الطريق يجب فيه العوادل والصعود والهبوط مقدار هذا ...

٢: مخطوطة برينستن ٤٦٥٧ (٤٩٨٣)، ق ٧١و - ٧٣و

... ولما فرغنا من ذكر التواريخ المشهورة أردنا أن نذكر طرفا من علم الهندسة في دور الأرض فصل السبب الباعث للمأمون على الرصد المنقول عن القاضي يحيى بن أكثم[1] إن أمير المؤمنين سمت همته إلى معرفة دور الأرض فسألني عن ذلك فقلت إن للمهندسين من أصحاب النجامة عندهم علم ذالك فأحضر خالد بن عبد الملك المرورودي ويحيى بن أبي منصور وعلي بن عيسى وأحمد بن البحتري[2] واختاروا[3] من الصفارين والنجارين جماعة فقال لهم المأمون كيف السبيل إلى ذلك فقالوا بأجمعهم هذا سهل وذلك أنهم توجهوا وقتا عند نزول الشمس رأس السرطان إذ ليس لها ميل هناك وقصدوا البرية الذي بين تدمر والرقة ورصدوا القطب حتى حققوا ارتفاعه ثم توجهوا على خط مستقيم من الشمال إلى الجنوب موازيا لدائرة نصف

[1]في الأصل: اكتم.

[2]في الأصل: التجري.

[3]في الأصل: واختارو.

النهار حتى تغير ارتفاع القطب مقدار درجة واحدة ثم مسحوا ما بين الموضعين فوجدوا مقدار درجة واحدة من وجه الأرض ستة وخمسين ميلا والميل أربعة ألف (!) ذراع بالسواد وهو ست قبضات القبضة أربع أصابع عرض الإصبع إلى الإصبع ست شعيرات مصفوفة بطون بعضها إلى بعض فضربوا نو في شس الذي هو دور السماء فخرج من الضرب عشرون ألف ميلا ومائة ميل وستون ميلا وهو مساحة دور الأرض فقسموا على ٣ وسبع فخرج قطر الأرض ٦٤١٤ ميلا و ٣ أخماس ميل فقال لهم المأمون إني أحب أن أعرف سمت القبلة وبعد ما بين بغداد ومكة فقالوا نعم إذلك (إذ ذلك) وقت خسوف القمر ثم انفذوا جماعة من أهل الرصد إلى مكة فحققوا وقت خسوف القمر ووقت تمام انجلائه وكذالك فعلوا (!) المقيمون ببغداد وقت خسوف القمر أيضا فوجدوا نصف نهار مكة غربيا عن نصف نهار بغداد بثلاث درج فعدلوا هذه الدرجات[4] بمطالع الفلك المستقيم فوجدوها درجتين و مح دقيقة وهو الطول بين مكة وبغداد معدلا ووجدوا ارتفاع القطب بمكة كا درجة وببغداد لج فعرض بغداد شمال عن مكة يب درجة ويه فإذا ضربوا ما بين الطولين في نفسه كان ز درج ون دقيقة وخمسا دقيقة وإذا ضربوا ما بين العرضين في نفسه كان ذلك قن جزء وثلث جزء وخمسون جزء فإذا جمعوهما وأخذوا جذر المجتمع يخرج يب درجة و ك بأقرب تقريب فإذا ضربوا ذلك في نو ميلا تجتمع (!) في ذ ميل و ٣ أميال ونصف وربع ميل وكان المأمون يحب أن يمتحن ما حسبه يحيى وأصحابه فوجه من ذرع طريق مكة فوجدوا الأميال بين بغداد ومكة على أقصر الطرق وأقربها استواء ذ ميل وي أميال فزاد الذرع على ما خرج بالحساب و أميال بالتقريب فقال المأمون الحساب أصدق وهذا التفاوت من استفال الأودية وارتفاع وجه الأرض فلما سكنت نفسه إلى ذلك سمت همته إلى أن يعرف أبعد بعد القمر عن وجه الأرض فسألهم عن ذلك فقالوا ...

[4]في الأصل: الدرجة.

Bibliography and bibliographical abbreviations

Ahmedov *et al.*, "Istanbul Manuscripts": A. A. Ahmedov, J. Ad-Dabbâgh and B. A. Rosenfeld, "Istanbul Manuscripts of Al-Khwârizmî's Treatises", *Erdem* (Atatürk Kültür Merkezi, Ankara) 3:7 (1987), pp. 163-186.

Ali, *al-Bīrūnī's Taḥdīd*: Jamil Ali, *The Determination of the Coordinates of Cities: al-Bīrūnī's Taḥdīd [nihāyāt] al-amākin*, Beirut: American University of Beirut Publications, 1966. [See also Bulgakov and Kennedy.]

Barani, "Muslim Geodesy": S. H. Barani, "Muslim Researches in Geodesy", *al-Bīrūnī Commemoration Volume*, Calcutta, 1951, pp. 1-52.

Berggren, *Episodes*: J. Lennart Berggren, *Episodes in the Mathematics of Medieval Islam*, New York, *etc.*: Springer, 1986.

al-Bīrūnī's Taḥdīd: See Bulgakov for the edition, Ali for the translation, and Kennedy for the commentary.

Britton, "Ptolemy on the Obliquity": John P. Britton, "Ptolemy's Determination of the Obliquity of the Ecliptic", *Centaurus* 14 (1969), pp. 29-41.

Bulgakov, ed., *al-Bīrūnī's Taḥdīd*: al-Bīrūnī, *Kitāb Taḥdīd nihāyāt al-amākin*, ed. P. Bulgakov, *Majallat Maʿhad al-Makhṭūṭāt al-ʿArabiyya* (Cairo) 8 (1962). [See also Ali and Kennedy.]

Cairo ENL Survey: David A. King, *A Survey of the Scientific Manuscripts in the Egyptian National Library*, (Publications of the American Research Center in Egypt, Catalog No. 5), Winona Lake, Ind.: Eisenbraun's, 1986.

Caussin, "Ibn Yūnus": Caussin de Perceval, "Le livre de la grande table Hakémite observée par le Sheikh ... ebn Iounis ... ", *Notices et extraits des manuscrits de la Bibliothèque Nationale* 7 (An VII = 1804), pp. 16-240. [The references here are to the separatum paginated 1-224.]

Debarnot, "Ḥabash's Zīj (Istanbul MS)": Marie-Thérèse Debarnot, "The *Zīj* of Ḥabash al-Ḥāsib: A Survey of MS Istanbul Yeni Cami 784/2", in *Kennedy Festschrift*, pp. 35-69.

DSB: *Dictionary of Scientific Biography*, 16 vols., New York, N.Y.: Charles Scribner's Sons, 1970-80.

EI[2]: *Encyclopaedia of Islam*, 2nd ed., 8 vols. to date, Leiden: E. J. Brill, 1960 to present.

Goldstein, "Ptolemy's *Planetary Hypotheses*": Bernard R. Goldstein, "The Arabic Version of Ptolemy's Planetary Hypotheses", *Transactions of the American Philosophical Society*, N.S., 57:4 (1967), pp. 1-55.

- , *Studies*: *idem*, *Theory and Observation in Ancient and Medieval Astronomy*, London: Variorum, 1985.

- , "Supernova of 1006": *idem*, "Evidence for a Supernova of A.D. 1006", *The Astronomical Journal* (New York, N.Y.) 70 (1965), pp. 105-114, reprinted in *idem*, *Studies*, XIII.

Herzfeld & Sarre, *Archäologische Reise*: E. Herzfeld and F. Sarre, *Archäologische Reise im Euphrat- und Tigris-Gebiet*, 3 vols., Berlin: Dietrich Reimer, 1911.

Hinz, *Maße und Gewichte*: Walther Hinz, *Islamische Maße und Gewichte*, (*Handbuch der Orientalistik*, Erg.-Bd. 1, Heft 1), Leiden: E. J. Brill, 1955.

History of Cartography, II:1: J. B. Harley and David Woodward, eds., *The History of Cartography*, vol. 2, book 1: *Cartography in the Traditional Islamic and South Asian Societies*, Chicago, Ill. & London: University of Chicago Press, 1992.

Irani, "Arabic Numeral Forms": Rida A. K. Irani, "Arabic Numeral Forms", *Centaurus* 4 (1955), pp. 1-12, reprinted in Kennedy *et al.*, *Studies*, pp. 710-721.

Juschkewitsch, *Mathematik im MA*: Adolf P. Juschkewitsch, *Geschichte der Mathematik im Mittelalter* (translated from the Moscow 1961 Russian edition), Leipzig: B. G. Teubner, 1964. [See also Youschkevitch.]

Kennedy, "Abū'l-Wafā' on the Distance Baghdad-Mecca": Edward S. Kennedy, "Applied Mathematics in the Tenth Century: Abū'l-Wafā' Calculates the Distance Baghdad-Mecca", *Historia Mathematica* 11 (1984), pp. 193-206.

- , *al-Bīrūnī's Taḥdīd*: *idem*, *A Commentary upon al-Bīrūnī's Kitāb Taḥdīd [nihāyāt] al-amākin*, Beirut: American University of Beirut, 1973. [See also Bulgakov and Ali.]

- , "Ibn al-Aᶜlam": *idem*, "The Astronomical Tables of Ibn al-Aᶜlam", *Journal for the History of Arabic Science* (Aleppo) 1 (1977), pp. 13-23.

- , "*Zīj* Survey": *idem*, "A Survey of Islamic Astronomical Tables", *Transactions of the American Philosophical Society*, N.S., 46:2 (1956), pp. 123-177, reprinted n.d. (*ca*. 1990).

Kennedy & Kennedy, *Islamic Geographical Coordinates*: E. S. and Mary Helen Kennedy, *Geographical Coordinates of Localities from Islamic Sources*, Frankfurt am Main: Institut für Geschichte der Arabisch-Islamischen Wissenschaften, 1987.

Kennedy *et al*., *Studies*: E. S. Kennedy, Colleagues and Former Students, *Studies in the Islamic Exact Sciences*, Beirut: American University of Beirut, 1983.

Kennedy Festschrift: David A. King and George Saliba, eds., *From Deferent to Equant: A Volume of Studies in the History of Science in the Ancient and Medieval Near East in Honor of E. S. Kennedy*, *Annals of the New York Academy of Science* 500 (1985).

al-Khaṭīb al-Baghdādī, *Ta'rīkh Baghdād*, XIV: Abū Bakr Aḥmad ibn ᶜAlī al-Khaṭīb al-Baghdādī, *Ta'rīkh Baghdād*, vol. XIV, Cairo: Maṭbaᶜat al-Saᶜāda, 1931.

King, *Astronomy in the Yemen*: David A. King, *Mathematical Astronomy in Medieval Yemen - A Bio-Bibliographical Survey*, (Publications of the American Research Center in Egypt, Catalogs, No. 4), Malibu, Ca.: Undena, 1983.

- , "Earliest Qibla Methods and Tables": *idem*, "The Earliest Islamic Mathematical Methods and Tables for Finding the Direction of Mecca", *Zeitschrift für Geschichte der Arabisch-Islamischen Wissenschaften* 3 (1986), pp. 82-149, with corrections listed *ibid*. 4 (1987/88), p. 270, reprinted in *idem*, *Studies*, C-XIV.

- , "al-Khwārizmī": *idem*, "al-Khwārizmī and New Trends in Mathematical Astronomy in the Ninth Century", *Occasional Papers on the Near East* (New York University, Hagop Kevorkian Center for Near Eastern Studies) 2 (1983).

- , "Orientations": *idem*, "The Orientation of Medieval Islamic Religious Architecture and Cities", *Journal for the History of Astronomy* 26 (1995), pp. 253-274.

- , "Sacred Direction": *idem*, "The Sacred Direction in Islam: A Study of the Interaction of Religion and Science in the Middle Ages",

Interdisciplinary Science Reviews 10:4 (1985), pp. 315-328.

- , "Sacred Geography": *idem*, "The Sacred Geography of Islam", to appear.

- , *Studies*, A-C: *idem*, *Islamic Mathematical Astronomy*, London: Variorum, 1986, 2nd rev. edn., Aldershot: Variorum, 1993; *Islamic Astronomical Instruments*, London: Variorum, 1987, reprinted Aldershot: Variorum, 1995; and *Astronomy in the Service of Islam*, Aldershot: Variorum, 1993.

- , *World-Maps*: *idem*, *World-Maps for Finding the Direction and Distance to Mecca: Examples of Innovation and Tradition in Islamic Science*, Leiden: E.J. Brill, and London: Al-Furqan Islamic Heritage Foundation, London, 1999. [A summary is in *Imago Mundi* 49 (1997), pp. 62-82.]

King & Lorch, "Qibla Charts": *idem* and Richard P. Lorch, "Qibla Charts, Qibla Maps, and Related Instruments", in *History of Cartography*, II:1, pp. 189-205.

Langermann, "Bibliographical Notes": Y. Tzvi Langermann, "Bibliographical Notes on Islamic Astronomy, the Results of a Study of the Exact Sciences among the Jews of Yemen", in G. Swarup, A. K. Bag and K. S. Shukla, eds., *History of Oriental Astronomy*, (Proceedings of International Astronomical Union Colloquium No. 91, New Delhi, India, 12-16 November, 1985), Cambridge, *etc.*: Cambridge University Press, 1987, pp. 203-206.

- , "Ḥabash on Distances": *idem*, "*The Book of Bodies and Distances* of Ḥabash al-Ḥāsib", *Centaurus* 28 (1985), pp. 108-128.

Mayer, *Islamic Astrolabists*: Leo A. Mayer, *Islamic Astrolabists and their Works*, Geneva: Albert Kundig, 1956.

Mercier, "Ibn al-Aᶜlam": Raymond P. Mercier, "The Parameters of the Zīj of Ibn al-Aᶜlam", *Archives Internationales d'Histoire des Sciences* 39 (1989), pp. 22-50.

- , "Muslim Geodesy": Raymond P. Mercier, "Geodesy", in *History of Cartography*, II:1, pp. 175-188.

Nallino, *Scritti*, V: Carlo A. Nallino, *Raccolta di scritti editi e inediti*, vol. V: *Astrologia - Astronomia - Geografia*, Roma: Istituto per l'Oriente, 1944.

Pingree, "al-Fazārī": David Pingree, "The Fragments of the Works of al-Fazārī", *Journal of Near Eastern Studies* 29 (1970), pp. 103-123.

Princeton Catalogue: Rudolf Mach, *Catalogue of Arabic Manuscripts (Yahuda Section) in the Garrett Collection, Princeton University Library*, Princeton, N.J.: Princeton University Press, 1977.

Ragep, *al-Ṭūsī's* Tadhkira: F. Jamil Ragep, *Naṣīr al-Dīn al-Ṭūsī's* Memoir on Astronomy (al-Tadhkira fī ᶜilm al-hay'a), 2 vols., New York, N.Y., *etc.*: Springer, 1993.

Sayılı, *The Observatory in Islam*: Aydın Sayılı, *The Observatory in Islam*, Ankara: Türk Tarih Kurumu Basımevi, 1960, reprinted New York, N.Y.: Arno, 1981.

Schoy, *Beiträge*: Fuat Sezgin *et al.*, eds., *Carl Schoy - Beiträge zur arabischen-islamischen Mathematik*, 2 vols., Frankfurt am Main: Institut für Geschichte der Arabisch-Islamischen Wissenschaften, 1988.

- , "Längenbestimmung": Carl Schoy, "Längenbestimmung und Zentralmeridian bei den älteren Völkern", *Mitteilungen der Kaiserlich-Königlichen Geographischen Gesellschaft* 12 (1915), pp. 27-62, reprinted in *idem*, *Beiträge*, I, pp. 95-130.

- , "Muslim Geography": *idem*, "The Geography of the Muslims of the Middle Ages", *The Geographical Review* 14 (1924), pp. 257-269, reprinted in *idem*, *Beiträge*, II, pp. 513-526.

- , "al-Nayrīzī über die Qibla": *idem*, "Abhandlung von al-Faḍl b. Ḥātim al-Nairīzī: Über die Richtung der Qibla", *Sitzungsberichte der Bayerischen Akademie der Wissenschaften, Math.-Phys. Klasse*, 1922, pp. 55-68, reprinted in *idem*, *Beiträge*, I, pp. 252-265.

Sezgin, *GAS*: Fuat Sezgin, *Geschichte des arabischen Schrifttums*, 9 vols. to date, Leiden: E. J. Brill, 1967 to present.

Tihon, "Alim": Anne Tihon, "Sur l'identité de l'astronome Alim", *Archives Internationales d'Histoire des Sciences* 39 (1989), pp. 3-21.

Varisco, *Yemeni Almanac*: Daniel M. Varisco, *Medieval Agriculture and Islamic Science - The Almanac of a Yemeni Sultan*, (University of Washington Publications on the Near East, no. 6), Seattle, Wa. & London: University of Washington Press, 1994.

van der Waerden, *Ancient Geometry and Algebra*: B. L. van der Waerden, *Geometry and Algebra in Ancient Civilisations*, Berlin, *etc.*: Springer, 1983.

Vernet, "Tabulae Probatae": Juan Vernet, "Las Tabulae Probatae", in

Homenaje a Millás-Vallicrosa, II, Barcelona, 1956, pp. 501-522, reprinted in *idem*, *Estudios sobre historia de la ciencia medieval*, Barcelona: Universidad de Barcelona & Bellaterra: Universidad Autónoma de Barcelona, 1979, pp. 191-212.

Youschkevitch, *Mathématiques arabes*: Adolf P. Youschkevitch, *Les mathématiques arabes (VIIIe - XVe siècles)*, translated by M. Cazenave and K. Jaouiche, Paris: J. Vrin, 1976. [See also Juschkevitch.]

XI

A WORLD-MAP IN THE TRADITION OF AL-BĪRŪNĪ (*CA.* 1040) AND AL-KHĀZINĪ (*CA.* 1120) PRESENTED BY SIRĀJ AL-DĪN AL-SAJĀWANDĪ (1210)

1 Introductory remarks

Muslim scientists were preoccupied with geographical coordinates – longitudes and latitudes – for over a thousand years. Their interest was essentially academic – knowledge for its own sake – and practical – these coordinates were necessary to calculate the qibla, that is, the sacred direction towards Mecca, or to modify astronomical tables from one meridian to another. (One should add that Muslim interest was often simply traditional, in that they copied geographical tables, often very carelessly, just for the sake of copying them; for the historian, even these corrupt copies can be useful in reconstructing the originals.) Altogether, some 200 lists of localities with their coordinates are known from Muslim sources.[1] The Muslims inherited the geographical tables of Ptolemy of Alexandria (*ca.* 125) and already in the early 9th century had greatly improved his longitude values by introducing a better equivalent for the length of the Mediterranean.[2] Particularly notable examples of their productions are the tables of al-Khwārizmī (Baghdad, *ca.* 825) for 545 localities,[3] al-Bīrūnī (Central Asia, *ca.* 1040) with 604 entries,[4] al-

[1] Kennedy & Kennedy, *Islamic Geographical Coordinates* (later abbreviated as K&K), contains information on some 2,500 localities and their coordinates from some 75 medieval Islamic sources arranged alphabetically according to place-name and again by source (each assigned a three-letter code), as well as by increasing longitudes and latitudes. This fundamental reference work is currently being updated by Mercè Comes of the University of Barcelona. See also King & Samsó, "Islamic Astronomical Handbooks and Tables", which provides the context for these tables, especially p. 27.

[2] For an overview of this activity see Kennedy, "Islamic Mathematical Geography".

[3] Kennedy & Kennedy, *Islamic Geographical Coordinates*, pp. xxii-xxiv and 399-408 (KHU, *etc.*).

Zayyāt (al-Andalus, *ca.* 1025) with 298 entries,[5] and the anonymous and highly influential, but still enigmatic *Kitāb al-aṭwāl wa-'l-'urūḍ* (Iran, *ca.* 1050?), featuring well over 450 localities.[6] In the sequel I use L and ϕ to denote longitude and latitude, and L/ϕ to denote a pair of coordinates or a position on a map.

It is well known to specialists that starting in the 9th century Muslim scientists prepared world-maps with localities marked on them according to their longitudes and latitudes.[7] My interest here is restricted to maps that bear either a longitude and latitude scale or a grid based on such scales, or which show signs of having been prepared using geographical tables so that the localities marked on them are in positions corresponding to the data in such tables; many other Islamic maps that are not of this kind do not concern us here. These maps might well be devoid of any physical features such as coastlines, rivers or mountains. The preparation of such maps, like much else in Islamic astronomy, mathematics and geography, started in Baghdad and continued in different centres for close to a millennium. As we shall see, these maps often suffered greatly at the hands of copyists who had no idea what they were doing.

Now not a single such world-map is known to have survived from the activities associated with the Caliph al-Ma'mūn (Baghdad, *ca.* 825),[8] the Fatimid Caliph al-Mu'izz (Cairo, 964),[9] or the universal scholar al- Bīrūnī (Central Asia, *ca.* 1040).[10] Two later maps have recently come to light, and some colleagues have seen these as possible remnants of the world-map of al-Ma'mūn:

(1) A 13th-century copy of an 11th-century world-map recently acquired by the Bodleian Library, sporting part of a carelessly-drawn longitude scale: cities are marked in addition to the coastal outlines and rivers, but not according to their positions in medieval geographical tables; thus, from the point of view

---

[4] *Ibid.*, pp. xviii and 450-464 (BIR).

[5] *Ibid.*, pp. xxxvii and 486-494. See also Comes, "Meridian of Water".

[6] Kennedy & Kennedy, *Islamic Geographical Coordinates*, pp. xvii and 422-430 (*sub* ATH FID, which I label ATW). On these particular tables, which were of fundamental importance in later mathematical geography in the Islamic East, see also King, *Mecca-Centred World-Maps*, pp. 33, 43, 157-161, and 596-597, and *idem*, *SATMI*, VIIc-4.

[7] See King, *Mecca-Centred World-Maps*, pp. 32-33, for a reconstruction by Joachim Lelewel (*ca.* 1850) of a world-map based on the coordinates of Naṣīr al-Dīn al-Ṭūsī (Iran, *ca.* 1250), and *ibid.*, p. 39, for a 15th-century German map based on the coordinates of the *Toledan Tables*.

[8] Kennedy, "Suhrāb and the World-Map of Ma'mūn"; and King, *Mecca-Centred World-Maps*, pp. 31-37.

[9] *Ibid.*, p. 35.

[10] On al-Bīrūnī see the *DSB* article by E. S. Kennedy. On his world-maps see King, *op. cit.*, pp. 41-42, 71-73, and 339-344.

of mathematical cartography, the map itself is something of a disaster.[11] The map is not necessarily corrupted by the copyist, and I do not underestimate its importance for the history of Islamic cartography generally. A careful analysis of the positions of the localities on the map would be worthwhile, if only to confirm that the scale was added as an afterthought.

(2) A 14th-century copy of a world-map showing mainly coastal outlines and rivers (but also a few cities and regions, albeit mainly without proper markers) which sports a not-unsophisticated grid based on an orthographic-type projection with meridians (as arcs of circles) for each 5° of longitude from 0° to 180° and parallels of latitude for each 5° from –20° to +90°. Since there are no cities marked on the map, the grid appears to be essentially superfluous.[12]

I strongly doubt the connection of either of these world-maps with al-Maʾmūn, although the former is at least of the right kind, showing the kind of localities one would expect, if not in their appropriate positions.

Some much more accurate maps of "Greater Iran" survive from the 15th and 16th centuries, as well as – until fairly recently – a few uninspired maps from the later period.[13] In the past 15 years, three remarkable Islamic world-maps from Safavid Iran (*ca.* 1675) have come to light: they are engraved on metal and are all related, in that they are all copies of an earlier proto-type. They bear a sophisticated cartographic grid so devised as to preserve direction and (the sine of) the (angular)

---

[11] Johns & Savage-Smith, "Newly Discovered Islamic Maps", pp. 11-13 and pl. 1. The authors recognize many of the problems of the map but are still tempted to suspect that it is a copy of something as important as the world-map of al-Ma'mūn. Yet the map is so problematic from the point of view of mathematical cartography that it is difficult to see how it could have had a serious ancestor. The longitude scale runs from [0°] to [180°] across half of the map; since it should extend to 180° across the entire map, it has been mislabelled. Also there are severe problems with the positioning of localities: for example, the Arabian Peninsula is situated half way across the right-hand side of the map, where the Indian subcontinent should be.

[12] On this, which Fuat Sezgin believes *is* the lost map of the al-Ma'mūn, see Sezgin, *GAS*, X-XII, and *idem*, "Arab Origin of European Maps", pp. 3-4, but also King, *Mecca-Centred World-Maps*, pp. 31 and 33-34. On this kind of grid in Islamic astronomical instrumentation see *idem*, *SATMI*, XIIb-16. Some of many problems surrounding the association of this map with al-Ma'mūn are the following:

First, the scale for the seven climates is absurd, beginning at the equator and running up to 70° with 10°-divisions for the climates (see already King, *Mecca-Centred World-Maps*, pp. 230-234, especially p. 234).

Second, there are very few cities featured on the map, and none of them is properly marked with a dot. The positions of the names are not helpful. Thus, for example, the name Mecca is written at *ca.* 76°/*ca.* 17° which corresponds neither to Ptolemy's 73;20°/22;0° (also 78;0°/23;0°) nor to the earliest Islamic values, namely, 67;0°/21;0°. The name Medina is at *ca.* 77°/*ca.* 18°, whereas Ptolemy has 71;40°/23;20° and the earliest Islamic sources have 65;20°/25;0°. What has happened here is simply that the hapless copyist has put both cities in the Yemen. For the coordinates cited here see Kennedy & Kennedy, *Islamic Geographical Coordinates*, pp. 225 (Mecca) and 202 (Madina).

[13] King, *Mecca-Centred World-Maps*, pp. 141-148.

distance to Mecca, positioned at the centre of the maps. I have shown that the geographical data used on the maps can be dated to early-15th-century central Asia, and that the underlying mathematics was known – in different manifestations – to various 9th-, 10th-, and 11th-century astronomers.[14] These Mecca-centred world-maps are therefore important in that they demonstrate how little we know about mathematical geography in the early Islamic period. But it is not these maps that concern us here: this paper deals with a world-map based on much simpler cartographical principles, and one that has suffered severely at the hands of a succession of copyists.

2 al-Khāzinī and a lost world-map by al- Bīrūnī

The geographical table of al-Bīrūnī, preserved in his astronomical handbook (*zīj*) called *al-Qānūn al-Mas'ūdī*, dedicated to the Ghaznawid sultan Mas'ūd, is masterly scientific production, and is fortunately available to us. The 604 localities are organized by region and numbered within each group.[15] No detailed analysis of this table has been conducted yet: this could yield a worthwhile doctoral dissertation. However, the ways in which al-Bīrūnī measured some of these coordinates himself and corrected and adjusted earlier values are outlined in his splendid work on mathematical geography, the *Taḥdīd nihāyāt al-amākin*, "The Determination of the Coordinates of Localities", which is the most important single work on the subject from the entire Middle Ages.[16] It is more than probable that al-Bīrūnī produced several world-maps. His writings on the accuracy of terrestrial coordinates, alas lost, filled over 300 folios, and one of these bore the title *Taḥdīd al-ma'mūra wa-taṣḥīḥuhā fī 'l-ṣūra*, which means something like "The Determination of (the Coordinates of Localities) in the Inhabited Parts of the World and their Correct Representation on a World-Map".[17] And why should he have written several treatises, fortunately still extant, on map projections, many originally thought to have been of later European provenance, if he had not tried to use them on a series of world-maps?[18]

I have already presented the evidence that al-Bīrūnī prepared a world-map featuring localities properly marked on an orthogonal grid,[19] which – in an original version or a copy – came into the hands of the scholar 'Abd al-Raḥmān al-Khāzinī

---

[14] *Ibid.*, pp. 197-371, on two examples, and *idem*, *SATMI*, VIIc, where a third example is presented, together with new manuscript sources and insights on the early Islamic (10th- and 11th-century) origin of the grids, based on information provided by Jan Hogendijk of Utrecht.

[15] al-Bīrūnī, *al-Qānūn*, II, pp. 547-579. The list is reproduced in alphabetical order of localities in Kennedy & Kennedy, *Islamic Geographical Coordinates*, pp. 450-460.

[16] Text, translation and commentary are listed under al-Bīrūnī, *Taḥdīd*.

[17] *Ibid.*, p. 344, notes 65-66, quoting a medieval bibliography of his works. See also n. 39.

[18] *Ibid.*, pp. 339-344, and the references there cited.

[19] See note 10 above.

(Marw, Central Asia, *ca.* 1120).[20] This evidence is simple to outline: in the list of about 250 localities in his major work, the *Sanjarı Zij*, al-Khāzinī presents a subgroup of the entries of al-Bīrūnī, but with coordinates generally rounded to the nearest ten minutes, and with an occasional jump of 1° or 2° in one or other of the coordinates. This can only have resulted from his reading the positions from a map.[21]

Now there are two known manuscripts of the *Sanjarı Zij*

– MSS London British Library Or. 6669 and Vatican arabo 761 – but the latter does not contain the geographical table. His own recension of his *zīj*, extant in MS Istanbul Süleymaniye Hamidiye 859, is alas missing a page of the geographical tables, but seems to represent more accurately his original table.[22] I have edited al-Khāzinī's geographical table[23] from the London manuscript, as well as this incomplete Istanbul copy of his recension and a few more relevant sources, such as the Persian *Ashrafī Zīj* (Shiraz, *ca.* 1310),[24] and the *Zīj* of Ibn al-Shāṭir (Damascus, *ca.* 1350).[25] Clearly, the table is only to be regarded as a kind of supplement to the much more extensive and orderly table of al-Bīrūnī.

Alas, the map of al-Bīrūnī used by al-Khāzinī is lost, and until recently we did not even know that any world-map based on their coordinates – precisely whose we may never know – had been preserved for us. This latter is the earliest known surviving Arabic world-map based on geographical coordinates. I begin by introducing the manuscript in which this map was found, explaining the evidence for attributing it to a well-known legal scholar *ca.* 1200 and how we know that it is based on al-Khāzinī, if not al-Bīrūnī, concluding with a commentary revealing, inevitably only partially, how a succession of copyists messed up the map.

---

[20] On al- Khāzinī see the *DSB* article by Robert E. Hall, and now King, *Mecca-Centred World-Maps*, pp. 42, 65-66, and 71-75. On the manuscripts of his *zīj* see *ibid.*, p. 65, n. 36.

[21] *Ibid.*, pp. 71-75. There are isolated exceptions where an occasional 5 minutes has been retained by al- Khāzinī: see, for examples, nos. 7 and 37 in Table 1. I have no explanation for this.

[22] I have no information on the date of copying of this manuscript or its cultural context. Also, I have not been able to consult MS Tehran Sipahsalar 71/1738, which is a copy of an(other?) abridgement of the *Sanjarī Zīj*. See further Pingree, "Editing the *Sanjarī Zīj*".

[23] King, *Mecca-Centred World-Maps*, pp. 564-585.

[24] *Ibid.*, p. 71.

[25] Kennedy & Kennedy, *Islamic Geographical Coordinates*, pp. xxxi and 541-545 (SHA), and King, *Mecca-Centred World-Maps*, p. 71.

3 The source: a treatise on folk astronomy by Sirāj al-Dīn al-Sajāwandī compiled in 1210

Our source is a treatise in 16 chapters on folk astronomy[26] attributed to one Sirāj al-Dunyā wa-'l-Dīn, compiled in the year 607 H [= 1210/11]. The treatise is preserved in fols. 40v-80r of MS Yahuda 4657 (Mach 4983) in the Special Collections Library of Princeton University, copied *ca.* 1350.[27] From the *naskhī* script it is clear only that the manuscript was copied somewhere between Egypt and Iraq. The map occurs on fols. 65v-66r, and is shown in Fig. 1. A related work is preserved in MS Ṭal'at *mīqāt* 127 in the Egyptian National Library, 59 fols., copied in 1255 H [= 1839/40]. This was compiled and copied in Egypt.[28]

A detailed study would be required to sort out the precise relationship of these two treatises. Both appear to be originally by the same individual, although the Cairo treatise contains some spurious 15th-century material on instruments. It must suffice here to stress the historical significance of some of the material preserved in both these two treatises. The Princeton treatise contains a new account of the first geodetic measurements and measurement of lunar distances made by Muslim scholars, this time by the judge Yaḥyā ibn Aktham, who was commissioned by the Caliph al-Ma'mūn to report back to him on the activities of the astronomers.[29] (The Bodleian manuscript mentioned above also contains an account of these geodetic measurements in addition to a world-map.[30]) Then the two treatises contain different information on the early development of the times of Muslim prayer.[31] Finally, although there is surely more awaiting us from a more careful investigation, the Cairo treatise contains a description of a very curious sundial based on the principles of folk astronomy.[32]

---

[26] On Islamic folk astronomy see Varisco, "Islamic Folk Astronomy", and King, "Folk Astronomy in the Service of Islam", A-B. On timekeeping by simple procedures involving shadow lengths see *idem*, "Islamic Shadow Schemes", reprinted in *SATMI*, III.

[27] *Princeton Catalogue*, no. 4983 (Yahuda 4657). The attribution is indicated only in the spurious title *Mukhtaṣar al-hay'a wa-'l-mafātīḥ fī 'ilm al-awqāt wa-'l-maṣābīḥ li-'l-Sirāj.*

[28] *Cairo ENL Survey*, no. C4. The treatise is entitled *Maṣābīḥ al-anwār wa-mafātīḥ al-asrār fī a'māl al-layl wa-'l-nahār* and is attributed to Sirāj al-Dunyā wa-'l-Dīn. See also Rosenfeld & Ihsanoglu, *MAIC*, no. 581, where the Princeton manuscript is not mentioned.

[29] King, "Earliest Muslim Geodetic Measurements", especially pp. 222-225.

[30] Johns & Savage-Smith, "Newly Discovered Islamic Maps", p. 9. This account needs to be compared with the others mentioned in the article cited in the previous note.

[31] King, *SATMI*, IV-3.3.

[32] King, "Shadow-Schemes", p. 233, and *idem*, *SATMI*, III-9.5.

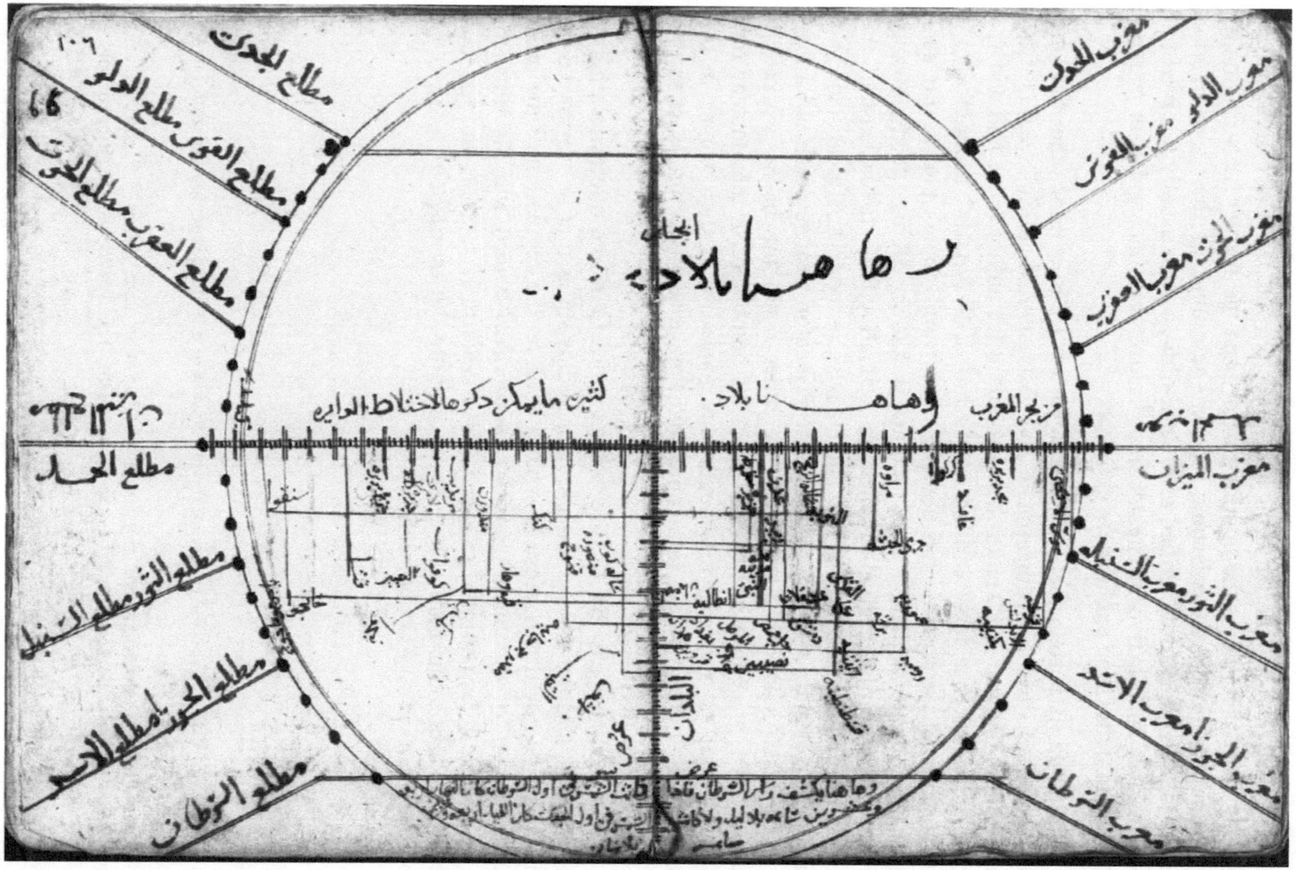

Fig. 1: The world-map in Sirāj's treatise on folk astronomy. [Courtesy of the Special Collections Library, Princeton University.]

I have previously labelled both the Princeton and the Cairo treatises as Egyptian, but I now incline to the opinion that at least the former hails originally from the eastern regions of the Islamic world. The author of the Cairo treatise attributes his sundial to an unidentified *al-ra'īs al-fāḍil*, a term usually reserved for the illustrious Ibn Sīnā, active in Greater Iran.[33] However, this sundial is marked for the *Coptic* months, used exclusively in Egypt.

It seems fairly certain that Sirāj al-Dunyā wa-'l-Dīn is none other than the legal scholar Sirāj al-Dīn Muḥammad ibn Muḥammad al-Sajāwandī, who flourished around the year 1200 and is best known for his authoritative and widely-used book on inheritance laws entitled *al-Farā'iḍ al-Sirājiyya*.[34] al-Sajāwandī's work on algebra and arithmetic has survived in several copies but, as far as I know, has never been studied.[35] Alas, nothing is known of his life. Sajāwand, whence he or his family originally hailed, is a small village to the east of the midway point of the road between Kabul and Ghazna, in modern Afghanistan. This attribution would explain why the author of the Princeton treatise presented a map based on Central Asian coordinates. On the other hand, maybe he worked in Baghdad or Cairo, or somewhere else far from his homeland. In the almanac that appears to be part of our text, the months are named with their names in the Western Islamic calendar before the Coptic and Persian equivalents, such as one would expect in an Andalusī source.

At the beginning of his treatise, after various Qur'ānic quotes relating to God, the heavens and the passage of time, our author states that he was asked by someone whom he could not refuse to write a book on the regulation of the lunar calendar by means of the lunar mansions (*manāzil al-ahilla*) and the times of prayer by simple observations (*al-awqāt bi-'l-adilla*). His work was in fact somewhat more ambitious than this, even mentioning the astrolabe (Ch. 8), but I shall not dwell further on its contents here.

---

[33] See King, *SATMI*, VIIb-1, especially n. 6, on the attribution of anything ingenious or useful to Ibn Sīnā.

[34] See the two articles "al-Sadjāwandī" by Rudolf Sellheim in *Enc. Islam*, 2nd edn., the first mentioning Sajāwand and the second devoted to Sirāj al-Dīn. The *Sirājiyya* is mentioned only in passing in Rebstock, *Rechnen im islamischen Orient*, p. 202.

When I first started in Islamic Studies I bought a little book from Pakistan at one of the Orientalist book-shops across from the British Museum: it was a reprint of the 1792 English translation of the *Sirājiyya* by Sir William Jones. Little did I think at the time that 30-odd years later I would publish Sirāj's world-map.

[35] See *Cairo ENL Survey*, no. G11, where the Cairo manuscript of his work on algebra and its relationship to his better-known work on arithmetic are discussed. Various other works on arithmetic, algebra and geometry are listed in Rosenfeld & Ihsanoglu, *MAIC*, no. 527.

4 The world-map presented by al-Sajāwandī

Although I had thrice published an illustration of the map, I confess that I never really looked at it properly, being the first time surprised that it was preserved in a treatise on folk astronomy,[36] the second time including it in amongst other charts and maps for finding the qibla,[37] and the third time being more concerned with the much more remarkable Safavid world-maps preserving direction and distance to Mecca at the centre.[38] Only in the Winter Semester of 2002-03 did I look closely at it, together with Mónica Herrera and Mohammed Abu Zayed in a seminar in Frankfurt dealing with Arabic scientific manuscripts. Investigating just a few of the coordinates of localities on the maps, we quickly established the connection with al-Khāzinī, if not to a corrupt version of al-Bīrūnī: see further §§4f and 4h.

a) The text on the world-map

al-Sajāwandī's chapter (no. 14) on latitudes and longitudes contains the world-map. He introduces it with the words: *hādhihi 'l-dā'ira tashtamil 'alā ṣifat aṭwāl al-buldān wa-'urūḍihā wa-jihātihā wa-samtihā wa-kull balad bi-smihi fī mawḍi' 'arḍihi ilā ṭūlihi*, "this circular diagram contains a map[39] showing the longitudes and latitudes of localities, as well as their directions and their azimuths, each locality being named at the position of its latitude and (lit., towards) its longitude". He states that he will mark only those localities that are easily markable (*wa-dhakartu fīhā mā tayassar mina 'l-bilād al-ma'rūfa bi-'l-ikhtiṣār*) and that he will not mark too many localities on the map, in order not to mess it up (*wa-law dhakarnā jamī' al-buldān la-khtilaṭat al-dā'ira*), a sentiment repeated on the world-map itself. He mentions the two base meridians of the Canary Islands and the western extremity of the Maghrib, some 10° apart;[40] he himself uses the former. Also he mentions the limit of the map being latitude 67°, and states what happens to the sun there, information also repeated on the map itself. He woffles on, devoting considerable ink to defining the quadrants of the horizon in which the qibla, the local direction of Mecca, is defined for the various situations depending on whether L is greater or less than or equal to the longitude of Mecca and ϕ is greater or less than or equal to the latitude of Mecca. He clearly has no idea of the complexity of the mathematical problem of determining the qibla properly,[41] and unfortunately there is no discussion of the qibla elsewhere in his treatise, which is

[36] King, "Astronomy for Navigators and Landlubbers", pp. 218-219.

[37] Lorch & King, "Qibla Charts and Maps", p. 199.

[38] King, *Mecca-Centred World-Maps*, pp. 90-91.

[39] The terms *ṣifa*, literally, "description", and *ṣūra*, literally, "image", were used for maps in the early Islamic period: see King, *op. cit.*, pp. 33 and 35, and also the text to n. 17 above.

[40] On this see, for example, Kennedy, "Islamic Mathematical Geography".

[41] On this see King, *Mecca-Centred World-Maps*, pp. 47-127, *etc.*, and, more recently, *idem*, *SATMI*, VII.

unusual for a treatise on folk astronomy. The Holy City is assigned the coordinates L: 67° and ϕ: 21;40°. The longitude value confirms my restoration of the reading from the map (see §4g). The latitude value is not that of al-Bīrūnī or al-Khāzinī (21;20°), already known from the 10th century, if not already the 9th,[42] but rather another one derived in the 10th century and adopted by the other main Eastern Islamic tradition, associated with the *Kitāb al-aṭwāl wa-'l-'urūḍ* (ATW).[43]

In general, his comments tend to confirm his lack of knowledge of the subject and invite the speculation that he simply copied the map from another source, which can only date from the period between *ca.* 1040 (the time of al-Bīrūnī) or *ca.* 1120 (the time of al- Khāzinī) and *ca.* 1200 (shortly before his own compilation). What al-Sajāwandī does not tell us is what we really want to know, namely, where he found this map, and who its original compiler was.

b) The frame for the world-map

The world-map is enclosed in an annular frame, which is divided around the outside as a horizon marked with the rising and setting points of the sun when it is at the boundaries between the zodiacal signs, with additional marks for each 10° of solar longitude(?) / azimuth(?). These have been inserted with appalling incompetence, judged by the high standards of medieval astronomical knowledge.[44] The rising- and setting-amplitudes are uniformly distributed around arcs of the horizon that bear no resemblance to their actual risings for any terrestrial latitude. (But it would be absurd if they did; in fact, the whole venture was doomed from the outset.) The span of the rising and setting amplitudes is determined by the scales on the map (see below), the limits for the solstices corresponding to latitudes ±67° on the latitude scale.

c) The inscriptions on the diagram

As we shall see, and as was standard in medieval cartography, north is at the bottom,[45] though this is nowhere stated. On the left of the diagram the rising points (*maṭla'*) of the signs are named, from the top: Capricorn (winter solstice, southern limit), Sagittarius/Aquarius, Scorpio/Pisces, Aries/Libra (equinoxes; due east), Taurus/Virgo, Gemini/Leo, and Cancer (summer solstice, northern limit). Similar information appears on the right for the setting points (*maghrib*). At the right end of the horizontal longitude scale are the words *min baḥr al-maghrib*, standing for "[longitude measured] from the Atlantic Ocean". At the bottom of the vertical

---

[42] King, "Earliest Qibla Methods and Tables", pp. 126-127.

[43] On values for the latitude of Mecca see, in addition to Kennedy & Kennedy, *Islamic Geographical Coordinates*, pp. 225-226, King, "Earliest Muslim Geodetic Measurements", pp. 225-228.

[44] Muslim astronomers tabulated to minutes of arc the rising amplitude of the sun for each degree of solar longitude for various latitudes, as well as the sine of that quantity for all latitudes as an auxiliary function for azimuth calculations: see King, *SATMI*, I-5.6-8 and 8.3 and 8.5*.

[45] See King, *Mecca-Centred World-Maps*, pp. 301-302.

latitude scale are the words *'arḍ al-buldān*, "latitude of localities". In large letters across the virtually empty top half of the circular frame there is an inscription that has been started but abandoned and then repeated in full along the top of the horizontal axis: *wa-hāhunā bilād kathīra mā yumkin dhikruhā li-khtilāṭ al-dā'ira*, meaning loosely: "there are many localities that we cannot feature here because this would mess up the map". The virtually empty upper semicircle is labelled *al-khālī*, "empty space". The lower of the two lines drawn perpendicular to the vertical axis through the rising and setting points of the solstices is labelled *'arḍ s-z*, "latitude 67°", and below this line is the following statement: *wa-hāhunā* [*yankashif*] (for *y-k-s-f*, the correct reading is in the text of Ch. 14) *ra's al-saraṭān fa-'idhā kānat al-shams fī awwal al-saraṭān kān al-nahār arba'a wa-'ishrīn sā'a bilā layl wa-*[*'idhā*] (text: *lā* !) *kānat al-shams fī awwal al-jady kāna al-layl arba'a wa-'ishrīn sā'a bilā nahār*, meaning: "Here (at latitude 67°) the sun does something weird (text unclear) at the summer solstice. If the sun is at the summer solstice, daylight will be 24 hours and there will be no night. If the sun is at the winter solstice, night will be 24 hours and there will be no day." As we have seen in §4a, these observations are found in the text of the treatise.

d) The scales on the world-map

The two perpendicular scales are divided for each 5° of longitude and latitude and subdivided for each 1°. No arguments are marked, which is a pity, because there is a problem at least with the longitude scale. Essentially, 5° is missing on each end of the scale. One would expect a scale to begin at longitude 0° (for the Western limit of the Maghrib) and run to 180° (the Eastern limit of the inhabited world in medieval geography), and it is not unreasonable to put the latitude scale at longitude 90°. If we assume that the central meridian is at 90° longitude, then the graduated longitude scale runs from 5° to 175°. One could live with this, but unfortunately the longitudes of some of the localities on the map are correct if one counts back or forward from the central meridian and others are correct only if one assume that the graduations begin at 0° rather than 5°. The latitude scale is unproblematic in comparison, but we shall encounter several latitude errors of ±5° or ±10°.

e) Aspects of the world-map

The choice of the localities appears to represent a fairly random selection from the table of al-Bīrūnī, which was skilfully organized according to region, with the entries sensibly numbered. Alternatively, it may be regarded as representing a subgroup of the entries of al-Khāzinī, itself a rather curious assortment, illogically ordered and jumping about from one region to a far one and then back. In any case, the localities on our map were not selected by virtue of their size or their religio-cultural importance.

There is no grid on the world-map. Localities that are marked properly – about one-fifth of the total – are joined by lines to the appropriate points on the longitude and latitude scales. The rest are marked only vaguely, with a single line drawn perpendicular to one of the axes, and often the intended length of the segment is unclear.

In the following investigation of the positions of the various localities, I have taken the liberty to adjust some of the longitudes by +5°, in order to demonstrate more clearly the correspondence with the values of al-Bīrūnī / al-Khāzinī. For nos. 1-13 the beginning of the longitude scale represented 0° rather than 5°. For some later entries with longitudes greater than 90°, this discrepancy occurs again, now surely for different reasons. Where a given measurement on the map, adjusted if necessary, corresponds reasonably well (within *ca.* 2°-3°) to al-Bīrūnī's / al-Khāzinī's values, I mark the measurement in bold font. Where a coordinate on the map clearly results from a misreading of a textual entry – a common feature of medieval Arabic tables (see below) – I have italicised the entry. In cases where the position of the locality is actually uncertain but the name is near a position that corresponds to the "correct" position, I have taken the liberty to assume that the name might originally have been associated with that position: see, for example, al-Tubbat (no. 38). Where the name cannot be securely associated with a coordinate, I have estimated the latter from the position and orientation of the name, indicating the readings with an asterisk: such values may be 2°-3° off what was intended.

f) Excursus: A comparison of coordinates in selected sources

In seeking a source of the geographical data used on this map, we need to identify a source that first features the localities shown on the map and second gives the same coordinates. This task is much facilitated by recourse to the lists of geographical data in over 75 Islamic sources compiled by Ted and Mary-Helen Kennedy.[46] It is further facilitated by the fact that our map features several obscure localities that do not often occur in medieval Islamic geographical tables. On the other hand, the whole operation is rendered more complicated by the fact that all manner of scribal errors affected the transmission of geographical coordinates as a result of the careless writing and copying of numbers in the *abjad* notation.[47]

I present here just five examples, with coordinates from our map (which I label SRJ), from al-Bīrūnī's table (BIR), al-Khāzinī's main table (SNB) and the derivative table of Ibn al-Shāṭir (SHA), and al-Khāzinī's table in the Istanbul recension (SNH), which perhaps represents more precisely his original table.[48] I occasionally use SNJ to designate the set of uncontested values of al-Khāzinī.

---

[46] See n. 1 above.

[47] On this phenomenon see King, *op. cit.*, pp. 161-163, and the references there cited.

[48] See n. 21 for the edition of SNB / SNH / ASH / SHA, also drawing on BIR.

1) The standard medieval value of the latitude of Cordova (no. 2) is 38° and some minutes (5, 20, 30, 50, depending on the source). BIR for reasons best known to himself introduced 35;0°, and this is reflected in the following values:

SRJ	BIR	SNB/SHA	SNH
14° ≈ **9° / 35°**	**8;40° / 35;0°**	**8;40° / 35;0°** (ASH: 35;5°)	**8;40° / 35;0°**

(The other Iranian tradition that used the same latitude had a different longitude base, and the longitudes were 10° higher.) Thus the values in SRJ are in the BIR / SNJ tradition, if one adjusts for the incorrect use of the longitude scale.

2) The locality Qata (*Q-t-'*) (no. 51), unidentified but apparently somewhere in China, is featured only in BIR, SNB/SHA, and SNH and a derivative of SNB. It is given the following coordinates:

SRJ	BIR	SNB/SHA	SNH
148° / *ca.* 27°	**148;40°** / 21;40°	108;30° / 21;40°	**148;30°** / 21;40°

What has happened here appears to be the following: al-Bīrūnī marked the longitude as 148;40° on a map, which al-Khāzinī measured and presumably registered in his table as 148;30°, but in the course of copying this was miscopied as 108;30° and as such was incorporated by Ibn al-Shāṭir. However, the Istanbul recension SNH presents al-Khāzinī's original value. Our map thus has the "correct" longitude. The latitude has jumped about 5°, which is precisely what one would expect on a map without arguments on the scales.

3) Kūkū (no. 6), that is, Gau in Mali, has latitude 10° or above in sources unrelated to BIR. We note the following coordinates:

SRJ	BIR	SNB/SHA	SNH
30° / *ca.* 5°	**30;0° / 5;0°**	**30;0° / 5;0°**	-

What has happened here appears to be that Kūkū was included in the original version of al-Khāzinī's table (so that it could be used on this map) but that it was featured on the missing page of the Istanbul recension.

4) The locality Taksīn (no. 43), somewhere in China, occurs only in BIR and SNB/SHA:

SRJ	BIR	SNB/SHA	SNH
125° ≈ **120° / 31°**	**120;15° / 32;50°**	**120;10° / 32;50°**	-

The connection to SHB/SHA is clear, and the original version of SNH probably had the same value.

5) Finally, Shanqū (no. 53), somewhere in China. The longitude shown on our map does not correspond to that of BIR or SNB/SHA, which are the only documented sources featuring this locality:

SRJ	BIR	SNB/SHA	SNH
165° / **15°**	115;10° / 40;20° (K&K have ϕ: 44;90° (*sic*))	115;30° / **15;0°**	-

Here we see that our map is using a reasonable longitude value for a locality in China, which was no doubt original to some source, although in BIR and the derivative SNB this has been miscopied as 115° and some minutes. In other words, SRJ is a witness to a lost tradition. The identity of Shanqu has likewise been forgotten.

In brief, it is already clear that either al-Bīrūnī's table or al-Khāzinī's table, if only in the incomplete Istanbul recension, might well be the source for the positions marked on our map. Further investigation will only confirm this, although the relationship is often obscured by the carelessness with which the map was prepared. "China" and three obscure localities in the Far East (nos. 49-50, 54-55) are not featured in the tables of al-Bīrūnī or al-Khāzinī, but the like of these localities are a law unto themselves in Islamic geographical tables.[49]

g) The positions of the localities marked on the map

In Table 1, the localities featured on the Princeton map are associated with their coordinates as far as this feasible. Values in bold correspond more or less with the values of al-Bīrūnī and/or al-Khāzinī (see §4d). Values in italics result from scribal errors using the *abjad* notation. References are given to K&K for the Kennedys' *Islamic Geographical Coordinates*, where the sources are introduced and where all known coordinates for each locality are listed (the place-names differ occasionally from those used here). References are to page numbers (Roman for the sources and Arabic for the coordinates), a + sign indicating that the entries extend to the next page. The coordinates read from the map can then be compared with the values in al-Bīrūnī's table, the B-nos. referring to the pages in the 1954-56 Hyderabad edition of Max Krause, and those in al-Khāzinī's tables, the K-nos. referring to their appearance in my edition of the latter table (*Mecca-Centred World-Maps*, pp. 564-585), in which I also included al-Bīrūnī's values in the notes. The B-nos. well reflect the order of the organization of al-Bīrūnī's table, and the K-nos. equally well demonstrate the chaotic arrangement of al-Khāzinī's entries. The entries for L/ϕ are those from al-Bīrūnī's table; for al-Khāzinī's original table, it there is a difference between SNB and SNH, this is noted in the column labelled "comments". If the values of al-Bīrūnī and al-Khāzinī are the same, they are underlined. If there is a discrepancy of up to 0;30° the values are underlined with dots. The various sources cited are here ordered chronologically:[50]

[49] A series of studies of the coordinates of localities in outlying regions of the Islamic world would not be without interest. See King, *Mecca-Centred World-Maps*, pp. 226-227, for some comments on problematic coordinates of localities in Central Asia.

[50] A full chronological overview of the sources used for Kennedy & Kennedy, *Islamic Geographical Coordinates*, is presented *ibid.*, pp. xxxviii-xxxix.

PTO The *Geography* of Ptolemy; geographical coordinates not readily accessible – see K&K xxviii-xxix / 395-398 (selection)

YAQ PTO attributed to Ptolemy by Yāqūt (*ca.* 1200); K&K xxxvi / 421-422

HTP: The *Handy Tables* of Theon, edited by N. Halma in 1822-25; K&K xxi / 389-394

KHU *Geography* of al-Khwārizmī (*ca.* 825), edited by Hans von Mzik in 1926; K&K xxii-xxiii / 399-408

KHZ a table attributed to al-Khwārizmī in a 13th-century manuscript; K&K xxiv / 409-412

SUH *Geography* of Suhrāb (ed. von Mzik); K&K xxxii-xxxiii / 412-421

ZAY *Geography* of al-Zayyāt (d. 1058), extant in an unpublished manuscript in Burdur, Turkey; K&K, xxxvii / 486-494

BIR al-Bīrūnī's *Zīj* (*al-Qānūn al-Mas'ūdī*), edited by [Max Krause] in 1954-56; K&K xviii / 450-460

SNB London copy of al-Khāzinī's *Sanjarī Zīj*; K&K xxxii / 476-478; used for the edition in King, *Mecca-Centred World-Maps*, pp. 564-585

SNH Istanbul recension of al-Khāzinī's *Sanjarī Zīj*; K&K xxxii / 478-480; used for the edition in King, *op. cit.*, pp. 564-585

SNJ uncontested values of al-Khāzinī, attested in all relevant sources: see King, *op. cit.*, pp. 564-585

SRJ coordinates derived from Sirāj al-Dīn al-Sajāwandī's map: see Table 1

SAA *Geography* of Ibn Sa'īd al-Maghribī (d. 1286), edited by Juan Vernet, 1958; K&K xxx / 486-494

ASH *Ashrafī Zīj* of Sanjar al-Kamālī (Shiraz, *ca.* 1310); coordinates listed in K&K xvi / 532-536; used for used for edition of al- Khāzinī's table in King, *op. cit.*

SHA *Zīj* of Ibn al-Shāṭir (Damascus, *ca.* 1350); K&K, xxxi / 541-545; used for edition of al-Khāzinī's table in King, *op. cit.*

Table 1 reveals that most of the positions of the various localities on the Princeton world-map correspond in some way or other to the al-Bīrūnī (BIR) and al-Khāzinī tradition (SNH or SNB, more often, in special cases, the former), unless substantial errors have crept in. See further §4h.

There are five localities that are not even listed in BIR / SNH or SNB, and these deserve special mention. See further §4i.

Table 1

Positions of localities in al-Sajāwandī's world-map

Notes: For nos. 1-13 the longitudes are measured assuming that the central meridian is at 90° and are then adjusted for the 5° error at the beginning of the longitude scale. For nos. 14-55 the longitudes are measured with respect to the central meridian, taken as 90°, but occasional entries are adjusted by –5°. For entries nos. 21-27 the longitudes on the map are 5° too high. There is an error of 10° in the longitudes of nos. 28 and 30-33. Notes on the orthography of the names that are asterisked are at the end of the table.

No.	Name	Equivalent	K&K	L	□	B/K	L/□ (BIR)	L/□ (SNJ)	Comments
1	Sūs al-aqṣā	Sus, W. Maghrib	335	9 → **4**	**20***	551/18	5;30/22;0	5;30/22;0	
2	Qurṭuba*	Cordova	95	14 → 9	35	563/24	8;40/35;0	8;40/35;0	
3	[Balansiyya]*	Valencia	372	25* → 20*	38	Ø/Ø	-	-	coordinates correspond to Maghribi sources only – see §4i
4	Bajā / Barbara	Beja (tribe)	58	20 → **15**	5*	548/11	55;0/12;0	55;[0]/12;30 (SNB) **15;0**/12;0 (ASH)	SNB has L: 55;08°; not in SNH; see §4h
5	Ghāna	Ghana	124	30 → **25**	7*	549/13	25;0/13;0	25;0/18;30 (SNB/ASH) 25;30/18;0 (SHA)	the 18° for ϕ is a copyist's error for the 13° in BIR
6	Kūkū	Gau (Mali)	191+	35 → 30	5*	547/8	30;0/5;0	30;0/5;0	
7	Rūmiyya	Rome	286+	41 → **36**	**41**	574/29	35;25/41;50	35;25/41;50	
8	Ammuriyya	Amorion (Gk.)	23+	41 → 36	**35**	564/28	33;0/38;0	33;0/33;0	the 33° for ϕ in SNH/ASH is a copyist's error for the better value 38° in BIR; it is not clear which latitude – 33° or 38° – was intended by the marker at 35°

9	[Zaghāwa]*	Zaghawa (tribe)	381	45 → 40	3*	547/3	56;0/-1;10	56;30/-1;10	L:45° could result from a copyist's error of 46 for 56
10	Jarmī al-Ḥabasha	Garama	157+	47 → **42**	**20**	549/14	41;40/9;40	41;40/**19;40**	the 19° for ϕ in SNB/ASH/SHA results either from a copyist's error for the 9° in BIR or from an error reading a map
11	Barqa	Barqa, Libya	70	45 → **40**	**35**	555/42	42;45/32;0	42;40/32;0	
12	Qusṭanṭiniyya	Constantinople	93+	55→ **50**	**46**	577/35	49;50/45;0	49;50/45;0	
13	Athīna	Athens	46+	55 → **50**	**41**	574/30	48;0/43;0	48;0/43;0	
14	ʿAkka	Acre	6	55	**32***	557/78	58;20/33;20	58;30/33;20	
15	al-Quds	Jerusalem	160+	**55**	26*	556/103	56;0/33;0	56;0/33;0	SNH/ASH have ϕ:33;0; SHA has 32;0°; the marker should be closer to that for no. 14
16	al-Nūba	Nuba (tribe)	109+	58	**14**	549/15	53;40/14;0	53;40/14;0	BIR/SNH have these values for Dongola
17	ʿAsqalān	Ascalon	42+	58	**33**	556/101	55;20/33;0	55;20/33;0	
18	Dimashq	Damascus	99+	**60***	**35**	557/79	60;0/33;30	60;0/32;30 (ASH) 60;0/33;25	not in SNB/SNH

								(SHA)	
19	Sufālat al-Zanj	?? in E. Africa	331	60	7*	547/1	50;0/-2;0	50;0/-2;0	
20	ʿAdan	Aden	7	**67***	8*	548/44	66;0/11;0	66;30/11;0	
21	al-Zabīd	Zabid	380	68* → **63***	**17***	549/46	63;20/14;10	63;20/15;10	not in SNB/SNH; the 15°in ASH/SHA for 14° in ϕ could result from a copyist's error or from a misreading of a map
22	Ḥalab	Aleppo	16+	68 → **63**	**35**	565/84	63;0/34;30	63;0/34;50	
23	Ḥims	Homs	143+	68 → **63**	**35**	565/82	61;0/33;40	61;0/33;40	
24	Niṣībīn	Nisibin	246	70*	45*	567/123	57;50/36;40	57;50/36;40	no comment !!
25	Makka	Mecca	225+	72 → **67**	**21**	551/86	67;0/21;20	67;0/21;20	
26	Madīnat al-Nabī	Medina	202+	72 → **67**	**27**	552/87	67;30/24;0	67;30/24;45	
27	[Suquṭra]* Island	Socotra	334	72 → **67**	14	548/12	67;0/9;0	67;30/9;0	
28	Khilāṭ	Ahlat	12+	75*	43*	575/133	64;50/39;40	64;50/39;40	error in L: 10°
29a	[Anṭākīya]* (?)	Antioch	29	75	33	565/Ø	61;35/34;10	-	see §4i
29b	Anṭālīya* (?)	Antalya	28	75	33	Ø/Ø	-	-	see §4i
30	Baghdād	Baghdad	55+	80*	38*	558/147	70;0/33;25	70;0/33;25	error in L: 10°, error in ϕ: 5°

31	Qazwīn	Qazwin	269+	83*	43*	568/176	75;0/37;0	75;0/37;0	errors roughly similar to those in previous entry
32	Hamadān	Hamadan	132+	85*	**37***	568/174	75;20/34;40	75;20/34;40	error in L: 10°
33	Iṣahān	Isfahan	149+	85*	**32***	560/159	77;20/33;30	77;20/32;30	L/ϕ from SNB/SNH; situation complicated – see my edition
34	Kāshghar	Kashgar	175	**94**	**47**	578/236	95;25/44;0	95;0/**47;0** (SNB1) 95;0/44;0 (SNB2) 95;0/45;0 (ASH)	situation complicated – see my edition
35	[Siyālkūt]*	Sialkot	295+320	95*	**31**	562/68	99;0/33;0	[99;0/33;0]	not in SNB or SNH, L/ϕ as in ASH/SHA
36	Manṣūra	Mansura	213+	101*	22*	552/75	95;0/26;40	95;0/25;40	
37	Qannawj	Kannauj	172	**105**	**25***	553/64	104;50/26;35	104;50/26;35	
38	al-Tubbat	Tibet	360+	**105**	**37**	573/242	92;0/37;0	**105;0**/37;0	the rationale behind the change in longitude between BIR and SNB/SNH is unclear
39	Jurjāniyya*	Gonbad-Ekavus	165+	*105*	**37**	570/185	80;10/38;10	82;10/38;10	L:105° possibly results from a misreading of 102 for 82

40	*K-n-k*	Kanaka in India	171+172	108	15	550/57	110;40/19;10	111;40/19;10	the 1° change in Lprobably results from a misreading of the position on a map
41	Qandahār	Qandahar	261	117*	**31/33(?)**	562/67	97;50/33;20	97;55/33;0 (ASH) 97;50/33;0 (SHA)	not in SNB/SNH
42	Mandūrat	?? in India	212	**120**	**15**	550/59	120;0/15;0	120;30/15;0	
43	Taksīn	?? in China	342	125 → **120**	**31**	563/246	120;15/32;50	120;10/32;50	note error of +5° in L
44	Banjū	?? in China	65	**125**	30*	554/243	125;0/22;0	125;0/22;30	
45	[Sarandīb]	Ceylon / Sri Lanka	304+	**127***	**10***	548/52	120;0/10;0	137;30/10;0 (SNH) **127;30**/10;0 ([ASH]/SHA)	ASH has L: 127;37, a copyist's error for 127;30 – see §4h
46	[Kazqū]*	?? in China	178+191	130* → **125**	25	554/244	127;0/21;0	127;30/21;0 (SNB) 124;30/21;0 (SHA)	
47	Lāmrī (Island)	Sumatra (?)	195	131* → **126***	**10***	548/53	127;0/9;0	126;30/9;0 (ASH) 127;30/9;0 (SHA)	note error of +5° in L

XI

48	*K-l-h* Island	?? in Indian Ocean	169	135 → **130**	10	548/55	130;0/8;0	130;0/5;0	error in L is less for BIR than SNJ
49	[al-Ṣīn]* (??)	"China"	91	136*	26*	Ø/Ø	-	-	see further §4i
50	[Sarīra]* Island	?? in Indian Ocean	306	**142***	7*	547/4	140;0/-1;0	140;30/-1;0	
51	Qata (*Q-t-æ*)	?? in China	272	148	27*	554/245	148;40/21;40	108;30/21;40 (SNB/SHA) 148;30/21;40 (SNH)	see §4f
52	Khānjū	Quazhou, China	180	**160**	*33*	550/248	162;0/13;0	162;30/13;0 (SNB) 152;30/13;0 (SHA)	ϕ: 33° on the map results from a misreading of 13° at some stage
53	Shanqū	?? in China	314	165	15	577/247	115;10/40;20	115;30/15;0	K&K have ϕ: 44;90 [*sic*] for BIR; L=115;30 in BIR is probably an original mistake for 165;30
54	Yājūj wa-Mājūj	Gog & Magog	377	165*	35*	Ø/Ø	-	-	see §4i
55	[Kattigara]*	Hanoi (?)	90	168*	-8*	Ø/Ø	-	-	see §4i

Notes on names: 2 - named *Qurṭubat al-Andalus*; 3 - written *b-k-n-y-h*; 9 - written *m-r-æ-w-h*; 27 - name garbled; 29 - written *Anṭālīyah*; 35 - written *s-æ-l-h k-w-b*; 39 - written: *m-q-d-r* Jurjāniyya, possibly a misrendering of *Qaṣbat Jurjān* as found in SNB; 46 - written *kūfūwān*, in SNB *kūfū*; 49 - name looks like *al-ḥ-y-y-z*; 50 - written *b-r-b-r-h*; 55 - written *k-x-l-d-r*

h) Arguments for and against a connection with al-Bīrūnī rather than al-Khāzinī

We have already seen in §4f that one of al-Sajāwandī's positions (namely, no. 2) agrees with al-Bīrūnī's coordinates and the Istanbul recension of al-Khāzinī's table, rather than the London version of al-Khāzinī's table. Perhaps his map was copied from a map in the tradition of al-Bīrūnī without any influence from al-Khāzinī, for whom the Istanbul manuscript simply preserves some values closer to al-Bīrūnī's table than the London manuscript?

But it is worthwhile to look at another locality for which the coordinates are simply wrong, when compared with al-Bīrūnī's original values. This is:

Beja (no. 4), the tribe in the Eastern Sudan, with L/ϕ: 20° ≈ [15°] / *ca.* 5°.

Whilst the adjusted longitude corresponds to the 15;0° in ASH / SHA, the 15 is a scribal error for the 55° in BIR. SNB happens to have 55;8°, the 8 perhaps resulting from a *ṭā'* misrepresenting an odd *jīm*, an abbreviation for *daraja*, "degree". (Several medieval coordinates have a 3 (*jīm*) or an 8 (*ṭā'*) for a 0 in the minutes, presumably for this reason.[51]) SNH is alas missing this entry. In brief, this position could surely *not* have been copied from a map due to al-Bīrūnī himself, for the master must have been aware of the location of the Beja.

Furthermore, another locality is marked on the map according to its position as given by al-Khāzinī, which is erroneous, with the latitude 10° higher than the value given by al-Bīrūnī. This is:

Jarmī or Garama (no. 10), with L/ϕ: 47° ≈ **42° / 20°.**

This position corresponds nicely to al-Khāzinī's **41;40°/19;40°**, but *not* to al-Bīrūnī's secure **41;40°** / 9;40°. The value 19;40° for the latitude of Garama goes back to al-Khwārizmī, so al-Bīrūnī must have thought that he had a good reason to reduce it by 10°.[52]

Then there is a locality for which al- Bīrūnī's longitude is too small, and the map shows it at the improved longitude of al-Khāzinī. It is not clear how the latter was derived. I refer to:

al-Tubbat (38), Tibet, with L/ϕ: 105° / 37°,

BIR has 92;0°/37;0°, and SNB/SNH has for some reason 105;0°/37;0°. A more realistic longitude value is 130;0°, found in al-Khwārizmī and several other early Islamic sources.[53]

---

[51] See, for example, Kennedy & Kennedy, *Islamic Geographical Coordinates*, pp. 10-11 (Ahwaz) and 177 (Kazirun).

[52] *Ibid.*, pp. 157-158.

[53] *Ibid.*, pp. 360-361.

Finally, there is one locality for which the longitude corresponds neither to BIR nor to SNH, but rather to ASH, which is based on SNB. This is:

Sarandīb (no. 45), Ceylon, with L/ϕ: *ca.* 127° / *ca.* 10°.

Now BIR has 120;0°/10;0°, whilst SNH has 137;30°/10;0°. It is conceivable that 120 could be misread as 137 in *abjad* notation. However, ASH/SHA have **127;30°**/10;0°, and it is these values, perhaps originally from SNB, that underlie the position on the map.

i) Five localities from a different textual tradition

First we consider:

3	Valencia (K&K 372)	L: *ca.* 25° ≈ ***ca.* 20°**	ϕ: **38°**

K&K gives only BIR: 14;0°/34;40°; **ZAY: 20;50°/37;30°; SAA: 20;0°/38;6°** for Valencia. Here we observe a most curious agreement with two Maghribi sources, ZAY and SAA. I have no explanation for this. I have shown elsewhere that ZAY was used in Egypt in the 14th century,[54] but our world map is apparently not an Egyptian production (although for many years, I thought it was). See also nos. 49 and 55 below.

Next we try to seek the identity of the locality marked on our map as *a-n-ṭ-ā-l-y-h*. There are two possibilities, of which only the first is viable:

29a	[Antioch] (K&K 29)	L: 75°	ϕ: 33°

For Antioch, K&K shows HTP: 69;0°/35;30°; PTO: 69;0°/35;30°; KHU: 61;35°/34;10°; KHZ: 61;35°/33;10°; SUH: 61;35°/35;0°; YAQ PTO: 69;0°/35;30°; SAA: 61;35/34;10°. No obvious correspondence is suggested. For the alternative:

29b	Antalya (K&K 28)	75°	33°

three non-relevant sources give longitudes of 65° and 52/54;30°, with latitudes *ca.* 37;30° and *ca.* 41°. Thus it is unlikely that Antalya was intended.

Muslim coordinates for localities in the Far East are problematic enough, those for regions even more so. Such is the case for "China", whose name our copyist could not even write legibly.

49	"China" (K&K 91)	L: *ca.* 136°	ϕ: ***ca.* 26°**

K&K gives over a dozen sets of coordinates, including KHU: 164;40°/18;30°; KHZ/SUH are similar but slightly different; ZAY has 177;12°/38;0°; and, remarkably, SAA has 167;0°/**26;0°**. It would be difficult to account for 136° as resulting from a scribal error for 167° or 166°.

---

[54] King, *Mecca-Centred World-Maps*, pp. 80-84.

In addition to various dubious localities in the Far East, we also find on our map the apocalyptic and mythical Gog and Magog, known from Biblical and Qur'ānic eschatology:[55]

54 Gog & Magog (K&K 377) L: *ca.* 165° ϕ: *ca.* 35°

Close to a dozen Islamic sources have L: *ca.* 170° with ϕ: *ca.* 43° for "Gog", and some half a dozen sources have L: *ca.* 172° with ϕ: 63° for "Gog & Magog".

Finally, for reasons that are unclear, our cartographer thought fit to mark Kattigara (= Hanoi?) on his map.

55 Kattigara (K&K 90) L: ***ca.* 168°** ϕ: ***ca.* -8°**

Only four sources listed in K&K feature this locality, namely, **HTP: 170;0°/-8;30°**; PTO: 177;0°/-8;30°; KHU: 161;30°/-6;0°; **SAA: 166;0°/-6;00**. The remarkable agreement with Ptolemy's *Handy Tables* is more likely to be significant than the agreement with SAA, though see no. 3 and also no. 49 above.

5 Concluding remarks

The world-map presented by Sirāj al-Dunyā wa-'l-Dīn in the form that we have it in the Princeton manuscript is surely a poor reflection of an original map based on the coordinates of al-Bīrūnī and/or al-Khāzinī. We can assume that on the original the localities were all properly as the intersection of a pair of perpendicular lines and that the names were clearly associated with these positions.[56] And we can be confident that the absurd horizon diagram surrounding the world-map was an addition of Sirāj al-Dunyā wa-'l-Dīn or one of his sources in the folk astronomical tradition.

Would that I could have presented a lost map of al-Bīrūnī to my close friend and distinguished colleague Hossam Elkhadem! It is with some regret that I have only been able to present only a rather miserable copy of a world-map based on the coordinates of al-Bīrūnī and/or al-Khāzinī. Already al-Khāzinī in the early 12th century was not capable of properly handling the geographical data of his illustrious predecessor,[57] and the compiler of the original version of this map was not more inspired than he.

---

[55] See the article "Yādjūdj wa-Mādjūdj" by Emory van Donzel and Claudia Ott in *Enc. Islam*, 2nd edn.

[56] This is the case on the Safavid world-maps, yet even on these some positions are confused: see *ibid.*, pp. 223-230. On the 13th-century copy of an 11th-century world-map recently illustrated by Johns and Savage-Smith (see n. 11), the localities are marked with red dots which appear to bear no relation to their positions as given in medieval Islamic geographical tables.

[57] al-Khāzinī also presented the qibla values for each locality in his table. These are sloppily calculated using an inaccurate table displaying the qibla for each degree of longitude and latitude distance from Mecca: see King, *Mecca-Centred World-Maps*, pp. 71-75. Accurate tables of this kind were available, but apparently not to al- Khāzinī.

In passing, we note that a world-map by Abū ʿAlī Qaṭṭān Marwazī (Central Asia, *ca.* 1125), bears two inscriptions along the equator and the Arctic circle, mentioning their lengths in *farsakhs*. For the latter circle, the altitude of the celestial pole there is stated to be 66°, reminiscent of al-Sajāwandī's map.[58] A task beyond the scope of the present study would be to investigate another world-map prepared about the same time as that of al-Sajāwandī by Muḥammad ibn Najīb Bakrān. His geographical work was dedicated in 605 Hijra [= 1208/09] to the Khwārazm-Shāh ʿAlā al-Dīn Muḥammad ibn Takash (*reg.* 596-617 Hijra [= 1200-1220]), extant in two manuscripts in St. Petersburg and Paris. This is now edited and translated into Russian, but is not currently available to me.[59] In Chapter 3, the author describes a world-map that he has made and states that he used a rectangular grid of longitude and latitude lines coloured in red. He mentions various early geographers, as well as al-Bīrūnī, and the 12th-century astronomer-mathematician Sharaf al-Dīn al-Ṭūsī, who is not otherwise known to have written on geography.[60]

The ultimate demise of Islamic science is well reflected in the fact that the highly sophisticated geographical achievements of al-Bīrūnī bore no virtually no fruit until they were rediscovered and published by orientalists in the 20th century.[61] In fact, the world-maps of Sirāj al-Dunyā wa-'l-Dīn al-Sajāwandī and Ibn Najīb Bakrān constitute the only known evidence of any practical influence of al-Bīrūnī's coordinates, and the second of these has apparently not survived. I am sure that al-Bīrūnī would have cried if he had seen the copy of al-Sajāwandī's map that we have before us, but, hopefully, Hossam Elkhadem will feel that my contribution to his *Festschrift*, essentially an attempt to make some sense of the map, is *aḥsan min balāsh*.[62]

---

58 Najafi, ed., *Qaṭṭān Marwazī - Gayhān Senāxt*, p. 187.

59 Persian text and Russian translation are listed in the bibliography as Ibn Najīb Bakrān, *Jihān-nāma*. See also King, *op. cit.*, p. 147.

60 On Sharaf al-Dīn al-Ṭūsī see the *DSB* article by Adel Anbouba, dealing predominantly with his mathematical works.

61 Fuat Sezgin, after outlining al-Bīrūnī's contributions, exaggerates his influence altogether: "(al-Bīrūnī's) data were used as the basis for the determination of coordinates which was continued unbroken for centuries in the eastern part of the Islamic world" ("Arab Origin of European Maps", p. 4). On the contrary, as stated by Kennedy & Kennedy, *Islamic Geographical Coordinates*, p. xviii: "Considering al-Bīrūnī's prestige, it is surprising that (his geographical table) was not more influential."

There was some serious activity in mathematical geography in Greater Iran during later centuries, but this was in the *other* tradition that was based on the mysterious *Kitāb al-Aṭwāl wa-'l-'urūḍ* (see the references in n. 6).

62 Egyptian Arabic for "better than nothing".

Acknowledgments: I am grateful to my colleagues Sonja Brentjes, Mercè Comes and Benno van Dalen, for their critical comments on a preliminary version of this paper, without implicating them in any remaining misinformation that it may contain.

Bibliography and bibliographical abbreviations

al-Bīrūnī, *al-* Bīrūnī: al- Bīrūnī, *al-* Bīrūnī *al-Mas'ūdī*, [Max Krause, ed.], 3 vols., Hyderabad, Deccan: Osmania Oriental Publications, 1954-1956.

— , *Taḥdīd*: Text: al- Bīrūnī, *Kitāb Taḥdīd nihāyat al-amākin*, ed. P. Bulgakov, Majallat Ma'had al-makhṭūṭāt al-'arabiyya (Cairo) 8 (1962). Translation: Jamil Ali, *The Determination of the Coordinates of Cities ... by al-Bīrūnī*, Beirut: American University of Beirut, 1967. Commentary: Edward Stewart Kennedy, *A Commentary upon Bīrūnī's Kitāb Taḥdīd [Nihāyāt] al-Amākin, An 11th Century Treatise on Mathematical Geography*, Beirut: American University of Beirut, 1973.

Cairo ENL Survey: David A. King, *A Survey of the Scientific Manuscripts in the Egyptian National Library*, (American Research Center in Egypt, Catalogs, vol. 5), Winona Lake, Ind.: Eisenbrauns, 1987.

Comes, "Meridian of Water": Mercè Comes, "Islamic Geographical Coordinates: al-Andalus' Contribution to the Correct Measurement of the Size of the Mediterranean", in *Science in Islamic Civilisation: Proceedings of (Two) International Symposia ...* , Ekmeleddin Ihsanoglu and Feza Günergun, eds., Istanbul: Research Centre for Islamic History, Art and Culture, 2000, pp. 123-138.

DSB: *Dictionary of Scientific Biography*, 14 vols. and 2 supp. vols., New York: Charles Scribner's Sons, 1970-80.

Enc. Islam, 2nd edn.: *The Encyclopaedia of Islam*, new edn., 11 vols. to date, Leiden: E. J. Brill, 1960 onwards. [Also available in French.]

Ibn Najīb Bakrān, *Jihān-nāma*: Muḥammad ibn Najīb Bakrān, *Jihān-nāma*, text edn. by Muḥammnad Amīn Riyāḥī, Tehran: 1342 HS [= 1963], Russian transl. by Y. E. Borchevski, Moscow: Institute of the Peoples of Asia, 1960.

Johns & Savage-Smith, "Newly Discovered Islamic Maps": Jeremy Johns and Emilie Savage-Smith, "*The Book of Curiosities*: A Newly Discovered Series of Islamic Maps", *Imago Mundi* 55 (2003), pp. 7-24.

Kennedy, "Islamic Mathematical Geography": E. S. Kennedy, "Mathematical Geography", in *Encyclopedia of the History of Arabic Science*, Roshdi Rashed, with Régis Morelon, eds., 3 vols., London: Routledge, 1996, I, pp. 185-201.

— , "Suhrāb and the World-Map of Ma'mūn": *idem*, "Suhrāb and the World-Map of Ma'mūn", in *From Ancient Omens to Statistical Mechanics: Essays on the Exact Sciences Presented to Asger Aaboe*, J. Lennart Berggren and Bernard R.

Goldstein, eds., *Acta Historica Scientiarum Naturalium et Medicinalium* (Copenhagen: University Library) 39 (1987), pp. 113-119.

Kennedy & Kennedy, *Islamic Geographical Coordinates*: *idem* and Mary Helen Kennedy, *Geographical Coordinates of Localities from Islamic Sources*, Frankfurt am Main: Institut für Geschichte der Arabisch-Islamischen Wissenschaften, 1987.

King, "Astronomy for Navigators and Landlubbers": David A. King, "Astronomy for Landlubbers and Navigators: The Case of the Islamic Middle Ages", *Revista da Universidade de Coimbra* 32 (1985), pp. 211-223.

—, "Earliest Muslim Geodetic Measurements": *idem*, "Too Many Cooks ... – A Newly-Rediscovered Account of the First Muslim Geodetic Measurements", *Suhayl – Journal for the History of the Exact and Natural Sciences in Islamic Civilisation* (Barcelona) 1 (2000), pp. 207-241.

— , "Earliest Qibla Methods and Tables": *idem*, "The Earliest Islamic Mathematical Methods and Tables for Finding the Direction of Mecca", *Zeitschrift für Geschichte der arabisch-islamischen Wissenschaften* (Frankfurt am Main) 3 (1986), pp. 82-149, repr. in *idem*, *Astronomy in the Service of Islam*, Aldershot (U.K.): Variorum, 1993, XIV.

—, "Folk Astronomy in the Service of Islam", A-B: *idem*, "Folk Astronomy in the Service of Religion: The Case of Islam", in *Astronomies and Cultures*, Clive L. N. Ruggles and Nicholas J. Saunders, eds., Niwot, Co.: University Press of Colorado, 1993 [published 1994], pp. 124-138 (A); and *idem*, "Applications of Folk Astronomy and Mathematical Astronomy to Aspects of Muslim Ritual", *The Arabist (Budapest Studies in Arabic),* 13-14 (1995) (*Proceedings of the XIVth UEAI Congress, Budapest, 1988*, Alexander Fodor, ed., Part I), pp. 251-274 (B).

—, *Mecca-Centred World-Maps*: *idem*, *World-Maps for Finding the Direction and Distance to Mecca: Innovation and Tradition in Islamic Science*, Leiden: E. J. Brill, and London: Al-Furqan Islamic Heritage Foundation, 1999.

— , *SATMI*: *idem*, *In Synchrony with the Heavens – Studies in Astronomical Timekeeping and Instrumentation in Islamic Civilization*, 2 vols., vol. 1: *The Call of the Muezzin. Studies I-IX*, and vol. 2: *Instruments of Mass Calculation. Studies X-XVIII*, Leiden: E. J. Brill, 2004, especially III: *A Survey of Arithmetical Shadow-Schemes for Time-Reckoning*, IV: *On the Times of Prayer in Islam*, and VIIc: *Safavid World-Maps Centred on Mecca: A third example and some new insights on their original inspiration.*

—, "Shadow-Schemes": *idem*, "A Survey of Medieval Islamic Shadow-Schemes for Simple Timereckoning", *Oriens* 32 (1990), pp. 191-249, and a new version in *SATMI*, III.

— & Samsó, "Islamic Astronomical Handbooks and Tables": *idem* and Julio Samsó, with a contribution by Bernard R. Goldstein, "Astronomical Handbooks and Tables from the Islamic World (750-1900): An Interim Report", *Suhayl – Journal for the History of the Exact and Natural Sciences in Islamic Civilisation* (Barcelona) 2 (2001), pp. 9-105.

Lorch & King, "Qibla Charts and Maps": David A. King and Richard P. Lorch, "Qibla Charts, Qibla Maps, and Related Instruments", in *The History of Cartography*, J. B. Harley and David Woodward, eds., vol. II, book 1: *Cartography in the Traditional Islamic and South Asian Societies*, Chicago, Ill., & London: The University of Chicago Press, 1992, pp. 189-205.

Najafī, ed., *Qaṭṭān Marwazī – Gayhān Senāxt*: *Gayhān Senāxt – 'Ayn al-zamān ... Abu-'ali Qattan Marwazi* ... (facsimile edn.), ed. Sayyid Mahmoud Mar'ashi Najafi, Qum, 1379 HS (= 199•).

Pingree, "Editing the *Sanjarī Zīj*": David Pingree, "Preliminary Assessment of the Problems of Editing the *Sanjarī Zīj* of al-Khāzinī, in *Editing Islamic Manuscripts on Science – Proceedings of the Fourth Conference of Al-Furqan Islamic Heritage foundation, 29th-30th Nov., 1997*, Yusuf Ibish, ed., London: Al-Furqan Islamic Heritage Foundation, 1999, pp. 105-113.

Pinto, *Ṣūrat al-arḍ*: Karen Pinto, *Ṣūrat al-arḍ – The World in the Medieval Muslim Cartographic Imagination*, unpublished Ph.D. thesis, Department of History, Columbia University, *ca.* 2000.

Princeton Catalogue: Rudolf Mach, *Catalogue of Arabic Manuscripts (Yahuda Section) in the Garrett Collection, Princeton University Library*, Princeton: Princeton University Press, 1977.

Rebstock, *Rechnen im islamischen Orient*: Ulrich Rebstock, *Rechnen im islamischen Orient – Die literarischen Spuren der praktischen Rechenkunst*, Darmstadt: Wissenschaftliche Buchgesellschaft, 1992.

Rosenfeld & Ihsanoglu, *MAIC*, Boris A. Rosenfeld and Ekmeleddin Ihsanoglu, *Mathematicians, Astronomers and other Scholars of Islamic Civilisation and their Works (7th-19th C.)*, Istanbul: Research Center for Islamic History, Art and Culture (IRCICA), 2003

al-Sarājiyya: *Al Sirajiyyah or the Mahommedan Law of Inheritance*, reprinted from the (Calcutta, 1792) translation of Sir William Jones, with notes and appendix by Almaric Rumsey, Lahore: Premier Book House, fourth impression, 1971.

Savage-Smith: see Johns & Savage-Smith.

Sezgin, "Arab Origin of European Maps": Fuat Sezgin, "Arab Origin of European Maps", *Zeitschrift für Geschichte der arabisch-islamischen Wissenschaften* (Frankfurt am Main) 15 (2002/03), pp. 1-23.

—, *GAS*: *idem*, *Geschichte des arabischen Schrifttums*, 12 vols. to date, Leiden: E. J. Brill, 1967 onwards, from 2000 Frankfurt am Main: Institut für Geschichte der Arabisch-Islamischen Wissenschaften, especially V: *Mathematik*, 1974, VI: *Astronomie*, 1978, Band VII: *Astrologie, Meteorologie und Verwandtes*, 1979;

X-XII: *Mathematische Geographie und Kartographie im Islam und ihr Fortleben im Abendland*, 2000.

Varisco, "Islamic Folk Astronomy": Daniel M. Varisco, "Islamic Folk Astronomy", in *Astrononomy across Cultures – The* [!] *History of Non-Western Astronomy*, Helaine Selin, ed., Dordrecht, *etc*.: Kluwer, 2000, pp. 615-650.

Mathematical Geography in 15th-Century Egypt: An Episode in the Decline of Islamic Science*

1. Introductory Remarks

Islamic mathematical geography deals with lists of longitudes and latitudes for numerous localities and the associated world-maps fitted with proper longitude and latitude grids and with localities properly marked according to their coordinates. The history of this subject is now, thanks mainly to the groundbreaking research of Ted and Mary Helen Kennedy and, more recently, Fuat Sezgin, a recognized discipline within the history of Islamic science.[1] Whilst some of the earliest tables of coordinates from the 9th and 10th centuries have survived, no such early world-maps are preserved for us.[2]

---

* This study first appeared in *Islamic Thought in the Middle Ages – Studies in Text, Transmission and Translation, in Honour of Hans Daiber*, edited by Anna Akasoy and Wim Raven, Leiden & Boston: Brill, 2008, pp. 319–44. It has here been reset, and the bibliographical references have been updated.

[1] Kennedy & Kennedy, *Coordinates* (sometimes abbreviated K&K): Edward Stewart Kennedy and Mary Helen Kennedy, *Geographical Coordinates of Localities from Islamic Sources* (Frankfurt am Main: Institut für Geschichte der Arabisch-Islamischen Wissenschaften (hereafter IGAIW), 1987). See also E.S. Kennedy, 'Mathematical Geography', in Roshdi Rashed, with Régis Morelon, eds., *Encyclopedia of the History of Arabic Science*, 3 vols. (London: Routledge, 1996), I, pp. 185–201.

The most recent volumes of Fuat Sezgin, *Geschichte des arabischen Schrifttums*, 13 vols. to date (Leiden: E.J. Brill, 1967 onwards, since 2000 Frankfurt am Main: IGAIW), are X–XIII: *Mathematische Geographie und Kartographie im Islam und ihr Fortleben im Abendland*, 2 vols. and 1 map vol., 2000. The volume of Sezgin's monumental work dealing with the sources for mathematical and descriptive geography – XIII: *Autoren*, published in 2007 – does not mention the work of Ibn al-ʿAṭṭār.

The standard bio-bibliographical sources for Muslim astronomers and mathematicians after the period covered by Sezgin (V: *Mathematik*, 1974, VI: *Astronomie*, 1978, VII: *Astrologie, Meteorologie und Verwandtes*, 1979) are:

- Suter, *MAA*: Heinrich Suter, 'Die Mathematiker und Astronomen der Araber und ihre Werke', *Abhandlungen zur Geschichte der mathematischen Wissenschaften* 10 (1900), and 'Nachträge und Berichtigungen', *ibid.* 14 (1902), pp. 157–85, repr. Amsterdam: The Oriental Press, 1982, and again in Fuat Sezgin *et al.*, eds., *Heinrich Suter, Beiträge zur Geschichte der Mathematik und Astronomie im Islam*, 2 vols. (Frankfurt am Main: IGAIW, 1986), I, pp. 1–285 and 286–314.
- Brockelmann, *GAL*: Carl Brockelmann, *Geschichte der arabischen Litteratur*, 2 vols., 2nd edn. (Leiden: E.J. Brill, 1943–49), and Supplementbände, 3 vols. (Leiden: E.J. Brill, 1937–42).
- Storey, *PL*: Charles A. Storey, *Persian Literature – A Bio-bibliographical Survey*, vol. II, pt.1: A – *Mathematics*, B – *Weights and Measures*, C – *Astronomy and Astrology*, D – *Geography*, London: Luzac & Co., 1972.
- King, *Cairo Survey*: *A Survey of the Scientific Manuscripts in the Egyptian National Library* (American Research Center in Egypt, Catalogs, vol. 5) (Winona Lake, Ind.: Eisenbrauns, 1987).
- Rosenfeld & İhsanoğlu, *MAIC*: Boris A. Rosenfeld and Ekmeleddin İhsanoğlu, *Mathematicians, Astronomers and Other Scholars of Islamic Civilisation and their Works (7th–19th C.)* (Istanbul: Research Centre for Islamic History, Art and Culture (IRCICA), 2003), with a supplement in *Suhayl – Journal for the History of the Exact and Natural Sciences in Islamic Civilisation* (Barcelona) 4 (2004), pp. 87–139.

[2] Kennedy, 'Suhrāb and the World-Map of Ma'mūn', in *From Ancient Omens to Statistical Mechanics: Essays on the Exact Sciences Presented to Asger Aaboe*, J. Lennart Berggren and Bernard R. Goldstein,

a) *Mathematical geography in Iran and Central Asia*

In some recent publications I have drawn attention to the two most influential traditions of mathematical geography in the Islamic East – Greater Iran and Central Asia – that lasted from the 11th century at least until the 17th century.[3] A driving force behind the activity in these two traditions, often overlooked in certain modern writings, was the determination of the qibla for all localities in the world.[4] These two different traditions of geographical coordinates and the associated world-maps, alas mainly lost,[5] and qibla-directions are the following:

First we have the tradition of al-Bīrūnī (Central Asia, *ca.* 1050), followed by ʻAbd al-Raḥmān al-Khāzinī (Marw, *ca.* 1125), Sanjar al-Kamālī (Yazd, *ca.* 1300), Ibn al-Shāṭir (Damascus, *ca.* 1350), and some lesser authorities. al-Bīrūnī surely prepared several world-maps, and I have shown that one of them was used by al-Khāzinī.[6] Alas, none of these maps has survived. A world-map based on the coordinates of al-Khāzinī, themselves derived from a world-map by al-Bīrūnī, is preserved in a treatise on folk astronomy by

---

eds., *Acta Historica Scientiarum Naturalium et Medicinalium* (Copenhagen: University Library) 39 (1987), pp. 113–19. This important study has been overlooked in some modern writings on 9th-century cartography.

[3] See King, *World-Maps for Finding the Direction and Distance to Mecca: Innovation and Tradition in Islamic Science* (Leiden: Brill, and London: Al-Furqan Islamic Heritage Foundation, 1999), and *idem*, *SATMI = In Synchrony with the Heavens – Studies in Astronomical Timekeeping and Instrumentation in Islamic Civilization*, 2 vols., vol. 1: *The Call of the Muezzin* (Leiden & Boston MA: Brill, 2004), and vol. 2: *Instruments of Mass Calculation* (Leiden & Boston MA: Brill, 2005), especially VIIc.

[4] On the qibla see the article ʻḲibla. i. Ritual and legal aspects' by A.J. Wensinck in *The Encyclopaedia of Islam*, new edn., 12 vols. and index vol. (Leiden: Brill, 1960–2007), hereafter EI_2. My article ʻḲibla, ii. Astronomical aspects' in EI_2 (1979) is long outdated. In it I expressed my surprise that in spite of all of the activity amongst Muslim scientists to determine the qibla on the basis of geographical coordinates and accurate trigonometric formulae, the majority of medieval mosques from one end of the Islamic world to the other are oriented in directions that do not correspond to the computed values. By the time the article ʻMakka, iv. As centre of the world' appeared in 1987, the basic documentation of medieval techniques for finding the qibla by the legal scholars, based on folk astronomy and using astronomical horizon phenomena to face an astronomically-aligned Kaʻba, had been achieved. For a new survey of medieval qibla determinations see King, *SATMI*, vol. 1, pp. 741–71; on the consequences for Cairo see *ibid.*, pp. 773–823. For the situation in al-Andalus and the Maghrib, as described in texts on the sacred law and folk astronomy, see Mònica Rius, *La Alquibla en al-Andalus y al-Magrib al-Aqṣà* (Barcelona: Institut 'Millás Vallicrosa' de Història de la Ciència Àrab, 2000). For a survey of over 20 different medieval schemes of sacred geography – the notion of the world divided into sectors around the Kaʻba and the associated non-mathematical qibla directions – see King, *The Sacred Geography of Islam*, forthcoming, with a summary in the EI_2 article 'Makka. iv'. For a more detailed analysis of some Yemeni schemes of sacred geography see Petra Schmidl, *Volkstümliche Astronomie im islamischen Mittelalter – Zur Bestimmung der Gebetszeiten und der Qibla bei al-Aṣbaḥī, Ibn Raḥīq und al-Fārisī*, 2 vols. (Leiden & Boston MA: Brill, 2007), and for the earliest schemes see the contribution of Mónica Herrera Casais and Petra Schmidl to the *Daiber Festschrift* (see the reference on p. 1 above), namely: 'The Earliest Known Schemes of Islamic Sacred Geography' on pp. 275–300.

[5] Much credit is due to Joachim Lelewel for his reconstruction of various world-maps. See his *La géographie du moyen âge*, with an *Atlas composé de cinquante planches*, Brussels, 1850–57 (repr. in Fuat Sezgin *et al.*, eds., *Islamic Geography*, vols. 129–33 (Frankfurt: IGAIW, 1993)), especially the atlas volume: figs. II (Ibn Yūnus), V (al-Bīrūnī), XIX (al-Ṭūsī), XX (*Kitāb al-Aṭwāl*), XXI (Ibn Saʻīd al-Maghribī), XXII (al-Marrākushī), and XXIII (Abu 'l-Fidā'), also King, *World-Maps*, pp. 32–3, for his map based on the coordinates of al-Ṭūsī.

[6] King, *World-Maps*, pp. 41–2 and 71–5, as well as App. D on pp. 564–85.

Sirāj al-Dīn al-Sajāwandī (1210): this is a singularly wretched copy that could serve no practical purpose.[7] It is possible that some of the numerous surviving maps of Greater Iran are based on one or other of these eastern Islamic traditions.[8]

Second we have the tradition of the anonymous and enigmatic *Kitāb al-Aṭwāl wa-'l-'urūḍ li-'l Furs* (Isfahan? 11th century?), Naṣīr al-Dīn al-Ṭūsī (Maragha, *ca.* 1260), Ghiyāth al-Dīn al-Kāshī and Ulugh Beg (both Samarqand, *ca.* 1425), and some anonymous scholars who compiled an enormous table that included qibla directions and distances to Mecca (Kish near Samarqand, *ca.* 1450).[9] For only one of these various sources have any traces of a serious world-map survived: the coordinates in the anonymous tables from Kish underlie three remarkable world-maps engraved on brass that were made in Isfahan in the late 17th century. These display the direction and distance to Mecca at the centre, and Jan Hogendijk has found the evidence to confirm my hypothesis that the mathematics underlying them was known in 10th-century Baghdad and 11th-century Isfahan.[10] It is perhaps too much to hope that a precursor to the Safavid world-maps will show up. The geographical data was used in gazetteers on Iranian astrolabes until the early 18th century.[11]

b) *Mathematical geography in Egypt and Syria*

In my study of mathematical geography in the Islamic East, I remarked that the magnificent world-map on silk prepared in Fustat in the year 964, at a reported cost of 22,000 *dīnārs*, which has not survived, had – as far as we know – no successor in Egypt, or, for that matter, in Syria either.[12] Although the celebrated astronomer Ibn Yūnus (Cairo-Fustat, *ca.* 990) in his geographical tables relied heavily on those of al-Khwārizmī, and in the 13th century the tables of the Andalusī Ibn al-Zayyāt (d. 1058) were clearly available in Egypt, there was, with one notable exception, no serious work done on coordinates and no serious world-maps prepared, this in spite of a colourful tradition of mathematical astronomy and instrumentation in Mamluk Egypt and Syria.[13] The only exception was

---

7 See King, 'A World-Map in the Tradition of al-Bīrūnī (*ca.* 1040) and al-Khāzinī (*ca.* 1120) Presented by Sirāj al-Dīn al-Sajāwandī (1210)', *Mélanges offerts à Hossam Elkhadem par ses amis*, Frank Daelemans, Jean-Marie Duvosquel, Robert Halleux & David Juste, eds., *Archives et bibliothèques de Belgique / Archief- en bibliotheekwezen in België*, numéro spécial / Extranummer 83 (2007), pp. 131–60, reprinted in *idem*, *Islamic Astronomy and Geography* (Farnham: Ashgate–Variorum, 2012), XI. This appears to be the only evidence that al-Sajāwandī (see Suter, *MAA*, pp. 192; Brockelmann, *GAL*, I, pp. 470–71, and SI, pp. 650–51; King, *Cairo Survey*, nos. C4 = G11, Rosenfeld & İhsanoğlu, *MAIC*, no. 581), well known in Ḥanafī legal circles, worked in Egypt, though this is still questionable. Apparently no biographical information is available on him.

8 For some examples see King, *World-Maps*, pp. 141–8.

9 *Ibid.*, pp. 42–3 and 149–68, and King, *SATMI*, I, p. 846.

10 On two of these maps see King, *World-Maps*, pp. 195–364, as well as App. A on pp. 456–77 and App. C on pp. 552–63, and on the third, as well as on Hogendijk's discoveries, see King, *SATMI*, I, pp. 825–46.

11 King, *World-Maps*, pp. 175–86.

12 *Ibid.*, p. 35.

13 For the context see the already outdated study, King, 'The Astronomy of the Mamluks', *ISIS* 74 (1983), pp. 531–55, repr. in *idem*, *Islamic Mathematical Astronomy* (London: Variorum, 1986, 2nd rev. edn., Aldershot: Variorum, 1993), III, and the new insights in François Charette, *Mathematical Instrumentation in Fourteenth-Century Egypt and Syria – The Illustrated Treatise of Najm al-Dīn al-Miṣrī* (Leiden: Brill, 2003), pp. 5–31.

the tables of al-Marrākushī (Cairo, *ca.* 1280), who presented his own measurements of the latitudes of various localities between the Maghrib and Cairo.[14] I have investigated several Egyptian and Syrian tables of geographical coordinates, showing them to be mainly of little initiative and, where calculation was involved for qibla-values, rather carelessly computed.[15] A world-map with a corrupt longitude scale (but no latitude scale) and numerous localities marked in defiance of any reconstruction of the scale in a 13th-century copy of an anonymous geographical work *Kitāb Gharā'ib al-funūn wa-mulaḥ al-'uyūn,* probably compiled in Egypt in the late 11th century.[16] A world-map presented by the historian-encyclopaedist Ibn Faḍlallāh in his encyclopedic *Masālik al-abṣār ca.* 1340 (see Fig. 2) bears a sophisticated looking grid, but this is completely out of place on a map of the simple *mappa mundi* tradition, and not a single locality indicated on the map is positioned according to any geographical tables.[17]

Two Syrians, however, stand out in their activities relating to geographical coordinates. The first is the scholar-prince Abu 'l-Fidā' (Hama, *ca.* 1325). He documented over 1200 pairs of coordinates from several earlier sources, some of which are no longer extant. His list is critical in the sense that he carefully indicated his sources, but uncritical in the sense that it is devoid of comment. Nevertheless, his heart was in the right place. His list has been particularly useful in reconstructing part of the lost *Kitāb al-Aṭwāl wa-'l-'urūḍ*, which underlies the main Eastern Islamic tradition mentioned above.[18] The second Syrian of merit is Shams al-Dīn al-Khalīlī, who *ca.* 1360 calculated the qiblas of 40-odd localities in Syria and Palestine, mainly accurately to the nearest minute. Even this achievement pales in comparison with his table of almost 3,000 entries displaying the qibla to minutes, usually accurately, for each degree of latitude and each degree of longitude difference from Mecca in the entire Muslim world.[19] But his qibla table was not widely used after his time, not least because it was simply too complicated for practical purposes.

An Egyptian scholar, some 150 years after Abu 'l-Fidā', tried to carry out the same exercise, and his endeavour is the topic of the present study. But first we need to introduce the notion of the climates, which was of fundamental importance in Islamic mathematical geography.

c) *The fate of the climates of Antiquity in medieval Islam*

The Greeks divided the inhabited earth into seven latitudinal 'climates' (*klímata*) whose centres and boundaries are defined in terms of the length of longest daylight – see Fig. 1. The climates were of paramount importance not only in ancient and medieval mathematical geography, but also in mathematical astronomy and astronomical

---

14 See Kennedy & Kennedy, *Coordinates*, p. xxv *sub* MAR.

15 King, *World-Maps*, pp. 76–86 and Tables F1–F8 on pp. 600–622.

16 See Jeremy Johns and Emilie Savage-Smith, '*The Book of Curiosities*: A Newly Discovered Series of Islamic Maps', *Imago Mundi* 55 (2003), pp. 7–24, especially pl. 1 and pp. 11–13.

17 King, *World-Maps*, pp. 34–7.

18 Kennedy & Kennedy, *Coordinates*, pp. xix–xx *sub* FID.

19 King, 'Al-Khalīlī's Qibla Table', *Journal of Near Eastern Studies* 34 (1975), pp. 81–122, repr. in King, *Islamic Astronomical Instruments* (see n. 13), XIII; also King, *SATMI*, I, pp. 386–93.

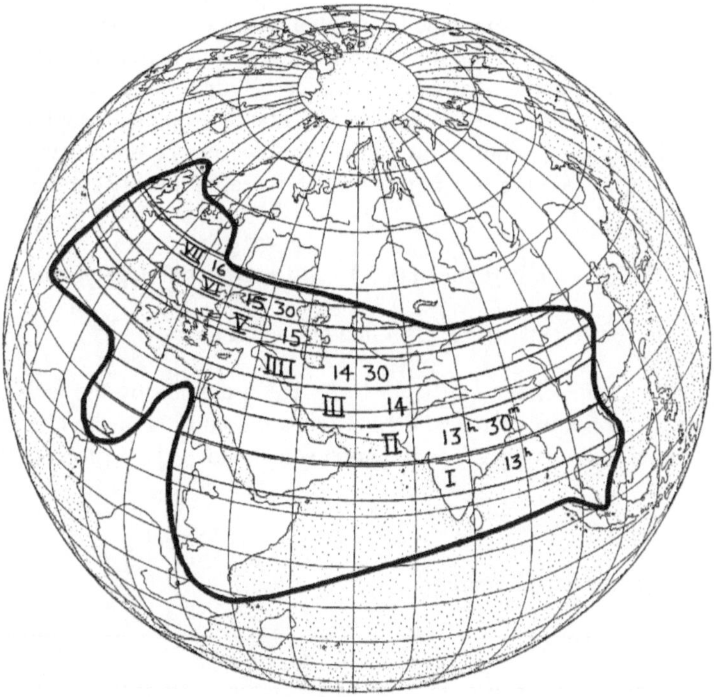

Fig. 1: The climates of Antiquity and the world known to Ptolemy. The world of Ibn al-ʿAṭṭār was the same, but he used a different definition of the climates. [From King, *World-Maps*, p. 24.]

instrumentation.[20] The problem with the climates defined by these criteria is that the latitudes of their centres and boundaries change by a few minutes over the centuries because the obliquity of the ecliptic changes slowly with time.[21]

[20] The standard work on the climates in Antiquity is Ernst Honigmann, *Die sieben Klimata und die πόλεις ἐπίσημοις – Eine Untersuchung zur Geschichte der Geographie und Astrologie im Altertum und Mittelalter*, (Heidelberg: Carl Winter's Universitätsbuchhandlung, 1929). Already in the Hellenistic world, new but unhappy definitions of the climates started to appear: see Burkhard Stautz, 'Die früheste bekannte Formgebung der Astrolabien', in Anton von Gotstedter, ed., *Ad radices – Festband zum fünfzigjährigen Bestehen des Instituts für Geschichte der Naturwissenschaften Frankfurt am Main* (Stuttgart: Franz Steiner, 1994), pp. 315–28, especially p. 318. For an introduction to the climates in Islamic descriptive geography see André Miquel's article 'Iḳlīm' in EI_2. On the importance of the climates in medieval Islamic and European science see King, *SATMI*, I, pp. 688–9, *etc.*, and *ibid.*, II, pp. 925–32.

[21] See the EI_2 articles 'Mayl' [= declination and obliquity] by D.A. King and 'Minṭaḳat al-burūdj' [= ecliptic and obliquity] by Paul Kunitzsch. The necessary technical background is provided in King, *SATMI*, I, pp. 27–38.

In this study we use the sexagesimal notation standard in the history of science; in the Islamic sources numbers are expressed in alphanumerical (*abjad*) notation.[22] Ptolemy had derived 23;51,20° for the obliquity and his latitudes for the climates are based on that parameter. Muslim astronomers, already in the early 9th century onwards, determined better values of this parameter, 23;33° and 23;35°, that of Ptolemy being long out of date. The maximum length of daylight *D* in equinoctial degrees or hours (where 360° = 24 hours, so that 1 hour = 15°) is a function of the terrestrial latitude (ϕ) and the obliquity of the ecliptic (ε):

$$D(\phi,\varepsilon) = 2\ \{\ 90^\circ + \sin^{-1}\ (\ \tan\varepsilon\ /\ \tan\phi\ )\ \} \quad \text{in equinoctial degrees, or}$$

$$2\ \{\ 90^\circ + \sin^{-1}\ (\ \tan\varepsilon\ /\ \tan\phi\ )\ \}\ /\ 15 \quad \text{in equinoctial hours}$$

Hence, for a specific climate C_n (n = 1, 2, ... , 7) defined by the length of longest daylight D_n, if the obliquity changes, then so does the latitude of the climate, $\phi_n(D_n,\varepsilon)$. The latitudes of the climates corresponding to the most widely used values of the obliquity are as follows:

Table 1 – The latitudes of the climates

	C_1	C_2	C_3	C_4	C_5	C_6	C_7
ε\max D	13^h	13;30	14	14;30	15	15;30	16
23;51°	16;27°	23;49	30;21	36; 1	40;53	45; 1	48;32
23;35°	16;39	24; 5	30;40	36;22	41;14	45;22	48;53

The climates presented no problem to most Muslim scientists. Thus, for example, al-Bīrūnī dealt with them in detail presenting a surplus of numerical information (based on ε = 23;35°) even on their areas.[23] In Greater Iran they seem to have remained as Ptolemy had defined them into the early modern period,[24] although different values based on different values of the obliquity were in circulation. However, the changing latitudes of the climates confused some scholars who were not completely in control of their subject

22 Sexagesimal numbers are written in the form p;q,r, representing p + q/60 + r/3600. On the Arabic *abjad* notation see Rida A.K. Irani, 'Arabic Numeral Forms', *Centaurus* 4 (1955), pp. 1–12, repr. in David A. King and Mary Helen Kennedy, eds., *E.S. Kennedy, Colleagues and Former Students: Studies in the Islamic Exact Sciences* (Beirut: American University of Beirut, 1983), pp. 710–21.

23 See Ahmad Dallal, 'Al-Bīrūnī on Climates', *Archives internationales d'Histoire des sciences* 34 (1984), pp. 3–18.

24 See King, *World-Maps*, pp. 230–34.

or who were bound by traditionalism.[25] The standard value of the obliquity used by Mamluk astronomers was 23;35°, derived by Ibn Yūnus *ca.* 990, but already used by his predecessors Ḥabash al-Ḥāsib and al-Battānī,[26] as well as al-Bīrūnī and Naṣīr al-Dīn al-Ṭūsī (Iran, *ca.* 1250), before he measured it properly. It is useful to compare the latitudes of the climates as presented by Ptolemy, and in two Islamic works, the *Tadhkira fī 'ilm al-hay'a* by al-Ṭūsī,[27] and the popular summary *al-Mulakhkhaṣ fi 'l-hay'a* by Maḥmūd ibn Muḥammad ibn 'Umar al-Jaghmīnī (*fl.* Khwarazm, *ca.* 1221/22),[28] whose complicated history has now been to some extent unravelled by Jamil Ragep.[29] In Table 2, adopted from Ragep, the latitudes of the middles of the climates are presented for Ptolemy (PT), based on $\varepsilon = 23;51°$, and al-Ṭūsī (TS), based on $\varepsilon = 23;35°$, and three different versions of al-Jaghmīnī: the 1246 Laleli copy (LL), the German translation (RH), and a commentary by Qāḍī Zāde al-Rūmī (QZ).[30]

In Egypt, some scholars proposed a different definition of the climates, to make them begin at the equator and embrace all latitudes up to the Arctic Circle. We find such aberrant definitions already in the geographical tables of Najm al-Dīn al-Miṣrī *ca.* 1325.[31] They are also implicit in the world-map of Ibn Faḍlallāh – see Fig. 2[32] – in which, however, the scale for the climates shows them to be of the same width. The centres of the Ptolemaic climates C_1, C_2, ... , C_6 become the beginnings of the Egyptian climates E_2, E_3, ... , E_7; also, E_1 begins at the equator, and E_7 ends at the Arctic Circle. It is perhaps significant that the historian Ibn Khaldūn *ca.* 1377 attributes this arrangement to Ptolemy himself, rather than recognizing it as a distortion of Ptolemy's scheme.[33]

---

25 This is one reason why, for example, many astrolabe-makers over the centuries preferred to keep on using, either explicitly or implicitly, Ptolemy's outdated value of the obliquity 23;51° in their presentations of the latitudes of the climates or the length of daylight for different latitudes. See King, *SATMI*, II, pp. 948–58.

26 See *ibid.*, I, pp. 44, 164 and 247.

27 F. Jamil Ragep, *Naṣīr al-Dīn al-Ṭūsī's Memoir on Astronomy* (al-Tadhkira fī 'ilm al-hay'a), 2 vols. (New York, *etc.*: Springer, 1993), I, pp. 250–53, and II, pp. 469–71. On his values for the climates see also King, *World-Maps*, p. 233.

28 Translated in G. Rudloff and A. Hochheim, 'Die Astronomie des Maḥmūd ibn Muḥammad ibn 'Omar al-Ǧagmīnī', *Zeitschrift der Deutschen Morgenländischen Gesellschaft* 47 (1893), pp. 213–75 (repr. in F. Sezgin, ed., *Islamic Mathematics and Astronomy*, vol. 77 (Frankfurt: IGAIW), 1998); see especially pp. 260–61. On the author see Suter, *MAA*, no. 403; Storey, *PL*, no. C88; King, *Cairo Survey*, no. G17; and Rosenfeld & İhsanoğlu, *MAIC*, no. 547. See also the next note and the text to n. 45 below.

29 See now Jamil Ragep, 'On Dating Jaghmīnī and his *Mulakhkhaṣ*', in M. Kaçar & Z. Durukal, eds., *Essays in Honour of Ekmeleddin İhsanoğlu*, 2 vols., Istanbul: IRCICA, 2006, pp. 461–6.

30 On Qāḍī Zāde see Suter, *MAA*, no. 430; Storey, *PL*, nos. A16 and C118; King, *Cairo Survey*, no. G50; and Rosenfeld & İhsanoğlu, *MAIC*, no. 808. Ragep cites an Iranian edition from *ca.* 1880 of his commentary on al-Jaghmīnī.

31 On his use of the climates see King, *World-Maps*, p. 79, where he states that the fifth climate runs from 31° to 36° (!!).

32 See *ibid.*, pp. 34–7. This map has nothing to do with the world-map of the Abbasid Caliph al-Ma'mūn, contrary to various claims in the modern literature. That map, which has not survived, is known to have had an orthogonal grid: see Kennedy, 'World-Map of Suhrāb' (n. 2 above). Besides, the sophisticated grid on Ibn Faḍlallāh's map is entirely superfluous, not least because no cities are marked on the map.

33 Franz Rosenthal, *Ibn Khaldūn – The Muqaddimah – An Introduction to History*, 3 vols. (Princeton, N.J.: Princeton University Press, 1958, 2nd edn., 1967), I, pp. 112–13.

Table 2 – The latitudes of the climates according to Ptolemy (ε=23;51°), al-Ṭūsī (ε =23;35°), and three versions of al-Jaghmīnī (after Ragep)

C	D max	PT	TS	LL	RH	QZ
C_1	13[h]	16;27° [0]	16;37,30° [a]	16;27	15;37 [!]	16;37
C_2	13;30	23;51 [+2]	24;5 [0]/24;40 [!] [b]	23;51	24;40 [!]	24;40 [!]
C_3	14	30;22 [+1]	30;40 [0]	30;22	30;40	30;40
C_4	14;30	36; 0 [-1]	36;22 [0]	36; 0	35;22 [!]	36;22
C_5	15	40;56 [+3]	41;15 [+1]	40;56	41;15	41;15
C_6	15;30	45; 1 [0]	45;21 [-1]	45; 1	45;21	45;21
C_7	16	48;32 [+1]	48;52,30 [c]	48;32	48;52	48;52
C_7 end	16;15	50; 4 [0]	50;20 [-5]	50;25 [!]	50;20	50;20

a: Accurately to seconds 16;38,48 (!!). b: Here the 40 minutes have probably been taken from the entry for C_3. c: Accurately to seconds 48;52,35°.

2. The *Geography* of Ibn al-ʿAṭṭār

Our source is a single manuscript, which shows how precarious and arbitrary our chances are of reconstructing history. MS HL 2520 (formerly 2469) in the Khoda Bakhsh Library in Patna is a collection of mainly astronomical treatises by a single author. The manuscript was catalogued in a masterful fashion in 1937 by Maulavi Abdul Hamid.[34] The author is the 15th-century Egyptian astronomer Muḥibb al-Dīn (Abū ʿAbdallāh) Muḥammad ibn Muḥammad and known as Ibn al-ʿAṭṭār, whose name is not new to the standard biobibliographical sources for late Islamic science.[35] Some of the treatises contained in it are known from other copies. However, fols. 11v–25r preserve for us a unique copy of

[34] Maulavi Abdul Hamid, *Catalogue of the Arabic and Persian Manuscripts in the Oriental Public Library at Bankipore*, vol. XXII (Arabic MSS): Science (Bihar, Patna: Government Printing, 1937), pp. 93–101 (no. 2469), especially pp. 94–5. These pages are full of information on Ibn al-ʿAṭṭār.

[35] The first mention of Ibn al-ʿAṭṭār in the modern literature (see n. 1) is in Suter, *MAA*, no. 431, listing his treatise on different types of quadrants and a set of astronomical tables in Oxford (see n. 37 below). Brockelmann, *GAL*, II, pp. 157–8, and SII, p. 158, mentions all of the relevant treatises in the Patna manuscript. King, *Cairo Survey*, no. C66, mentions his treatise on quadrants, his treatise on the construction of curves on astrolabe plates, and a lost treatise on astronomical timekeeping entitled *Jawharat al-yawāqīt*, albeit associated with Shams al-Dīn Muḥammad ibn al-ʿAṭṭār al-Bakrī al-Shāfiʿī (see further *ibid.*, no. C124 *ad* 3.1.28). Maybe

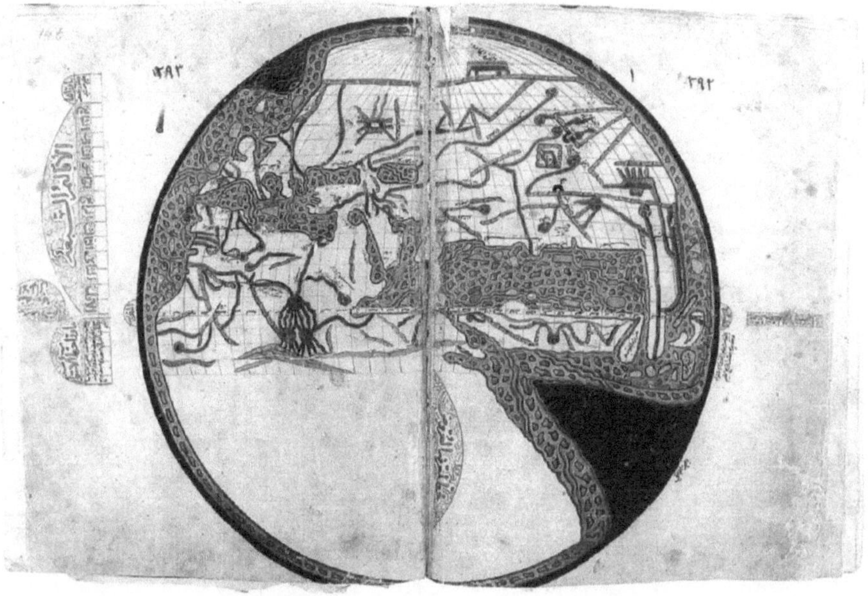

Fig. 2: The climates in the world-map presented by Ibn Faḍlallāh are defined in a different fashion from those in Fig. 1. The unhappy scale to the left of the map divides the climates more or less equally between the equator and the Arctic Circle. [From King, *World-Maps*, p. 34.]

Ibn al-ʿAṭṭār's geographical work which has no title, but begins: *hādhhi ʿurūḍ al-bilād wa-aṭwāluhā wa-nḥirāfuhā ʿani 'l-jihāt fī 'l-aqālīm al-sabʿa* ... , 'These are the latitudes and longitudes of localities in the seven climates and their qibla directions with respect to the cardinal directions'. See Fig. 3. This text was copied in 874 H [= 1469/70] – see the Appendix for the colophons. Ibn al-ʿAṭṭār has already attracted our attention as the author of a treatise entitled *Kashf al-qināʿ fī rasm al-arbāʿ*, which deals with different kinds of quadrants, and reveals that he was in touch with the impressive developments in earlier Mamluk instrumentation; indeed, he had studied with some of the leading astronomers of Cairo.[36] Nevertheless, the works attributed to him in the Patna manuscript need to be investigated more thoroughly, not least to ascertain whether or not it is the same individual who also wrote treatises in Damascus.[37]

---

this is the father of our author? Rosenfeld & İhsanoğlu, *MAIC*, no. 813, has no 'new' materials, but conveniently lists all of the 'old' ones.

36 On his activities within the context of Mamluk astronomy see Charette, *Mathematical Instrumentation* (cited in n. 13), pp. 7, 19–20, *etc*. On his treatise on quadrants see *idem* & King, *The Universal Astrolabe of Ibn al-Sarrāj*, forthcoming. Charette has located a detailed commentary on this treatise by Ibn al-ʿAṭṭār himself, which remains to be studied: see his 'The Locales of Islamic Astronomical Instrumentation', *History of Science* 44 (2006), pp. 123–38.

37 MS Oxford Bodleian Laud. Or. 221 (Uri I.974) contains an extensive set of planetary tables, stated to be for the longitude of Damascus, said to be based on those of the celebrated 14th-century astronomer Ibn al-Shāṭir and attributed to (and copied by?) Muḥammad ibn al-ʿAṭṭār. The date mentioned in the instructions on the first page is 930 H. The mean-motion tables begin with the date 870 H and have entries for each 30 years.

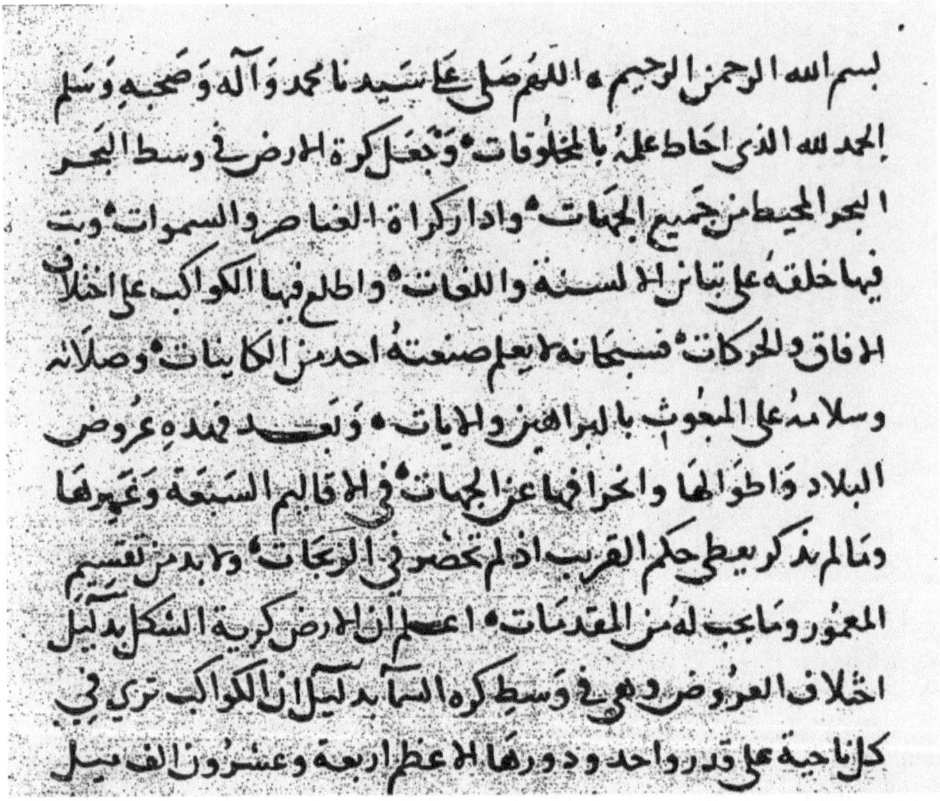

Fig. 3 The beginning of the treatise [Figs 3–8 are courtesy of the Khoda Bakhsh Library.]

The copy of Ibn al-ʻAṭṭār's *Geography*, and various copies of other works of his in the Patna manuscript are in the hand of another Egyptian astronomer (Shihāb al-Dīn) Aḥmad ibn Aḥmad ibn Timurbāy al-Ḥanafī al-Qādirī, a student (*tilmīdh*) of Ibn al-ʻAṭṭār, who read it back to his teacher.[38] Ibn Timurbāy's elegant *naskhī* script sometimes degenerates

For each of the five planets there are double-argument equation tables. Dr Benno van Dalen kindly showed me a microfilm of the Oxford manuscript.

38 King, *Cairo Survey*, no. C94, and Rosenfeld & İhsanoğlu, *MAIC*, no. 908, mention *al-Barq al-sāṭiʻ fī mukhtaṣar al-Bāriʻ*, a commentary on the treatise of the astrologer Ibn Abi 'l-Rijāl (also Sezgin 1979, VII, p. 187); a second astrological treatise *Risāla fī Maʻrifat al-ḥawādith al-sufliyya min dalālāt al-ashkhāṣ al-ʻulwiyya*; an astrological treatise copied in 894 H in his own hand; and some solar and lunar tables in his hand. See also the next note on another manuscript copied by him. Perhaps our man was a grandson of the Mamluk *amīr* Tīmūrbughā Mintāsh. See also the text to note 68 below.

The vowelling of the Turkic name T-m-r-Bāy or T-m-r-Bughā has caused some problems. In the *Patna Catalogue* (n. 34) our author becomes 'Aḥmad Tamirtāshī' (p. 95) and 'Aḥmad ibn Muḥammad al-Tamirtāshī' (p. 96). In Sezgin 1979, VII, p. 187, he becomes 'Aḥmad ibn Tīmūrbughā (or Tamrīta)". In Rosenfeld & İhsanoğlu, *MAIC*, no. 908, he becomes Aḥmad ibn Tamirbuqā. In the *Encyclopedia of Islam* we encounter the Mamluk sultan (al-Ẓāhir) Temirboghā (*reg.* 1467-68) (IV, col. 462b), and the Mamluk *amīr* Tīmūrbughā Mintāsh *ca.* 1390 (VI, cols. 548a and 580b), but in the Index these become Tīmūrbughā. It seems that in Arabic

completely in his marginalia (see Fig. 8), and we have encountered worse examples of his handwriting elsewhere.[39] In the treatise and tables all numerals are written in the standard Arabic alphanumerical notation,[40] and Ibn Timurbāy has sometimes succumbed to all of the attendant dangers of possible copyists' mistakes because of his careless handwriting.[41]

The potential historical importance of Ibn al-'Aṭṭār's geographical work was announced in 1999.[42] It attracted our attention again more recently because it is cited by the 16th-century Tunisian scholar 'Alī al-Sharafī al-Ṣafāqusī in the text accompanying his sea atlas.[43] The work does little to enhance Ibn al-'Aṭṭār's reputation, for it is naïve and ill informed. Furthermore, Ibn Timurbāy distinguishes himself by his occasional dreadful and careless handwriting. What concerns us here are, firstly, Ibn al-'Aṭṭār's statements about the latitudes of the climates and Ibn Timurbāy's marginalia, which may also be due to the author, since Ibn Timurbāy read the text back to his teacher, and secondly, their joint presentation of a substantial corpus of geographical data.

3. Ibn al-'Aṭṭār on the climates

Ibn al-'Aṭṭār uses the climates as defined by, say, Ibn Khaldūn.[44] When he writes 'the nth climate: its latitude is from the latitude of the previous climate to ϕ_n degrees', the values ϕ_n are in fact *the middles* of each of the Ptolemaic climates. Also he writes 'the latitude of the first climate is from the equator to ϕ_1 degrees', and 'the limit of (the seventh climate) is latitude 66;25°'. So the climates have been stretched to fill all the space between the equator and the Arctic Circle. In addition, since the Arctic Circle is at latitude 66;25° = 90° – 23;35°, we might expect that our author would use obliquity 23;35° throughout.

In any case, the latitudes associated with each of the climates are shown in Table 3, with some marginal additions labelled Mg. Other marginal insertions, which are labelled in the text *(al-)Mulakhkhaṣ*, here abbreviated Mu, indicate that our astronomers have compared the values with those in the very popular summary of Ptolemaic astronomy entitled *al-Mulakhkhaṣ fī 'l-hay'a* by al-Jaghmīnī.[45] We here distinguish between the Laleli copy, LL, the German translation, RH, and Qāḍī Zāde's commentary, QZ. We sometimes refer to Ptolemy's values, PT, based on ε = 23;51°, which are essentially as found in LL, whereas al-Ṭūsī's values, TS, are based on ε = 23;35°, and recur in RH and QZ (see Table 2 above). In some cases, values for the climates are given in the geographical tables that follow (§4), here abbreviated GT. In the accompanying notes, the reader should bear in

---

the long vowels of Tīmūr (meaning "iron or steel" in Turkish) could be shortened when followed by such appendages, as in Timurtāsh and even Timurtash. In modern Turkish the orthography Temür is preferred.

39 See King, *Cairo Survey*, pl. LXXXIII on p. 303, illustrating a fragment of an astrological history, as yet unstudied, of the campaigns between the Mamluks and Ottomans around Adana during the period 1470–73.

40 See n. 22.

41 On copyists' errors in geographical tables see King, *World-Maps*, pp. 161–3.

42 *Ibid.*, p. 83.

43 This is the topic of a doctoral dissertation by Mónica Herrera Casais at the Institute of History of Science in Frankfurt and the Universidad de La Laguna (Tenerife).

44 See the text to n. 33 above.

45 See n. 28 above.

mind that the phrase 'has been computed for a certain value of the parameter ε' means 'lifted from a source in which the latitudes were based on that value for ε':

Table 3 – The latitudes of the climates given by Ibn al-'Aṭṭār

E_1 begins at latitude 0°	This confirms that we are dealing with the Egyptian definition of the climates.
Beginning of E_1 (middle of C_1): 16;25°, confirmed in GT, Mg: (16);35°, Mu: 16;37°	16;25° has been computed for ε = 23;51° (accurately 16;27°, as in PT/LL) or is a copyist's error for 16;35°; 16;35° has been computed for ε = 23;35° (accurately 16;39°). 16;37°, probably originally rounded from TS 16;37,30, agrees with QZ.
Beginning of E_2 (middle of C_2): 24;40°, Mg: 23;50°, Mu: 24;40° or 23;50° (unclear)	24;40° is an aberrant error – perhaps resulting from some copyist combining the degrees of 24;5° of C_2 in TS and the minutes of 30;44 for C_3 – which is found in copies of TS and recurs in RH/QZ. 23;50° is probably rounded from 23;51°, which is found in PT/LL.
Beginning of E_3 (middle of C_3): 30;40°, Mu: 30;40°	30;40° has been accurately computed for ε = 23;35°, and is found in TS/RH/QZ.
Beginning of E_4 (middle of C_4): 36;22°, Mu: 36;22°; GT: 36;0°	36;22° agrees with TS/QZ and has been accurately computed for ε = 23;35°. 36;0° is computed for ε = 23;51° (accurately 36;1°) and agrees with PT/LL.
Beginning of E_5 (middle of C_5): 40;15°, Mu: 41;15°, GT: 40;55°	40;15° is probably an error for 40;55°, perhaps rounded from the 40;56° in PT/LL; 41;15° is computed for ε = 23;35° and is found in TS/RH/QZ.
Beginning of E_6 (middle of C_6): 45;0°, confirmed in GT, Mg: (45);21°; Mu: 45;21°	45;0° has been computed for ε = 23;51° or rounded (the accurate value is 45;1°, as in PT/LL); 45;21°, as in TS/RH/QZ, is computed for ε = 23;35° (accurately 45;22°).
Beginning of E_7 (middle of C_7): 48;30°, confirmed in GT, Mu: (48);52°, Mg: 50;25°	48;30° has been computed for ε = 23;51° (accurately 48;31°) or miscopied from 48;32° as in PT/LL. 48;52°, as in TS/RH/QZ, was computed for ε = 23;35° (accurately 48;53°). 50;25° is the upper limit of C_7 according to the Ptolemaic definition and is accurately computed for ε = 23;35°: TS/RH/QZ have 50;20°, but LL has 50;25°.
E_7 ends at latitude 66;25°	The end of E_7 at 66;25° implies an ε of 23;35°, still using the Egyptian definition of the climates.

In other words, the information presented here is a hopeless jumble (by the higher standards of Islamic mathematical geography in contemporaneous Iran and Central Asia).

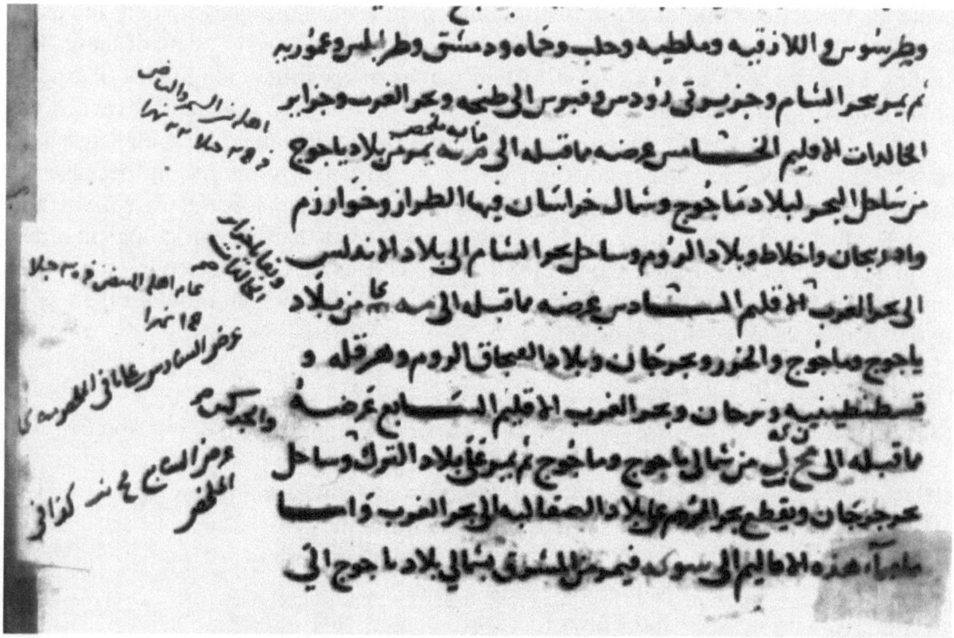

Fig. 4: The information on the 5th, 6th and 7th climates.

The latitudes in the text and in the marginalia have been taken from at least two separate sources and have been shuffled a few times.

The numerical information on the climates is followed by lists of localities in each one – see Fig. 4; these will not concern us here, save to note that they show no advance over the previous 600 years.[46] All this does not reflect well on either the author or the copyist, and it does not augur well for the geographical tables that follow.

4. Ibn al-ʿAṭṭār's geographical data

The text is followed by a table of longitudes, latitudes and qiblas of about 425 localities, in which some of the entries are left blank and others are given in duplicate and more, in one case, even quintuplicate. There are 12 folios of carefully-framed tables with 30 entries on each folio, and the other entries are given in the margins. To edit this properly would be a nightmare and there are far more urgent tasks awaiting us.[47] As an example

[46] We present one set of data here, namely, the localities in the 'third climate' on fol. 12v: 'Northern China and India, including Qandahar, Sind, then Kabul, Kirman, Sijistan, Mahdiyya, coast of al-Basra, including Istakhr (Persepolis), Siraf, Shiraz, Ahwaz, Sapur, then al-ʿIrāq, including Wasit, Anbar, south of Kufa and Basra, Baghdad, Syria or Damascus (*al-Shām*), Acre, Jerusalem, Gaza, then Madyan [= Madyan Shuʿayb], Qulzum, Cairo, and the south of Alexandria, Barqa, south of Ifriqiya, Kairouan, to the Maghrib and the encompassing Ocean (*al-baḥr al-muḥīṭ*).'

[47] Most other late medieval Egyptian geographical tables are full of errors of one sort or another: see King, *World-Maps*, pp. 76–84, and various tables edited on pp. 600–622.

of the challenge awaiting anybody who would venture into the geographical tables of Ibn al-ʿAṭṭār, we mention just seven sets of entries: see Figs. 5–6 for some of these. The heading indicates that we can expect to find names of localities, longitudes, latitudes, qiblas (*inḥirāf*), and their general directions (*jiha*), that is, the quadrants between the cardinal points. We use L for longitude, ϕ for latitude as before, and q for qibla, measured from the local meridian. The values for Mecca are particularly interesting because for the former their sources are identified, if cryptically, with the four labels *Qānūn*, Ibn Saʿīd, Kūshyār and *Rasm*. These, and other relevant sources, here in chronological order, are abbreviated as follows, in the tradition of the corpus of Islamic geographical data compiled by Ted and Mary Helen Kennedy (hereafter K&K), which provides the key to any investigations of Islamic coordinates.[48]

PTO	Ptolemy (*ca.* 140)
RSM FID	*Rasm al-rubʿ al-maʿmūr* (*ca.* 820), cited by Abu 'l-Fidā' (FID) (K&K: RES)
KHU	the *Kitāb Ṣūrat al-arḍ* by al-Khwārizmī (*ca.* 830)
KHW	other sources for al-Khwārizmī
KHZ	a related list in an Istanbul manuscript
SUH	Suhrāb (*ca.* 930)
YUN	Ibn Yūnus (*ca.* 990)
KUS	Kūshyār ibn Labbān (*ca.* 1000)
BIR	*al-Qānūn al-Masʿūdī* of al-Bīrūnī (*ca.* 1025)
BIR FID	values attributed to BIR by Abu 'l-Fidā' (FID)
ATW	The anonymous *Kitāb al-Aṭwāl wa-'l-ʿurūḍ li-'l-Furs*, compiled in Isfahan (?) in the 11th (?) century (K&K: ATH)
ATW FID	values attributed to ATW by Abu 'l-Fidā' (FID)
SNJ	the reconstructed values in the *Sanjarī Zīj* of ʿAbd al-Raḥmān al-Khāzinī (see King, *World-Maps*, p. 71–5 and 564–85)
ZAY	the *Kitāb Aḥkām al-marjān* ... by Isḥāq ibn al-Ḥasan al-Zayyāt (*ca.* 1040)
SAA	the *Kitāb Basṭ al-arḍ fi 'l-ṭūl wa-'l-ʿarḍ* of Ibn Saʿīd al-Maghribī (d. 1286)
SAA FID	values attributed to SAA by Abu 'l-Fidā' (FID)
MAR	al-Marrākushī (*ca.* 1280)
BNA	Ibn al-Bannā' (*ca.* 1300)
LYD	a 14th-century Egyptian list, being essentially a selection from ZAY, but with a list of the Egyptian pilgrim-stations appended
FID	Abu 'l-Fidā' (*ca.* 1325), cites RSM FID, ATW, BIR, SAA, amongst others
SHA	Ibn al-Shāṭir (*ca.* 1350), derived from SNJ
AIN	the *Āʾīn-i Akbarī* of Abu 'l-Faḍl ʿAllāmī (India, *ca.* 1580), derived from ATW and its derivatives

[48] Kennedy & Kennedy, *Coordinates* (see n. 1 above). The abbreviations are on pp. xv–xxxvii.

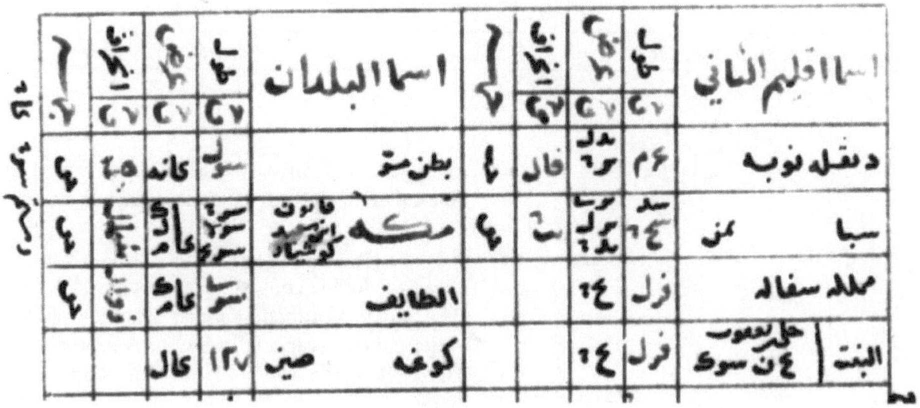

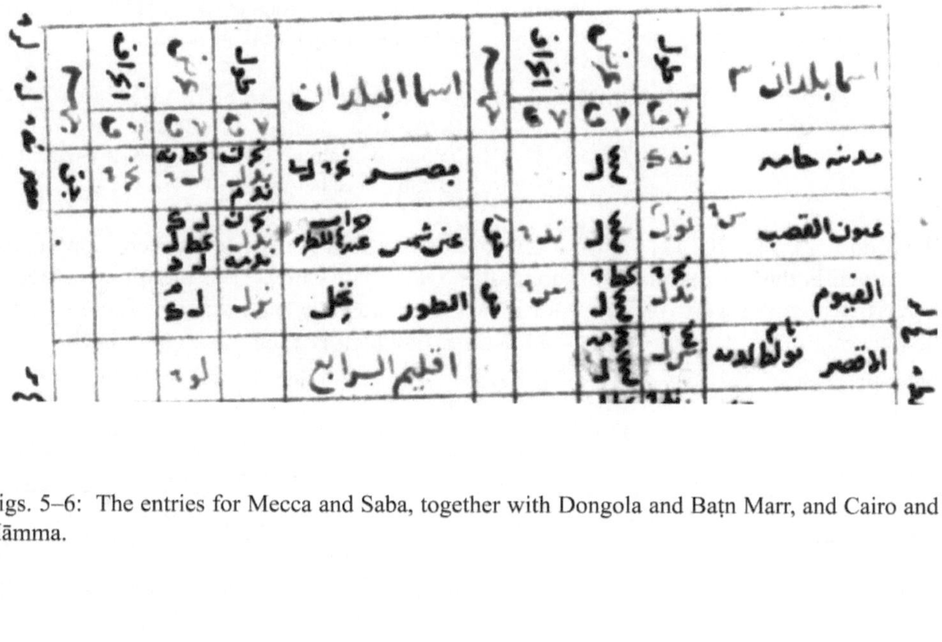

Figs. 5–6: The entries for Mecca and Saba, together with Dongola and Baṭn Marr, and Cairo and Ḥāmma.

Table 4 – The latitudes of various localities given by Ibn al-'Aṭṭār

Locality, L, ϕ, q, stated source				*Comments*
Mecca[49]				
67; 0°	21;20°	N[!]SE[!]	BIR	as in BIR, confirmed by BIR FID, also in SNJ and SHA. On the entry for the qibla see below.
67;31	21;30	–	SAA	SAA & SAA FID have 61;31°/21;31°, which is clearly a double copyist's error for 61;0°/21;0°.[50] The latter coordinates are as found in several other sources, especially the derivative SAA & BNA. It is the value 21;30° that surprises us here.
67;13	21;40	–	KUS	KUS has 67;10°/21;40°, which is repeated in some dependent sources. ϕ = 21;40° is an early value, also found in ATW: see below.
67; 0	21; 0	–	RSM	as in RSM FID, KHU, SUH, YUN, also ZAY & BNA. These are the coordinates used by Ibn Yūnus to compute the qibla for Cairo-Fustat at 55°/30° – see below.

The values 21°, 21;20°, 21;30° and 21;40° for the latitude of Mecca were apparently all derived in the 9th century, by whom, alas, we do not know.[51] For the qibla at Mecca there is written the word *shamāl*, 'north', which is absurd,[52] and this is associated with the quadrant of south-east (*j-q* for *janūbī-sharqī*), which is even more absurd. The same quadrant is associated with the next entry, for Taif, which has *zawāl*, 'meridian', for the qibla.

Locality, L, ϕ, q			*Comments*
Cairo/Fustat[53]			
53; 5°	30;10°	13; 0° SE	written by the name *Miṣr*; no other source has these values.

[49] Kennedy & Kennedy, *Coordinates*, pp. 225–6.

[50] The symbols for '0', of Greek origin, have been mistaken for the vaguely similar ligature for *lām-alif*: see Irani , 'Arabic Numeral Forms' (n. 40), pp. 11–12.

[51] See King, 'Too Many Cooks ... – A Newly-Rediscovered Account of the First Muslim Geodetic Measurements', *Suhayl – Journal for the History of the Exact and Natural Sciences in Islamic Civilisation* (Barcelona) 1 (2000), pp. 207–41 (reprinted in *idem*, *Islamic Astronomy and Geography* (Farnham: Ashgate–Variorum, 2012, X), especially pp. 225–6.

[52] One would expect a phrase such as *qiblat al-'ālam*, 'qibla of the whole world".

[53] Kennedy & Kennedy, *Coordinates*, pp. 111–12.

Locality, L, ϕ, q			*Comments*
[6]4;50	29;15	–	MAR has 64;50° / 29;55°; no other sources have longitudes with 50 minutes; KHZ has ϕ = 29;55° and SUH has 29;15°. On ϕ = 29;15° and 29;55° see below.
-;30	30; 0	–	no other sources have longitudes with 30 minutes; ϕ = 30° common after YUN.
34;40	[30;0°]	–	no ϕ is given; KHZ has L = 54;40°, which explains the absurd copyists' error; several later sources have 54;40°, with ϕ = 30;0°.
55; 0	30; 0	37; 0° SE	YUN, BNA, and various Yemeni sources, have these coordinates. On the qibla value see below. What could be taken as a source may simply be the word *Miṣr* repeated.

The value 29;15° for the latitude of Cairo seems to a careless mistake for the 29;55° used by al-Marrākushī.[54] Furthermore, two different values for the qibla at Cairo are presented without comment: they are 13°, a careless scribal error for 53°, and 37°, the former being measured from the meridian and the latter from the east west line. The single direction is that calculated by Ibn Yūnus *ca.* 990 (for coordinates 55°/30° for Cairo and 67°/21° for Mecca), which was generally favoured by the Mamluk astronomers.[55]

Locality, L, ϕ, q			*Comments*
Damascus[56]			
60; 0	33;30	31;10 / 29; 9	BIR has these coordinates, as well as various late sources. On the two qibla values see below.

The qibla value of 29;9° appears to be a scribal error for 29;4° (*ṭā'* for *dāl*!), the value of al-Khalīlī for these coordinates and 67;0°/21;30° for Mecca.[57] The other value does not correspond to any known medieval values.

Two lesser localities are featured in the extract in Fig. 5, namely, Dongola in Nubia and Baṭn Marr near Mecca. Also, as was the case with some earlier Mamluk Egyptian tables, all manner of irrelevant localities – places that had been destroyed or abandoned centuries before – are listed. Modest amounts of information are given for two such localities of

[54] Although it may be that the 29;15° is an older value: see King, *SATMI*, II, p. 932, n. 25.

[55] *Ibid*.,, I, pp. 268–70 and 804.

[56] Kennedy & Kennedy, *Coordinates*, pp. 99–100.

[57] King, *World-Maps*, entries F4-93 on p. 615 and F7-14 on p. 620, and *idem*, *SATMI*, I, entry 14 on p. 392.

this kind – Saba' in the Yemen (Sheba, modern Marib) and Madīnat Ḥāmma in Southern Ifrīqiya – featured in Figs. 5–6, but the details are of considerable historical interest for other reasons, not least for identifying the source of the data:

Locality, L, ϕ, q			*Comments*
Dongola[58]			
53;40°	14;30°	–	as in SUH and probably ATW FID (K&K give *L* = 43;40°); BIR has 53;40°/14;0°
53;40	17; 0	81;30 SW[!]	ϕ as in ZAY and MAR; ZAY has *L* = 53;0°. On the qibla value see below.
Baṭn Marr[59]			
66;30	21;55	5; 0 SE	only as in ATW FID, as in AIN, a late Indian source relying on ATW (now with 77;0°/21;55° and a meridian 10° to the west). On the qibla value see below.
Saba'[60]			
64°	17;10°	–	as in KHU & RSM FID; PTO has 73;40°/17;10°.
68; 0	17;30	12; 0 SE[!]	no sources have ϕ = 17;30°. On the qibla value see below.
68;0	14; 0	–	no value given for *L*; ATW FID & BIR have 68;0°/14;0°.
Ḥāmma[61]			
10;20°	28;30°	–	this locality occurs only in PTO (33;40°/28;15°), KHU (34;30°/28;30°) and SUH (34;20°/28;30°).

The qibla-value for Baṭn Marr is attested in an anonymous 14th-century list of stations on the pilgrim road from Cairo to Mecca (LYD), where it is associated with other coordinates (to which it does not correspond).[62] The qibla of Dongola is between 70° and 75° in other Egyptian sources.[63] The qibla of Saba' is absurd, and bears no relation to medieval

58 Kennedy & Kennedy, *Coordinates*, pp. 109–10 (two entries).

59 *Ibid.*, p. 87.

60 *Ibid.*, pp. 289–90.

61 *Ibid.*, p. 133 (confused?).

62 King, *World-Maps*, entry F4-M11 on p. 618.

63 *Ibid.*, entries F2-51, F3-84, and F4-20, on pp. 604, 609 and 613.

Egyptian values.[64] The longitude given for Ḥāmma shows that our two astronomers had no idea where Ptolemy intended it to be.

Finally, we consider the entries at the bottom of the last page of tables: see Fig. 8. These are:

Locality, L, ϕ, q			*Comments*
Hiraqla (Ereğli)[65]			
57;20°	46;35°	29;20°	ATW has 57;22°/46;30°. No other early sources have *L* between 57° and 58°. ϕ = 46;35° attested in KHU, KHZ, SUH, BIR, *etc*. The qibla value is lifted from SNJ (or one of its derivatives): there, however, it is associated with coordinates 53;25°/46;35°.[66]
Yājūj *al-dākhila* ('Inner Gog')[67]			
172;30°	63;0°	–	KHU has these coordinates for Yājūj *al-dākhila*, which locality does not occur elsewhere. SUH and various later sources have these coordinates for Yājūj and Mājūj.

In brief, the information presented is confused, and more of it would need to be evaluated in order to identify all of the various sources that have been used. However, this would probably be an exercise with ever diminishing returns. With but one exception, Mecca, no sources are named, but these are not the sources used for other localities. The fact that there is not even a clear statement of the qibla at Cairo, and the fact that al-Khalīlī's qibla value for Damascus has been miscopied, are poor reflections on Ibn al-'Aṭṭār and his student.

5. Ibn Timurbāy on distances between cities

By his colophon on fol. 20v, Ibn Timurbāy has penned the following remark (see fig. 8):

> "If you multiply the number of degrees and fractions thereof between two localities by fifty six and two thirds, the result will be the number of miles between them."

This conversion factor was attributed by al-Farghānī (Baghdad, *ca.* 850) to the earliest Muslim geodetic measurements in the early 9th century. al-Bīrūnī regarded it with

---

64 *Ibid.*, entry F2-52 on p. 604.

65 Kennedy & Kennedy, *Coordinates*, pp. 138–9.

66 King, *World-Maps*, entry 36 on p. 567.

67 Kennedy & Kennedy, *Coordinates*, p. 377.

some suspicion since it is slightly different from the values in the actual reports of these measurements.[68] It would be interesting to know Ibn Timurbāy's source for the value. Inevitably, he does not present the complicated formula for finding the distance in degrees between two localities, which had been known for centuries.

6. Concluding remarks

Almost 500 years had passed since al-Bīrūnī had written on the divergences between the coordinates in different sets of geographical tables that were in circulation in his time. Alas, his writings on this topic have not survived.[69] Almost 150 years had passed since the scholar-prince Abu 'l-Fidā' had noted different sets of coordinates in his *Geography*, properly identifying his sources. But one cannot conduct such research seriously if one is using and / or producing manuscripts in which the numbers are not uniquely identifiable or already hopelessly corrupt anyway. Scholars like Ibn al-'Aṭṭār and Ibn Timurbāy were amongst the leading scholars of their time, but they fell victim to the manuscript tradition. Certainly, their hearts were in the right place, but circumstances beyond their control rendered their efforts little more than an exercise in futility. And although they surely did not realize it, they were contributing to the decline of Islamic science. In this case, we are witness to an uncritical handling of a superfluity of uncontrolled data, without regard to the reliability of the sources, which could not be checked anyway. Not even the two apparently inconsistent values for the qibla at Cairo, one measured from the south (and miscopied) and the other measured from the east, received a comment.

Mathematical geography in medieval Europe was not much better, because there the coordinates in the *Toledan Tables*, a careless mix of values from al-Khwārizmī and al-Battānī, reigned supreme until the recovery of Ptolemy's original coordinates in the 15th century.[70] These then had to be corrected, which was a long and complicated procedure that sometimes involved recourse to Islamic lists. In the mid 18th century the German scientist Tobias Mayer drew maps of Europe based on the different sets of coordinates available to him, graphically demonstrating the need for new geodetic measurements. Such important undertakings were very much in the spirit of al-Bīrūnī and the compilers of the enigmatic *Kitāb al-Aṭwāl wa-'l-'urūḍ li-'l-Furs* in the 11th century, as well as poor Ibn al-'Aṭṭār in the 15th.

*****

---

68 King, 'Earliest Geodetic Measurements' (n. 51), pp. 216 and 219.

69 They are listed in King, *World-Maps*, p. 344, n. 65 (after Boilot).

70 See, for example, Noel Swerdlow, 'The Recovery of the Exact Sciences of Antiquity: Mathematics, Astronomy, Geography', in Anthony Grafton, ed., *Rome Reborn – The Vatican Library and Renaissance Culture* (Washington, D.C.: Library of Congress, 1993), pp. 125–68.

Appendix: More information on Ibn al-ʿAṭṭār

The colophons of the text of the treatise and the geographical tables (fols. 17v and 20v – see Figs. 7–8) are of some historical interest. They read:

> The copyist completed the reading of this text before its author (*balagha kātibuhu qirā'ata*[*hu*?] *'alā mu'allifihi*), the *shaykh*, *imām*, and scholar (*'ālim*) and luminary (*'allāma*) Muḥibb al-Dīn Muḥammad ibn Muḥammad ibn Aḥmad ibn Aḥmad ibn Muḥammad ibn Mushrif, whose noble ancestor was known as Ibn al-ʿAṭṭār – may God treat him with kindness and with the blessing of His chosen Prophet and (all) the Muslims – in the months of the year 874 Hijra [= 1469/70]. This was said and written (? *qālahu wa-katabahu* ?) by Aḥmad ibn Aḥmad ibn Timurbāy – may light clothe his sleep (? *w-l-b-s* ? *nawmahu al-nūr* ?)[71] – al-Ḥanafī al-Qādirī – may God forgive him and his parents and the Best of the Prophets and His Family. God knows best.

> Completed by the hand of the needy servant of God, Aḥmad ibn Aḥmad ibn Timurbāy al-Ḥanafī al-Qādirī, student of the scholar Muḥibb al-Dīn Ibn al-ʿAṭṭār, may God treat both me and him with His hidden (? *khafī*) kindness, [and] His favour and generosity, in Rajab *al-fard* of the year 874 (Hijra) [= January, 1470].

Elsewhere in the Patna manuscript, at the end of a work labelled *al-Fuṣūl* on the coordinates for marking astrolabes and quadrants, Ibn Timurbāy writes:[72]

> This copy was checked against the manuscript of the author, then its coyist read it back to the author, who is the *shaykh*, the *imām*, the scholar and luminary Muḥibb al-Dīn Muḥammad ibn Muḥammad ibn Aḥmad ibn Aḥmad Ibn Muḥammad ibn Mushrif, whose ancestor was known as Ibn al-ʿAṭṭār al-Bakrī by lineage, al-Shāfiʿī by rite, al-Wafā'ī by (Ṣūfī) order … . Written by Aḥmad ibn Aḥmad al-Timurbāy al-Ḥanafī al-Qādirī.

On fol. 78r of the Patna manuscript, Ibn al-ʿAṭṭār informs us:[73]

> I spent about thirty years working on this subject (of astronomy) from eight hundred thirty Hijra [= 1426/27] to the year 859 [= 1454/55] then until this date, which is the year 871 [= 1466/67].

---

71 On the problems of identification of our copyist's grandfather see n. 38 above. The phrase used here is curious because Timurbāy must have been about 100 years old or deceased.

72 *Patna Catalogue* (see n. 34), p. 96.

73 *Ibid.*, p. 94. The date is given as 849 in the reproduction of the Arabic text but 859 in the English translation.

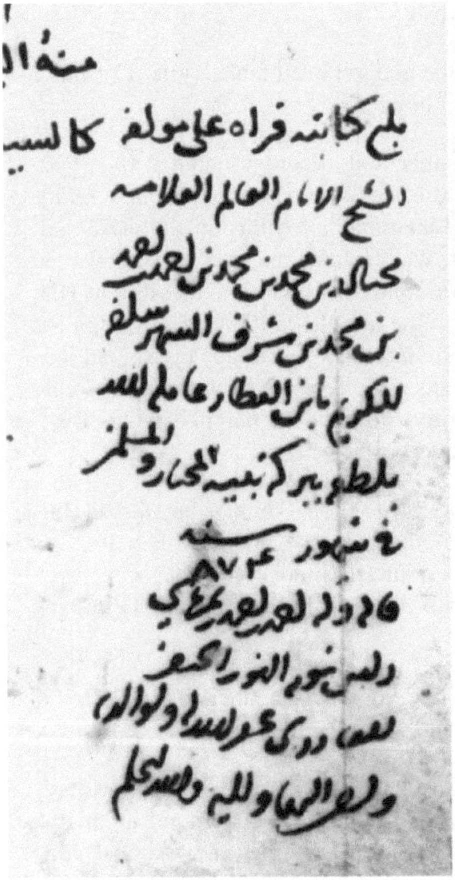

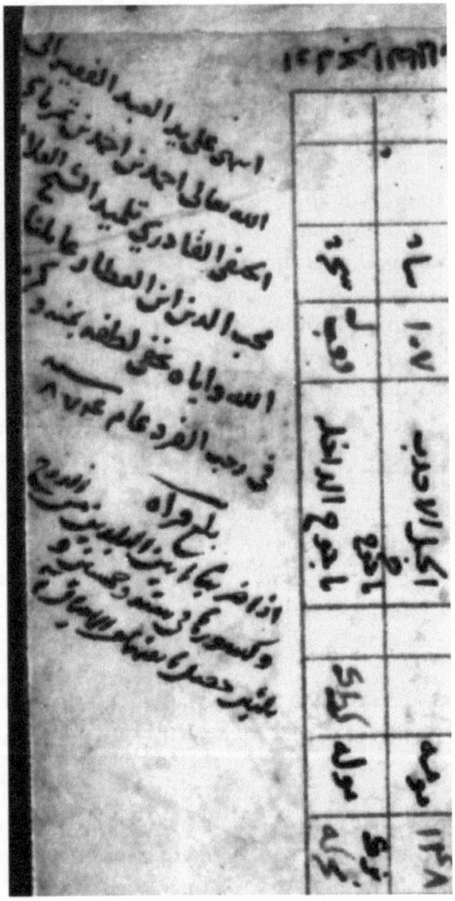

Figs. 7–8: The colophons on fols. 17v and 20v.

At the end of some copies of his treatise on quadrants, we find the following note by Ibn al-ʻAṭṭār:[74]

> All of this is from the useful information (*fawā'id*) of our *shaykh*, the *ustādh*, the luminary (*'allāma*) Nūr al-Dīn al-Naqqāsh ibn ʻAbd al-Qādir in the year eight hundred and thirty (Hijra) [= 1426/27]. He, through his modesty and humanity, when he saw that I excelled over my peers, he took me and passed me on to the unique teacher (*ustādh*) in the(se) difficult (*'aziza*) subjects, our *shaykh*, the teacher, the luminary Shihāb al-Dīn al-Majdī al-Shāfiʻī, may God make his knowledge useful to us. I studied carefully under him (*ḥaqqaqnā 'alayhi*) the principles of this science, and I read under him about the different

[74] Recorded with numerous errors in King, *Fihris al-makhṭūṭāt al-'ilmiyya al-maḥfūẓa bi-Dār al-Kutub al-Miṣriyya*, 2 vols. (Cairo: General Egyptian Book Organization, 1981 and 1986), II, pp. 368–9.

kinds of sundials, horizontal and vertical, and cosmology, and other subjects relating to the sciences of religion in the sacred law and other aspects, at the Azhar Mosque and at his home in the next to (the Mosque) and at the Madrasa of Jānī Beg. I accompanied him for close to twenty years. In God is success and from Him we ask for help. This was reported by the luminary Muḥibb al-Dīn Muḥammad ibn Muḥammad known as Ibn al-ʻAṭṭār ibn Aḥmad ibn Aḥmad ibn Muḥammad ibn Mushrif ibn Badr al-Dīn ʻAbd al-Khāliq ibn al-Ḥasan ibn Dā'ūd ibn Idrīs ibn ʻAbd al-Jalīl ibn ʻAbd al-Kāfī ibn Ṭalḥa ibn ʻAbdallāh ibn ʻAbd al-Raḥmān ibn Abī Bakr al-Ṣiddīq, may God be pleased with him and make him content, and may He make Paradise his resting place. Amen. Praise be to God, Lord of the Worlds.

This cannot be a complete genealogy back to the Caliph Abū Bakr, because it does not span six centuries. It is of course naïve to expect that it would. His teachers Ibn al-Majdī and Nūr al-Dīn (ʻAlī ibn Muḥammad) al-Naqqāsh are well known.[75]

*****

Acknowledgements: In 1998 Dr Sonja Brentjes kindly furnished me with a set of incomplete and poor-quality photocopies of parts of the Patna manuscript. During the winter semester of 2004–2005, I read the text of Ibn al-ʻAṭṭār's *Geography* with Petra Schmidl and Mónica Herrera Casais in my Arabic Scientific Manuscripts seminar. In December, 2005, I received a microfilm of the complete text of the *Geography* thanks to the kindness of Dr Imtiaz Ahmad, director of the Khoda Bakhsh Library. In January, 2006, Prof. Jamil Ragep kindly sent me a preprint of his paper on al-Jaghmīnī, which proved to be of considerable relevance to the present study and saved me not least from the error of thinking that all was well with the climates in the Islamic East.

[75] See King, *Cairo Survey*, nos. C62 and C74, also Rosenfeld & İhsanoğlu, *MAIC*, nos. 815 and 847, as well as Charette, *Mathematical Instrumentation* (n. 13), pp. 19–20.

INDEX

"Never index your own book."
Kurt Vonnegut, Jr., in *Cat's Cradle* (1963)

The Arabic article 'al-' as well as 'b.' for 'ibn' are ignored in the order of the entries. 'Ibn' with a capital 'I' and 'Abū' are not. The numerous astrolabe-makers from 9th and 10th century Baghdad listed in II, pp. 115–17 (nn. 50–52) are not included in this index; all their surviving productions are published in *SATMI*, vol. 2. Also not included in this index are the numerous Maghribi manuscripts, authors and instrument-makers listed in VI, pp. 198–206.